W9-CHL-525

CIGR Handbook of Agricultural Engineering Volume I

CIGR Handbook of Agricultural Engineering

Volume I
Land and Water Engineering

Edited by CIGR–The International Commission of Agricultural Engineering

Volume Editor:
H. N. van Lier
Wageningen Agricultural University, The Netherlands

Co-Editors:
L. S. Pereira
Instituto Superior de Agronomia, Portugal
F. R. Steiner
Arizona State University, USA

Published by the American Society of Agricultural Engineers

LCCN 98-93767 ISBN 1-892769-01-8

For Information, contact:

Manufactured in the United States of America

Editors and Authors

Volume Editor
H. N. van Lier
Wageningen Agricultural University, Department of Environmental Sciences, Laboratory for Spatial Analysis, Planning, and Design, 13 Gen. Foulkesweg, Wageningen 6703 BJ, The Netherlands

Co-Editors
L. S. Pereira
Instituto Superior de Agronomia, Departamento De Engenharia Rural, Tapada da Ajuda, Lisbon 1399, Portugal

F. R. Steiner
School of Planning and Landscape Architecture, College of Architecture and Environmental Design, Arizona State University, Tempe, AZ 85287-2005, USA

Authors
R. G. Allen
Department of Agricultural Irrigation Engineering, Utah State University, Logan UT 84322-4105, USA

J. Amsler
Swiss Federal Office of Agriculture, Section Land Improvement, 5 Mattenhof Strasse Berne CH-3003, Switzerland

V. Bagarello
Universita di Palermo, Dipartimento di Ingegneria e Tecnologie Agro-Forestali, Sezione Idraulica, Viale delle Scienze, Palermo 90128, Italy

A. Brasa Ramos
Escuela Tecnica Superior de Ingenieros Agrónomos, Departamento de Producción Vegetal y Tecnología Agraria, Universidad de Castilla–La Mancha, Campus Universitario, Albacete 02071, Spain

G. Chisci
Universita di Firenze, Dipartimento di Scienze Agronomiche e Gestione del Territorio Agroforestale, 18 Piazza delle Cascine, Firenze 50144, Italy

H. Depeweg
Section Land and Water Development IHE; International Institute for Infrastructural, Hydraulic and Environmental Engineering, Westvest 7, P.O. Box 3015, 2601 DA Delft, The Netherlands

D. De Wrachien
Università di Milano, Istituto di Idraulica Agraria, 2 Via Celoria, Milan 20133, Italy

M. W. van Dongen
c/o Wageningen Agricultural University, Department of Environmental Sciences, Laboratory for Spatial Analysis, Planning, and Design, 13 Gen. Foulkesweg, Wageningen 6703 BJ, The Netherlands

C. Fabeiro Cortés
Escuela Tecnica Superior de Ingenieros Agrónomos, Departamento de Producción Vegetal y Tecnología Agraria, Universidad de Castilla–La Mancha, Campus Universitario, Albacete 02071, Spain

V. Ferro
Universita di Palermo, Dipartimento di Ingegneria e Tecnologie Agro-Forestali, Sezione idraulica, Viale delle Scienze, Palermo 90128, Italy

M. Greppi
Università di Milano, Istituto di Idraulica Agraria, 2 Via Celoria, Milan 20133, Italy

M. B. A. Hes
c/o Wageningen Agricultural University, Department of Environmental Sciences, Laboratory for Spatial Analysis, Planning, and Design, 13 Gen. Foulkesweg, Wageningen 6703 BJ, The Netherlands

A. Hoogeveen
Optifield, 197 Boeckstaetedreef, Nijmegen 6543 JH, The Netherlands

C. F. Jaarsma
Wageningen Agricultural University, Department of Environmental Sciences, Laboratory for Spatial Analysis, Planning, and Design, 13 Gen. Foulkesweg, Wageningen 6703 BJ, The Netherlands

J. Arturo de Juan
Escuela Tecnica Superior de Ingenieros Agrónomos, Departamento de Producción Vegetal y Tecnología Agraria, Universidad de Castilla–La Mancha, Campus Universitario, Albacete 02071, Spain

A. Legorburo Serra
Escuela Tecnica Superior de Ingenieros Agronomos, Departamento de Produccion Vegetal y Tecnologia Agraria, Universidad de Castilla–La Mancha, Campus Universitario, Albacete 02071, Spain

F. Martin de Santa Olalla
Escuela Tecnica Superior de Ingenieros Agronomos, Departamento de Produccion Vegetal y Tecnologia Agraria, Universidad de Castilla–La Mancha, Campus Universitario, Albacete 02071, Spain

J. Martinez Beltran
Land and Water Development Division, Food and Agriculture Organization of the United Nations, Room: B 725, Viale delle Terme di Caracalla, Rome 00100, Italy

F. Preti
Istituto di Genio Rurale, Facolta di Agraria, Universita della Tuscia, via S. Camillo de lellis, 01100 Viterbo, Italy

N. Romano
*Università di Napoli, Federico II, Dipartimento di Ingegneria Agraria,
100 Via Università, Portici (Naples) 80055, Italy*

J. M. Tarjuelo
*Centro Regional de Estudios del Agua, Instituto de Desarrollo Regional (IDR),
Universidad de Castilla–La Mancha, Campus Universitario, Albacete 02071, Spain*

T. J. Trout
*USDA/ARS/Water Management Research Laboratory, 2021 S. Peach Avenue, Fresno
CA 93727-5951, USA*

A. P. Wolleswinkel
*c/o Wageningen Agricultural University, Department of Environmental Sciences,
Laboratory for Spatial Analysis, Planning, and Design, 13 Gen. Foulkesweg,
Wageningen 6703 BJ, The Netherlands*

Editorial Board

Contents

Foreword

This handbook has been edited and published as a contribution to world agriculture at present as well as for the coming century. More than half of the world's population is engaged in agriculture to meet total world food demand. In developed countries, the economic weight of agriculture has been decreasing. However, a global view indicates that agriculture is still the largest industry and will remain so in the coming century.

Agriculture is one of the few industries that creates resources continuously from nature in a sustainable way because it creates organic matter and its derivatives by utilizing solar energy and other material cycles in nature. Continuity or sustainability is the very basis for securing global prosperity over many generations—the common objective of humankind.

Agricultural engineering has been applying scientific principles for the optimal conversion of natural resources into agricultural land, machinery, structure, processes, and systems for the benefit of man. Machinery, for example, multiplies the tiny power (about 0.07 kW) of a farmer into the 70 kW power of a tractor which makes possible the production of food several hundred times more than what a farmen can produce manually. Processing technology reduces food loss and adds much more nutritional values to agricultural products than they originally had.

The role of agricultural engineering is increasing with the dawning of a new century. Agriculture will have to supply not only food, but also other materials such as bio-fuels, organic feedstocks for secondary industries of destruction, and even medical ingredients. Furthermore, new agricultural technology is also expected to help *reduce* environmental destruction.

This handbook is designed to cover the major fields of agricultural engineering such as soil and water, machinery and its management, farm structures and processing agricultural, as well as other emerging fields. Information on technology for rural planning and farming systems, aquaculture, environmental technology for plant and animal production, energy and biomass engineering is also incorporated in this handbook. These emerging technologies will play more and more important roles in the future as both traditional and new technologies are used to supply food for an increasing world population and to manage decreasing fossil resources. Agricultural technologies are especially important in developing regions of the world where the demand for food and feedstocks will need boosting in parallel with the population growth and the rise of living standards.

It is not easy to cover all of the important topics in agricultural engineering in a limited number of pages. We regretfully had to drop some topics during the planning and editorial processes. There will be other requests from the readers in due course. We would like to make a continuous effort to improve the contents of the handbook and, in the near future, to issue the next edition.

This handbook will be useful to many agricultural engineers and students as well as to those who are working in relevant fields. It is my sincere desire that this handbook will be used worldwide to promote agricultural production and related industrial activities.

Osamu Kitani
Editor-in-Chief

Acknowledgments

At the World Congress in Milan, the CIGR Handbook project was formally started under the initiative of Prof. Giuseppe Pellizzi, the President of CIGR at that time. Deep gratitude is expressed for his strong initiative to promote this project.

To the members of the Editorial Board, co-editors, and to all the authors of the handbook, my sincerest thanks for the great endeavors and contributions to this handbook.

To support the CIGR Handbook project, the following organizations have made generous donations. Without their support, this handbook would not have been edited and published.

Iseki & Co., Ltd.
Japan Tabacco Incorporation
The Kajima Foundation
Kubota Corporation
Nihon Kaken Co., Ltd.
Satake Mfg. Corporation
The Tokyo Electric Power Co., Inc.
Yanmar Agricultural Equipment Co., Ltd.

Last but not least, sincere gratitude is expressed to the publisher, ASAE; especially to Mrs. Donna M. Hull, Director of Publication, and Ms. Sandy Nalepa for their great effort in publishing and distributing this handbook.

Osamu Kitani
CIGR President of 1997–98

Preface and Acknowledgments

Land and Water Engineering are becoming increasingly important globally for the future of humankind. There are at least two main reasons for this growing significance: First, it is well understood that, aside from several other means, the wise use of land and water will play a key role in the provision of enough good food for future generations. Despite all types of programs and policies, the global population still increases and most probably will continue to do so for the next half century or so. This means that more food and fiber will be necessary. At the same time the demand for food changes in terms of types and quality. There is undoubtedly an increase with regard to the demand for high-quality food with a larger variety in daily nutrition. Both effects mean an increased concern for better land use and improved agricultural water management to provide for future food requirements.

Second, the demand for different land uses in the rural countryside (often referred to as the green space) is increasing tremendously, especially in the developed world. However, each day, it is becoming clearer that the developing world also needs to pay more attention to this aspect. Land used for housing, industries, infrastructure, outdoor recreation, landscape, and nature is in high demand. The increased uses of land for urban development are a direct result of the increasing population, urbanization, welfare, etc. The growing concerns for landscape and nature are a result of a better understanding of its vital role in creating a sustainable countryside. Farming, as an important and mostly dominant user of space in rural areas, has to change in at least two ways. The way in which farming is performed, especially the high-intensive mechanized farming, has to change to farming methods in which the natural resources, soils, and water are safeguarded from depletion. Methods that keep or improve the natural qualities are considered to be sustainable and therefore important for future generations. At the same time, there is an increasing understanding that landscape or nature has an important meaning for earth and humankind in the long run. The protection and re-creation of nature areas, the planning of new nature areas, the design of ecological corridors, the redesign and improvement of (often small-scale) landscapes with their value for living and enjoyment are at stake. Finding new balances in green spaces among these different demands is the ultimate challenge for land and water engineers and related professions.

Volume I of this CIGR Handbook attempts to address this challenge by first focusing on the changing role of agriculture within society and within the rural countryside. Today's and future farming systems have to face the challenge of finding a balance between further development in terms of increased volumes and productivity, diversification in food, and improved qualities of food on the one hand and the establishment of farming methods that safeguard the environment, the natural resources, and the ecosystem, nature, and landscape on the other.

Although many types of action are necessary to achieve these goals, there is certainly a special role for land- and water-use planning. This creates possibilities for rearranging farms, fields, and rural roads and for improving soils and the water management systems. The opening up of the countryside is not only important for farming but also for other functions, such as for outdoor recreation, living, nature education, forestry, and others. Chapters 1 through 3 describe these possibilities in more detail.

Soil, together with water, provides the basis for our life on Earth. After humankind in distant history changed from hunters and nomads to settlers, soil was worked to improve its productivity. The reclaiming and conservation of soils through regeneration, improvement, erosion control, and so forth was, is, and will be extremely important for our species to survive and prosper. Chapter 4 gives the most important of today's knowledge and practice in this regard.

Together with our care for good soils is the establishment of good water management systems, both in its quantitative and qualitative meanings. Crop production, the relation of water to soils, the providing of good and sufficient water for crop protection, the drainage of agricultural lands and regions, and finally the quality of water, especially related to drainage, are important topics in this regard. Chapter 5 gives the latest and most important knowledge and practice in these fields.

To conceptualize a book on Land and Water Engineering is not an easy undertaking. This undertaking is practically impossible for a single individual. I am therefore indebted to many others. First, I would like to thank the many authors who set aside time not only to produce the first drafts of their paragraphs or chapters but also to do the corrections and improvements after one or two review processes. This book would not have been possible at all without them, and the workers in the field. Second, I owe very much to the "Wageningen crew," N. Berkhout, A. Hoogeveen, C. Jacobs and M. Riksen, for their substantial help in writing Chapter 2 and reviewing the first drafts of Chapters 4 and 5. Their willingness to do this critical reading and improvement was the beginning of the last and long effort to complete the book.

I am also very much indebted to our official reviewers from abroad, who read and commented on the different chapters in their final phase. W. Schmid (ORL, ETHZ, Switzerland) who, with his team, took the responsibility for Chapters 1–3 and M. Fritsch (LWM, ETHZ, Switzerland) and his colleagues, who was responsible for Chapters 4 and 5, did a marvellous work in reading and commenting on the many paragraphs and chapters.

The whole project would have failed without the enormous help of my two co-editors. L. Santos Pereira not only co-authored two main sections, but also reviewed, corrected, and improved the contents of the Chapters 4 and 5. For this reviewing he thankfully was assisted by D. Raes (Leuven, Belgium), M. Smith (FAO, Rome), D. Kincaid and D. Bjomberg (Idaho, USA), F. Lamm (Kansas, USA), and F. Morissey (California, USA). It proved to be a time-consuming process for which I owe him very much. Special thanks goes also to W. H. van der Molen and W. Ochs for reviewing the Land Drainage Chapter F. R. Steiner spent very much of his time not only in reviewing and commenting on Chapters 1–3 but also in correcting and improving the English for the large majority of all sections and chapters. Without his performance in this regard, the book could not have been published.

A last but specific thanks goes to Andreas Hoogeveen, who from the very beginning to the last minute helped not only to co-author one of the sections, but also to prepare all intermediate and final layouts of each paragraph and chapter. Without this heavy and time-consuming work, the book would not be in its present state. Andreas, thank you.

Hubert N. van Lier
Editor of the Volume I

1 Balancing Agriculture Between Development and Conservation

M. W. van Dongen and H. N. van Lier

Increasing agricultural production has always been the main goal of agricultural research. Because of the continuing growth of the world's population and the important share of agricultural products in the world economy, this increase of agricultural production will continue to be important, but under different conditions. The more or less uncontrolled growth in agricultural production during the past few decades, in industrial as well as developing countries, has pushed agricultural production to and, in many cases, over the edge of sustainability. This means that the traditional ways to increase production by, for example, land and water engineering are meeting a new challenge: to find a new balance between agricultural development and the conservation of the natural resources.

1.1 Agriculture: a Description

1.1.1 Scope of Activities

Agriculture is the practice of cultivating the soil, harvesting crops, and raising livestock to produce plants and animals useful to human beings and, in varying degrees, it is the preparation of these products for humans use and their disposal [1]. According to this definition, agriculture includes horticulture, seed production, dairy farming and livestock production (animal husbandry), the management of land that supports these activities [2]. Forestry and fisheries are excluded.

The main activity of agriculture is the production of food. Grigg [3] gives as an example the United States, where 90% of the value of farm products is consumed as food, and 93% of farmland is sown with food crops. Besides food, agriculture provides raw materials such as cotton and wool for clothing, stimulants such as tobacco, coffee, and tea, as well as flowers and rubber.

In many countries, agriculture is not only the main source of food and raw materials, it is the main source of income. For the period 1990–1992, about 48% of the total world labor force was working in agriculture [4]. In the developing countries, this was about 58% and in the industrial countries about 10% of the total labor force [4]. However, the progressive industrialization of agriculture in developing countries will lead to a smaller labor force in these countries in the (near) future. This will have a major impact on the socioeconomic development of rural areas as well as of their agglomerations.

Nevertheless, the figures make clear that agriculture is still the most important human (economic) activity in the world.

1.1.2 Description of Agricultural Systems

Agriculture is practiced in many different ways, resulting in a great diversity of agricultural types all over the world. These agricultural types can be roughly divided into eight major groups [5]:

- subsistence farming and shifting cultivation,
- pastoral nomadism,
- wet-rice cultivation,
- Mediterranean agriculture,
- mixed farming systems and dairying,
- plantations,
- ranching, and
- large-scale grain production.

Subsistence Farming and Shifting Cultivation

Subsistence farming is the production of enough food and fiber for the needs of the farmer's family [6]. This form of agriculture is quite common in the developing countries.

Shifting cultivation is a system in which a relatively short cultivation period alternates with a long fallowing period. In the traditional system, a cultivation period of two to three years is followed by a fallowing period of 20 years or more. In this fallowing period the original vegetation (mostly forest) regenerates and soil fertility is restored. Because such a system is only possible where the population density is very low, shifting cultivation is being replaced by other agricultural systems. In Africa, for example, the original shifting cultivation is replaced by *bush fallowing* in which the fallowing periods are only long enough for grass or bush to regenerate [3]. If more than 30% of the arable and temporarily used land is cultivated annually, we no longer speak of shifting cultivation but of *semipermanent farming* [7].

Pastoral Nomadism

Pastoral nomadism is a system in which farmers (nomads) and their households are more or less continuously moving with the herd, on which they depend for food, fuel, clothing, and cash. Pastoral nomadism is found in the arid and semiarid tropics of Asia and Africa. In these areas the grass production is very low and seasonal. This means that the livestock cannot stay at one place for a long period [5]–[7]. Ruthenberg [7] distinguishes two main types of nomadism:

- total nomadism, in which the livestock owners do not live in permanent settlements, practice no regular cultivation, and move their families with their herds;
- seminomadism, in which the livestock owners have a permanent place of residence (which is kept for several years) and combine some kind of regular cultivation with long periods of travel with their herds.

Wet-Rice Cultivation

Wet-rice cultivation is a system in which rice is grown in slowly moving water to an average height of 100–150 mm for three-quarters of its growing period [5]. Wet-rice cultivation is found mainly in the river deltas and the lower reaches of rivers of the Far East. It supports a majority of the rural population in that part of the world [5].

Mediterranean Agriculture

Mediterranean agriculture is found in the areas surrounding the Mediterranean Sea and in areas with a similar climate, characterized by mild, wet winters and hot, dry summers [5]. There are four land-use patterns that characterize the traditional Mediterranean farming system [5]:

- extensive wheat cultivation;
- extensive grazing by sheep and goats;
- cultivation of tree crops such as olives and figs, grapes, and date palms;
- cultivation of fruits and vegetables, with apples, peaches, and pears as most important fruit crops and potatoes, tomatoes, lettuce, onions, cauliflower, and peas as most important vegetables.

Intensification and specialization has partly changed the Mediterranean agricultural system, with olive and grapes becoming the most important crops.

Mixed Farming Systems and Dairying

Mixed farming systems integrate crop and livestock production [6]. Mixed farming was originally a typical agricultural system for northwestern Europe and the easter United States [5]. Some characteristics are [5]:

- high level of commercialization,
- a declining agricultural labor force,
- ownership and operation by families (family farm),
- use of a large part of cereal crops to feed livestock.

A better access to inputs and services, cheap chemical fertilizers, and the legislation and price interventions of the European Union have resulted in specialization and intensification of the mixed farming systems in Europe. The number of traditional mixed farms has strongly declined in the past few decades. They have been replaced with specialized farms that produce crops or livestock.

Dairying is the production of milk and milk products and is most common within the farming systems of Europe, North America, and Australasia [5]. Table 1.1 shows the dominant role of Europe in milk and cheese production worldwide.

Plantations

Plantations are large-scale tropical and subtropical crop production systems, specializing in one or two crops. Typical crops are rubber, coconuts, oil palm, sisal, cacao, coffee, bananas, tea, cotton, jute, tobacco and groundnuts [5]. The plantation system origined during early European colonization of North and South America and was exploited with slaves. After abolition of slavery, many large plantations broke down or were split up. Now, most plantation crops are grown by smallholders or corporations.

Ranching

Ranching is livestock production primarily through extensive commercial grazing. The main products are beef cattle and sheep for mutton and wool. The major ranching areas are [5]:

Table 1.1. Milk and cheese production in 1992

Area	Milk Production (1,000 MT*)	Cheese Production (MT*)
World	453,733	13,532,455
Africa	14,686	483,544
North and Central America	86,481	3,730,897
South America	34,175	3,268,500
Asia	66,792	866,180
Europe	153,392	6,846,694
Oceania	15,989	356,227
USSR[a] (former)	100,921	1,686,250

[a] Data from 1991.
Compiled from [8].
* MT = metric tons.

- North America, including the western United States and adjacent parts of Canada and Mexico;
- South America, including Brazil, Argentina, Venezuela, and Uruguay;
- South Africa;
- Australia and New Zealand.

Large-Scale Grain Production

The major cash crop grown in the *large-scale grain production system* is wheat. This system is found in North America, Argentina, the former Soviet Union, and Australia.

1.1.3 Characteristics of Agricultural Systems

To characterize an agricultural type in a certain area more precisely, a lot of variables should be taken into account. Grigg [3] gives a good example of a list of such variables (Table 1.2). This list was compiled by a commission of the International Geographical Union.

1.2 Dynamics in Agriculture

Agricultural development is a dynamic process and is highly affected by external conditions. These external conditions encompass the natural environment as well as socioeconomic and political factors. In other words, agricultural development is caught between the conditions for growing crops and raising livestock and the demand for agricultural products.

1.2.1 Limiting Conditions for Agriculture

Crop and livestock production depend on biological, physical, and spatial conditions and access to inputs and services.

Biological Conditions

Biological conditions refer to the nature of crops and animals and the pests and diseases that threaten them. The nature of crops and animals can be described in terms

Table 1.2. Characteristics of agricultural systems

A	**Social attributes**
1	Percentage of land held in common
2	Percentage of land in labor or share tenancy
3	Percentage of land in private ownership
4	Percentage of land in state or collective ownership
5	Size of holding according to numbers employed
6	Size of holding according to area of agricultural land
7	Size of holding according to value of output
B	**Operational attributes**
8	Labor intensity: number of employees per hectare of agricultural land
9	Inputs of animal power: draught units per hectare of agricultural land
10	Inputs of mechanical power: tractors, harvesters, etc. per hectare of agricultural land
11	Chemical fertilizers: nitrogen, phosphorus, and potassium per hectare of cultivated land
12	Irrigation, irrigated land as a percentage of all cultivated land
13	Intensity of cropland use, ratio of harvested to total arable land
14	Intensity of livestock breeding, animal units per hectare of agricultural land
C	**Production attributes**
15	Land productivity: gross agricultural output per hectare of agricultural land
16	Labor productivity: gross agricultural output per employee in agriculture
17	Degree of commercialization: proportion of output sold off farm
18	Commercial production: commercial output per hectare of agricultural land
D	**Structural characteristics**
19	Percentage of land in perennial and semiperennial crops
20	Percentage of total agricultural land in permanent grass
21	Percentage of total agricultural land in food crops
22	Percentage of total agricultural output of animal origin
23	Animal production as percentage of total commercial output
24	Industrial crops (sugar, fiber, rubber, beverages) as percentage of total agricultural land

Source: J. Kostrowicki (1976) as cited in [3] p. 3.

of nutritive value (or usefulness in the case of nonfood products), productivity (yield), and growth characteristics (perennials or annuals, planting season, bearing time, etc.). Besides being useful to man, crops and livestock are part of the natural food chain. This means that other living creatures, such as bacteria, viruses, parasites, fungi, insects, birds, and beasts of prey, also are interested in the crops and livestock. To the farmer, these other living creatures are pests that can threaten the production of crops, livestock, and the products derived from them and thus food security and the farmer's income.

Physical and Spatial Conditions

The *physical and spatial conditions* refer to land and water, which are the limited resources necessary for farming. This means that to produce agricultural goods and gain an income, the farmer depends on the quantity and quality of land and water.

Land quantity is the amount of land available for agriculture,which is dependent on geographic and socioeconomic factors. The geographic factors refer to landform, altitude, and the ratio of land to open water. The socioeconomic factors are population density (competition for land among farmers), development rate of the country (competition among land uses such as for housing, infrastructure, outdoor recreation, and nature), and the amount of people dependent on agriculture.

Land quality refers to both soil quality and the spatial configuration of farmed areas. Soil quality depends on soil type, structure and fertility (workability of the land, water-holding capacity, erosion sensitivity). The spatial configuration of farmed areas is determined by the fragmentation of property, the distance between farm buildings and fields, and the quality of the rural infrastructure, and is in many cases a reflection of the social structure in an area or the political situation in the country.

Water quantity depends on geographic, socioeconomic, and political factors. Geographic factors include climate (rainfall, temperature), altitude, steepness (water runoff), soil (water-holding capacity, permeability), and the presence of groundwater and surface water (rivers, lakes, brooks). The socio economic factors are population size and density and development rate of the country (amount of water for drinking and industrial uses). Political considerations include, for example, the relationship of a country with other (surrounding) countries that are using the same source for water. Table 1.3 shows the agricultural share of world water use.

Water quality depends on geographic and socioeconomic factors. Geographic factors include the influence of the sea and other natural sources of water not suited for agricultural use. Socioeconomic factors are related to water pollution.

Table 1.3. World water use

	Water Withdrawals						
	Annual	Domestic		Industry		Agriculture	
Area	(km^3)	(km^3)	(%)[a]	(km^3)	(%)	(km^3)	(%)
Africa	144	10	7	7	5	127	88
Asia	1,531	92	6	122	8	1,317	86
North & Central America	697	63	9	293	42	342	49
South America	133	24	18	31	23	78	59
Europe	359	47	13	194	54	118	33
USSR (former)	358	25	7	97	27	233	65
Oceania	23	15	64	1/2	2	8	34
World	3,240[a]	259[a]	8	745[a]	23	2,235[a]	69

[a] Percentage of total water withdrawals.

[b] Numbers are not column totals.

Compiled from [9].

Access to Inputs and Services

The *access to inputs and services* determines to a large degree the possibilities of agricultural development in a country or area. With good access to, for example, cheap fertilizers, good seeds, bank credit, and the results of agricultural research, a farmer has a greater chance of increasing the farm's profitability. A good example is the Green Revolution in Asia. When farmers in that part of the world got access to high-yield varieties of rice and wheat, the production of these crops increased enormously.

1.2.2 Improving the Limiting Conditions for Agriculture

Improving the conditions for growing crops and raising livestock makes it possible to sustain or even increase, agricultural production.

Biological Conditions

The *biological conditions* can be improved by manipulating the nature of crops and animals and improving the production environment. The most important way to manipulate the nature of crops and animals is by developing new varieties that give a higher yield or production, are more resistant against pests and diseases, more tolerant of drought, etc. The development of new varieties is possible through crossbreeding of plants and animals and biotechnology.

The production environment can be improved by animal and plant production engineering (see Vols. II and III of this Handbook). The production environment also can be improved by fighting pests and diseases through animal and plant production engineering and management, development of new varieties, and the use of pesticides and vaccinations.

Physical and Spatial Conditions

The *physical and spatial conditions* can be improved by manipulation of the quantity and quality of land and water.

Land quantity can be increased through, for example, land reclamation (see Chapter 4).

Land quality can be improved by improving the soil structure and fertility (see Chapter 4), land-use planning (see Chapter 2), and rural road development (see Chapter 3).

Water quantity can be increased through drainage (in the case of water surplus), irrigation and inundation (in the case of water shortage), and water management in general (see Chapter 5).

Water quality can be improved through water purification and management (see Chapter 5).

Which conditions should be improved and by what method depends on the local land and water quantity and quality and socioeconomic and political conditions as well.

Access to Inputs and Services

The *access to inputs and services* can be increased by improving rural infrastructure and distribution of agricultural inputs, forming cooperatives, subsidizing agricultural investments, funding agricultural research, among others. Governments can, with or without development aid, start programs to improve infrastructure, subsidize investments, or

Table 1.4. Effects of measures to improve agricultural conditions

Measure	Positive Effects	Negative Effects
Planning		
Land reclamation	Increasing the land availability for agriculture	Destruction of wetlands, mangrove forest, and riverine systems
		Disturbance of natural river dynamics, causing floods and droughts
Land redevelopment	Improving the spatial conditions for growing crops and working the land	Loss and disturbance of ecological systems
		Habitat fragmentation
		Degradation of landscape
Irrigation	Increasing the water availability for agriculture	Salinization of soils, what causes land degradation and a collapse of agricultural production
		Overconsumption of unrestorable water sources
Management		
Fertilizer use	Higher fertility of soils; higher yields (in the short term)	Ground water contamination by nitrogen and phosphates
		Mining of nutrients other than nitrogen and phosphate, what causes fertility problems in the long term
		Decline in biodiversity
Pesticide use	Higher yields (in the short term)	Increased pest resistance to pesticides
		Ground water contamination

fund research. Farmers can form cooperatives to get cheaper inputs or share expensive machinery such as tractors and harvesters.

Conclusion

In general, the conditions for agriculture can be improved through planning and management of the limiting resources. Unfortunately, produce although many of these measures positive short-term effects in agriculture, they can have negative long-term effects on the environment or human health. A few examples are given in Table 1.4.

1.2.3 Demand for Agricultural Products

The *demand for agricultural products* determines what a farmer is grows or raises and in which amounts. Because the main activity of agriculture is food production, human food requirements generally determine agricultural production and development. Of course, some exceptions exist for areas where farmers produce crops for certain industrial products (e.g., fiber for clothes).

Food Requirements

To remain healthy, the human body has minimum nutritional requirements. Although these requirements depend on gender, climate, age, and occupation, certain intakes of energy, vitamins, minerals, and proteins, among other nutrients are needed [3]. Energy

Table 1.5. Population growth trends

	Number of People (million)			Growth Rate (%)		
	1990	2000	2010	1980–1990	1990–2000	2000–2010
World	5,296.8	6,265.0	7,208.6	1.8	1.7	1.4
Developing countries	4,045.9	4,946.9	5,835.2	2.1	2.0	1.7
Developed countries	1,248.9	1,314.7	1,369.7	0.7	0.5	0.4

Compiled from [10].

is needed "to maintain the metabolic rate and to allow the body to work" [3]. Protein foods contain amino acids, of which 10 or 11 are essential. They are found in both animal products (eggs and meat) and plant foods [3]. Although eggs and most meats provide all of the essential amino acids, they are expensive and thus are unavailable to many people in adequate amounts. Plant foods are a less efficient protein source; they contain lower levels of protein and no crop contains all the essential amino acids. This drawback can be overcome by eating a combination of food crops, for example, cereals and legumes, which are present in most traditional diets [3].

Demand for Food

The demand for food is determined by population growth and changes in income [3].

Population Growth

World population is still growing. Even with a declining growth rate, about 90 million people are being added to the population each year [9]. The United Nations predicts an increase in world population from 5.8 billion in 1995 [9] to 7.0 billion in 2010 [10].

Population growth means a growing demand for food. This is especially true in developing countries, where the largest population growth is expected (Table 1.5). The growth in demand for food leads to intensification of agricultural production and an increasing pressure on existing and potential agricultural lands. Theoretically, this intensification is possible in most developing countries [10], but this increase in food production will certainly strain an already fragile ecological balance.

Changes in Income

Changes in income mean changes in composition of diet and the expenditures on food. In low-income countries the demand is for food products that give the most calories for the least money, which are cereals and root crops such as potatoes and yams. These crops give a high yield of calories (and protein) per hectare [3]. This explains the predominance of crops such as rice, maize, sorghum, or millet in many developing countries. With increasing income, the food pattern changes. Most important is the shift from plant food to the less efficient and more expensive animal products.

1.2.4 Major Trends in World Agriculture

Agricultural Production Growth

Together with the ongoing population growth, the rate of agricultural is declining. Table 1.6 shows the trends over the past few decades and the prognoses for the next

Table 1.6. Agricultural production growth rates

Period	Growth Rates (%)
1960–1969	3.0
1970–1979	2.3
1980–1992	2.0
1993–2010	1.8

Source: [10].

decade. There are a few explanations for this development:

- Urban areas are located mostly in parts of a country that also have the highest agricultural potential; when a city expands it often takes the best agricultural lands. The loss of agricultural production is not always (totally) restored elsewhere.
- Intensification of agricultural production and mismanagement of agricultural lands leads to land degradation and loss of agricultural lands and thus loss of agricultural production.
- The countries with the best possibilities for production growth are not the countries where demand for food is increasing most (i.e., developing countries). So, the possibilities for keeping up with the increasing demand for food are not fully used.

Mechanization

Since the Industrial Revolution, replacement of human labor and draught animals in agriculture with machinery has occurred worldwide, but there is a big gap between industrialized and developing countries in this respect.

The first application of power to agricultural production in the industrialized countries took place at the beginning of the nineteenth century. At first this had little impact on the overall agriculture production, because the machinery was cumbersome and costly. In North America and Australia, tractors became a significant factor after World War I, but in Europe, this was not the case until after the World War II. The mechanization in the rest of the world stayed far behind [5]. Table 1.7 shows some recent developments in mechanization and the regional differences.

Fertilizer Use

Soil fertility is one of the most important conditions for agricultural production. In traditional farming systems, several methods are in use to maintain and improve soil fertility, the most common being livestock manure, fallowing, and planting of legumes. Although agricultural output can be maintained with these methods, they are insufficient for keeping up with the food demands of the increasing world population.

The introduction of chemical fertilizers in the middle of the nineteenth century made it possible to increase yields per hectare in substantial amounts. Untill the 1950s, however, intensive use of chemical fertilizers was limited mainly to northwestern Europe. Since then, their use has increased enormously worldwide, their use is especially significant in areas with a high population density.

It is expected that fertilizer use will continue to increase, especially in the highly populated, developing countries. However, the effects of fertilize use on agricultural

Table 1.7. Mechanization

Area	Population Economically Active in Agriculture in 1992 (×1000)	Tractors			Harvester/Threshers		
		Number of Tractors in 1992	Change Since 1979–1981 (%)	Persons per Tractor	No. of Harvesters/ Threshers in 1992	Change Since 1979–1981 (%)	Persons per Harvester
World	1,116,057	26,137,136	16.4	43	3,861,239	8.8	289
Africa	158,025	554,349	20.4	285	71,797	33.6	2,201
North and Central America	19,878	5,843,151	3.2	34	849,812	−1.3	23
South America	24,080	11,52,142	23.0	21	121,655	18.2	198
Asia	874,897	5,670,108	40.1	154	1,316,934	30.2	664
Europe	20,055	9,864,083	14.4	2	781,663	−4.2	26
Oceania	2,003	401,399	−5.8	5	60,088	−3.7	33
USSR (former)	17,933[a]	2,580,000[a]	−0.6	7	675,300[a]	−10.9	27

Source: Compiled from [8].

[a] Data from 1991.

Table 1.8. Fertilizer use

	Nitrogen use (1,000 metric tons)		
Area	1991–1992	1994–1995	Change (%)
World	75,481	73,599	−2.5
Africa	2,112	2,022	−4.3
North and Central America	13,376	13,534	1.2
South America	1,670	2,360	29.2
Asia	37,908	40,348	6.0
Europe	12,093	11,819	−2.3
Oceania	544	717	24.1

Compiled from [11].

production are not unlimited and, eventually, its use will decline. Table 1.8 shows some figures.

Pesticide Use

Although traditional plant protection methods are still important in developing countries, pesticides have become widely used in the past few decades. There are three types: herbicides (weed killers) fungicides, and insecticides. Their use is influenced by socioeconomic as well as the agroecological factors [10].

The developing countries have a relatively small share in overall pesticide use (about 20% [10], but a very big share in insecticide use (about 50% [10]). For the first figure, there are two explanations: the relatively high costs of pesticides and the relatively low costs for labor. The latter figure can be explained by the higher incidence of insects in the humid tropics [10].

It is expected that pesticide use in the developing countries will still be increasing in the coming decade. This will occur because of the rising labor costs in some countries and the intensification and expansion of agriculture [10]. Through a combination of technological change, improved management and incentives, and increasing application of integrated pest management (IPM), this growth could be contained at fairly low rates [10]. In the industrial countries the declining growth in agricultural production, improved legislation, and a further spread of IPM could lead to a absolute decline in the total use of pesticides [10].

1.3 Agriculture and the Countryside

1.3.1 Description of the Countryside

The countryside or the rural area can be defined roughly as the land outside the cities. More specific definitions are given by, for example, van Lier [12]:

- "rural areas are the areas composed of the (open) fields";
- "rural areas is all the land outside the urban areas, with a low population density."

About 37% of the total land area in the world is in use by agriculture as cropland and permanent pasture [9]. From a global point of view, this means that agricultural activities determine the landscape structure of the countryside. The *Agriculture Dictionary* [13] gives a definition of rural land that shows this direct relationship between agriculture and the countryside. "Land which is occupied by farmers or used for agricultural purposes as distinguished from urban land, park or recreational land, and wilderness."

Although agriculture is the major land-use type in the countryside, it is not the only one. Outdoor recreation, infrastructure, rural housing, and nature, for example, have a certain share in the spatial structure of the countryside. In general, that share is determined by the geographic and socioeconomic situation of a country.

1.3.2 Countryside Values

The human race depends for its existence on the natural environment. The natural environment provides us with air to breathe and the potential to grow food, gather resources for industry, and recover from work. The natural environment can be characterized by *countryside values*: nature, landscape, resources, and ecology.

Nature

A clear and objective definition of nature as a countryside value is difficult. It could be defined as that part of our environment that is not affected by any anthropogenic influences. This definition would limit nature only to the largest jungle, the deepest sea, and the highest mountain and even this would be disputable. A more workable definition is the following: *Nature is that part of the environment that is not dominated by anthropogenic influences and where natural processes are more or less in balance.* This definition excludes, for example, cropland and includes, for example, forest, which also is used for production of wood.

An other approach is not to speak of nature as a countryside value as such, but to speak of *land with a certain nature-value*. This approach is most popular in highly developed and densily populated countries, which have almost no land that is not under more or less total human control.

Landscape

As a countryside value, landscape can be defined as *the spatial cohesion of the environment*. The landscape can be approached in two ways:

- In the anthropogenic approach, landscape has a certain experience value. The importance of this value depends on the place of the countryside in society. In industrial societies and in agricultural societies, the experience value will be different.
- In the ecological approach, landscape has a certain ecological value. This value refers to the diversity of the landscape and thus the presence of different macro- and microenvironments. These environments are interrelated with biodiversity and thus with the ecological health of the countryside (see also "Resources," below).

Resources

As a countryside value, resources can be defined simply as everything useful to man, but not produced by man. This includes water, soil, minerals, food (not produced by farming), knowledge, and gene pools. Some of the resources are unrenewable (minerals and some water sources) or easily destroyed (the tropical rain forest as a gene pool for medical research). In general, resources are more or less sensitive to mismanagement by man and should be managed carefully.

Ecology

As a countryside value, ecology can be defined as the interrelationship between plants and animals and their environment. These interrelationships keep the natural processes running and thus are important for our survival. The ecological health of the countryside is directly connected to the management of nature, landscape, and resources. Mismanagement of these countryside values disturbs the ecological relationships and unbalances the natural processes, with sometimes unrecoverable damage to our natural environment.

1.3.3 Countryside Values and Agriculture

Worldwide, agriculture is the main "user" of the countryside. This means that agricultural development has a large influence on the countryside values.

Since World War II, agriculture has been very successful in increasing production. This development has taken place mainly in the industrialized countries, but many developing countries also have had their share of success.

The increase in output can be attributed to increased output per hectare, increased output per capita, and in some countries, expansion of the cultivated area. These developments were made possible through the introduction of inputs such as chemical fertilizers, pesticides, high-yield varieties of crops, and machinery; management systems such as irrigation and dry farming; and planning systems such as land realotment.

But this is not the end of the story. The increased output has put great pressure on the countryside values and thus on our natural environment. In many areas in the world, natural processes have been seriously disturbed and unrenewable resources used up. The following list gives an impression of the damage done:

- soil erosion,
- soil nutrient mining,
- salinization of soils (waterlogging),
- desertification,
- water contamination,
- eutrophication
- acidification,
- deforestation,
- habitat fragmentation.

To achieve more sustainable agricultural development, good management of the countryside values is necessary.

1.4 Farming: a Balance

The farmer's first interest is to provide for the family livelihood. Farming, though, is no longer just growing crops, raising livestock, and gaining an income. Population growth and environmental hazards ask for solutions relating to food security and management of the natural environment. These are issues directly related to agriculture on both global and local levels.

To meet future challenges of food security, further development of agriculture is necessary. This is not only development in the sense of increasing the agricultural output, but also in the responsible use of natural resources. A responsible use of the natural resources is important because of the dependence of agriculture on these resources. Human innovations such as chemical fertilizers cannot totally replace the natural basis. This means that the natural environment should be treated or managed in such way that the future of farming is secured. Food security is not only a matter of quantity, but also of continuity.

Agriculture thus is forced to find a balance between development and conservation. On a global level, this means the gathering and distribution of knowledge about how to reach this balance. The support of developing countries by industrialized countries is important to achieve this goal. On the local level, this means that existing possibilities for improving farming conditions should be used to achieve a responsible management of the countryside values.

1.5 Sustainability in Farming and the Countryside

The responsible use of natural resources, as mentioned in Section 1.4, also can be described in terms of sustainable development: "Sustainable development is development that meets the needs of the present without compromising the ability of future generations to meet their own needs" [14].

More specifically connecting it with agricultural development, Herdt and Steiner [15] define sustainability as "the result of the relationship between technologies, inputs and management, used on a particular resource base within a given socio-economic context." Following Lynam and Herdt [16], Herdt and Steiner [15] argue for a system approach to sustainability.

An approach to sustainable rural systems from a land-use planner's perspective is given in [12]. In this approach the term sustainability can be viewed, or be defined, from several angles. In most cases, its notion is based upon the protection of our natural resources because of their production and reproduction qualities for now and forever, if properly managed or used. There are, however, more dimensions in the term. Bryden [17] distinguishes at least into three meanings:

- Sustainability in the meaning of husbandry. In this sense, it is related to terms such as continuity, durability, and exploitation of natural resources over long periods of time. It also refers to certain methods by which land is managed—crop rotation systems, fallowing, etc.—all meant to make it possibile to restore the quality and abundance of soil and water systems. This meaning actually refers strongly to the long-run physical and economic sustainability.
- Sustainability in terms of interdependence. As described by Bryden, this meaning is strongly related to the spatial dimension of sustainability. It refers to such aspects as fragmentation (which has contradictory meanings for farming, nature, and outdoor recreation and is therefore an important land-use planning aspect), and relations

between different land uses (e.g. cropped areas and seminatural vegetations). It is this meaning of sustainability that gets a great deal of attention in land-use planning studies, because there is still a great lack of knowledge, there are many uncertainties, and there clear policies often are lacking in this regard.

- Sustainability in terms of ethical obligations to future generations. This refers to the many observed losses and depletions of natural resources in combination with the expected increase in population. In particular, are depletion of fossil fuels and forests, soil losses, water and air pollution, losses of nature areas and of old landscapes, etc. It is clear that, in the field of better management and of restorations, much needs to be done to ensure the future of mankind.

The term rural or rural area is already described in Section 1.3.1. Besides this description in terms of land use, rural areas also are considered to consist of specific local economies and to bear a specific social living pattern. It is than considered to be a specific way of living.

The term system most probably comes from systems analysis, a scientific field that was developed after World War II. In its most elementary definition, in the beginning, it was defined as "a collection of objects, having mutual relations." Because many systems have a relationship with their environment, a distinction was made between open and closed systems. Closed systems are considered to operate outside other systems, whereas open systems depend on other systems. In this case the output of other systems is often an input into another system (and vice versa). Later definitions of a system described it as "a collection of objects, having mutual relations, and so forming an autonomous unity" [18]. In this sense, rural areas can be considered as open systems: They are composed of several objects that are related to one another and form a unity, but often undergo strong influences from the outside world.

Bringing the three words together leads to the following description: *sustainable rural systems are areas outside the urban areas that form a unit and that are composed of specific land uses, in which the activities are performed in such way that a durable situation results regarding the social, economic, and natural properties of the area.*

Scientists interested in the planning and management of land often have to struggle with two, seemingly contradictory, dimensions of sustainability: ecological conservation and economic existence.

The first form of sustainability refers strongly to conservation: to conserve the natural resources (clear water, air, and soils), to preserve plants and animals (biodiversity; gene sources), etc. In many cases, it goes even further than just conservation: it seeks a re-creation of lost values. Examples are the creation of nature areas out of farmland or reforestation of pieces of land. Other examples are: restoration of high water tables in formerly drained lands and finding less intensive uses for meadows, thus restoring bird areas. Many more examples can be given for many parts of the world. Generally, this approach is a clear one, especially in terms of spatial consequences. Sustainability in terms of conservation is focused either on halting certain autonomous developments, retracking on past developments, or a combination of both. It can conflict with the other meaning of sustainability, but does not necessarily do so.

The second meaning, that of a durable socioeconomic existence, is often argued as a very important goal to achieve in order to create a sustainable rural system. In many places across the world, local economies are under strong pressure, notably so in farming. Surplus production, low quality outputs, worsening production conditions (lack of water or other important means), and rising production costs make it even more difficult for many people to survive at a reasonable standard of living in rural areas. This results in such things as outmigration among other effects. This dimension of rural sustainability often is felt when activities concerning land-use planning and management are at stake. A very important task for land-use planning has always been to improve the socioeconomic situation of the rural population.

These views seem to be contradictory. They are to certain degree, but they are also a challenge to mankind. Would it be possible to achieve both ecological and socioeconomic sustainability all at one time? And if so, what strategies would be needed for that?

Several planning and management instruments that can be important factors in sustaining rural systems development are:

- land-use planning (see Section 2.2);
- IPM,
- integrated plant nutrition systems,
- legislation on use of fertilizers (organic and chemical) and pesticides.

It is outside the scope of this Handbook to discuss the whole sustainability concept, but it should be clear that it is the assignment of this generation to strive for sustaining the agricultural and rural system.

References

1. Gove, P. B. (ed.) 1981. *Webster's Third New International Dictionary*, p. 44. Springfield, Mass.: Merrian-Webster.
2. Dalal-Clayton, D. B. 1985. In *Black's Agricultural Dictionary*, 2nd ed., p. 10. London: A&C Black.
3. Grigg, D. B. 1995. *An Introduction to Agricultural Geography*, 2nd ed. London: Routledge.
4. United Nations Development Programme. 1995. Human Development Report, pp. 150–230. New York: Oxford University Press.
5. Grigg, D. B. 1974. *The Agricultural Systems of the World: an Evolutionary Approach*, Cambridge Geographical Studies, No. 5, pp. 1–284. London: Cambridge University Press.
6. Spedding, C. R. W. 1988. *An Introduction to Agricultural Systems*, 2nd ed., pp. 101–129. London: Elsevier Applied Science.
7. Ruthenberg, H. 1971. *Farming Systems in the Tropics*, pp. 252–281. London: Oxford University Press.
8. Food and Agriculture Organization of the United Nations. 1994. Production Yearbook 1993, pp. 215–221, 233–238. Rome: FAO.
9. World Resources Institute. 1994. *World Resources 1994–95*, pp. 285–384. New York: Oxford University Press.
10. Alexandratos, N. (ed.) 1995. *World Agriculture: Towards 2010. An FAO Study*, pp. 1–34, 421–425. Chichester: Food and Agriculture Organization of the United Nations.

11. Food and Agriculture Organization of the United Nations. 1996. *Fertilizer Yearbook 1995*, p. 4. Rome: FAO.
12. van Lier, H. N. 1996. Sustainable rural systems: Concepts from a land use planner's perspective. In: *Geographical Perspectives on Sustainable Rural Systems*, pp. 14–23. Proceedings of the Tsukuba International Conference on the Sustainability of Rural Systems. Kasai Publications, Tokyo, Japan.
13. Herren, R. V., and R. L. Donahue. 1991. *The Agriculture Dictionary*. New York: Delmar.
14. World Commission on Environment and Development. 1987. Our Common Future. Oxford, England: Oxford University Press.
15. Herdt, R. W., and R. A. Steiner. 1995. Agricultural sustainability: Concepts and conundrums. In: *Agricultural Sustainability. Economic, Environmental and Statistical Considerations*, eds. V. Barnett, R. Payne, and R. Steiner, pp. 1–13. Wiley: Chichester.
16. Lynam, J. K., and R. W. Herdt. 1989. Sense and sustainability: Sustainability as an objective in international agriculture research. J. Agric. Econ. 3:381–398.
17. Bryden, J. M. 1994. Some preliminary perspectives on sustainable rural communities. In: Bryden et al (ed.), Towards Sustainable Rural Communities, eds. J. M. Bryden et al., pp. 41–50. The Guelph Seminar Series, Guelph, Ontario: Arkleton Trust and University of Guelph.
18. Hanken, A. F. G., and H. A. Reuver. 1973. Inleiding tot de systeemleer (Introduction to System Analysis). Leiden, The Netherlands: St. Kroese.

2 Land- and Water-Use Planning

2.1 The Planning Issue

M. B. A. Hes and H. N. van Lier

2.1.1 Why Land- and Water-Use Planning Are Important

In Chapter 1, it is stated that a traditional way of increasing agricultural production is by means of land and water engineering. After a few decades of improving the conditions in favor of people and agriculture, a new view of this matter has arisen, namely the use of the resources of this earth in a sustainable way [1]. If we continue to produce in the same way as we have since the start of modern agriculture, the soil will be exhausted and then it will be impossible to produce food, resources for shelter, and other products that are necessary to sustain human health, safety, and welfare. It is, therefore, necessary to look well at how we use this earth and its resources, if we do not want to get ourselves (further) into trouble.

One way to do this, is to watch closely how we use the earth. Good agricultural soils are scarce, and so, we have to use them in the best, most sustainable way. We have to determine how intensively a soil can be used, not only agriculturally, but also for other land uses such as housing and recreation, and how much we will gain from that specific land use, economically and socially. A combination of different land uses should be made in a way that provides for all desired products, and whereby it will still possible to use the land in the same or any other way in the future. This can be achieved through careful planning of the uses of land and water. This chapter explains land- and water-use planning, what the goals are, and the planning process.

First, it is important to explain land- and water-use planning. The phrase can be divided into three parts: land use, water use, and planning. To start with the latter part, "planning" is derived from the verb "to plan" and the noun "plan." The *Oxford Dictionary* [2] gives several distinct meanings of "plan" (n&v): "*1a) a formulated and esp. detailed method by which a thing is to be done; a design or scheme, b) an intention or proposed proceeding, 2) a drawing or diagram made by projection on a horizontal plane, 3) a large-scale detailed map of a town or district, 4a) a table etc. indicating times, places, etc. of intended proceedings, b) a diagram of an arrangement, 5a) an imaginary plane perpendicular to the line of vision and containing the objects shown in a picture.*" The list shows that "plan" refers both to something on paper (1a, 3, 4a, 4b, 5) and to a more abstract concept: an idea that exists in the minds of people who are involved in the planning and that is expressed by them (1a, 1b). In land- and water-use planning, both meanings are involved. The planning process is recorded on paper, both the steps in the process and the proposed development for an area, the land-use plan. The land-use plan consists of two parts:

- a map of the area for which the development zones and other proposed changes and developments are represented;
- an accompanying text that explains the symbols used on the map.

This is discussed in more detail in Section 2.1.3. The wishes of the people involved are not always clear and may not be presented in the actual land-use plan. These ideas are necessary to start the land-use planning process and continue to the end. Without these ideas and thoughts, the process will stop quickly after it has started. The thoughts and wishes are needed to make decisions and to proceed to the next step in the planning process. To be able to make right decisions, as part of the planning process, requires experience in planning.

The dictonary [2] gives for "planned, planning" (-v) the following definitions: "*1) arrange beforehand; form a plan, 2a) design, b) make a plan of (an existing buiding, an area, etc), 3) in accordance with a plan (his planned arrival)* and *4) make plans*." Meanings 1, 2, and 4 refer to the planning itself, the whole process of thinking of solutions to a problem. Meaning 3 refers to planned; the past tense shows that this meaning only exists after the planning is completed. In land- and water-use planning, meaning 3 is a very important part of the whole planning process. After the planning is completed, after the development trends are determined and presented in land-use plans and the physical changes in the area are made, one has to check whether the proposed and intended improvements have taken place, if the changes have been made in accordance with the plan. This is called "evaluation." More can be read about this in Section 2.1.3, Step 10.

In the preceding part of this section, much is said about land use, but what is understood by it? Looking in the dictionary again, "to use" means "*to employ for a purpose, put into action or service*." The noun "use" means "*the act or way of using or fact of being used, the ability or right to use something, the purpose or reason for using something*." These meanings intend that "use" is something introduced by people. Something can only be called "use" when people have meant it to be that way, when some action has been taken to let it be like it is now, to be able to use it in the way for which it was designed and for which it is of any purpose for people. In land use, this intends that only if people have accomplished some action in the land use, can it be called land use. In that sense, agricultural, living, and recreational grounds that are specifically designed for those uses, are examples of land uses. Looking at it in a broader perspective, natural grounds also are a form of land use. They also are of use for people, both economically (e.g., wood) and socially (e.g., enjoyment of the beauty of nature). In this way, natural ground is used, and is therefore an example of land use, although people have not choosen at first hand to use the land in this way. The vegetation arose spontaniously, without intervention by

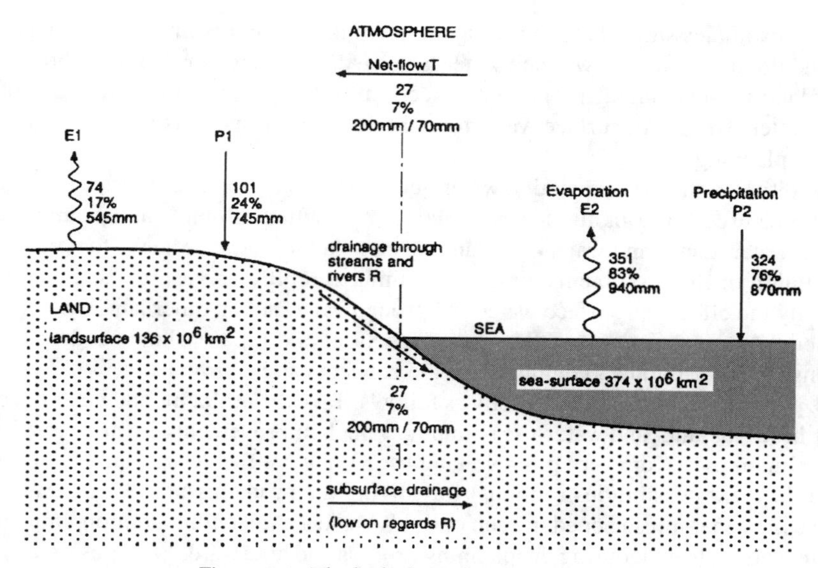

Figure 2.1. The hydrological cycle. ***Source:*** **[4].**

people. It became the land use "nature," by the people's choice to preserve the vegetation and enjoy it.

All activities on the surface of the earth can be called land use. Almost everything can be used by people, socially or ecomomically, and has been given a name. People have chosen to preserve it in a natural state or to try to change it for themselves. As a result, the surface of the earth is assigned a land use.

Water use is a little different from land use. The surface water is sometimes there naturally (oceans, seas, rivers) and is sometimes created by people. Humans are always a part of the hydrological cycle (Fig. 2.1) and therefore have a function or use. A main function of streams is water drainage. Streams discharge the superfluous water from an area and this is stored in the lakes, seas, and oceans. In this way, mankind does not have an active role; it happens by nature. What people can do is interfere and make an area drier or wetter (by artificial drainage or irrigation). In this sense, it can be called water use because a choice has to be made about what to do with this water. People depend a great deal on water: The use it for drinking, irrigation, and in industries. Bigger streams, seas, oceans, and lakes are used for transport and fisheries. It is also in limited supply, especially freshwater. Therefore, it is important to think about what we do with our water and how we do it. In many areas, there is a water shortage, which means that the area cannot function to its best abilities, in both socioeconomic and technical senses. The area cannot produce the products needed for survival of people, such as food, fuel, and shelter. This is the case in deserts and in many semiarid environments. In other areas, there is too much water, which also causes problems. For example, roots of plants drown in groundwater, which causes production loses.

These examples show that the shortage or surplus of water is linked to the land use. The land use determines how much water is needed. This is why water also should be a part of land-use planning [3]. The first concern of water-use planning is the state of the groundwater. The use of surface water for fisheries and transport also can be a topic for land-use planning.

In addition to the water quantity, water quality is also important. If water is polluted, it cannot be used for drinking, for example. Also, a natural ground can be damaged by polluted water. Land- and water-use planning can play important roles in the prevention and solution of these problems. One has to think carefully about proposed land uses, especially the effects on surface water and groundwater and on other land uses.

After these explanations of the main aspects of the phrase "land- and water-use planning," a description and definition of the concept can be provided.

The process can be described briefly as follows: Take the problems and potential of an area as bases for land- and water-use planning. Which improvements are necessary to get the maximum profit for every possible land use? What are the desired land uses? What are the different possibilities (options) for solving the problems in the area? What are the necessary measures needed in each option? Make a choice for an option, implement the plan, and evaluate it during the planning process and afterward. With this description a definition can be made: Land-use planning is the process of systematically describing the problems in an defined area, the way in which the problems can be solved, the combination of these solutions into plan options, and the weighing of these options to come finally to the economically and socially optimal use of the land and its resources.

"Systematically" in the definition means that, in every situation, in principal, the same planning process is followed. In Section 2.1.3, a description of the land-use planning process is given. This cannot be adopted gratuitously in every situation. A land-use planner always has to decide, for every situation, if all of the described steps are necessary, how they must be filled in, and if an extra step is needed. The political situation in a country with its existing laws also will influence the planning process. Some countries have very extensive planning laws, for example, several countries in northwestern Europe and in Asia. The planning laws can order that everyone who has any interest in the land use be consulted or be given the opportunity to participate in the process.

2.1.2 Goals

Goals are important in every planning situation. Often called common goals, they are stated in the planning procedure adopted by the specific nation or lower jurisdiction, such as state, province, or region. If a country has not adopted a planning procedure and a bureau is involved with the planning of the uses of the land, the bureau will have to describe the goals itself. The goals cannot be the same for each country or be stated generally for land-use planning. They also depend on the specific situation and earlier interventions, done by planning or caused by a lack of planning. How land is being used determines the goals of planning. The vision on the development and direction of the land uses is not a topic of the land-use planning process or the bureau in charge of it. Other levels of policy should develop a vision on desirable future land uses. Land-use planning is the way in which the proposed vision can be realized.

Goals can be further divided into objectives and targets [5]. *Objectives* are the more detailed goals of the planning process. They allow the judging of different solutions of a concrete problem in detail in the planning area. They can be clarified after an analysis of the planning area and its problems. Usually the objectives are known beforehand because, after one has seen problems in a certain area, a proposition for land-use planning will be made. This is often even the incentive to start the planning process. In fact, usually it is the effects of the stated problems that make one realize that problems and/or opportunities exist in an area. In the analysis of the area, the cause of the problems, which is what must actually be solved through land-use planning, becomes clear. The objective then is to take away the effects of the stated problems.

The actual problem is the topic of the *targets*, the most detailed goals of land-use planning. They lead to the actual measures that have to be taken in an area to solve the problems, to take away the effects in such a way that it is in accordance with the vision of development. An example may clarify this.

At a certain level, it is stated that the agricultural situation in a specific area of the country should be improved. The crop production is not as high as it should be. This means that a part of farming should be improved. A way to do this is through land-use planning. In planning area X, there is a shortage of water. The production in this area can be improved by irrigation. Perhaps there is no irrigation system or the existing irrigation system does not work well enough. The goal of land-use planning is to improve the crop production. The objective is to improve the water supply. The target is either building an irrigation system or improving the existing one.

If the water used in the irrigation system comes from a stream and this stream also is used for transport of the crops to the market, there should remain enough water in the stream that this latter function will not be lost. With this example, it is shown that goals, objectives, and targets can be conflicting. Therefore, it is necessary to look at alternatives for solving the problems. In this example, it could be that transport of the products is possible through another mode, for example, by building a road to the market. If the costs of this road are less than the loss in production, and if the water from the stream is not used for irrigation, it is more profitable to build the road.

Goals and objectives also identify the best use of land. If two different forms of land use give exactly the same profit (economically and socially, which in practice is hardly ever the case), the goal will determine which of the two land uses should be implemented. For example, an analysis may show that a dairy farm can produce as much as an arable farm on the same area. If the goal is to improve the dairy production, the best land use will be dairy production. If the goal is to improve crop production, the best land use is crop production.

The possible goals of land-use planning depend on the present situation, the earlier measures, and the developments proposed or desired by bureaus (or agencies) of several levels of government. The following describes a few possible goals of land-use planning.

The United Nations Food and Agriculture Organization (FAO) [6] distinguishes three different goals of land-use planning: efficiency, equity and acceptability, and sustainability. Efficiency refers to the economic viability of the land-use plan: The plan should yield more than it costs. However, it is not always clear which land use is the most profitable one; this depends on the point of view. A farmer, for instance, has a different point of view than the government, so they do not necessarily have to agree about which land-use plan is best.

Equity and acceptibility represent the social part of land-use planning. The plan must be accepted by the local population; otherwise, the proposed changes will not take place. People will not cooperate if they do not agree with the plan. Equity refers to the leveling of the standards of living by land-use planning. People living in the planning area have to gain from the land-use plan even if they do not own a farm. Others state that a plan should be a fair and just consideration and treatment for all those affected by a plan or course of action [3].

Sustainability is, as stated before, an important part of land-use planning. It meets the needs of the present while conserving resources for future generations. One should use the present resources to meet today's needs but also reserve the resources to be able to use the land in a different way if needed in the future. Other goals could be [3]:

- Livability. After the land-use plan is implemented, the area should still be a suitable place to live for the people who were already there.
- Amenity. The land-use plan should have provisions for making life pleasant.
- Flexibility and choice. The plan should leave options for individuals to fill in by themselves. The plan should not dictate to people what to do. If this is the case the plan probably will not be accepted, and so, this goal is basically the same as acceptability.
- Public involvement in the planning process. Every group or individual with an interest in the plan should be able to participate in the process. They should be able to defend their interests in the land-use plan, to keep their land use from disappearing through the plan, or to be offered a new land use as part of the plan. This both helps the planners (who could have forgotten a possible land use, or have not thought of the right motives to include the particular land use in the plan) and the interest groups or individuals because they can exert an influence on the plan and are not restricted to only the "hope that everything will be all right."

2.1.3 The Planning Process

The planning process can be divided into several steps. Not every step is equally necessary in all processes and not all steps need equal attention in every new plan. Sometimes, several steps can be integrated into one large step, depending on the situation. The FAO [6] describes the following steps:

- Step 1: Establish goals and terms of reference.
- Step 2: Organize the work.
- Step 3: Analyze the problems.
- Step 4: Identify opportunities for change.
- Step 5: Evaluate land suitability.
- Step 6: Appraise the alternatives.
- Step 7: Choose the best option.
- Step 8: Prepare the land-use plan.

- Step 9: Implement the plan.
- Step 10: Monitor and revise the plan.

Step 1: Establish Goals and Terms of Reference

This step includes more than the title indicates. To start a planning process, the boundaries of the area for which the plan is to be made should be defined. After this is done the analysis of the present situation begins. Not all information needs to be very detailed, but it should give a basic idea of the situation, whether more detailed information is available, and where more information can be obtained. The basic information that is needed includes

- Land resources, i.e., climate, hydrology, geology, landforms, soils, vegetation, fauna, pests, and diseases;
- Present land use, i.e., farming systems, forestry, production levels, and trends;
- Infrastructure, i.e., transport, communication and services to agriculture, livestock management, and forestry;
- Population, i.e., numbers, demographic trends, location of settlements, the role of women, ethnic groups, class structure, leadership;
- Land tenure, i.e., legal and traditional ownership and user rights to land, trees, grazing, forest reserves, national parks;
- Social structure and traditional practices, because the current land use is a result of the history of the area and the culture of the people and when proposing changes, understanding of the current situation is a prerequisite;
- Government, i.e., administrative structure, the agencies that are involved in planning, and the laws and rules that exist;
- Nongovernmental organizations, i.e., farming and marketing cooperatives or others that may have a role in planning or implementing the land-use plan;
- Commercial organizations, i.e., companies that may be affected by the planning.

With the basic information, it becomes clear which (groups of) people can be affected by the land-use planning. All these people or their representatives should be contacted in order to make them aware of the upcoming changes and to obtain a view on the area and its problems from the inside.

Next, the goals of the land-use plan can be established. In Section 2.1.2, the different levels of goals are described. The goals meant here are the objectives of the land-use plan, for which the problems are to be solved. Sometimes it is necessary to establish the goals for the land-use plan, depending on the planning situation of a country or in the area. It can also be very useful to record the goals of the planning in the land-use plan because the goals may not have been recorded previously. Different planning agencies may have prepared different views for the development of the planning area. In the land-use plan, these different goals can be united. This helps to clarify the differences between the views and whether they are conflicting or not.

The terms of reference dictate the limits of the land-use plan. They state what may change, how it may change, and, more specifically sometimes, how it may not change. These limits may be legal, economic, institutional, social, or environmental. Specific conditions also can be included, such as the amount of nature areas in the plan or how much the local population average should benefit, or how much specific changes may cost. These may be derived from the goals set out by the different planning agencies. The total budget for the land-use plan and the implementation period (the length of time for which the plan will operate) are also terms of reference. The terms of reference limit the possible alternative land-use plans. If in the planning process the terms of reference are excluded, many options may be considered that are not feasible. To limit the number of alternatives and to save time when creating them, the terms of reference are described in the plan.

Step 2: Organize the Work

The work must be organized to coordinate the different activities needed. Some activities have a long lead time and should be done in time in order to be able to proceed with the next step. When, for example, some specific information is needed in a step, the garthering of this information should be finished on time in the previous step. If this is done too late, then it will cause a delay in the planning process.

A list of needed activities and planning tasks should be made in order to organize the work. This list should include the organizations and people responsible, skilled personnel and other resources needed, and the estimated time needed for each task. The tasks and activities should be placed on the list in the order necessary to complete the plan. Now that all tasks and activities are clear, the planning team can be completed (if this has not yet been done). The planning process and work plan can be drawn up as a whole. This can be done in different formats: a planning table (Table 2.1), a bar chart (Fig. 2.2), or a critical-path chart (Fig. 2.3).

Table 2.1. Example of a planning table: Land-use plan for District X

	Planning Step	Task	Resources	Responsibility	Due date
1	First meeting	Identify Participants	Director, Decisionmaker	J. Cruz	09/01/97
		Assemble materials	Agency library, five-year plan, national database	E. J. Evans	09/15/97
		Arrange venue, support staff, transport	Administration unit, motor pool	M. Wong	09/30/97
3	Structure problems and opportunities	Develop questionnaire	Regional statistican, consultant on public involvement	S. Moe (with E. J. Hoover)	02/01/98
3.1	Problem statements	Identify and interview key people	Contact list, interview forms, team vehicle field assistant	T. F. Guy	03/20/98
		Prepare problem statements	Interview data	T. F. Guy S. Moe	04/01/98
3.2	Find options for change	Set benchmarks	Land resources survey (1995), district agronomist, team vehicle	S. Moe (with M. Wong)	05/05/98
		Summarize regulations	Agency code book, law clerk	F. Sims	05/30/98

Source: [6].

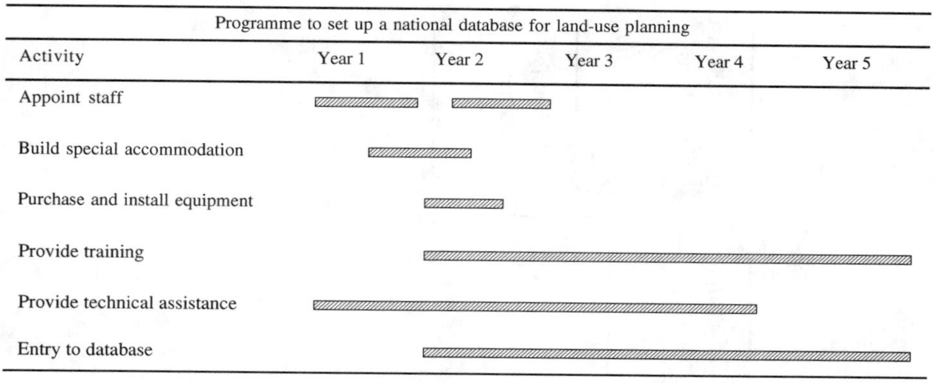

Programme to set up a national database for land-use planning

Activity	Year 1	Year 2	Year 3	Year 4	Year 5
Appoint staff					
Build special accommodation					
Purchase and install equipment					
Provide training					
Provide technical assistance					
Entry to database					

Figure 2.2. Example of a bar chart. ***Source:*** **[6].**

Each activity and task should be assigned to a member of the planning team to make sure all activities and tasks are done and to make sure every member knows his or her responsibility. This assignment of activities and tasks should be drawn up in individual, personal work plans. The work plan for the whole project should also include when meetings of the planning team need to be held. In this way it becomes clear whether all scheduled activities and tasks are done and if there are any delays. When these meetings point out that the work plan needs to be changed, the individual work plans need to be changed as well.

In the work plan the money and equipment should be allocated. Each activity gets a budget and the resources needed to complete the task are drawn up. Administrative matters and logistics also should be arranged, such as transport, equipment and office facilities for the planning team, technical support (e.g., inputs from other agencies, field assistance, laboratory and secretarial workers). Seasonal differences, holidays, contingencies, and iteration should be kept in mind while making the work plan.

Step 3: Analyze the Problems

In this step, the current situation is studied carefully. The information gathered is in much more detail than the basic facts gathered in step 1. It encompasses all information that will be needed in subsequent steps, up to implementation. Data should be collected on population (numbers, age, gender, trends, and distribution); land resources (any data relevant to the planning task, e.g., landforms, climate, agroclimatic regions, soils, vegetation, pasture resources, forests, and wildlife); employment and income (summarized by area, age, social and ethnic groups); current land use, production, and trends (tabulated production data, graphic production trends, and economic projections for the planning period, as quantitative as possible); and infrastructure (roads, markets, and service centers). Maps should be made where possible.

Step 1 should have clarified what information is already available and what should be gathered by surveys. Surveys take more time than the gathering of available information, and allowance for this time should be included in the work plan.

The planning area can be split into land units, that is, areas that are relatively homogeneous with respect to climate, landforms, soils, and vegetation. Each land unit presents

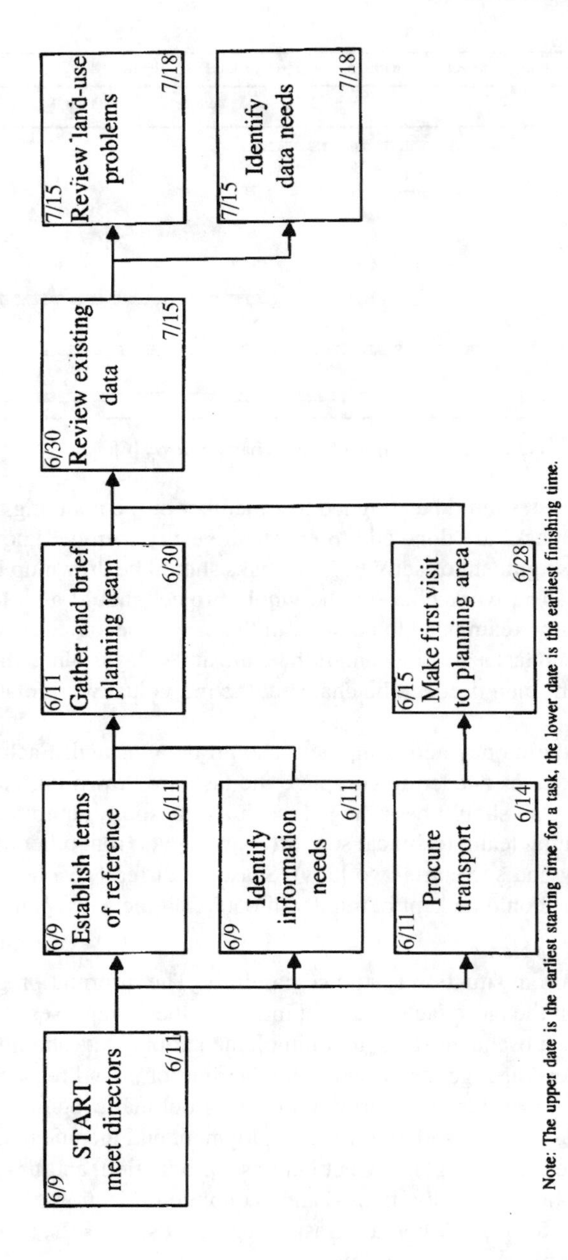

Note: The upper date is the earliest starting time for a task, the lower date is the earliest finishing time.

Figure 2.3. Example of a critical-path chart. ***Source:*** **[6].**

similar problems and opportunities and will respond in similar ways to management [6]. Then, land-use systems can be identified: areas with similar land use and economy, based on farming systems, the dominant crops, size of the farms, or the presence of lifestock.

The identification of the problems of land use is necessary to be able to improve the situation. There are several methods of identifying the problems, such as farming systems analysis (described, for example, in [7] and [8]), diagnosis and design (D&D, as described in [9] and [10]), and rapid rural appraisal (see [11], [12], and [13]). They are all based on interviewing a sample of rural land users, preferably stratified according to identified classes of farming systems. However, they are centered on different aims. Farming systems analysis is used to identify problems at the farm level in order to adapt technologies for specific farming systems. In D&D, the problems with land-use systems are identified and their causes analyzed (diagnosis), after which new agroforestry land-use types are designed to solve the problems. In rapid rural appraisal, the existing land-use systems are analyzed in a short period of time (a number of weeks), including the problems of the current systems.

The last part of the problem analysis is to prepare problem statements. For each problem, they describe its nature and severity and its short-term and long-term effects, and provide a summary of its causes.

Step 4: Identify Opportunities for Change

In this step, possible ways to solve or ameliorate the problems analyzed in step 3 are formulated. All possibilities should be included in this step; then, the most promising and most feasible solutions should be analyzed further. When making a range of options for solving the problems, possible changes include

- People. They can contribute labor, skills, and culture. Cooperation of the people in the planning area is necessary for success of the land-use plan.
- Land resources. Regions may be underdeveloped, or resources may be unexploited. Land often has the ability to produce more or new crops with a change in management. Also new crops or land-use systems can be implemented to improve the current situation.
- Technology. New or improved technologies can increase production, e.g., new fertilizers, pesticides and irrigation and drainage techniques.
- Economic measures. New sources of capital, new or improved markets, changes to price structures, or improvement of transport and communications may offer opportunities for change.
- Government action. Possibilities include reform of land tenure and administrative structure, taxation policies, pricing policies, subsidies, and investment.

Not all of the above mentioned solutions are part of land-use planning, but they make it possible to isolate problems that can be better solved through other means of action. Also, different land-use strategies can be followed: no change, maximum production, minimum investment, maximum conservation, or maximum equity. Options for solving problems also can be generated in terms of different kinds of production, the role of conservation, and self-reliance versus external investment. What is important when designing the solutions is to keep all interested parties informed and to seek their views.

When all possibilities have been identified, realistic options that best meet the needs of production, conservation, and sustainability and that minimize conflicts in land use can be developed. The number of options can be limited by social imperatives, budgetary and administrative constraints, the demands of competing land uses, and an initial assessment of land suitability.

The problem statements and the alternatives for change should be presented in terms suitable for public and executive discussion: clear, brief summaries but with detailed evidence available for scrutiny. Now, the question is whether the original goals still appear to be attainable. If this is the case, the decision has to be made about which problems are to be given priority and which are the most promising alternatives for a feasibility study. It is possible that action is needed at other levels of land-use planning or outside the scope of land-use planning. After making these decisions, targets for this subsequent work must be specified. Subsequent steps may be more specifically planned than before. This is in fact a partial reiteration of step 2. It also could be necessary to prepare an additional or a revised budget and time schedule.

Step 5: Evaluate Land Suitability

For each of the promising land uses identified in step 4, the land requirements have to be established and matched with the properties of the land in the planning area to establish physical land suitability. To do this, the land-use types should be described in terms of their products and management practices. Depending on the geopolitical level, the descriptions should be more or less detailed; that is, at district and local levels, more detail is necessary than at national levels.

Next, the land-use requirements are described by the land qualities (e.g., availability of water and nutrients) necessary for sustained optimal production. Most land qualities are determined by the interaction of several land characteristics, that is, measurable attributes of the land.

The land units identified in step 3 should be mapped in more detail when necessary. Land units are choosen because they are expected to respond to management in a relatively similar ways at similar scales of study. The need to carry out original surveys depends on the need for and availability of these data.

Now, the requirements of the land-use types can be compared to the properties of the land units. Each land unit can be put into a land suitability class for each land-use type. Table 2.2 gives the structure of the standard land suitability classification used by the FAO. The limiting values of land quality or a land characteristic determine the class limits of land suitability for a certain land use. First, a determination has to be made as to whether the land unit is suitable or not suitable. The important criteria used in this decision are sustainability and ratio of benefits and costs.

After the land suitability classes of the different land units are determined, the matching of land use with the land quality starts. Compare the requirements of each land-use

Table 2.2. Structure of FAO land suitability classification

Code	Classification	Description
S	SUITABLE	The land can support the land use indefinitely and benefits justify inputs.
S1	Highly suitable	Land without significant limitations. Include the best 20%–30% of suitable land as S1. This land is not perfect but is the best that can be hoped for.
S2	Moderately suitable	Land that is clearly suitable but that has limitations that either reduce productivity or increase the inputs needed to sustain productivity compared with those needed on S1 land.
S2e[a]		Land assessed as S2 on account of limitation of erosion hazard.
S2w		Land assessed as S2 on account of inadequate availability of water.
S3	Marginally suitable	Land with limitations so severe that benefits are reduced and/or the inputs needed to sustain production are increased so that this cost is only marginally justified.
N	NOT SUITABLE	Land that cannot support the land use on a sustained basis, or land on which benefits do not justify necessary inputs.
N1	Currently not suitable	Land with limitations to sustained use that cannot be overcome at a currently acceptable cost.
N2	Permanently not suitable	Land with limitations to sustained use that cannot be overcome.
N2e[a]		Land assessed as N2 on account of limitation of erosion hazard.

[a] There is no standard system for letter designations of limitations; first-letter reminders should be used where possible.
Source: [6].

type with the qualities of each land unit. Check measured values of quality or characteristics against the class limits and allocate each land unit to its land suitability class according to the severest limitation. Consider which modifications to the land-use type will be most suitable. Also, consider which land improvements could make the land better suited for the type of land use. Land can be made physically suitable for many types of land use. Sometimes, however, these technical changes are so intense that the land is no longer sustainable or the change is not economically feasible. In those cases, the proposed land use is not possible on that specific piece of land. One also can argue about which interferences are still contributing to a sustainable environment and which are not.

The last part of this step concerns the mapping of the land suitability, which shows the suitability of each land unit for each land-use type.

Step 6: Appraise the Alternatives Through Environmental, Economic, and Social Analysis

In this step, a number of studies are carried out. They refer first to individual combinations of land uses to which land units, classified as suitable, can be put and, second, to alternative combinations of land uses that are being considered in the plan. These proposed combinations can be considered as the alternatives among which the choices should be made. The following types of analyses are made:

- Environmental impact assessment. An assessment will help to determine what will happen under each alternative system of management in terms of the quality of the whole community. It also should consider effects both within and beyond the planning area. Examples of environmental considerations are soil and water resources, pasture and forest resources, quality of wildlife habitat, and scenic and recreational values.
- Financial analysis. This analysis considers the profitability from the point of view of the farmer or other private investor, by comparing the producers' revenues with their costs. The proposed land-use types should be profitable for the farmer or other land users.
- Economic analysis. This analysis estimates the value of a system of land use to the community as a whole. In addition, the monetary value of clear economic consequences of environmental effects can be estimated and included in the economic analysis. This also holds true for the unintended side-effects, such as damage done to the environment.
- Social impact. An assessment studies the effects of proposed changes on different groups of people. Examples of social factors are population, basic needs of the people, employment and income opportunities, land tenure and customary rights, administrative structure and legislation, and community stability.
- Strategic planning. A strategic study will help to determine how the proposed changes in land use affect wider aspects of rural development planning, including national goals. The critical importance of land for specified uses has to be clear. This means that, besides the economic and physical suitability, the use of a specific area in a particular way has to be taken into account.

Step 7: Choose the Best Option

First, a series of options for the allocation or recommendation of land-use types to land units should be set out. Their evaluation in terms of land suitability and environmental, financial, economic, and social analysis should be stated. For each alternative, all other consequences should be listed, including the advantages and disadvantages of every possible combination of land uses. Advantages and disadvantages are not always very easy to list: some effects might be favorable for one and unfavorable for another. It is not really necessary to work out every alternative in great detail; this is a lot of work and does not really make a difference when comparing them. Enough data should be available to make a fair comparison. The alternatives should be in the same level of detail.

All proposed alternatives should meet the goals and objectives set earlier and the terms of reference. If an alternative does not correspond to those, the proposal cannot be an alternative.

One of the alternatives should be a description of what will happen if nothing is done at this point in time. This alternative, usually called the zero, no-action, or steady-state alternative, can be used as a reference. The steady-state alternative is actually a wrong name. It implies that nothing will happen in the future, but it actually describes what will happen if no planning is done now. Therefore, it is also called the autonomous development alternative. Usually, this zero-alternative is not a realistic option but helps the planning team to make clear what is actually gained by carrying out a land-use plan. It points out the relative difference between carrying out and not carrying out a plan.

Next, the options or alternatives and their effects should be presented in a way that is appropriate for review. Make arrangements for public and executive discussions of the viable options and their effects. These are necessary to safeguard the public involvement in the planning process. The people who should be able to take part in these discussions are from the communities affected and the implementing agencies. These people now have the opportunity to find out in detail what the plan is designed to achieve and how it will affect them. Planners should allow adequate time for reviews and comments, and should obtain views about feasibility and acceptability. Sometimes the discussions have as a result a new alternative, being a combination of two existing ones, or a completely new one. This new alternative also should have a fair chance in the comparison and, therefore, it is necessary to go back to step 5 (and sometimes even further if not all data are available to work out the new alternative into the same level of detail as the other alternatives).

With the comments from the public, any necessary changes to the options have to be made. Now the real decision about the options or alternatives can be made. Sometimes it is completely clear which alternative should be chosen. If this is not the case, an objective method of weighing alternatives, such as multicriteria analysis (MCA), can be used (this method is described in more detail in Section 2.1.5).

Finally, the subsequent steps must be authorized. The source of this authorization depends on the level at which the planning is being done, e.g., local or national. At the local level, it might be an executive decision; at national level, it might require a decision at the highest level of government.

Step 8: Prepare the Land-Use Plan

After an alternative is selected, the land-use plan has to be worked out in great detail. The plan consists of two parts: the map of the planning area with the location of the proposed land uses and several supporting maps; and a report describing what these different land uses look like, how the needed changes will have to be made, and when and how all of this will be put into practice. The report also contains a summary of all results from the previous steps. Next to presenting the plan, the report also has the function of preparing the plan for implementation. Therefore, the report consists of three elements: what should be done, how it should be done, and the reasons for the decisions.

An important part for those who need to know what is to be done next is the description of the land-use allocations or recommendations, in summary form and then in more detail. Here, the selected option is set out without confusing the reader by references to rejected alternatives. The selected land-use types, including their management specifications and the land units for which they are recommended, are described.

Then, the reasons for choices and decisions must be given, in outline and in some detail. Funding agencies need these explanations if they want to review the soundness of the proposals from technical, economic, or other points of view.

Next, practical details for implemention must be considered: deciding the means, assigning responsibility for getting the job done, and making a timetable for implementation. In large plans, it might be wise to divide the plan into phases. In this case, a map is made for the first phase, for the second, and so on. The needs for land improvements are itemized, including supporting services, physical infrastructure, and credit and other internal financial services. The needed inputs are based on the phases and the management specifications for the land-use types. Land improvements, such as engineering works, are ranked. Extension programs and incentives are planned. Responsible parties are identified for each activity. Adequate arrangements for financing staff costs, inputs, and credit are ensured. Particular attention must be paid to providing for maintenance of all capital works. Details of the arrangements to be discussed with the decisionmaker and relevant agency staff in terms of feasibility and acceptability and the availability of advisory staff, logistic support, and supervision. The need for staff training is assessed. The necessary arrangements for research, within the plan or through outside agencies, are made. A procedure for reviewing the plan's progress is established. The financing needed for each operation and the sources of funds are determined. Policy guidelines and any necessary legislation are drafted.

Because of the wide range of readership and what they want to read in the report of the land-use plan, the report is usually divided in the folowing sections:

- Executive summary. Written for nontechnical decisionmakers, it is a summary of the land-use situation, its problems and opportunities, and the recommendations for action. Reasons for decisions are given briefly. Clear concise writing is of the highest importance. This section should include at least one key map, the (master) land-use plan and possibly other maps at small scales. It is typically 20 to 50 pages long at the most.
- Main report. It explains the methods, findings, and factual basis of the plan. Written for technical and planning staff who want to know details, it includes reasons for decisions and often is 5 to 10 times longer than the executive summary.
- Maps volume. This is an integral part of the main report, presented seperately for convenience of binding.
- Appendices. These provide the technical data that support the main report. These may run to several volumes. They include the results from the original surveys conducted as part of the plan, e.g., soil surveys, forest inventories, records of river flow.

Table 2.3 gives an example of the contents of a report for a land-use plan.

Because not all of the people that may need to be informed may (be able to) read the full land-use plan report, a range of public information support documents may have to be created. The support documents will inform interested parties about the plan, its relevance, the benefits to the community as a whole and the participation needed from different sections of the community.

Step 9: Implement the Plan

Implementation is a step of a totally different nature than the other steps, but because the objective of the land-use planning process is to identify and put into practice beneficial land-use changes, it is considered a part of the whole process. Implementation involves a

Table 2.3. Example of contents of a land-use plan report

Title[a]
- Land-use plan for. . .

Summary
- Highlights of problems, recommendations, and the main reason for these recommendations.

Introduction
- Long-term goals for the planning area and the purpose of the plan.
- Relationship to other documents; briefly describes legislation and any higher-level plans as well as local plans that are related to this plan.
- Description of the planning area; brief overview of location, area, population, land resources, current land use and production.

Management problems and opportunities
- Statement of land-use problem and opportunities.
- Rationale for the selected option.
- Summary of the changes the plan will bring about, by subject area or geographical area.

Direction
- List of land-use types standards that apply to the whole planning area and to individual planning units.
- Identification of projects, illustrated with maps and diagrams.
- Timescale for action.

Monitoring and revision
- Description of procedure for reviewing progress and revising the plan.

Work plan for implementation
- List of individual project with details of location, time, resources required, and responsibility for implementation.

Appendices
- supporting information:
  - physical environment, planning units, agroclimate, and soil data
  - population, settlement, infrastructure, tenure
  - present land use
  - land-use types and land requirements
  - land suitability
  - economic projections

[a] Until the plan has been approved by the decisionmaker, it is a "proposed land-use plan."
Source: [6].

wide range of practical activities, many of which lie beyond the scope of this overview of the land-use planning steps. The following implementation strategies refer specially to the roles that the planning team may undertake. Depending on the level of planning, the team has different roles. On the national level, it supplies information to the government as a basis for decisions. At the local level, the planning team may draw up detailed plans for implementation while leaving other agencies to put the plan into action. The focus of the planning team could be:

- Ensure that the changes and measures recommended in the plan are correctly understood and put into practice. Be available for technical consultations. Discuss with implementing agencies any suggested modifications.

- Help to maintain communications among all people and institutions participating in or affected by the plan, i.e., land users, sectoral agencies, governments, nongovernmental organizations, commercial organizations. A part of this is also the explanation of the land-use situation and plan to the media, at public meetings, and in schools.
- Assist in coordination of the activities of the implementing agencies.
- Assist in institution building by strenghening links between existing institutions, forming new bodies where necessary, and strengthening cooperation.
- Focus on the participation of the land users. Ensure adequate incentives.
- Organize research in association with the plan. Ensure that results from research are communicated and, where appropriate, incorporated into the plan.
- Arrange for education and training of project staff and land users.

Step 10: Monitor and Revise the Plan

In this step, it becomes clear how well the plan is being implemented and whether it is succeeding. The implementing agencies can still modify the "strategy for implementing the plan" or just "modify the plan," if necessary, before the full plan has been applied. The planning team may learn from experience and respond to changing conditions. When monitoring the plan, implementation data are collected to discern whether the land-use activities are being carried out as planned, if the effects and costs are as predicted, whether the assumptions on which the plan has been based have proven to be correct, whether the goals are still valid, and how far a long the goals are toward being achieved. One should keep in mind that the analysis and the action are more important than the gathering of the data. The more time spent on gathering data, the less time there is for adjustments, and the greater the loss or the failure may be. Monitoring also may involve observations at key sites, regular extension visits, and discussions with officials and land users. These periodic checkups make clear whether the goals are being met also in the long term.

There are many possible reasons for failure of the plan: It could have been based on the wrong assumptions, there may be changes in economic circumstances, the logistics of the implementation could be failing, and there could be a problem of communication and participation. Try to find solutions for these problems. Initiate modifications or revisions of the plan: Minor modifications can be made through action by implementing agencies, and larger revisions can be made by the preparation of proposals and reference back to decisionmakers. Continuous minor revisions are preferrable because more substantial changes can lead to delays.

The focus monitoring will change with the passing of years. In the first period, during and immediately after implementation of the plan, results will become visible, for example, new roads, water supplies, job opportunities. The second stage, consisting of extension and maintenance and operation of capital works, is harder to monitor. This transition is difficult and so, the latter phase calls for even more effective and willing cooperation between implementing agencies and land users.

In several of the above steps throughout the planning process, contact must be established with local people in the planning area. This is done to make sure that they are involved in the planning process. Without their support and involvement, the plans likely will not succeed [6]. This is because many changes will have to be made voluntarily by the people, such as changing crops. For example, through analysis, the planning team, determines that wheat will provide more income for farmers than potatoes in a certain area. In the land-use plan, it is stated that the crop to be grown will be wheat, not potatoes. The individual farmers in the planning area have to grow wheat to make the plan succeed. When the farmers are not willing to do that, their income will not increase and the planning will have failed. In addition, the farmers may have had a very good reason not to grow wheat; they may have tried it before and their yield and/or income dropped and so, they switched back to potatoes. If the planning team had asked the farmers about their ideas, they would have found this out earlier so that another way to increase the farmers' income could have been explored .

Too often in land-use planning, a top-down approach is followed. This means that the government starts the planning process and fails to integrate the local people into it ([14], [15]). Thus, local people have no opportunity to participate in the development of their area. There is no link between the people and the planning agencies. The local people usually are the most knowledgeable about major problems and constraints as well as opportunities, in their area. Therefore for more successful planning, the more holistic, bottom-up approach should be followed. Not only will local people be involved in the planning process, but it is more likely that all relevant issues will be integrated in the planning.

More about the importance of people's participation can be found in an FAO report [16].

2.1.4 Levels of Planning

In the description of the steps, it seems that there is only one level on which land-use planning can take place, that of a relatively small area. This is not really the case; there are some levels above and beneath the local level. The following seven major levels are distinguished by the United Nations Economic and Social Council [15]:

- global,
- regional,
- national,
- provincal/district,
- local,
- municipal/village,
- household/farm.

The following five are distinguished by the FAO [14]:

- international,
- national,
- district/local/government/subnational,
- local community/watershed/ecosystem,
- primary land user.

Global is joined with regional and called international and local is partly joined with provincial/district and partly with municipal/village. It is clear that the level of land-use planning is not always easy to establish.

How rigidly the steps that describe the land-use planning can be followed also depends on the level. On the global or international level, for example, not all people involved can be asked about their development ideas. Planning also is not as detailed at higher levels as at lower levels. At higher levels, a global direction of development is given for large areas; for example, world-scale parts of South America are designated as tropical rainforest and parts of Africa as desert. This does not mean that other land uses are not allowed; it only indicates the importance of the conservation of these types of land use in these areas or, in the case of the deserts, that very little can be done to change the present situation. Also, the boundaries of the areas or not fixed; the patches on the map only give an indication of the location of the rainforests and deserts.

On the regional or national level, the planning area is already a lot smaller and therefore the planning is more detailed. The edges of the patches better indicate what kind of land use is planned for which area, although the boundaries still are not fixed on the map. The steps of the planning procedure cannot be followed in every detail but participation is more important than on a higher level.

On the lowest level of land-use planning—the household, farm, or primary land user—planning the use of land generally is called "management of the land." Whereas the use of the land is determined at a higher level, on the lowest level, only decisions on how to implement the plan have to be made.

Very little can be said about the exact scale of the maps that can be used at the different levels. On a global or international level, the maps will be of a smaller scale; conversely, maps that show a small area are of a larger scale.

On the different levels, the land uses are not described in the same detail. Going from a small scale (global/international level) to a larger scale (local level), the land uses will be more specific. For example, agriculture can be divided into arable land and pasture. Arable land use can be specified as crop and pasture by cattle. It is not always possible, desirable, or even necessary to describe the land uses in much detail on higher levels. The data that are required to make a more detailed distinction within a land use are not always available on smaller scales. But even if these data are available, making such a detailed plan allows the planners on a lower level little space in further detailing the land use.

2.1.5 Knowledge Systems

Knowledge systems are techniques or methods that support the land-use planning process. Sometimes, they can be used in different steps of the process but some are especially designed for one particular step. Not all knowledge systems or available methods are described in this book. Also, a general description is given of the knowledge systems, but there are substantial handbooks available for each knowledge system, sometimes even a different one for each possible application. The most important knowledge systems that are used in land-use planning are described shortly.

Geographical Information Systems (GIS)

GIS are very important knowledge systems in land-use planning. They are used widely for multiple purposes in the process. A GIS can be described in many different ways but all descriptions deal with georeferenced or spatial data. Standard (commercial) software or specific applications are available for dealing with these spatial data [17]. Within GIS the real world and its elements are transformed and formalized into spatial data, which then are processed into information from which results are presented. A GIS can be cell based or vector based. In the cell-based systems, all data are stored in grids. The grids together form one layer in which an area of the real world is represented. In different layers (maps) different kinds of data can be stored. The size of the grids defines the accuracy of the representation, smaller grids (grids that represent a smaller piece of the modeled area) give a better representation of reality than larger grids.

In the vector-based systems the data are stored as points, polygons, and lines. In this format the real world can be represented more accurately. A real point can be depicted as a point and a line as a line. This is not possible in a grid-based format. Points and lines then are depicted as a small area of a certain size, namely the size of one (point) or more cells (lines). Figure 2.4 shows how points, lines, and polygons are represented in grid-based and vector-based formats. Just as in the grid-based format, different sorts of information can be stored in different maps in vector-based formats. This structure of the GIS allows different kinds of information to be recalled and combined.

A GIS can be very useful when a large amount of data must be gathered and stored. With a GIS, this information can be retrieved and processed more easily than when it has to be done manually. It is especially helpful in performing land suitability analyses, described in step 5 of the planning process.

Systems Analysis

The use of system analysis in land-use planning is described in [18]. In systems analysis, interrelated processes are analyzed and modeled. In [19] and other references, it has been defined as a general framework of thoughts that creates the possibility to project specific problems as seen from a general background. This description is rather general. It shows that system analysis is applicable to many disiplines and that an interdisciplinary approach is possible as well. Land-use planning deals with concrete systems. A concrete system is a limited and coherent part of the real world (e.g., a technical instrument, a farm, a local community, or a land-use planning area). Such a system is, by definition, limited. The type of system determines whether it is open or closed. An open system interacts with its surroundings; a closed system does not interact. Land-use planning areas are (predominantly) open systems: They have an input (influence of their surroundings on them) and an output (their influence on their surroundings).

There are at least three reasons to apply systems analysis in land-use planning:

- to encourage a more holistic perspective, rather than the reductionistic view from which science usually works;
- to try to learn more about the most important aspects of a system and to focus further research upon these aspects;
- to learn about the interaction of the different aspects and, where applicable, to promote the study of these interactions.

An objective of a systems analysis approach in land-use planning is "to learn about the meaning of planning in the functioning of rural areas so that priorities can be given

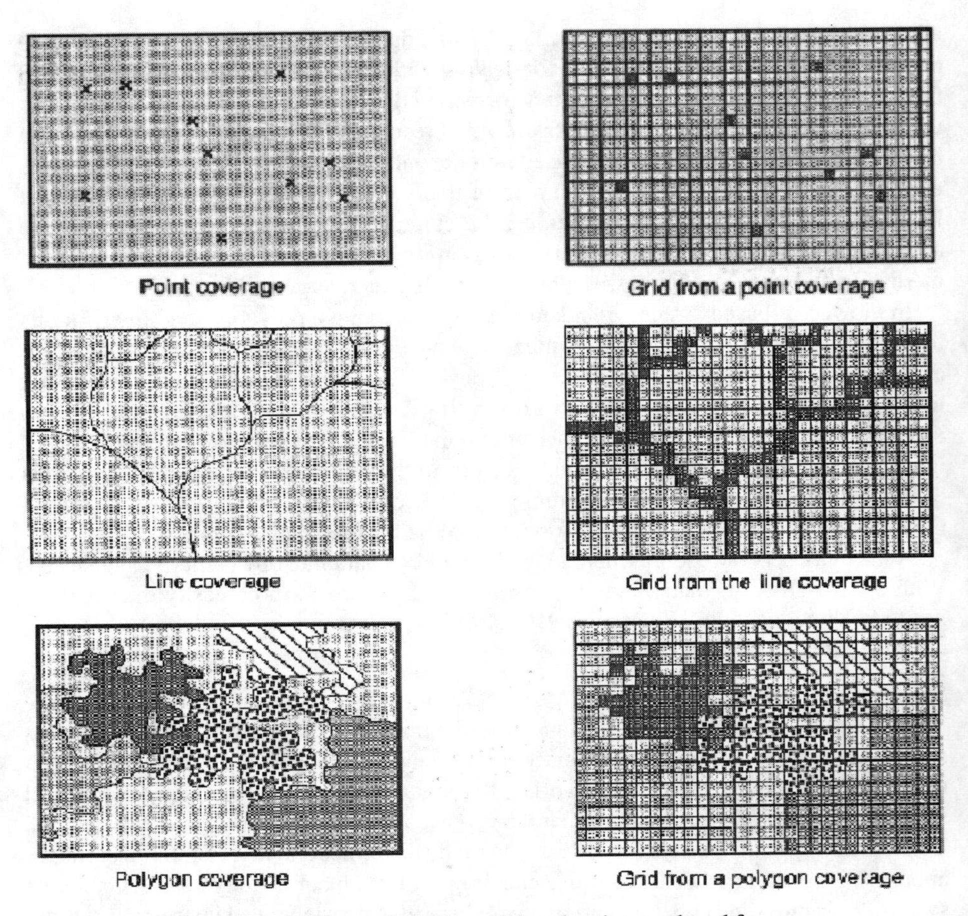

Figure 2.4. Difference between grid-based and vector-based formats.

for research as well as that the interaction between different studies can be clarified." Systems analysis in land-use planning should lead to knowledge about interactions, for example, between the different land-use types; knowledge about the most striking problems in the system; a determination of what knowledge is lacking; and the setting of priorities and promotion of new research. Research in land-use planning is based almost entirely on field research. A systems analysis approach to land-use planning in specific rural areas, therefore, is predominantly based upon properties, land use, and problems of that specific areas. Systems analysis of a specific planning area may study the different land-use types in such a way that the long-term effects of changes on area properties, often based upon spatial concepts, can be given for the different land uses.

Systems analysis is not new. It has been used in many fields throughout the past 50 years. It also has been used in land-use planning. An example of this is the research described by Jorjani [20]. The effects of changes in a rural water management system were analyzed from a total systems approach, as Fig. 2.5 shows. The approach clarifies many things. First, it demonstrates that working with systems is a complicated approach. Figure 2.5 shows, for example, that answering the question "What are the effects of changing the water (management) system?" leads to 33 knowledge fields. Second, it makes clear that teamwork is needed. Third, it probably is impossible, despite teamwork and an interdisciplinary approach, to cover everything: priorities have to be set and choices made. In the case of the drainage system study [20], which took four years, the researcher could study only line 1-2-3-4-5-6-7-8 (see numbered boxes in Fig. 2.5), with some references to 27-28 and 14-15-16.

Cost-Benefit Analysis (CBA)

Financial and economic analyses are important parts of the land-use planning process. A land-use project, like any other project, can be implemented only if the total benefits exceed the total costs. This has to be true for the project as a whole (economic analysis) as well as for the individual land users in the project area (financial analysis). It is not always possible or necessary to perform a financial analysis for every individual land user but it has to be clear that most people—not just a few—will benefit from the project. The problem with financial and economic analyses is that the total benefits and total costs cannot simply be added. The costs usually occur in the first year(s) of the project whereas the benefits are spread out over a longer period of time and usually occur later in the project (e.g., when the project is halfway through implementation or after it is completed). The amounts have to be discounted to a standard year.

After this is done, there are several methods for comparing the costs and benefits. These methods, as well as the other aspects of CBA in an agricultural project, are explained in the literature [21].

Multicriteria Analysis

The decision to implement a particular land-use plan or a certain alternative that is designed within a project is not based only on the economic and financial analyses. Other criteria, which have to be established earlier in the project, play an important role. Each criterion usually has a specific format and therefore cannot just be compared to each other criteria or added to obtain "score" of a plan or alternative. To make a fair comparison of criteria and alternatives, MCA can be used. It consists of a few steps [22]: A table is made with the value of each criterion for every alternative. This is called the criterion score matrix. Each criterion has different units in which it is measured. To be able to compare them, the values have to be transformed and standardized. To transform means to express the qualitative criteria in a figure (by valuing the qualitative items). To standardize means to express the criteria in equal units with the help of the same formula. All scores are expressed on the same scale so that the criteria are comparible among themselves. The converted criterion scores are put into an effectivity matrix. In the next step, called priority standing, the criteria are weighted. Then it is possible to state which criteria are considered more important than the others. The distribution of weights is represented in the priority matrix. The last step is to multiply the converted criterion scores by the weights and add all values to calculate a total score for each alternative.

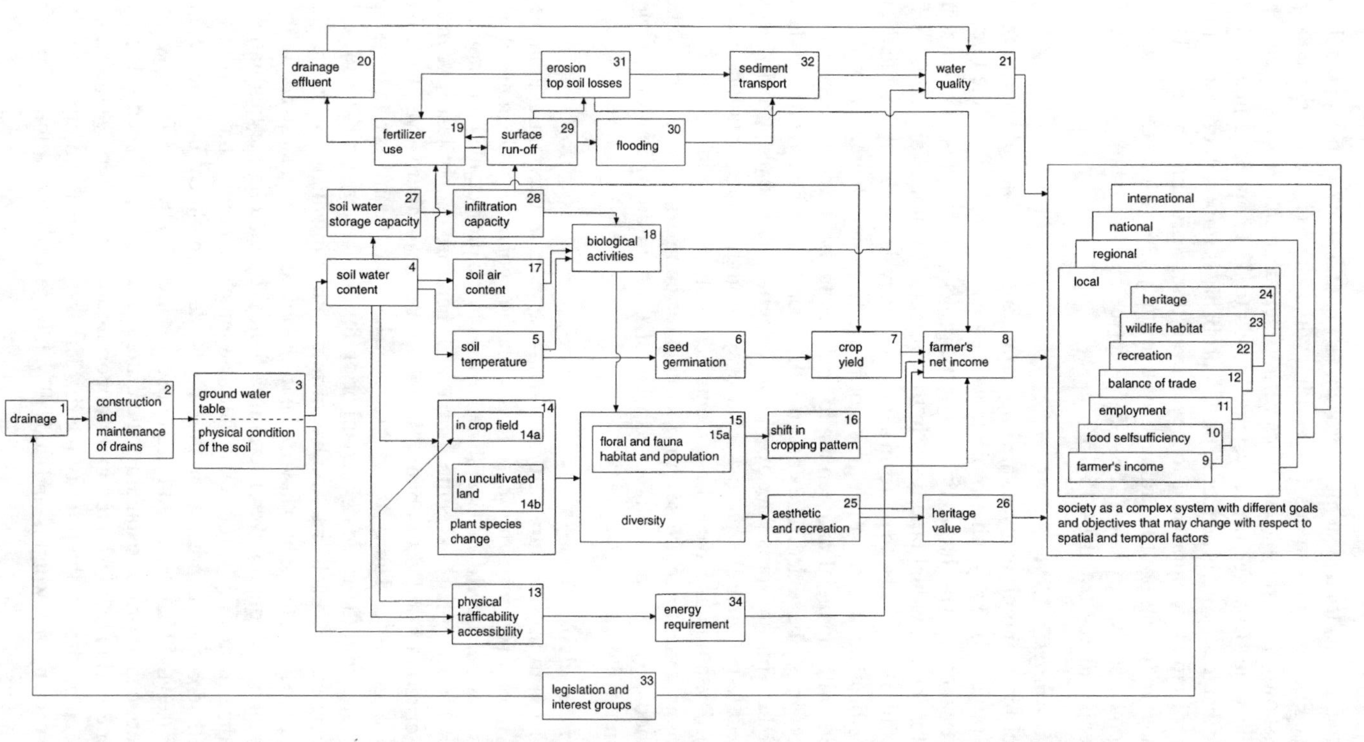

Figure 2.5. A systems approach to the study of the cause-effect relationship through drainage systems. *Source:* [20].

There are several ways to transform and standardize the values. Each way gives a different total score. Also, the weighting is of great influence on the total score. The distribution of the weights is influenced by the point of view from which the comparison is made. Someone who thinks the conservation of natural beauty is more important than the growth of agricultural output will distribute the weights differently than someone who thinks otherwise. These two people most likely will not have a preference for the same alternative and that is shown objectively by the total scores of each alternative. By using MCA, the personal preferences of the planning team for an alternative will be ruled out (or at least clarified and made public).

2.2 Land-Use Planning for Farming

A. Hoogeveen and H. N. van Lier

2.2.1 Land-Use Planning

In Section 2.1, a general outline of land-use planning is provided. This section focuses on land-use planning for farming. Land-use planning can be defined as an idea for the future land uses, expressed in a map and a written text, presenting the development zones and other proposed changes and developments (see also Section 2.1). In this case, land-use planning can be considered as a form of (regional) agricultural planning. Its target is to establish the best use of land, within certain environmental and societal conditions, and with certain objectives [7]. A way to establish the best use of land is to develop the right spatial structures. These spatial structures set the right conditions for farming. Apart from the actual making of a plan, the implementation of the plan also is considered to be a part of land-use planning. Usually, the achievement of the desired use of land takes place in the form of projects or programs.

When it comes to farming, the development of spatial structures is regarded from a point of view of the modernization of agriculture. Modernization involves the increase of agricultural productivity, that is, more output with less input. During the past few decades, however, a general sense of responsibility toward ecosystems, of which humankind is a part, has resulted in some additional targets. The notion of sustainable development, used by the Brundtland Commission, has become an important issue in agricultural science as well as in most communities throughout the world. It encompasses the idea that humankind has certain responsibilities for future generations, who will need to provide a way of living with the same (or more) possibilities and natural resources as the present generations. The principle of sustainable development also identifies a certain intrinsic value in the natural environment [1].

Agriculture can contribute to sustainable development. This is in the interest of agriculture because production is dependent on the resources provided by the natural environment. Land-use planning is able to provide agriculture with some of the conditions that make sustainable development possible.

The first objectives of land-use planning for farming—increasing production and reducing production costs—can be achieved through changes in spatial conditions and

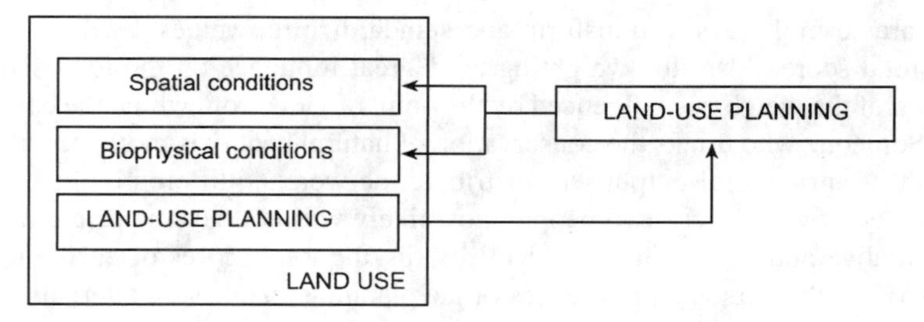

Figure 2.6. Land-use planning and its conditions.

biophysical processes. The second target, which can be summarized as sustainability, also can be achieved by modifying these two conditions. At the same time, sustainability sets the limits for these modifications. These limits are a part of socioeconomic conditions, which cannot be altered by land-use planning alone.

To a certain extent, the spatial and biophysical conditions are the subject of land-use planning for farming and the socioeconomic conditions set the limits for this planning. These relationships between land-use planning and the two conditions are illustrated in Fig. 2.6.

2.2.2 Spatial Conditions

Land-use planning for farming has an impact on agricultural spatial conditions. It is a means to improve these conditions and provide better production. Spatial conditions can be divided into four aspects: access, location, shape, and dimensions. These aspects can be used to describe the way in which spatial conditions influence agricultural production. See Fig. 2.7 and Table 2.4 for examples of these aspects.

Various kinds of access are necessary for the farm to operate competitively. Farm products have to be transported from the fields to the farm buildings and to the market. Animal feed supplements must be taken to the stables, fertilizers must be transported from the (local) market to the fields. In agricultural systems, with high mechanization, it is also neccesary that the fields be accessible to heavy mechanical equipment.

Spatial parameters vary from one location to another. It is very important to fix the right location for the right activity. The best location for storage facilities is near a road. Conversely, an isolated field in the middle of a forest area will be of little use to a farmer. On a higher level, the right location will depend on regional types of land use and on the suitability of the location for farming.

At different levels, the dimensions indicate the economic and, sometimes, social survivability of a farm or a farming system. Almost always, a minimum size is required. For example, the use of mechanical equipment is tied up with the size of the area (the field) that has to be worked. When the field is too small, the investment in a harvester is too high for a single farm or group of farms.

The shape is less important as a spatial condition. Its meaning as a factor to determine the suitability of a farm or a field for agricultural production is limited almost entirely to mechanized farming systems. In these systems, the shape of the fields influences the yields and the productivity of the equipment used.

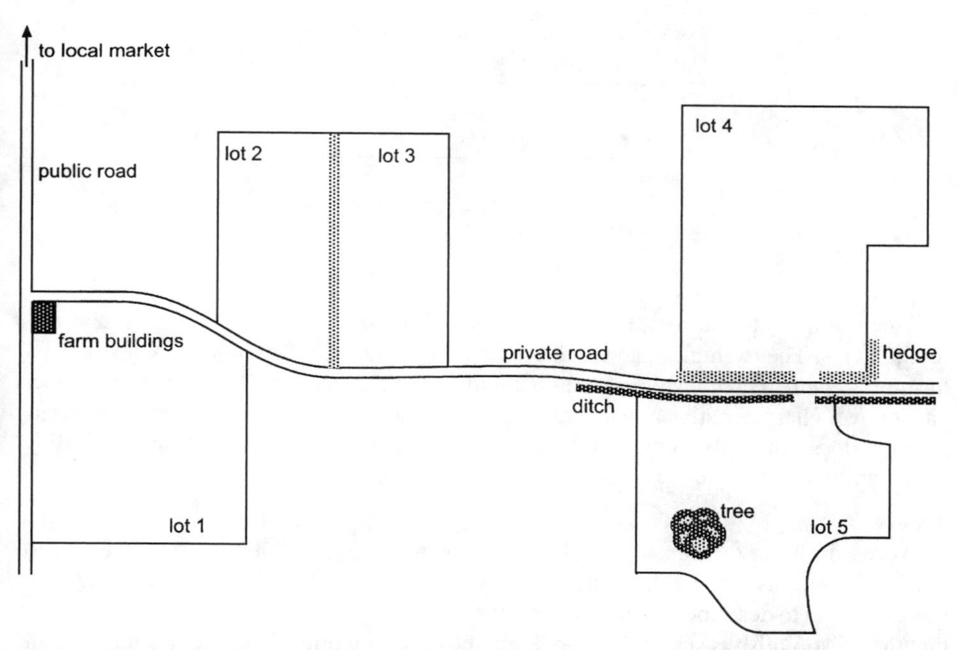

Figure 2.7. Example of the four aspects of spatial conditions (see also Table 2.5). Three of the four aspects can be illustrated by a dairy farm that is not too distant from a town, where the local market is situated. The farm buildings are situated along a public road. A private road leads from the buildings to five lots. The local market at which all products from the farm are sold, also is situated along the public road. The text in Table 2.4 reviews the aspects of the spatial conditions. The regional level is left out, because of the scale of the example. The dimension aspect also is left out because it was not possible to incorporate it in the figure.

Table 2.4. Aspects of spatial conditions using a dairy farm example

	Level	
Aspect	Farm	Field
Access	Excellent. The buildings are situated along the public road. Trucks and other vehicles have direct access to the farm.	Moderate. Lots 1, 2, and 3 are easily accessible, but lots 4 and 5 have restrictions: a hedge and a ditch, respectively.
Location	Fairly good. The local market, which is very important to the farm, is not too far away.	Could be better. The distance from the buildings to the lots, especially lots 4 and 5 are restricting the farming method. The animals (in this example cows) cannot graze on these two lots, because they have to be milked twice or three times a day.
Shape		Good, with the exception of lot 5, which is irregularly shaped and is in the shadow of a tree.

Table 2.5. Levels and aspects of spatial conditions

Aspect	Level		
	Regional	Farm	Field
Access	x	x	x
Location	x	x	
Dimensions	x	x	x
Shape			x

The four aspects—access, location, shape, and dimensions—can be reviewed at different levels. The examples show that access to a field is very different from access to a farming system. Three levels can be distinguished: field, farm, and region. Each aspect has its own characteristics at each level. Shape applies only to the field level, whereas location does not apply to the field level. The four aspects and the levels at which they occur are given in Table 2.5.

Access

Accessibility is defined by Mitchell and Town (in [3]) as "the ability of people to reach destinations at which they can carry out a given activity." Moseley et al. [23] use this definition to describe a form of physical accessibility. The other form of accessibility that they distinguish is social. It refers to the fact that sometimes individuals must fulfill certain requirements to reach something they want.

In this Handbook, the first form of accessibility is used, as far as it concerns agriculture. The central idea of accessibility is the capacity to overcome distance. Hence it refers to the term "ability" [23]. The interaction between the destinations and the people is an important factor of accessibility. Accessibility is determined by the nature of the destination, but also by the people who (desire to) reach the destination to carry out a certain activity. As far as accessibility is concerned, both aspects have to be taken into account.

In land-use planning, the accessibility of a certain location (e.g., a field with cattle) can be improved, according to the wishes of the people who need to have access to the location. Improvement can take place in many different ways, ranking from the removal of physical barriers to the relocation of the land.

For farming, accessibility is very important. At the field level, all lots have to be accessible, either for small-scale activities or for large-scale mechanized activities. The activity carried out on the field determines the requirements with regard to accessibility. In practical terms, this means, for example, that for a lot situated near a road, it is possible to get from the road onto the field without having to cross barriers such as ditches or fences.

At the farm level, the transport of raw materials and finished products determines economic survivability. The importance of accessibility at the farm level depends on the farming system. In modern farming systems, the trucks that transport cattle fodder to the farm and milk from the farm must have access to the right location on the farm. In less developed systems, with extremes such as autarkic systems, accessibility is less important for economic survivability. Land-use planning at the farm level has roughly the same tasks as at the field level. The difference is the scale of the measures. A road from a farm to a local market is more complicated and has more side effects than a road from the farm buildings to a field. On top of that, regional development also applies to this level of planning and land-use planning has to interact with regional economic development strategies (see Section 2.2.4).

The access to a farm from a local market (and vice versa) is very important for the economic development of an area. First, there have to be roads and waterways before the potential of an area can be fully realized. In areas with less potential, but better transportation systems, farming may be a more attractive enterprise [24].

Large-scale economic principles on the regional level lead to more complexity. Land-use planning is one of the many instruments that determine the use of land on this scale. Factors such as the services provided in a certain area and the number of food-producing factories in the region are important. Access to these facilities is crucial for farming in developed countries. On a regional level, access is a key factor to provide farming with the means to play its role in the economic system. However, access on this level is more a matter of regional development and large-scale land-use planning than of land-use planning for farming.

Location

The location of certain elements, such as fields at a farm or farms within a community, can be defined as the relative geographic position of each element with regard to other elements. Access and location interact when planning the use of land for agriculture. A field that is located at a great distance from the farm is not very accessible. Access is also poor when a field is situated in an extensive woodland area.

At the farm level, the location of each field determines the time needed to move from one field to another and from the fields to the farm building. Location is an important factor for reducing the costs in mechanized agricultural systems. For less mechanized systems, the losses will not be counted in money but in time. Walking to the fields takes even longer than driving, and so, every reduction of the distance will be welcome.

At the regional level, the location of the local market is of great importance. Also at this level, the interaction between accessibility and location is eminent (see "*Access*").

Land uses may conflict in their way of using natural resources. Some forms of industry threaten vulnerable ecological systems and a wrongly designed water management system in an agricultural area may drain the water from valuable wetlands. In most cases, land uses that are not economically powerful are vulnerable. Usually, nature is the most vulnerable use of land, industry, agriculture, and urbanization are the most threatening uses. In densely populated areas, agriculture also can be one of the vulnerable uses of land. The right location of each land use, and of agriculture in particular, can prevent either agriculture from threatening nature or agriculture from being threatened by industry or urbanization. These considerations take place at the regional level. In Fig. 2.8, three examples are given of how different land uses can intertwine. In the three examples, the scale at which the mix of land-uses takes place, varies.

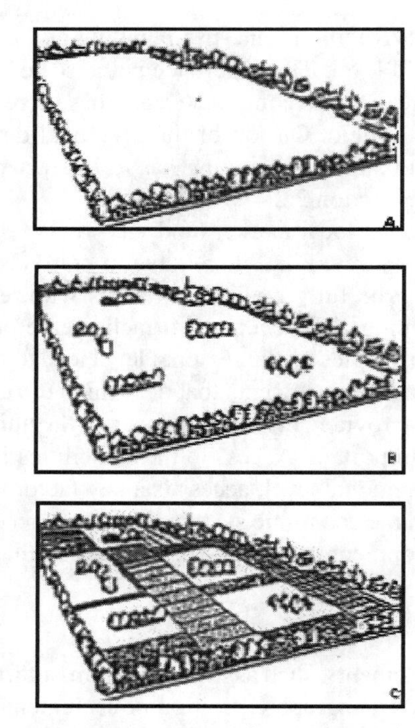

Figure 2.8. Separated or intertwined land uses [25].

Dimensions

Land-use planning can be a means to enlarge farms. Lots that belong to farmers who quit can be sold to the remaining farmers. Land-use planning can provide strategies for the exchange of lots, in order to reduce the distance from the farm buildings to each lot.

Mechanized agriculture requires optimized production conditions. Only when the circumstances are optimized, can the production take place as efficiently as possible. In most processes, economies of scale occur when the amount processed increases. This is also the case in agriculture. The fields have to have a certain area for the machinery to be able to produce efficiently. The smaller the field, the more time that is spent on turning and the more area that is lost at the edges (in relation to the yields), as is shown in Fig. 2.9.

However, farmers of all size of operations tend to enlarge their farms, more for the increase in income than for the decrease in costs per unit [26]. This applies to the field level as well as to the farm level.

Farm size is influenced by a number of factors. These factors have been given for California [27], but they apply to most other developed countries. They include government politics, taxation, the product marketing system, labor costs, energy use, mechanization, and the rural community. If land-use planning is used to enlarge farms, these factors have to be taken into account to ascertain the mutual influence of these factors and farm size.

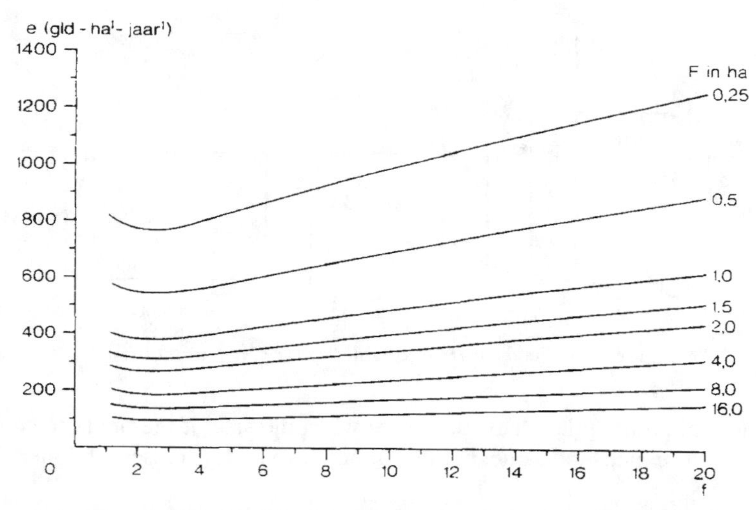

Figure 2.9. Losses on edges and corners (jaar = year; gld = dutch guilder = approx. U.S. $0.5). ***Source:*** **[31].**

Shape

The aspect of shape is important at the field level. It influences the time spent on each field in the case of mechanized farming systems. On irregular-shaped fields, the number of turns and the area that is not cultivated at the end of the bouts increase. This consumes time and money [28]. In Table 2.6, some figures are given to indicate the loss of time on irregular-shaped fields compared to regular-shaped ones [28].

In Fig. 2.10, the field shapes that are used in Table 2.6 are visualized. All examples have an area of 10 ha (25 acres).

As with dimensions, the shape is only important in mechanized agricultural systems. The shape of a field is not very important to the farmer in less-developed systems, who does all the work by hand.

Table 2.6. Effect of field shape on the time to cultivate 10 ha[a]

Field Shape[b]	Cultivation Time (Min ha)	Index
Square	56.6	100
Rectangle (2:1)	54.0	95
Rectangle (4:1)	52.4	93
Re-entrant side	59.1	104
Building plots	60.5	107
Obstacles in field	62.0	109

[a] See also Fig. 2.9.
[b] See Fig. 2.10.
Source: [28].

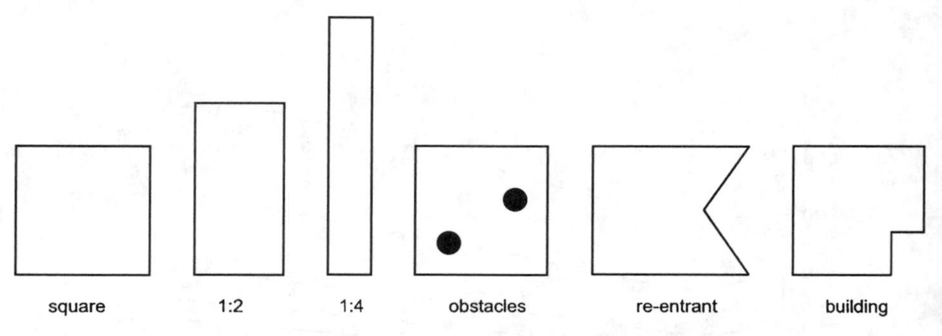

Figure 2.10. Field shapes (see Table 2.6 for cultivation times).

The effect of an irregular shape decreases when the size of the field increases. The time that is lost remains the same, but the amount of yield increases, and therefore the efficiency rises.

In Fig. 2.9, the relation between size and shape is shown in the context of losses at the edges. The horizontal axis represents the ratio of the length to the width of a field. The various lines indicate the field size, ranging from 0.25 ha to 16 ha. The graph shows that, as the acreage increases, the losses decrease. It is also clear that the effect of shape decreases as the acreage increases. On small fields, the shape is more important than the size. On big fields, it's the other way around. The optimum is at a ratio of approximately 2.5.

2.2.3 Biophysical Conditions

The biophysical conditions are qualities of the land. Water, soil, and air are the components of the biophysical surroundings of any farm system. The nature and extension of natural biotopes is determined by biophysical conditions [29]. Agriculture is heavily dependent on natural biotopes because they form the main production factor of farming. Throughout the years, the components of the biophysical surroundings are changed by various land uses, resulting sometimes in near-perfect production conditions.

The quality of these components is an important factor for growing any crop or keeping most animals. Some sorts of farming can function without their natural environment (such as greenhouse systems), but they are not reviewed here because they are hardly affected by land-use planning.

The FAO has defined a land quality as "a complex attribute of land which acts in a distinct manner in its influence on the suitability of land for a specific kind of use" [30]. If the suitability of land for farming has to be increased with land-use planning, the qualities of the land have to be altered through land-use planning. The land qualities that apply to rainfed agriculture are listed in Table 2.7.

Some of these land qualities cannot be influenced by land-use planning. For example, the radiation regime is a factor that is unchangeable. The same goes for temperature regime and climatic hazards. Other land qualities are influenced by normal farming procedures. Rooting conditions are improved by plowing, moisture availability is improved

Table 2.7. Land qualities for rainfed agriculture

No.	Quality
1	Radiation regime
2	Temperature regime
3	Moisture availability
4	Oxygen availability to roots (drainage)
5	Nutrient availability
6	Nutrient retention capacity
7	Rooting conditions
8	Conditions affecting germination and establishment
9	Air humidity as affecting growth
10	Conditions for ripening
11	Flood hazard
12	Climatic hazards
13	Excess of salts
14	Soil toxicities
15	Pests and diseases
16	Soil workability
17	Potential for mechanization
18	Land preparation and clearance requirements
19	Conditions for storage and processing
20	Conditions affecting timing of production
21	Access within the production unit
22	Size of potential management units
23	Location: accessibility
24	Erosion hazard
25	Soil degradation hazard

Source: [32].

by sprinklers, and nutrient availability is improved by application of manure. The improvement of these qualities is not land-use planning.

Typical land qualities that can be ranked under land-use planning are access (Nos. 21 and 23 in Table 2.7), size (No. 22), location, and shape of the farms and fields. These land qualities are reviewed in Section 2.2.2, "Spatial Conditions." Other land qualities that can be improved through land-use planning are structural soil improvements (Nos. 4, 7, 16, and 17), erosion control (Nos. 24 and 25), and flood hazard (l.q. 11). The nature of these qualities and the way to improve them are dealt with in other sections of this Handbook. In this chapter, the way land-use planning affects these qualities is reviewed.

Structural soil improvements can be achieved outside land-use planning but nearly all land development plans contain measures to improve the soils. The conditions of the soil are important for crops and for mechanization.

The rooting conditions, that is, the conditions for the development of an effective root system [31], refer to the ability to keep the plant in place and the plant's ability to extract moisture and nutrients. If the volume of the root system is limited, the parts of the plant that are aboveground will suffer [32]. The characteristic by which the rooting conditions are measured is the effective soil depth. Additional characteristics include soil structure, consistency, and texture. Land-use planning involves measures to improve rooting conditions, such as deep ploughing, adding chalk or sand to clay or peat, removing of the topsoil, or applying drainage. If the land-use plan contains uses of land that require perfect rooting conditions, the measures should be part of the land-use plan.

Almost all crops need to take up oxygen through their roots, although plants vary in their tolerance of short periods of waterlogging. Oxygen availability to roots is therefore very important. Because oxygen is available above the water table and not below it, the depth of the water table is the key factor for this land quality. Drainage is a way to improve oxygen availability. Most land-use plans involve interventions in the water management system. Ditches, dikes, pumps, and field drainage, are ways to influence the amount of water in an area. Because water management is of a regional nature, land-use planning is a preeminent way to achieve the right level of availability of water.

Water management is also very important for soil workability and mechanization. The moisture content of the soil is particularly important for the workability. Some soils are always easy to work, whereas other soils have strict limitations with regard to the water content. Generally, sandy soils are easier to work than clayey soils and well-textured soils are easier to work than massive soils [32]. The measures in a land-use plan that can be applied to improve soil workability are the same as with land quality 7, rooting conditions.

Apart from soil workability, the texture of soil is also important for mechanization. This land quality applies only to agricultural systems that are highly mechanized. Characteristics of land that define potential for mechanization are slope angle, rock hindrances, stoniness or extreme shallowness of the soil, and the presence of heavy clays [32]. The potential for mechanization can be improved by lessening the slope angle and by improving the soil characteristics. The slope angle can be lessened by terracing the lots (Fig. 2.11). These improvements, like soil improvements, can very well take place in connection with land-use planning schemes.

Erosion is one of the main problems for agriculture across the world. It causes fertile soils to be washed or blown away, leaving poor, nonproductive land. The many aspects of erosion are discussed in other parts of this book. The ways in which land-use planning can reduce erosion are reviewed here.

Land-use planning can prevent erosion by assigning uses of the land that protect sensitive soils (such as dense woods). Less sensitive soils can be used for agriculture. Where this is not possible, a small-scale mix of certain land uses provides a way to prevent erosion: For example, when agriculture is combined with forestry or nature reserves, the forest or the reserve functions as a buffer, preventing the water or the wind from eroding the nearby fields. Other ways to prevent erosion include terracing fields, building dikes, planting grass strips for water flows and to trap sediments, and planting

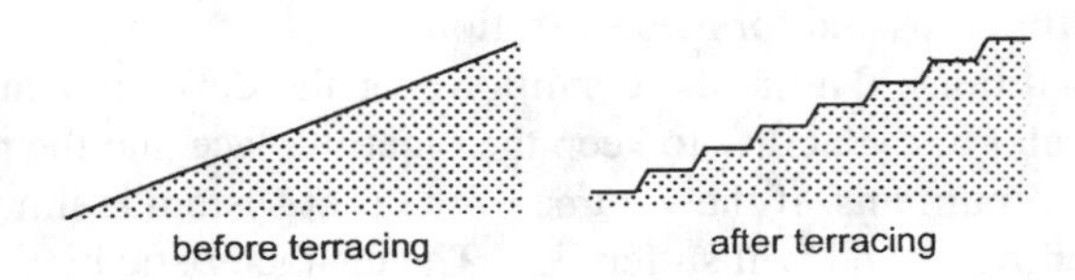

Figure 2.11. Terracing of fields.

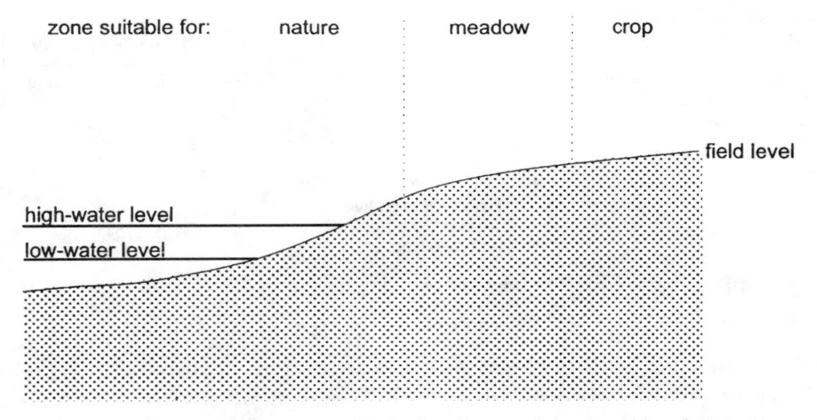

Figure 2.12. Damage resulting from flooding can be reduced by planning the use of the land.

trees as a protection against wind. These measures can also be taken outside land-use plans and they are discussed elsewhere in this volume (see Chapter 4).

Flood hazard is a problem that requires more than just land-use planning. However, planning can mitigate the effects of flooding by making sure that land threatened by flooding is used appropriately (see Fig. 2.12). Some uses of land even benefit from incidental or permanent inundation.

2.2.4 Socioeconomic Conditions

Although land-use planning changes the spatial and physical aspects of a rural countryside, it also has impact on and is influenced by social and economic conditions. The status of these conditions can hardly be changed by land-use planning on a farm-level scale. Only through nationwide or statewide land-use planning can certain changes be achieved. This is especially so for multiple land-use planning. In this section, some issues between land-use planning for farming and socioeconomic conditions are presented.

The first issue is sustainability, the general outlines of which are given in Chapter 1. The concept of sustainable land use embraces two goals: optimal use and protection of natural resources for the long term (environmental sustainability) and meeting the needs and aspirations of the present generation (socioeconomic sustainability) [33]. Natural resources are the main production factor for farming, and so, farming has, by definition, an impact on these resources. To provide the right conditions for sustainability within farming systems, land-use planning must improve the spatial and biophysical conditions of the agricultural area, so that an optimal and sustainable use of the natural resources is possible. Land-use planning provides the tools to do this. Some examples of these tools are

- the spatial separation of conflicting land uses (e.g., agricultural and urban land use);
- legislation and regulation with regard to the agricultural use of land (e.g., nitrate application criteria to protect water resources);
- promotion of and provision of the means for nature management by farmers.

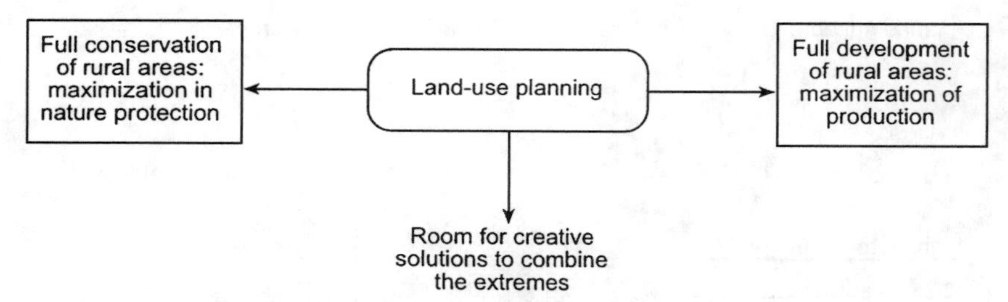

Figure 2.13. The land-use planning spectrum. ***Source:*** **[34].**

Land-use planning has to struggle with two dimensions of sustainability [34]: ecological conservation and economic existence. Conservation is a striving to restore and preserve; Development is a striving to create and "improve." Figure 2.13 illustrates this.

Land-use planning, however, can go further than the above mentioned protection of the natural resources; it can help to restore lost values. This is the case when, for example, farmland is turned into nature areas [35].

The second issue is multifunctional land-use planning, which occurs on a larger scale than one or a few farms. More types of land uses have to be taken into account when planning on this scale; it cannot be restricted to farming alone. The most suitable use of land has to be determined with regard to the land qualities and the requirements that each type of land use has. This may not be agriculture, even in an agricultural area. Sometimes mixed forms of land use are possible and, in fact, have even been encouraged [33] where they take into account the interrelationships among the land uses (see Fig. 2.14).

The third issue concerns the market principle and economic conditions. The agricultural interest in certain kinds of land is determined by the situation in the agricultural market [36]. The market has more influence in some political systems than in others, depending on the extent of freedom of pricing. If land prices are low, many farmers will want to buy land, so that the pressure on the market for land increases. Land-use planning has to deal with this problem, for example, by making more land suitable for farming.

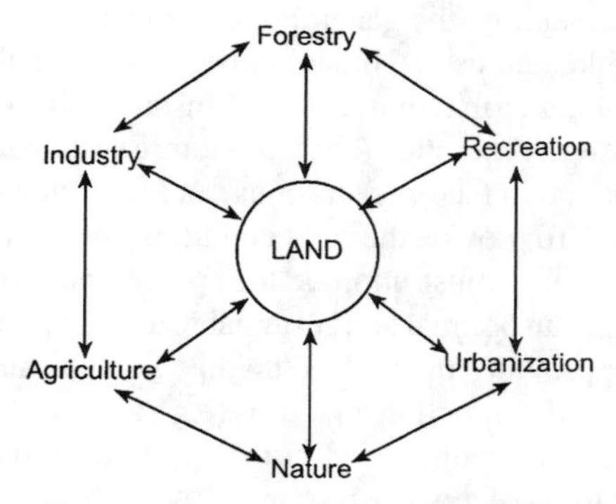

Figure 2.14. Multiple land-use planning.

When the demand for farmland is low, land-use planning can transform agricultural land into nature reserves or recreational sites.

Agriculture always has been an important sector in almost every economic system in the world. Most nations have been and are dependent on agricultural exports [37]. Governments have treated agriculture according to its importance to the national economy and, in many cases, have made the improvement of agricultural conditions the single goal of land-use planning.

In developing countries, land-use planning and rural development may be used to increase exports consisting of primary agricultural products, such as cocoa butter, raw timber, and soybeans. These products are processed, after being imported, in developed countries [38]. In this situation, land-use planning can be used in two ways: The first is to help increase the production level of the primary production systems. The second is to help to develop ways to process the primary products in their country of origin, thus stimulating the national economy. Land-use planning is only one of the many instruments needed to enhance these complex, international relationships.

The fourth issue is rural development. When the agricultural system in a certain area is improved by land-use planning, the rural area, which usually is connected closely with agriculture, benefits from this improvement. In most cases, land-use planning will improve not only the conditions for farming, but also for other land uses (see the preceding discussion of multifunctional land-use planning). Conditions in the area as a whole will improve, in most cases causing a higher level of economic activity. When this happens, land-use planning can provide some of the conditions for increased economic activity, for example, through improvement of the infrastructure.

The improvement of a rural area with regard to its economic possibilities and its social and environmental "health," is called rural development. Land-use planning and rural development go hand in hand: one needs the other, and there is no clear boundary between them [6].

However, there are, two sides to rural development. The most important is the above-mentioned development of the area. The other side is the increase in property taxes and the general cost of living [37]. The increase in property taxes is a result of the increased value of the land arising from the demand for more land to facilitate the economic activities. This is even more the case when the increased economic activity leads to urbanization of the countryside. Speculation and the market principle lead to higher land prices. A higher service level for the inhabitants of the area leads to higher cost of the primary services.

The fifth, and final, issue is knowledge. Generally, much knowledge and information are required about what, how, where, and why planning should take place. For farming, land-use planning involves information about the land and about farming. Information about the land provides insight into the opportunities and constraints of each type of land for agriculture. The land qualities mentioned in Section 2.2.3 can be a good guide when gathering information. The spatial requirements of agriculture can be found in Section 2.2.2. It is essential that all information be integrated in an information system. Such a system makes it possible to assess biophysical and spatial effects of the alternative land-use options and to make decisions as to the best alternative [39].

Aside from the information about land and agriculture, some knowledge about the planning process is necessary. This is reviewed in Section 2.1.

References

1. World Commission on Environment and Development. 1987. *Our Common Future*. Oxford, UK: Oxford University Press.
2. Allen, R. E. (ed.). 1990. *The Concise Oxford Dictionary of Current English*. Oxford: Clarendon Press.
3. Beatty, M. T., G. W. Peterson, and L. D. Swindale (eds.). 1979. *Planning the Uses and Management of Land*. Madison, WI: American Society of Agronomy, Crop Science Society of America, Soil Science Society of America, Inc.
4. Werkgroep Herziening Cultuurtechnisch Vademecum. 1988. *Agricultural Engineering Handbook* (Cultuurtechnisch Vademecum). Utrecht: Cultuurtechnische Vereniging.
5. Hall, P. 1975. *Urban and Regional Planning*. London: Newton Abbot.
6. Food and Agriculture Organization. 1993. Guidelines for Land-Use Planning. Rome: FAO.
7. Fresco, L. O., H. Huizing, H. van Keulen, H. Luning, and R. Schipper. 1989. Land Evaluation and Farming Systems Analysis for Land Use Planning. FAO Guidelines: Second Draft. Rome: FAO.
8. Food and Agriculture Organization. 1991. Land use planning applications. *Proceedings of the FAO Expert Consultation, 10–14 December 1990, Rome, Italy*. World Soil Resources Report 68. Rome: FAO.
9. Raintree, J. B. 1987. *D&D User's Manual: An Introduction Agroforestry Diagnosis and Design*. Nairobi, Kenya: International Council for Research in Agroforestry.
10. Young, A. 1986. Land evaluation and diagnosis and design: Towards a reconciliation of procedures. *Soil Surv. Land Eval.* 5:61–76.
11. Abel, N. O. J., M. J. Drinkwater, J. Ingram, J. Okafor, and R. T. Prinsley. 1989. Guidelines for Training in Rapid Appraisal for Agroforestry Research and Extension. Amelioration of Soils by Trees. London: Commonwealth Science Council and Harare, Zimbabwe: Forestry Commision.
12. Food and Agriculture Organization. 1989. Community Forestry Rapid Appraisal. Community Forestry Note No. 3. Rome: FAO.
13. McCracken, J. A., J. N. Pretty, and G. R. Conway. 1988. An Introduction to Rapid Rural Appraisal for Agricultural Development. London: International Institute for Environment and Development.
14. Sims, D. 1996. Integrating Land Resources Management. Increasing Production and Achieving Conservation Through People's Participation, More Logical Decision Making, and Improved Institutional Structures. Rome: FAO.
15. United Nations Economic and Social Council. 1995. Integrated Approach to the Planning and Management of Land Resources. UN, Economic and Social Council, Geneva.
16. Food and Agriculture Organization. 1990. Participation in Practice: Lessons from the FAO People's Participation Programme. Rome: FAO.

17. van der Knaap, W. 1997. The Tourist's Drives. GIS Oriented Methods for Analysis Tourist Recreation Complexes. Ph.D. Thesis. Wageningen Agricultural University, Wageningen.
18. van Lier, H. N. 1996. "Sustainable rural systems: A challenge also for land use planners. *Second Workshop on Sustainable Land Use Planning with Special Regard to Central and Eastern European Countries*. Gödöllö, Hungary: Gödöllö University of Agricultural Sciences and National Committee of CIGR.
19. Hanken, A. F. G., and H. A. Reuver. 1973. *Introduction to Systems Analysis (Inleiding tot de Systeemleer)*. Leiden: Stanford Kroese.
20. Jorjani, H. 1990. Analysis of Subsurface Drainage for Land Use Planning. Ph.D. Thesis. Wageningen Agricultural University, Wageningen.
21. Gittinger, J. P. 1982. *Economic Analysis of Agricultural projects*. 2nd ed. Baltimore. Johns Hopkins University Press.
22. Voogd, J. H., C. Middendorp, B. Udink, and A. van Setten. 1980. *Multicriteria Methods for Spacial Evaluation Research (Multikriteria-Methoden voor Ruimtelijk Evaluatieonderzoek: Verslag van een Onderzoekuitgevoerd in Opdracht van de Rijksplanologische Dienst)*. Delft: Planologisch Studiecentrum TNO.
23. Moseley, M. J., R. G. Harman, O. B. Coles, and M. B. Spencer. 1977. Rural Transport and Accessibility. Vol. I: Main Report. Norwich, UK: Centre of East Anglian Studies, University of East Anglia.
24. Held, R. B., and D. W. Visser. 1984. *Rural Land Uses and Planning. A Comparative Study of the Netherlands and the United States*. ISOMUL (International Studygroup on Multiple Uses of Land) Amsterdam: Elsevier.
25. Buitenhuis, A., C. E. M. van de Kerkhof, and I. J. van Randen. 1986. *Schaal* van het *Landschap*: *Opbouw* en *Gebruik* van een *Geografisch Informatiesysteem* van *Schaalkenmerken* van het *Landschap* van Nederland, met *Landelijke Kaarten* 1:400.000. (Landscape Scale: Uses of a Geographical Information System for Scale Properties of the Dutch Landscape, with Maps). Wageningen: Stiboka. (in Dutch.)
26. Miller, T. A., G. E. Rodewald, and R. G. McElroy. 1981. Economies of Size in US Field Crop Farming. Washington, DC: U.S. Department of Agriculture, Economics and Statistics Service.
27. Carter, H. O., and W. E. Johnsten (eds.). 1980. Farm-Size Relationships, with an Emphasis on California. A Review of What Is Known About the Diverse Forces Affecting Farm Size, and Additional Research Considerations. Davis: University of California, Department of Agricultural Economics.
28. Sturrock, F. G., J. Cathie, and T. A. Payne. 1977. Economies of Scale in Farm Mechanisation. A Study of Costs on Large and Small Farms. Cambridge, UK: Agricultural Economics Unit, Department of Land Economy.
29. Brouwer, F. M., A. J. Thomas, and M. J. Chadwick (eds.). 1991. *Land Use Changes in Europe. Processes of Change, Environmental Transformations and Future Patterns*. Dordrecht: Kluwer.
30. Food and Agriculture Organization. 1976. A Framework for Land Evaluation. Wageningen: International Institute for Land Reclamation and Irrigation.
31. Sprik, J. B., and J. A. Kester. 1972. Kantverliezen op Rechthoekige en Onregelmatig Gevormde Akkerbouwpercelen (Edge Losses on Rectangular and Irregular

Shaped Fields). Wageningen: Instituut voor Cultuurtechniek en Waterhuishouding. (in Dutch.)

32. Food and Agriculture Organization. 1984. Guidelines: Land Evaluation for Rainfed Agriculture. FAO Soils Bulletin 52. Rome: FAO.
33. van Lier, H. N., C. F. Jaarsma, C. R. Jurgens, and A. J. de Buck (eds.). 1994. *Sustainable Landuse Planning. Proceedings of an International Workshop, 2–4 September 1992, Wageningen.* ISOMUL, CIGR. Amsterdam: Elsevier.
34. Sasaki, H., I. Saito, A. Tabayashi, and T. Morimoto (eds.). 1996. *Geographical Perspectives on Sustainable Rural Systems. Proceedings of the Tsukuba International Conference on the Sustainability of Rural Systems.* Tokyo: Kaisei Publications.
35. Kneib, W. D. 1996. A landscape development plan as a tool for sustainable ecological planning. *Second Workshop on Sustainable Land Use Planning with Special Regard to* Central and Eastern European *Countries.* Gödöllö, Hungary: Gödöllö University of Agricultural Sciences and National Committee of CIGR.
36. Organisation For Economic Cooperation and Development. 1976. Land Use Politics and Agriculture. Paris: OECD.
37. Cloke, P. J. (ed.). 1989. *Rural Land-Use Planning in Developed Nations.* London: Unwin Hyman.
38. Food and Agriculture Organization. 1995. Planning for a Sustainable Use of Land Resources: Towards a New Approach (Draft). Rome: FAO.
39. Fresco, L. O., L. Stroosnijder, J. Bouma, and H. van Keulen. 1994. *The Future of the Land.* Mobilising and *Integrating Knowledge for Land Use Options.* Chichester: Wiley.

3 Rural Roads

3.1 Overview

A. P. Wolleswinkel and C. F. Jaarsma

A rural road network creates many contradictions. The presence of a well-developed road network in a region is a *conditio sine qua non* for economic development and efficient access to and use of land resources. Accessibility of rural areas and mobility for the rural people are also social aims. Simultaneously, the presence of the road network and its traffic flows often can have harmful effects. The problems that are encountered in this context may be approached in two ways: quantitatively and qualitatively. Both aspects occur everywhere, but may have especially large effects in less developed countries. Remoteness, isolation, and inaccessibility are the key characteristics of many rural regions in Africa, Asia, Central America, and South America. A lack of sufficient roads, both qualitatively and quantitatively, results in a bad transportation system, leading to economic and social losses. Another problem in these countries constitutes the impacts due to the disturbed balance between the road functions and maintenance. In industrialized countries, the quantitative aspect is less important than the qualitative aspect. Emissions and noise affect local people, flora, and fauna. There also is the problem of traffic safety on rural roads with low capacities, because of high traffic speeds.

Terms such as remoteness and isolation vary with national living standards. In Australia, the extensive use of air transport and radio links brings the most distant farm into easy communication with essential services. On the other hand, in much of Southeast Asia, even the bicycle is beyond the means of many villagers. Thus, these terms need to be handled with caution. It is too complex to restrict these terms to developing nations alone [1].

In the next sections, a substantial distinction is drawn between industrialized and developing countries. The latter nations must deal with mobility problems that differ from those in the industrialized countries. In most developing countries, at least two-thirds of the population still can be classified as rural, although densities vary considerably according to levels of economic activity. In Nigeria, for example, rural densities of 400 persons per km^2 are recorded in the southeast region, whereas the drier interior savannahs support densities of only 20–30 persons per km^2. If rural isolation is interpreted in terms of the absence of roads for cars, then 196 million village dwellers in India belong in this category. A survey of 16 provinces in Indonesia indicated that 30% of all villages had no link to the road network [1].

Many nations, however, find themselves in an intermediate position between the two extremes of being industrialized or developing. Within these countries, the levels of isolation and remoteness in rural areas can vary substantially. In nations in the Middle East and Latin America, for example, the mobility is relatively high in some rural areas but inaccessibility is still a severe problem in other regions [1]. Eastern European countries, with economies in transition, face the problem of rapidly growing car ownership. At the same time, the existing road network needs structural changes to serve the new forms of land use and the related land layout changes.

Where this chapter deals with roads, it should be realized that a road is not a goal unto itself. Roads are only constructed to serve traffic and transportation. Both traffic and transportation are derived functions. They strongly depend on local land use (location and type of activities). Simultaneously, all human activities are strongly dependent on the road network. In this way, there is a narrow relationship between spatial planning, land-use planning, and the planning, construction, and maintenance of the road network. This holds for both urban and rural areas.

This chapter is divided into three main parts. Section 3.2 addresses the fundamentals of rural roads. It explains rural road structure and the density of networks. Then traffic problems are discussed in the context of the need for traffic planning. Section 3.3 discusses in greater depth the motives and perspectives for rural road planning. Examples of planning systems and cases are described to give an adequate view into the international planning for rural roads. In Sections 3.4 and 3.5, respectively, basic principles of road construction and maintenance are presented and discussed.

3.2 Rural Road Development in a Wider Context

A. P. Wolleswinkel and C. F. Jaarsma

3.2.1 Introduction

A well-designed regional road network consists of several mutually related hierarchical networks. First, there is the network of high-quality roads, such as motorways, to provide long-distance connections with high-speed travel. This is mostly a national network. Second, the network of regional highways connects regions with each other and with the first network. In countries of the Organization for Economic Cooperation and Development (OECD), these major roads only represent some 20% to 40% of the overall stock. In developing countries, this is much lower—some 5% [2]. So, the network of high-quality trunk roads only includes a relatively small part of a country's total road network. The majority of this network is the so-called low-traffic network, embracing the low-traffic roads (LTRs). Other terms, such as low-volume roads and minor rural roads, are also in use. In the following sections, LTR is used to mean roads with lower functions than the trunk roads. Both one-lane and two-lane LTRs and paved as well as unpaved LTRs appear. In developing countries, unpaved roads represent about two-thirds of the

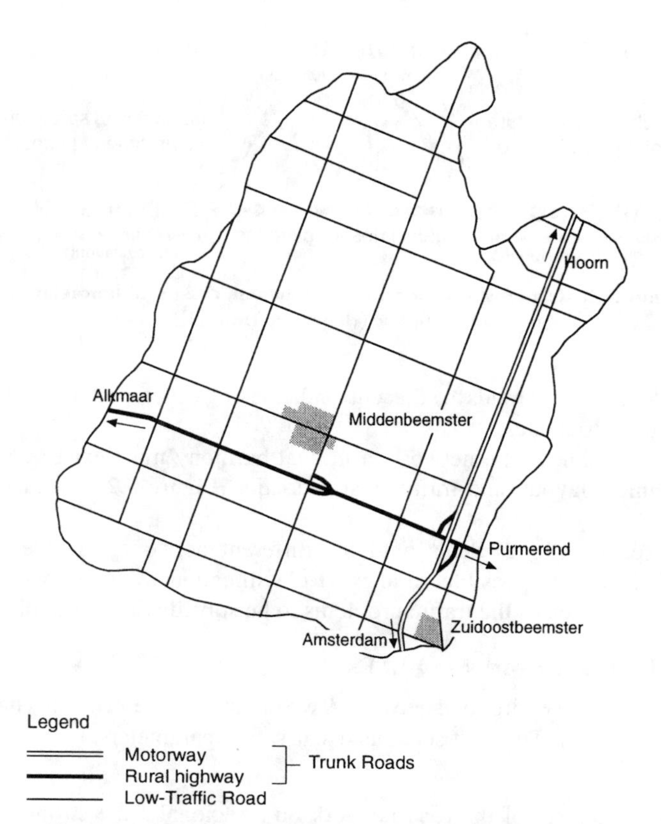

Figure 3.1. Coherence between networks of trunk roads and LTRs. *Source:* Reprinted by permission of the Wageningen Agricultural University.

network, excluding tracks that cannot be defined as roads [2]. Figure 3.1 illustrates the coherence between the networks of trunk roads and LTRs.

LTRs play a major role in the industrialized countries in promoting and facilitating the proper development of land and natural resources, serving the needs of local industry and promoting commercial, social, and cultural activity at both local and regional levels. In developing countries, roads are designed essentially to open up new resources, develop agriculture and mining, promote industry and commerce, and stimulate the social and cultural life of the regions [2].

In defining an LTR, several criteria may be used, including traffic volume, road function, administrative classification, and management and financing arrangements. LTRs are mainly local collector and access roads but they might also be of a higher road class. Traffic volumes on these roads is less than 1,500 (2,000 in certain exceptional cases) motor vehicles per day [2]. Among the network of LTRs, there are mainly local

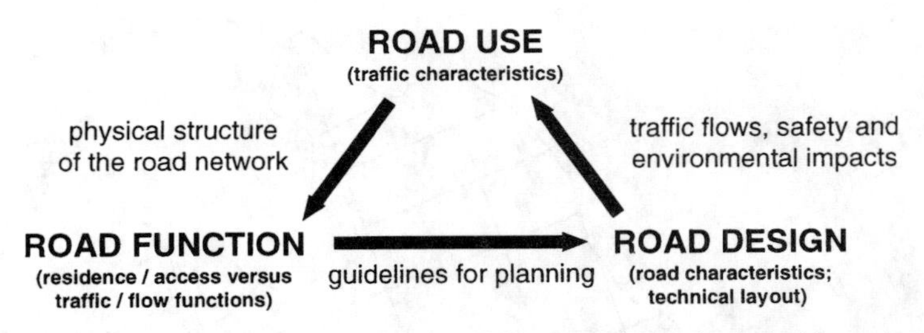

Figure 3.2. Relationships between road function, design (technical layout), and traffic use (elaborated from [3]).

collector roads and access roads to adjacent land, housing developments, and agricultural, forestry, and industrial sites.

For every road link in the network, a mutual harmony must exist between desired function, technical layout, and traffic characteristics. Figure 3.2 illustrates these relationships.

The next section (3.2.2) deals with the different networks that generally occur. Section 3.2.3 discusses possible rural road classification systems, and finally, Section 3.2.4 briefly indicates the traffic problems to be solved with traffic planning.

3.2.2 Rural Road Networks and LTRs

LTRs are part of the entire road network. Two parameters describe this network: form (structure) and density. The next sections discuss both parameters.

Network Form

Tradionally, the form of the road network on a regional scale strongly depends on natural constraints and/or the historical rural occupation patterns. For example, in the Dutch province of Drenthe, several features of the triangular network can be observed through the old Sachsian settlement pattern. Principally, there are two basic shapes: radial and tangential (the somewhat confusing terms functional and geometric network are also in use). In a radial network, the roads start from the center village or town and go into the surrounding rural area. This system is found in areas with natural barriers such as hills, mountains, or rivers and/or in areas where the occupation lasted relatively long (Fig. 3.3A). Adequately expanded radial networks provide access to the formerly fairly inaccessible mountainous areas. In contrast, a tangential network is applied in more or less homogeneous areas, occupied during a relatively short period. They are characterized by a clear, easily expanded structure. This type of network is suitable in areas with intensive agriculture and forestry. Its chessboard structure is well known in the United States. It also occurs in both historically (Fig. 3.3B) and more recently constructed reclamation areas (e.g., The Netherlands), in extensive agricultural regions (e.g., eastern Germany), and in lowlands of mountain regions (e.g., Switzerland).

In practice, adapted forms of these basics appear. Radial networks are shaped either as a branched structure or an "antichessboard." A grid structure introduces hierarchy into the basically unhierarchical tangential chessboard network [4]. Figure 3.4 shows the most important advantages and disadvantages.

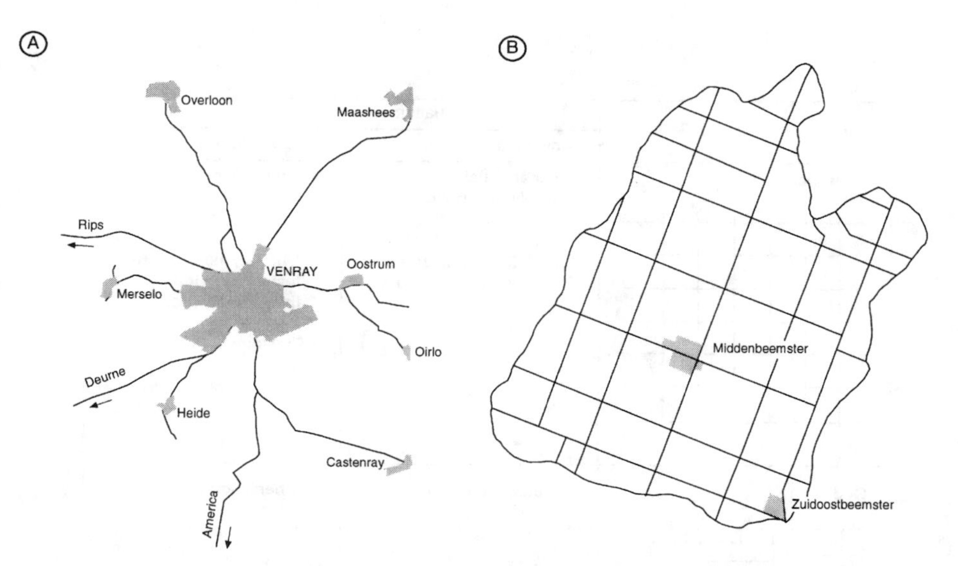

Figure 3.3. Basic network forms: (A) radial and (B) tangential (original pattern after reclamation). ***Source:*** **Reprinted by permission of the Wageningen Agricultural University.**

Network Density

An easily interpretable statistic for the network density is the so-called road density (symbol W, dimension km/km^2). This statistic is calculated by dividing the road length by the acreage of the study area. In this calculation, only roads with an access function for the study area are considered. A road density of 1.5 to 6 often is used for farmed areas. The desired value strongly depends on the type of agricultural land use. The lower values (1.5 to 2.0) count for arable land and pasture land in flat areas. For horticulture, a density of 2 to 3 is used, whereas glasshouse districts need a density of 3 to 6 km/km^2. Furthermore, the values depend on farm sizes, the number of parcels per farm, form and acreage of the parcels, and the topographical structure of the landscape. In newly reclaimed land, with large arable farms and the farmland concentrated around the farm buildings, the road density is only 0.6 to 0.8 km/km^2 for public roads. This value roughly doubles when the frequently constructed, paved private farm roads are included.

In developing countries the road density is much lower. Overall data for selected countries in Africa from the International Road Federation show values from 0.02 km/km^2 (in Botswana and Ethiopia) to 0.4 km/km^2 (in Rwanda). However, these data are on a national scale. They cannot be compared directly with the regional data for a specific form of agricultural land use, as presented above.

It might be more illustrative to recalculate the road density in a mesh size of the network. This can be done easily if a standard representation of reality is dividing all roads outside built-up areas in a regular rectangle (the chessboard structure). The relationship between road length and road density can be calculated by considering the total road length around a rectangle as $4L$ km; the surface amounts L^2 km^2. To determine the road density, each length needs to be halved because each road serves the opening up of two

Form of network	Evaluation	
	Advantages	Disadvantages
Chessboard 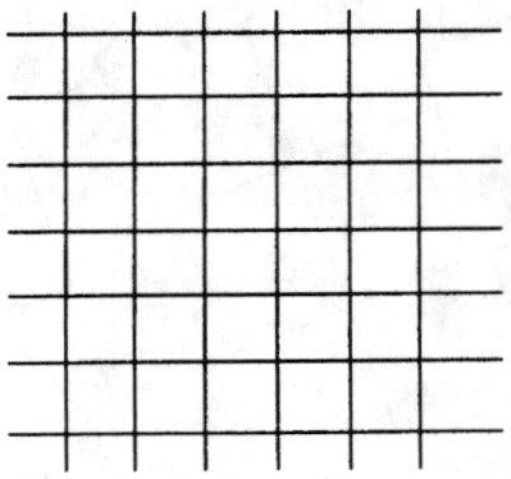	- Clear structure, easily expanded - Economical utilization and division of land (uniform road network)	- Orientation (center?) - Hierarchy not evident - Unclear, monotonous → fast through-traffic, tiring (safety problem) - Problems with diagonals - Capacity not evenly used
Grid 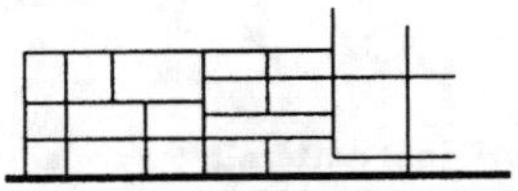	- Better hierarchy possible compared with 'chessboard'	- Orientation (center?) - Unclear, tedious

Form of network	Evaluation	
	Advantages	Disadvantages
Branched 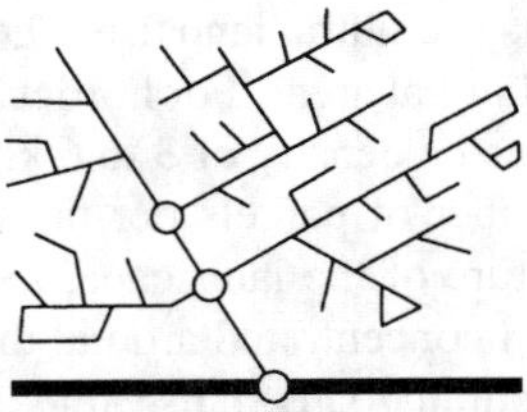	- Clear structure - Complex - Can be 'exactly' dimensioned because of the lack of external traffic - Good for opening up residential areas	- Supply and disposal - Road cleaning - Relatively large traffic area
Grid 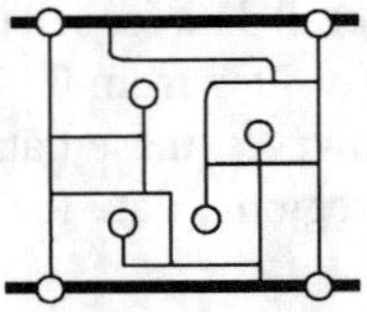	- High level of safety - Parceling situation	- Unclear structure - Hierarchy - Poor orientation

Figure 3.4. Adapted forms of networks with their advantages and disadvantages.
***Source:* [4], reprinted by permission of PUDOC-DLO, Wageningen.**

adjacent blocks. The road density then amounts to

$$W = 1/2 \cdot 4L/L^2 = 2/L, \tag{3.1}$$

where W = road density (km/km^2) and L = mesh size (km).

This mesh-size approach may give more insight into the real network situation. It is a good starting point for calculation of accessibility measures of a network. These can be defined as the travel time necessary within the network to reach a destination in the rural area. The kind of land uses in the rural area determine the desirable mesh size (km), just as it was for road density (km/km^2).

Most regions in industrialized countries have, from a quantitative point of view, a sufficient rural road network. With regard to quality, however, increasing volumes cause increasing problems. On the contrary, in developing countries the main problem is a proportionally more expensive infrastructural network over large distances, dispersed population centers, and low traffic flows. Large, sometimes oversized, networks often mean that maintenance is a major financial problem for these countries [5].

3.2.3 Classification of (Rural) Roads

Roads can be categorized in many different ways. In relationship to Fig. 3.2, a classification may be based on road design (construction), use (traffic characteristics), or function. For planning purposes, a classification by function is most useful. This is elaborated in the next sections, for both industrialized and developing countries. Classifications by road design are commonly based on pavement width and type. The distinction "paved/unpaved" is the simplest classification. Road surfaces can be paved with seales or oil-based materials as well as asphalt or concrete. Unpaved roads are characterized as earth, sand, or gravel. Traffic characteristics embody features such as traffic volume and composition by mode. The narrow relationships between the three classifications are self-evident. The more important the function of the road, the higher the volumes, the bigger the construction of the road, and the wider the pavement. Table 3.1 gives an idea of such relationships.

Divisions into categories have been developed with the intention of forcing the adjustment of the driving behavior to conditions of road and traffic. To enable road users to do this, differences among the categories should be clear to them. The clearest road feature is the width of the pavement. Therefore, in each country, standard sizes are handled and recommended.

Industrialized Countries

In the introduction, LTRs are defined as roads with lower functions than the trunk roads. Their most common function is giving access to rural areas. In addition to this function, LTRs may serve functions specific to certain countries [2], such as

- providing access to remote communities, e.g., in Australia and Canada;
- providing access to and circulation within recreational areas;
- providing long-distance connections between remote centers, e.g., in Australia, Canada, and Finland.

Table 3.1. Relationships among possible road classificiations in rural areas

Pavement Width (m)	Access Functions	Traffic Modes Allowed	Traffic Composition	Management Authority
3.0–3.5	Local access to parcels and farms	All (mixed traffic)	Destination traffic: agricultural vehicles	Private, municipality
4.5–6.0	Local/regional access to villages; (collecting and distributing local traffic to higher-level networks)	Mixed traffic; sometimes bicycles separated	Destination traffic: agricultural traffic, recreational traffic	Municipality, county
6.0–7.5	Regional access (opening up of regions, connection of regional centers)	Bicycles and sometimes agricultural vehicles separated	Through and destination traffic (cars and commercial vehicles such as trucks, tractors, and buses)	County, region, district
>7.5	Flow function: through traffic on long distances (no access)	Motor vehicles only	Through traffic (fast vehicles)	Federal, national

This also illustrates the enormous differences between countries, resulting in completely different purposes of the LTR network. However, even within the LTRs for rural access, considerable differences appear. In the Netherlands, three categories of LTRs are distinguished. As shown in Fig. 3.5, trunk roads are neglected. The three categories of LTRs imply different levels of access. The lowest category (VIII) gives access to some parcels or a single farm. The next category (VII) gives access to farms, whereas the highest category (VI) gives access to the villages (local distributor road). The figure also shows that both categories VII and VIII are subdivided into two types with different pavement width. So, in practice five types of LTR appear for three access functions.

Table 3.2 presents an indication of the differences between the three access functions of LTRs. Traffic volumes, characteristics, and speeds and road characteristics are compared. The table also indicates an important characteristic of LTRs: the mixed composition of traffic by mode. Heavy and light vehicles, slow and fast vehicles, and cars and bicycles occur together. In terms of numbers, passenger cars are dominant. Remarkable is the small portion of agricultural vehicles even on agricultural access roads. Because of general developments in transport, the portion of trucks slightly decreased during the past decade but truck dimensions (including axle loads) clearly increased [6].

A recent development is a desired reduction of the number of categories of roads. For LTRs, one category with only two types is proposed. One type will be provided with a separate bicycle path. Further technical layout of both types is still open to discussion.

For example, to obtain a policy framework to be able to manage, improve, and maintain the Devon Rural Road Network (United Kingdom), the network was categorized in a special way. An appraisal of the entire network was carried out by classifying routes under a number of generalized functional headings. These headings were related to the settlement pattern but also recognized the requirements of industry and the volumes of

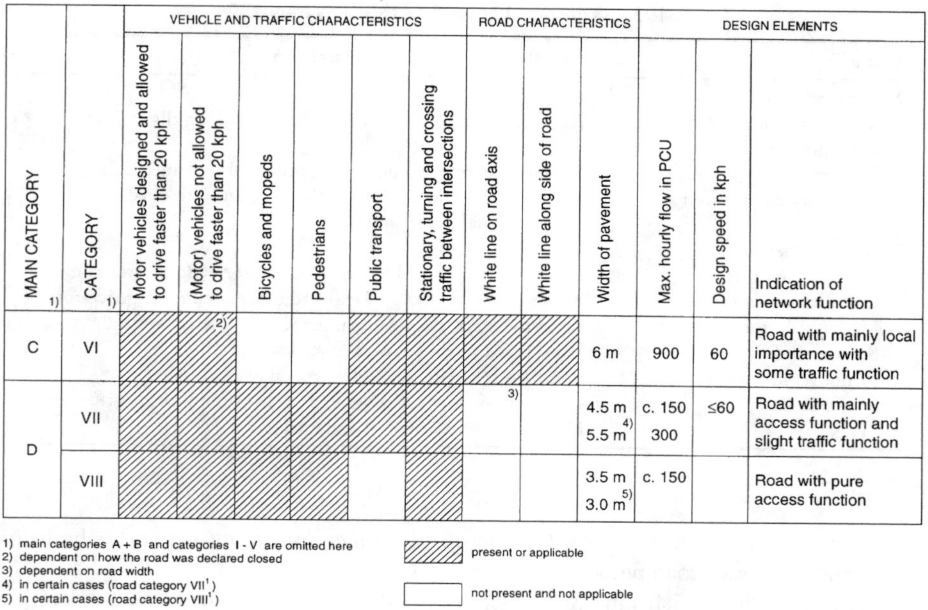

MAIN CATEGORY [1)]	CATEGORY [1)]	VEHICLE AND TRAFFIC CHARACTERISTICS: Motor vehicles designed and allowed to drive faster than 20 kph	(Motor) vehicles not allowed to drive faster than 20 kph	Bicycles and mopeds	Pedestrians	Public transport	Stationary, turning and crossing traffic between intersections	ROAD CHARACTERISTICS: White line on road axis	White line along side of road	Width of pavement	DESIGN ELEMENTS: Max. hourly flow in PCU	Design speed in kph	Indication of network function
C	VI	▨	▨ [2)]			▨	▨	▨	▨	6 m	900	60	Road with mainly local importance with some traffic function
D	VII	▨	▨	▨	▨	▨	▨	[3)]		4.5 m 5.5 m [4)]	c. 150 300	≤60	Road with mainly access function and slight traffic function
	VIII	▨	▨	▨	▨		▨			3.5 m 3.0 m [5)]	c. 150		Road with pure access function

1) main categories A + B and categories I - V are omitted here
2) dependent on how the road was declared closed
3) dependent on road width
4) in certain cases (road category VII')
5) in certain cases (road category VIII')

▨ present or applicable

☐ not present and not applicable

Figure 3.5. Classifications of roads with a lower hierarchical function [6].

Table 3.2. Comparison of traffic features on different access roads

Traffic Characteristics	Access Road to: Parcel	Farm	Village
Volumes (motor vehicles/day)	20–100	50–500	500–3,000
Portion of (%)			
Passenger cars	65	69	78
Heavy-goods vehicles	3	4	5
Agricultural vehicles	5	3	2
Bicycles and mopeds	27	24	15
Average speeds (kph)			
Heavy-goods vehicle	44.9	50.1	57.9^{a}
Motor vehicle	47.1	57.6	66.5^{a}
Exceeding speed limit (%)	2.6	10.3	26.9
Road type	VIII	VII	VI
Road characteristics			
Pavement width (m)	3.5	4.5	6.0
Design speed (kph)	—	≤ 60	60
Capacity (motor vehicles/hour)	50	150	900

^{a} Pavement width 5.5 m.
Source: Adapted from [6].

Table 3.3. Functional network in Devon County, UK

Route Type	Function
Major road network	
Motorway and primary routes (national routes)	National strategic routes for through and long-distance traffic
Primary county routes	Main county routes connecting principal settlements
Secondary county routes	Main access routes to large settlements and principal recreational attractions
Minor road network	
Local distributors	Access routes to small settlements and recreational attractions
Collector roads	Access routes to small villages and other significant generators
Minor collector roads	Local roads serving small hamlets and scattered communities
Service roads	Local roads serving a few properties
Minor service roads	Local roads serving only one property
Minor lanes	Other minor roads serving fields only or duplicating other routes
Tracks	Not normally used by vehicular traffic

Source: [7].

traffic, particularly trucks. Table 3.3 sets out the functional route network—both the major and the minor road network—for Devon County [7].

The Swiss distinguish among connection roads (point-to-point traffic), collection roads (area-related traffic), and access and minor access roads (parcel-related traffic) [4]. The German subdivision roughly consists of community roads, farm roads, forest roads, and rural roads for other purposes [8].

Developing Countries

The most common classification system used by developing countries can be given in terms of motorways and primary, secondary, and tertiary roads. However, major differences exist among countries in their breakdown of groups, or types of traffic, terrain, and road surfaces [2]. Rearranging the various data leads to three useful main categories, outlined in Table 3.4.

Category 3 includes the LTRs. It may be further divided into two subclasses: local roads and special service roads. This distinction is not meant to be a strict one; there is

Table 3.4. Classification of roads in developing countries

Category	Road Function	Surface Type	Kind of Management
1	High-volume traffic lanes	Bituminous concrete surfacing	Central government
2	All primary and secondary roads with low or medium traffic	All	Central government or, in the case of a federal structure, by the state
3	Tertiary roads with low or very low traffic volumes	Mainly nonsurfaced	Local or central government; in many cases, ministry for agriculture or development

Source: Based on [2].

a wide range of intermediate situations. Local roads are designed primarily to serve the rural population. In many cases, they are constructed using local human resources. A lot of these roads started out as a trails or village tracks used by people traveling on foot or on local animal-drawn vehicles. Such roads usually are not inventoried or classified and do not fall under any particular authority.

Special service roads are known as feeder roads. They serve to span the distance between the source of the product transported, which is generally a farm or a forest, and the place of use. In contrast with the local network, these roads are planned mostly from scratch, frequently by foreign consultants, and integrated into national defense or development projects [2].

In India, LTRs can be defined as "the roads primarily serving a group of villages, passing through mainly agriculture areas, and having relatively low volume of traffic, often slow moving." These roads are generally shorter, sometimes even as short as a kilometer. Rural roads consist of other district roads (ODRs) and village roads (VRs). ODRs serve the rural areas of production and provide them with an outlet to market centers or other main roads. VRs connect a group of villages with each other and to the market centers, or with the nearest road of a higher category [9].

Comparative analyses of LTRs in developing countries are difficult. There are no commonly accepted definitions of route status, and the three general categories of seasonal road, track, and footpath frequently coincide. A fundamental distinction can be drawn between all-weather and dry-weather routes, and also between tracks capable of carrying wheeled vehicles and those open only to pack animals and pedestrians. Figure 3.6 shows a road map of a part of Zimbabwe. It categorizes the qualitatively insufficient routes as being "unsuited to use by tourists." This also gives a realistic impression of the increasing importance of international tourism for the national economy [1].

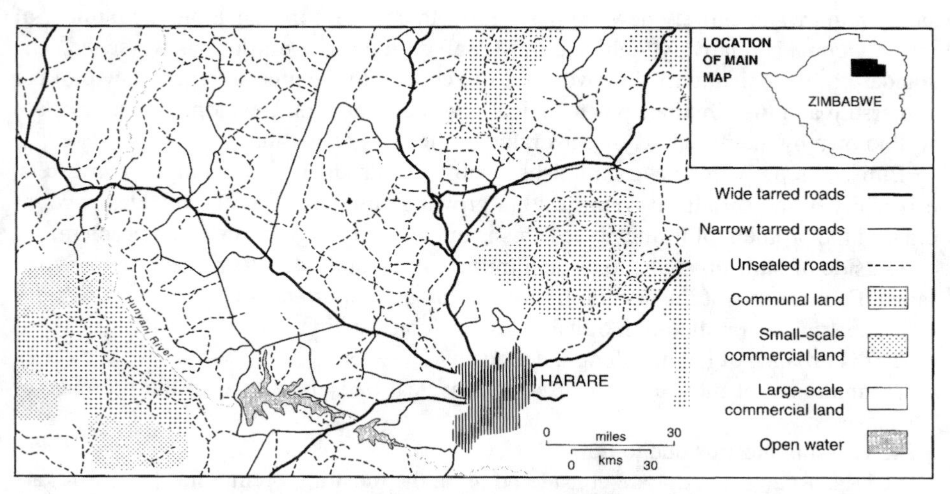

Figure 3.6. Road categories west of Harare, Zimbabwe. *Source:* [1], Reprinted by permission of Addison Wesley Longman Ltd.

3.2.4 Rural Traffic Problems and the Need for Planning

Traffic Problems in a Wider Context

Worldwide traffic problems are becoming manifest in different forms. Densely populated countries regularly have to struggle with large traffic volumes on roads with capacities that are too low, causing large traffic jams. On the other hand, problems of mobility and inaccessibility are created because of low population densities and different settlement patterns in remote areas. In African and Asian countries, road planning is usually a part of an integrated development program in order to advance the common welfare and living conditions. For this purpose, plans for mobility within and accessibility of the countryside are integrated in the development program.

Traditionally, road planning focused on the construction of new roads and on the modernization of existing ones, in order to satisfy traffic demands. However, gradually it became clear that this way of planning insufficiently considers side effects of roads and traffic. Technological, economic, and social developments have caused new trends in the planning of roads [2], such as

- the importance of safety and consistency in design in each road category;
- the harmonization of the service conditions offered by different networks and road categories;
- the inclusion of side effects of traffic for local people (livability, which is negatively affected by noise and pollution) and for flora and fauna (habitat fragmentation by roads and their traffic).

Today, safety generally is considered to be one of the most important traffic issues. In the United States and in western European countries, traffic safety has increased significantly over the past decades, despite a considerable growth of traffic volume. However, this increase has slowed recently. The number of victims is still much too high. Differences between countries are considerable. A substantial further increase in safety requires an entirely new approach [10]. In countries in transition, the number of traffic victims is relatively high, and traffic safety in these countries has worsened. The standard of the existing road network fails to cope with the present rapid growth of car ownership and the simultaneous technical upgrading of cars. Developing countries are well known by their relatively unfavorable traffic safety situation.

Emissions problems have diminished with the introduction of catalytic converters, especially in industrialized countries, but growing volumes of motor vehicles have increased the problem of traffic noise. This is mainly, but not only, an urban problem.

Outside of built-up areas, roads and traffic cause habitat fragmentation for flora and fauna. Four aspects of this habitat fragmentation are important [11]:

1. destruction or alteration of habitat due to construction works;
2. disturbance of habitat along the roads by noise, etc.;
3. hindrance of movements caused by physical barriers related to the presence of roads;
4. crossing hazards due to traffic.

For LTRs, aspects 3 and 4 are considered to be the most important ones [11]. Both are barrier effects, especially affecting fauna. They force a separation of functional areas, such as living and reproduction areas or rest and food areas as well as kill the crossing animals. Both an increase of the overall stock of roads and growing traffic volumes intensify habitat fragmentation.

Traffic Problems in Western European Countries

Traffic problems on LTRs are caused by changes in the use of the road infrastructure (growing volumes of vehicles in general and of rat-run traffic (defined below), larger farm equipment, and larger and heavier trucks). Through these changes, the actual usage of some roads is not in accordance with their originally planned functions and designs. Therefore, usage should be restricted to what is desirable and appropriate for the layout of the road. Attention should be given to the three major types of traffic problems on LTRs:

1. mixed composition of traffic by mode;
2. high speeds and large differences in speeds;
3. rat-run traffic.

Rat-run traffic is through traffic that uses roads with a lower hierarchical function in order to avoid a longer travel time (e.g., caused by traffic jams) and/or a longer distance on the functional route (along roads with a higher hierarchical function). This can cause several problems on misused LTRs: capacity (especially for LTRs with an access function only), safety (especially for nonmotorized road users such as pedestrians and cyclists, related to higher car speeds), and annoyance and emissions for residents. Relative to the traffic performance, the number of personal injury accidents between fast and slow traffic on LTRs is two-to threefold the number on trunk roads. These problems are still increasing because of the increased volume (and congestion) on trunk roads. The introduction of tolls on the major roads can cause a further shift of traffic to minor roads.

Road administration is used to respond to such growing volumes through an adaption of the road. At first view, this seems reasonable. Damage to the road may result in decreased safety and must be repaired. For management, a broadening of the pavement may be cheaper than a continuous repair of damaged verges. However, the disadvantages of this "following" approach become clear:

- Road improvements frequently attract "new" traffic, quickly consuming the newly reconstructed road capacity.
- The increase in traffic safety, if any, is very modest.
- Diffuse traffic flows are spread over the rural area, resulting in emissions and noise for residents and habitat fragmentation for flora and fauna.

To overcome these shortcomings, another approach needs to be followed [12]: the integral approach to regional rural traffic planning. See Section 3.3.3 for elaboration.

Mobility Problems in Developing Countries

The minimum socially acceptable levels of mobility identified in the developed world may be irrelevant in most developing rural societies, given the great differences in overall living standards. In developing countries, the quality of the infrastructure and levels of personal and household income are often the most important factors. This is unlike the industrialized world, where levels of mobility and access are closely related to the availability of public transport and private car ownership.

In most developing countries, the road is the principal mode available. Rail and inland waterway transport generally play a less important role within rural areas. However, for the developing world, the physical condition for the rural road network is one of the main conditions for economic expansion and the upgrading of social facilities. In the mid-1960s the road networks of most African states, apart from South Africa, had on sealed surfaces less than 5% of their total length. A large proportion of the minor feeder roads were, and still are, tracks beaten out by walkers and animal carts [1].

Agricultural improvement programs change the land use and increase crop yields. They also result in a need for longer trips by farmworkers to farming areas. When time spent on walking increases, the period available for farming activities decreases and the effectiveness of farm labor is even further reduced. In that case, farmers may decide to abandon their remotest fields to concentrate their efforts on the more accessible food crops. The need for regular supplies of drinking water and firewood in rural areas also can involve long trips. This can take up to four hours each day as, for example, in the eastern part of Africa. In parts of Tanzania, for example, water sources are often found at distances of at least 5 km from a village. Almost all of these trips have to be undertaken on foot. These long trips for water can be avoided if wells or boreholes are created closer to the villages [1].

Table 3.5 displays information on the small distances and the modest loads transported for the majority of trips in developing countries. To provide for these kinds of trips, an adapted network for local available modes may be preferable.

Traffic Problems: A Solution?

The foregoing leads to somewhat confusing and contradictory conclusions. On the one hand, in many parts of the world (such as developing countries, countries in transition) the primary goal still is to design and construct new roads. This is a main condition for economic expansion and the upgrading of social facilities. On the other hand, in other parts of the world (notably in densely populated, developed countries) disadvantagous side effects of the growing traffic flows become more and more apparent. It is also quite clear that neither the side effects nor the growing demand can be neglected: No country in the world is rich enough to permit a continuously congested traffic system.

Table 3.5. Rural transport trips by length and loads transported in some developing countries

	Distance of Transport (Km)			
Country	Typical	Average On-Farm	Average Off-Farm	Load (kg)
Kenya	90% of trips <7	0.8		70% of trips <25
Malaysia	75% of trips <7	1	10	
India	90% of trips <5	1.5	8.3	
Bangladesh	Most trips <12			Most trips <50
Western Samoa	Most trips <5			Most trips <80
Republic of Korea	Most trips <10			30–80

Source: [1].

Therefore, all over the world, physical planning and improvement plans for the rural road network are necessary to provide for people's needs in the future. Simultaneously, however, harmful effects of this network conflict with the principles of environmental sustainability. In practice, planners must search for solutions. It is quite clear that the kinds of solutions will differ with the circumstances, such as the land uses to be served, population density, possible alternative solutions, other lands uses that are affected, opportunities for mitigation, and compensation for side effects. Obviously, these are great challenges for planners [13].

3.3 Planning of Rural Road Networks

C. F. Jaarsma

This section presents the motives for rural road planning.

Traffic and transportation should be conceived as derived functions rather than as objectives in themselves. They originate from social activities, the volumes and characteristics of traffic depending on both location and type of activity. The location of the various land uses is a task of spatial planning. As a result, spatial planning forms an important determinant of traffic volumes and characteristics.

Many rural areas around the world are used as farmland. In the early days, the opening up of the farms to the fields and to the villages and towns was the most essential traffic aspect in these areas. It is still very important. However, the extent of importance varies from nation to nation, depending on the level of development.

The function of traffic is characterized by continuing development in the rural areas. In densely populated and well-developed countries, space has always been scarce. So, much attention has been paid recently to having a diverse, liveable, and sustainable rural area. Apart from the dynamic agricultural activities, many societal demands have arisen (e.g., for outdoor leisure activities) as well as demands for nature conservation and development outside the cities. These developments have had a clear impact on the rural road traffic. These other uses have increased the impacts of traffic on the rural landscape, especially the ecological system. Consequently, a good traffic system depends on a careful land-use planning system. This section deals with such a system. Some examples of planning systems and cases are described to give an adequate view into the international planning for rural roads.

First, the specific road planning for farming is elaborated. Then, a more integrated approach of planning for rural roads in developing countries is outlined. A description of the planning for multiple uses of rural areas is the third part of this section. In that part, ecological habitat fragmentation and different concepts for traffic planning are presented, illustrated with case studies for solving traffic problems through a regional approach.

3.3.1 Planning for the Opening Up of Agricultural Areas

Several studies are available on so-called optimal agricultural field length. Boss and Flury [4] cite a Swiss study from the early 1980s. This study shows that fields suitable for arable farming require a road on both sides, whereas one road is sufficient for grassland.

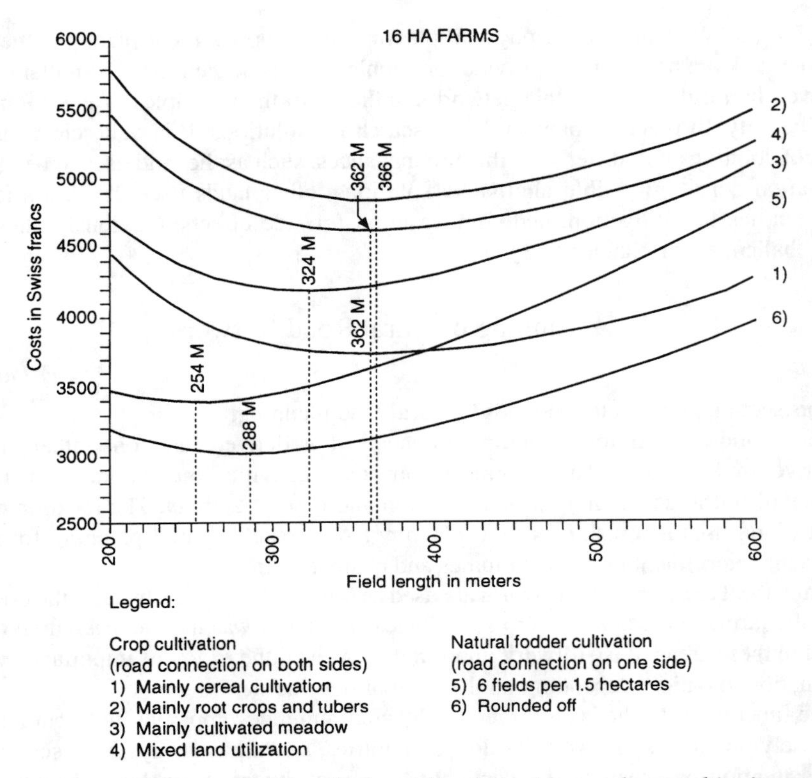

Figure 3.7. Optimum field lengths for a 16-ha farm with several crop cultivations. *Source:* [4], reprinted by permission of PUDOC-DLO, Wageningen.

This difference is explained by the burden of the harvest. Instead of a second public road, private farm roads on the field may be an alternative in arable farming. Further, Boss and Flury [4] state that the field lengths are dependent on the cultivated area per farm, the degree of mechanization, the soil utilization system, and the crop rotation. Figure 3.7 demonstrates the various effects of these factors on a 16-ha farm. The optimum field length is based upon the cost of turning (related to reduced yields at field edges and in the turning area), "empty" trips (due to additional filling and emptying during fertilizer application, spraying, and harvesting), double work during turning at field edges, and the additional expenditure of seed and fertilizer during an entire crop rotation. In addition to the 16-ha farms in the figure, it can be summarized that for 9-ha farms and 25-ha farms, respectively:

- the optimum field lengths are about 250 m and 450 m, respectively;
- the associated road densities are 3.5 and 2.5 km/km^2, respectively;
- the optimum ratio of field width to field length is 1:5 or 1:4, respectively.

It is clear that the optimum field length increases and the optimum road density decreases with increasing farm size.

To calculate the optimum road density in agricultural areas, it is advisable to take more factors into account than the agricultural ones mentioned above. Surely, factors such as costs of construction and maintenance of roads and ditches, as well as costs of drainage systems, should be considered. Then, an optimum size for arable farms appears to be 300 × 1,000 m for a 30-ha farm, 500 × 1,200 m for a 60-ha farm and 500 × 1,700 m for a 85-ha farm [14].

Generally, it can be stated that the larger the farms, the larger the fields and the lower the road density. This holds for public roads. Depending on the bearing power of the soil, private paved farm roads are advisable for larger farms in order to avoid long distances for internal transports on bad subsoil.

3.3.2 Planning for the Opening Up of Developing Rural Areas

Today, planning issues regarding the opening up of areas that are solely agricultural are confined mainly to developing countries. Especially in these countries remoteness, isolation, and inaccessibility are key characteristics. Moreover, the economic and social deprivation that these areas suffer is often largely due to inadequate transportation services. Investment in transport facilities is usually just one element of an integrated development program. This is designed to improve standards of living in these rural areas. Commonly, the main objectives are to increase the quality and quantity of commercial farming and to ensure a better distribution of educational, health, and other social services [1].

Although economic motives are reasons enough for road improvements, the same road also can serve the needs of farmers and provide connections to shops, schools, and villages, to vehicular or foot traffic. Road building and upgrading projects are essential if access to local markets, schools, and clinics are to be increased. Because of the high costs of road-building schemes, most plans are preceded by feasibility studies that investigate the potential benefits to be gained from road upgrading and the introduction of mechanized transport (if appropriate).

Rural communities, however, are just one of many sectors that depend on funds from the national level for road improvements. Even when an appropriate allocation of these funds has been made, the problem of how the optimal distribution of investments within individual rural districts, to secure the most effective return in social and economic terms, remains. The specific transport needs of each area must be identified, followed by the selection of the most effective technology in terms of road surface standard and vehicle type. Finally, the extent to which the opportunities, offered by new or improved roads, for the advancement of living standards, can be realized must be assessed wherever possible. Specific surveys usually are limited to rural areas with an existing or potential agricultural value, as well as to areas that are judged most likely to benefit from the investments. Given the costs of road construction, the issue of cost-effectiveness is often a major criterion [1].

Early studies of the effectiveness of road improvement programs in Africa and Southeast Asia were concerned primarily with improvements in vehicle operating costs and changes in agricultural output and prosperity. In areas with low crop yields, typical for remote areas as well as regions in the developing countries and countries in transition, it is likely that additional investments in road upgrading and new feeder road construction will continue to be highly selective and confined to areas producing the more profitable and higher-yielding crops. In the Ashanti region of Ghana, for example, 98% of the population are within 2 km of all-weather roads or tracks suitable for motorized vehicles. There, additional road improvements would have little effect on either farm-gate or village market prices here, because nontransport factors such as marketing practices are more significant in the local agricultural economy. However, schemes to upgrade footpaths to allow vehicles would be effective here, because improvements in the field–village linkages could lead to more efficient farm practices [1].

Today, "intermediate technology" is often accepted as being the most effective approach to satisfying immediate needs for better accessibility. It matches the limited financial means of isolated farming communities with the demands for more efficient but affordable means of transportation.

Traffic densities in many parts of rural Africa and Asia are insufficient to justify the extension of all-weather road networks. The ultimate objective of a more prosperous rural society with higher levels of accessibility may be obtained by means of a less sophisticated traffic and transport system. Moreover, this can be reached in a more rapid and cost-effective manner. Individual local authorities, rather than regional or national agencies, would be capable of initiating and operating these intermediate technology improvements, which then could be adjusted more closely to the transport needs of specific settlements. If commercial agriculture then prospered to a level where motorized transportation for marketing were feasible, a more ambitious program of road improvement could be introduced.

In many rural areas, therefore, current short-term investments are best directed at the lower end of the road hierarchy, converting simple tracks into serviceable roads to benefit local communities. As soon as information on the existing transportation requirements has been collected, a decision can be made about what level of transport technology is most appropriate for the situation.

In developing countries, the agricultural output from rural areas is still a very meaningful component of the national economy. Therefore, it is important to realize that the small-scale transport systems serving those areas are, in turn, a substantial part of the national transportation system. The systems require as much attention from transport planners as nonrural interurban transport [1].

Guidelines are published for the European Union, aiming to provide a comprehensive overview of the issues that are important for moving toward more sustainable transport infrastructure in developing countries [5]. Furthermore, the guidelines provide a sectoral framework in which project proposals and requests for European Union assistance to the sector can be examined. The guidelines comprise three parts: The first part provides an insight into the key issues in transport infrastructure in developing countries and the emerging solutions of a sectoral approach. The second part provides the means to apply a sectoral approach to examine proposals and requests for assistance in financing transport infrastructure projects. It is organized according the phases in Project-Cycle

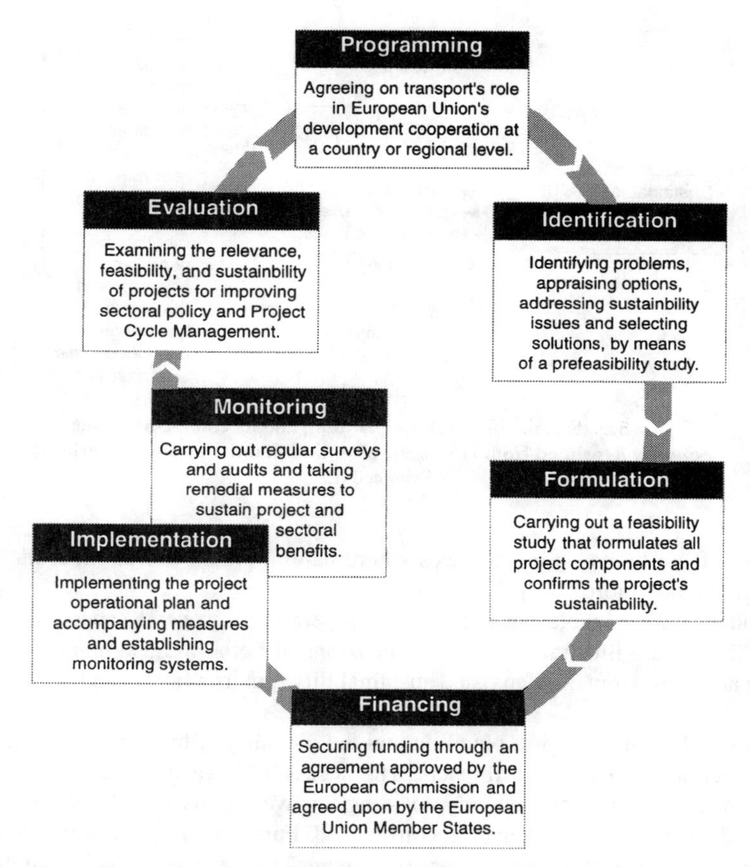

Figure 3.8. The six phases of the project cycle, as accepted by the European Commission [5].

Management, as adopted by the European Commission (Fig. 3.8). For each of these phases, the issues affecting project sustainability are raised in a series of key questions. This list of questions should be used as the starting point to raise further questions in examining the underlying problems and causes. Finally, the third part provides tools for developing and monitoring projects within a sectoral framework. These are standard formats for terms of reference and reports of studies in different phases of the project cycle.

3.3.3 Sustainable Planning for Multiple-Use Rural Areas

"Land-use planning deals with an active planning of land to be used in the (near) future by people to provide for their needs. These needs are diverse: from food products to places to live; from industrial production sites to places to relax and to enjoy beautiful

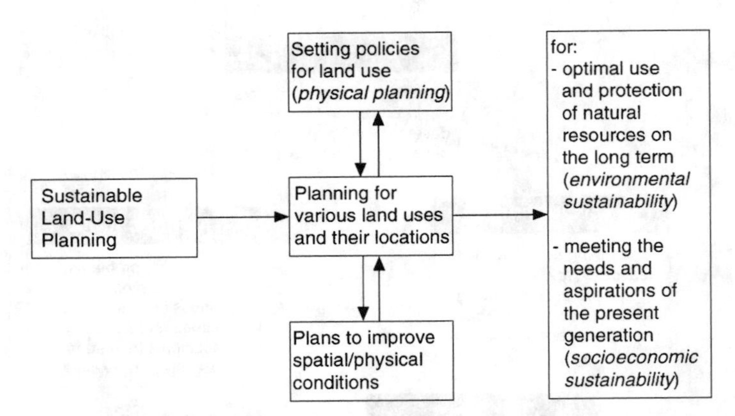

Figure 3.9. Sustainable land-use planning and its embracing aspects
***Source:* Reprinted from [15, p. 9], with kind permission from Elsevier Science NL.**

landscapes; from human uses to places where natural plants and animals can live and survive; and many more" [15].

The phrase "sustainable land-use planning" embraces several aspects, as given in Fig. 3.9. The figure illustrates the two dimensions of both land-use planning (physical planning and improvement plans) and sustainability (environmental and socioeconomic sustainability).

Traditionally, both land-use planning and the planning of the rural road network focused on economic and social impacts. The accessibility of the rural area used to be the first consideration. Consequently, existing unpaved roads were paved and existing paved roads were widened or even reconstructed. Furthermore, new paved roads were constructed. In practice, many minor roads were constructed to such a high technical standard that they competed with major roads. This is a follow traffic approach, which is based on road links, not on the whole network. In most industrialized countries, this policy was abandoned during the 1970s. Mainly because of lack of financing, (re)construction of the network of LTRs came to a standstill. In 1986, the OECD [2] established a policy, primarily designed to conserve the network of high-quality major roads. According to the OECD [2], the minor road network had become somewhat run down and a sizeable effort in terms of investment and energy would be needed to bring them up to acceptable standards. Furthermore, the OECD stated that design standards were not static but were changing in the light of technological, economic, and social developments. New criteria in geometric design, involving such considerations as energy consumption and side effects of traffic (noise, pollution, unsafety), were being included [2].

The inclusion of new criteria marked a transition to road planning with a wider perspective. It also marked a transition from planning for road links to planning for road networks. So, the question arises as to how such planning can be realized. For this purpose, Jaarsma [13] suggests a planning system for a network of rural LTRs in three stages. This system is based on existing or planned traffic flows.

To obtain a safe and efficient traffic system, first a functional classification of the network should be made. The assignment of functions to the regional road links should be based on an inventory of the present situation (function, traffic volumes and speeds, traffic accidents, acceptable road capacity, geometric features), regional transportation plans, possible changes in land uses, and traffic volumes. A proper distinction between minor roads with mainly a residential access function and minor roads with mainly a traffic flow function should result from this functional classification. For example, five classes of LTRs are distinguished for a densely populated country such as the Netherlands. The main difference in road design is the pavement width, for which 3, 3.5, 4.5, 5.5, and 6 m as standards are advised (Fig. 3.5).

The next step in the planning is a mutual harmony among desired function, technical layout, and traffic characteristics, for every road link in the regional network in question. This is schematized in Fig. 3.2. In this second step, the actual technical layout of each road link in the network is compared with its desired layout, concluded from the road function assigned in the first step. Discrepancies, if any, should be removed by introducing the desired technical layout [3]. This may result in a downgrading of some roads.

In the third step, traffic characteristics on each road link, such as volumes and speeds of cars, are compared with the characteristics belonging to a road with the function assigned in the first step. If discrepances appear, the roadlink is classified as a bottleneck. For these bottlenecks, a choice should be made between two fundamentally different solutions:

- upgrading a road from a lower to a higher category,
- reduction of traffic and/or adaptation of the traffic characteristics to the road.

The first solution is obvious when the desired function of the road is higher than the function belonging to the present layout of the road. In general this means reconstruction of an access road to a road with mainly a traffic function. Widening of roads, construction of bicycle paths, or even construction of new roads are alternatives within this solution (new construction).

Unlike the foregoing, the desired function may be lower than the function of the present layout of the road. Then, the second solution comes into the picture. This is also the case when actual traffic characteristics do not match with the traffic characteristics belonging to the desired function of the road. In general, this means refuse admittance to through traffic and/or enforce the desired driving behavior on a road with mainly an access function. There are two ways to realize this solution: structural regulations and incidental regulations. This approach to the problem has a small-scale character. It can be called amelioration [6].

Structural spatial changes to the road network influence rural access definitively, for example by removing a rural road or by transforming a rural road into a bicycle path. These changes can be very effective but they are very interfering for local residents and firms, and so, they should considered carefully. Thus, in practice, spatial changes are difficult to realize [6].

Incidental changes in the network imply either legal regulations (traffic signs) or small-scale adaptations of the roads in the field of traffic engineering. A detailed description of 16 civil technical regulations and five legal regulations, applicable for LTRs, is given by the CROW [16]. The suitability of these incidental regulations depends on the type of problem and the function of the road. This is illustrated in Table 3.6.

The functional network approach, as described earlier, differs fundamentally from the traditional follow traffic approach. This holds for both the regional scale and the leading role of the road function, resulting in an adjustment of traffic characteristics instead of the road characteristics in cases where discrepancies appear. It also means that the accepted traffic characteristics are embedded. However, for the road users the differences between the classes may seem too small, and so, they do not sufficiently modify their behavior (traffic characteristics, such as speed). Therefore, a clarification of the functional classification for LTRs for road users (especially car drivers) requires further research. The concept of traffic calming brings new ideas into this discussion.

In the functional network approach, the assignment of road functions is implicitly or explicitly based on existing or planned traffic flows. A starting point is a more or less stressed traffic function for each link in the network of LTRs. In built-up areas, this idea was abandoned with the introduction of residential areas. The approach of urban traffic calming started with speed reduction. Urban traffic calming became a wider idea with the Dutch concept *woonerf* in the 1970s, followed by the German *Verkehrsberuhigung* in the 1980s [17]. The basic principle is an integration of traffic in residential areas, but based on priority to the needs of *people*, not the needs of *traffic*. In the residential space, traffic is unavoidable but it should be subsidiary to other spatial functions. Traffic is concentrated on distributor roads that are able to cope with the flow [13]. This approach represents a fundamental change in rural transportation planning: from following actual traffic flows to *regulation* of it. Its application creates bridges between contradictory demands for rural road networks.

The suitability of this new concept of urban traffic calming as a principle for rural traffic planning is currently under investigation. For this purpose, the concept of a traffic-calmed rural area (TCRA) was established [13]. A starting point in the TCRA is the desired spatial function of the rural area, rather than the existing traffic flows. Aside from the demands of the population, the need for nature conservation also is considered. Principally, in the concept, *residence functions* (for inhabitants and recreationists as well as for local fauna) refer to the *flow function* for through traffic. With the modest technical design for this function, roads within the region mainly will have an access function. With a flow function, the region is surrounded by and accessible from rural highways. The underlying idea is a regulation of rural traffic flows with a clear separation of space for living and staying and space for traffic flows.

Although this underlying idea is the same in rural and built-up areas, large differences occur in desired traffic characteristics and in measures to realize them. In the TCRA, traffic speeds should be modest, for example, 40 to 60 kph. However, it is not desirable that the network be planned in such a way that cars are forced to drive too long on these roads. Therefore, a second type of road, with somewhat higher technical standards and speeds of,

Table 3.6. Legal and civil technical means and their applicability in relationship with the type of problem (rat-run traffic, traffic speeds, or mixed traffic) and road type[a]

		Rat-Run Traffic					High Speeds or Large Differences in Speeds					Mixed Traffic				
Incidental Regulations		VI	VII′	VII	VIII	VIII′	VI	VII′	VII	VIII	VIII′	VI	VII′	VII	VIII	VIII′
L	***Legal regulations***															
1	Closed for motorized traffic			+	+	+								+	+	+
2[b]	Road signalization	+	+	+								+	+	+		
3[b]	Reducing maximum speed	+	+	+	+	+	+	+	+	+	+					
4	Restrictions on width, length, and/or weight				+	+									+	+
5	Left and/or right turning not allowed	+	+													
C	***Civil technical regulations***															
1	Visual narrowings						+	+	+							
2	Physical narrowings						+	+								
3	Central rumble strips			+				+	+				+	+		
4	Design of pedestrian or cycle paths						+	+				+	+			
5	Cycle suggestion lanes											+	+	+		
6	Roundabout						+	+	+							
7[b]	Central line deflections			+	+			+	+	+						
8	Rumble strips						+	+	+	+	+					
9	Service roads											+				
10	Blocked, with side lane for turning around				+	+									+	+
11	Median painted or raised traffic-guiding islands, with traffic lanes bending outward						+	+	+							
12	Idem, in curves						+	+	+							
13	Barricades and curve attention signs						+	+	+							
14	Traffic-guiding earth walls in outside curves						+	+	+							
15[b]	Emphasizing entrance of urban areas			+	+	+	+	+	+	+	+					
16	Stressing the crossings at intersections						+	+	+							

[a] Road types VI–VIII; see Fig. 3.5 for more details. Pavement widths of types VI, VII′, VII, VIII, and VIII′ are 6.0, 5.5, 4.5, 3.5, and 3.0 m respectively.
[b] Only applicable as a component of other regulations.
Source: Elaborated from [16].

for example, 60 to 80 kph is planned. Only this second type of LTR is connected with the network of rural highways and trunk roads. This part of the implementation of the TCRA strongly differs from the usual classification as given in Fig. 3.5, with five categories. However, it refers to another recent development, the concept of sustainable safe road traffic [10, 18]. This concept introduces a preventive approach to traffic safety, instead of the traditional curative approach. The reason for this new approach is the decreasing efficiency of the curative approach, especially in most western European countries. The curative approach can be successful as long as a systematic improvement of the so-called "black spots" in the road network is possible. The remaining safety concerns are more complex and more diffuse and are too difficult for the curative approach.

The sustainable safe-road traffic concept describes a durable safe (DS) traffic system. Instead of the car, the human being with all his constraints is the starting point of DS. This concept consists of three components: traffic participants, vehicles, and infrastructure. For the infrastructure, DS suggests a classification of the complete road network into only three categories, based on three traffic functions:

1. a flow function, for a fast and comfortable service for through traffic on long distances;
2. an access function for regions (opening up of regions); and
3. an access function on the local level (accessibility of destinations).

The technical design of the roads in the DS concept depends on the function assigned, in which an optimal safety is guaranteed. Three safety principles are systematically and consistently applied:

1. to prevent unintended use;
2. to ensure that speed differences are small on roads where traffic is in both directions on one lane and/or where there are large differences in vehicle mass; and
3. to promote the predictability of road path and traffic behavior through a maximum standardization of the road design.

LTRs will have the third function: access on the local level. For such roads, the safety principles lead to

- prevention of through traffic and rat-run traffic,
- enforcement of low speeds (e.g., maximum 40 kph),
- standardization of design (as much as possible).

Low speeds on LTRs are necessary in the DS concept because of the presence of slow vehicles, traffic flows in two directions, and a mix of light and heavy vehicles on only one lane. For cars, the speed limit is reduced (40 kph e.g., to). The acceptability of the low speeds, from the point of view of the drivers and the perspective of regional accessibility, strongly depends on trip duration on these roads. A maximum trip duration of 3 minutes on roads with this low speed limit has been proposed [18].

Realization of the DS principles strongly depends on the mesh sizes of the regional and local access roads. The larger the area to be opened up with local access roads, the greater the travel time. This demands that small regions be opened up by local access roads. On the other hand, small regions need a high density of roads with a regional access function. This is an expensive solution that requires much space. Therefore, large regions with only access roads on the local level seem to be preferable. However, this leads to another problem. The larger the region with local access roads only, the greater the traffic volumes and the wider the needed roads. For that reason, the current planning distinguishes five categories of LTRs (Fig. 3.5). In other words: the technical layout of LTRs cannot be similar. This conflicts with the DS aim of standardization. As yet, two types of DS LTRs are proposed. For modest traffic volumes (average annual daily traffic [AADT], less then 1,000 motor vehicles per day), a maximum pavement width of 4.5 m is proposed. This type of DS LTR only services local inhabitants and firms. It has an access function only. A second type services local land uses with a higher traffic volume, such as larger villages. For this type a pavement width between 4.2 m and 5.4 m is proposed. To guide the traffic, either median or side lines are painted. Further, because of the higher speed level on this type (60 kph is proposed), a separate path (minimum size 2.25 m) is provided for bicyclists. Only this second type of DS LTR is directly connected with the rural highways. To reduce speed, especially at the intersections, four-way intersections (full crossings) are avoided as much as possible. Miniroundabouts and T-crossings offer a safer solution. To reduce speed, especially at potential conflict points such as exits, the civil technical regulations presented in Table 3.6 may be applied.

Figure 3.10 illustrates a possible solution for the functional classification of a rural road network. Within the rural area between the two rural highways, two types of LTRs are distinguishable. To provide for the access to farms and fields, the modest type of DS LTR will satisfy (pavement width 3.5 or 4.5 m; see A-A and B-B in Fig. 3.10). The connection with the rural highways is made by the second type of DS LTR (pavement width 5.4 m; see C-C in Fig. 3.10). Note that direct connections between the modest type of DS LTR and the rural highway are avoided. If necessary for access, a parallel road is applied. Crossings between the DS LTRs are T-crossings wherever possible.

Implementation of the DS concept still needs further investigation into the classification of the network of LTRs and the technical design belonging to it. This is also so for the acceptable travel time on roads with low speed limits. The DS concept only considers traffic safety and does not include the other side effects of traffic. Nevertheless, the approach has given new perspectives to the principles of regional road network planning based on regulation of rural traffic flows.

Based on recent case studies, a procedure for sustainable rural traffic planning is elaborated in Fig. 3.11. A structural regional approach starts with determination of the boundaries of the area to be considered (step 1). It is important that a coherent network exist within the area included in the study. Natural boundaries, such as rivers and canals, may be much more useful than boundaries of management and administration. In step 2, data on road and traffic characteristics are sampled. Information about present and planned land uses, as well as policy documents on physical planning, is also important. Based on these data, a functional classification of the road network is made (step 3). In step 3, it is important to check the future accessibility especially of the land uses with high traffic volumes. After the functional classification, for which options may be elaborated, a traffic forecast is made in step 4. Also, the main impacts of the flows are calculated. In step 5, for each link in the network, the calculated flows and their impacts

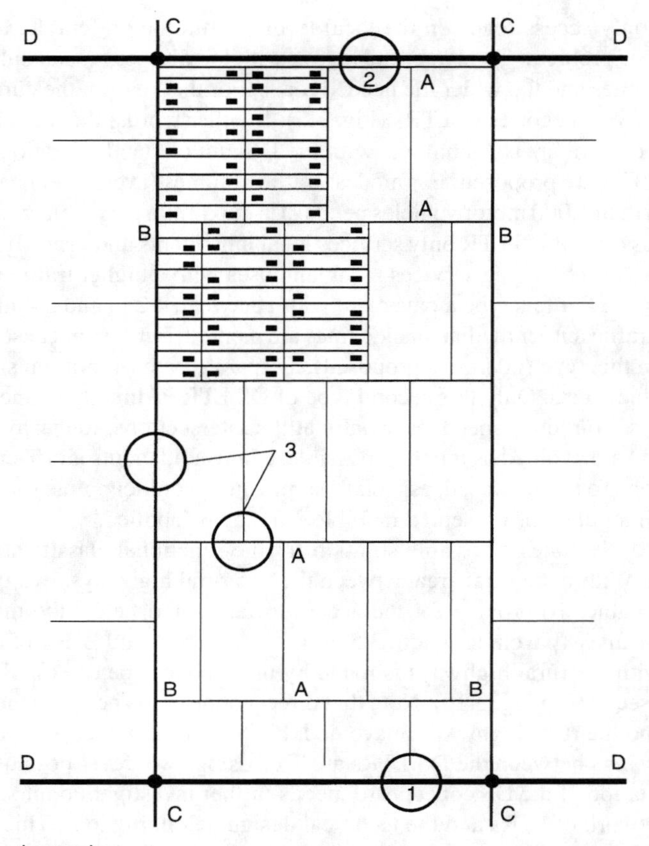

Legend:

A - A } access roads to parcels and
B - B } farms only (DS LTRs of the lowest category)

C - C access roads to villages (DS LTRs of the higher category)

D - D rural highways

▬ farm ● roundabout

(1) direct connection between rural highway and DS LTR of the lowest category is missing

(2) as in 1; parallel road gives access to farms

(3) T-crossings instead of four-way intersection

Figure 3.10. A possible solution for the functional classification of the road network, within the context of rural traffic calming.

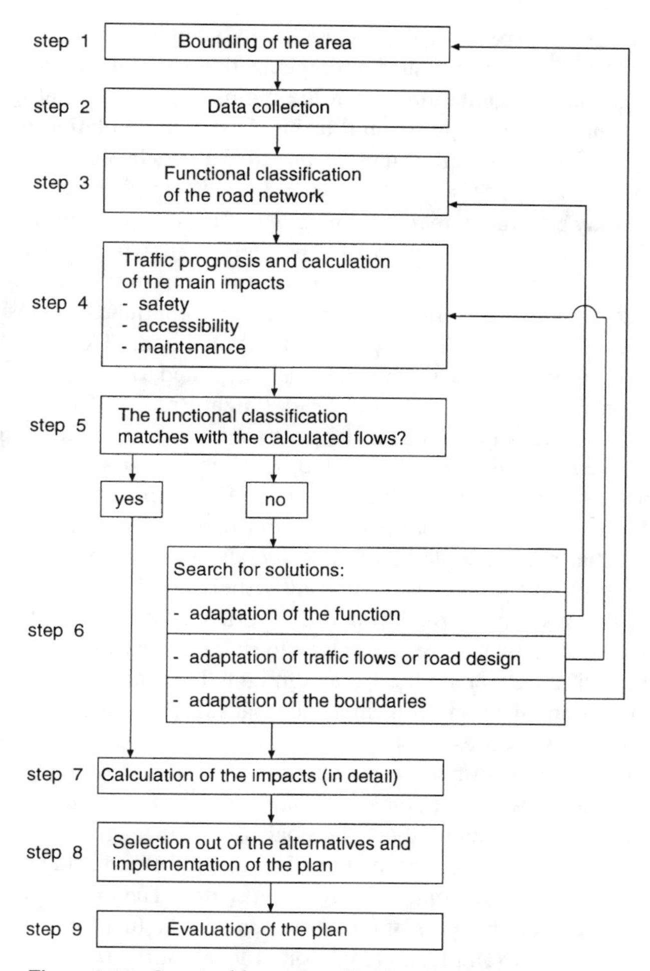

Figure 3.11. Sustainable rural traffic planning, step by step [19].

are compared with the flows that are characteristic for the function assigned to that link. If discrepancies appear, step 6 suggests three different types of solutions. When the road and traffic characteristics from the functional classification match with the calculated flows, impacts can be calculated in more detail (step 7). Step 8 implies a selection from the possible options and implementation of the plan. Step 9, evaluation of the plan, is strongly recommended [19].

In conclusion, the concept of the TCRA decreases traffic speeds on LTRs. With a well-designed network, this may result in a reorganization of traffic flows:

- Diffuse volumes at the minor rural roads will be concentrated at a few trunk roads.
- Remaining traffic flows at minor rural roads will be rural bound (origin and/or destination along a minor road).

- Traffic volumes and speeds within the region will decrease.

Case studies [19, 20] show that such a reorganization will counter both traffic safety problems and habitat fragmentation within the region. These cases also illustrate the surplus value of integral planning of rural traffic, based on regulation of traffic flows. Even other impacts of rural traffic may be included in such planning. For example, emissions of exhaust gases, noise levels, or energy use [21] may be considered.

In practice, it may be difficult to realize integral traffic planning. Many rural areas currently do not suffer from the same car-related problems with which urban environments are faced. Therefore, the need for rural traffic planning and traffic management is not widely recognized. There is a feeling that rural areas can continue to absorb large traffic increases [22]. Cullinane et al. [22] report on a Devon County Council and Dartmoor National Park Authority proposal to implement an integrated traffic management scheme at Burrator Reservoir in Dartmoor National Park (Great Britain). This scheme proposes the closing of some roads in the area (either permanently or on summer Sundays only), the running of a frequent minibus service on days when these roads are closed, and the extension of bus services and an existing cycle path. However, criticisms of the scheme, promulgated by a minority of local residents, mushroomed into a major problem. It became obvious that many people had failed to understand the motivation behind the proposals. As a result, the scheme was reluctantly shelved [22].

What are the lessons to be learned from this experience? It is generally accepted that, to be successful, a traffic management plan should be integrated and composed of carrot-and-stick elements [22]. Other studies have highlighted the role of effective marketing. In our opinion, communication with interested parties early in the planning process is very important. These parties must have the opportunity to share their visions on the problems. The next step, finding solutions, should not be started too early because nearly every proposed measure will have some disadvantages for some of the parties involved. Therefore, a complete agreement about the goals of the management scheme and the problems to be tackled should be reached in advance. In a second stage, measures may be developed. Sometimes, even options may be presented. The impacts of the proposed measures can be tested on the basis of the agreements in the first step. This allows for a check of agreements, for example, on reduction of local traffic flows and traffic victims and on a minimum level for local accessibility.

The final conclusion is that, throughout the ongoing growth of motorized traffic, there never has been a greater need for the successful implementation of integral rural traffic planning. To realize this, early communication with interested parties and effective communication play major roles. Main difficulties met during the planning process may be of a social, not a technical type.

3.4 Construction of Rural Roads

J. Amsler

Access roads are generally the first infrastructural features to be installed in a rural or remote area. Such roads are designed to meet the needs of agriculture or forestry management. It is usually difficult to predict how an area will be developed. This will depend on several factors:

- the area's attractiveness for nonagricultural and nonforestry activities (tourism, residential use and leisure activities, quarrying, landfills, commercial fishing, transportation facilities, and so on);
- activities of the resident population (interest in the marketing of advantages, amenities);
- political planning preferences oriented either toward protection or utilization (landscape planning);
- natural physical requirements, including danger zones (avalanches, landslides, floods, etc.), water protection zones, zones with few human activities, species diversity (occurrence of rare species).

In areas where agricultural and forestry activities remain dominant, rural roads usually manage to satisfy the requirements for a number of decades. If, on the other hand, an area is developed for tourism (e.g., winter sports) or for quarrying (heavy transport), the access requirements will increase dramatically. In such cases, the existing roads have to be widened, reinforced, made safer for traffic, or rebuilt entirely, depending on the extent of the new requirements.

In several countries, minor roads are built by private individuals or cooperatives. Such roads probably receive state subsidies but must themselves pay for the upkeep, financing this with the help of the users and, where one exists, through a cooperative. However, it should be clear that, in the case of changed requirements, a new maintenance model will be necessary.

In cases where the planning requirements are to be determined as part of a political process prior to opening an area to a variety of activities, attention also must be paid to such questions as the ownership and the responsibility for operating and maintaining the roads concerned. It may be appropriate to take these roads into public ownership.

Therefore, it must be concluded that the concept of minor rural roads is not a clear-cut one, which makes it more difficult to prescribe general guidelines for their construction. The following guidelines apply only to roads with a width of 3.0–4.0 m for single-track roads and 5.0–6.0 m for two-track roads with a layout of 100,000 to 150,000 standard single-axle loads (8.2 t). See Table 3.7 for definitions of technical terms.

It is particularly worth noting that, where paths and roads have been well laid out, with appropriate geometry, they generally are easy to adapt to more demanding requirements. However, steeper gradients and the crossing of sliding areas in particular represent almost insurmountable obstacles to enlargement. It is difficult to alter a gradient once it has been laid out.

3.4.1 Framework Conditions

Those involved in the planning and construction of minor rural roads often find that the financial means of the main contractor (rural commune, cooperative, individual) are limited and the project will have low priority among competing government activities. Moreover, these roads have to be planned for slow-moving, infrequent traffic, and, at the

Table 3.7. Definitions of some important technical terms

Term	
Design load	Definition standard single-axle load or equivalent standard-axle load: 8.2 t or about 80 kN (US: 18 kip [1 kip = 4.4 kN])
HMT	Heissmischtragschicht: Hot-mix asphalt (mixed in plant), used as bearing bed and pavement
GIS	Geographic information system: a system for capturing, storing, checking, integrating, analyzing, and displaying data about the Earth that is spatially referenced. It is normally taken to include a spatially referenced database on appropriate applications software.
LIS	Land information system: a system for acquiring, processing, storing, and distributing information about land.
CBR	California Bearing Ratio
CBR%	$\frac{P}{P_s} \cdot 100$ P : resistance of the material in place P_s : resistance of standard material
ME	Disk standard procedure
ME	$\frac{\Delta P}{\Delta S} \cdot D$ [N/mm^2] ΔP : change of change (pressure) ΔS : corresponding change of settlement or [kg/cm^2] D : Diameter of disk

same time, designed to bear heavy loads. They should also be easy to drive on, should conform to appropriate standards in terms of safety and consolidation, and should offer sufficient comfort.

Although in many countries a complete set of standards for high-performance roads is readily available, these standards simply are not applicable to minor rural roads for a variety of technical and economic reasons. Generally speaking, there are considerably fewer regulations intended for minor rural roads than for the high-performance type. This allows greater freedom to those concerned in the modelling of such works (variants and selection criteria). Despite this fact, more assessments and decisions are required. Two main criteria are worth mentioning: utility and cost-effectiveness.

Conspicuous examples are found in the work of the pioneering Swiss bridge builder Robert Maillart. His celebrated bridges often are found on country roads and are fine examples of the conservative use of materials, which at the time was an important consideration in view of the higher relative costs of both materials and transport, as well as for the superlative and even daring way in which they are laid out [25].

For the following remarks we make reference to Hirt [24] and Kuonen [23].

3.4.2 Concepts

The standard concept of a minor rural road is given by its cross section, as demonstrated in Fig. 3.12.

3.4.3 Substratum and Foundation

The optimum requirements will have been met if the subsoil has excellent bearing capacity and also can be used as building material, which can reduce the construction costs drastically. Unfortunately, such conditions are rare. Most sites either have been developed already or lack the appropriate dimensions. Geology is differentiated only

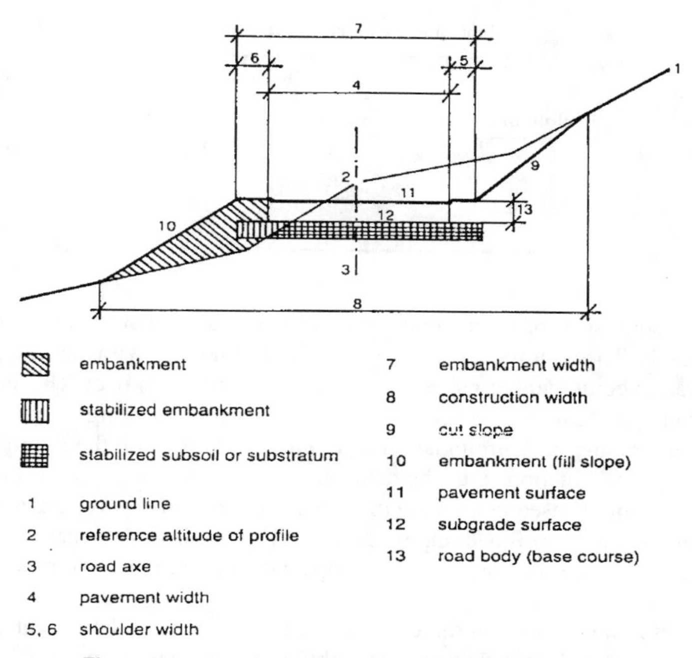

Figure 3.12. Standard cross section of a minor rural road.
***Source:* Adapted from [23].**

in wide areas. However, development areas are usually smaller. It is improbable that a subsoil with sufficient bearing capacity can be found only by changing the alignment of the road; subsoil treatment measures will be required in any case. Local zones with bad subsoil should be upturned.

The following substrata are more or less suitable for the building of minor rural roads (suitability of other materials can be determined from Table 3.10, as discussed later in Section 3.4.7):

- good substrata with the requisite bearing capacity include limestone, moraine, ballast, alluvium or alluvial deposit, conglomerate, granite, gneiss, schist, or slate (solid rock);
- substrata lacking both sufficient bearing capacity and suitability as building material include opalinus stone or clay, marl, marl slate, decomposed moraine, slope wash, loess loam, molasse debris or soft sandstone, alluvial soil or silt, flysch debris, decomposed sandstone, clay slate, marl slate, and peat.

Cost-conscious construction requires as little displacement of earth (earthwork volume) as possible, from excavation to deposition. The aim is to achieve balance. It is with this in mind that the choice of machines for transport and spreading should be made. If earth displacement over long distances by means of trucks, dumpers, or other transport is unavoidable, preliminary steps should be taken to stabilize the subsoil or to provide the necessary foundation layer. This can result in a considerable increase in cost, however.

Table 3.8. Stability of earthslope

Bottoming	Slope Gradient	
	Excavation	Embankment
Noncohesive rock	1:1	4:5
	4:5	2:3
Solid rock	5:1	—
	10:1	—

For cost reasons, state-of-the-art bottoming (filling) in accordance with the regulations is rarely possible. The bottoming can be done makeshift fashion with the caterpillar track of the excavation equipment or using pressure from the power shovel. This often results in long-lasting problems of uneven settlement. Therefore, waiting one to two years before fine subgrading of the foundation layer and surfacing with the bituminous layer is recommended. Distinctions must be made among the following types of bottoming:

- Side bottoming is used in the case of mixed profiles with excavation and embankment immediately to full dumping height. To prevent any danger of slipping or shifting it may be necessary to form a slope base and stepping. No real compacting will be necessary.
- Head bottoming is used in the case of short embankments built over gullies and ditches immediately to full dumping height. Again, no real compacting is necessary. Better results will be achieved with the help of excavation equipment.
- Layered bottoming is used only in the case of long embankments. The bottoming should be done in layers, with subsequent compacting.

The compaction requirements for the subgrade surface are contained in several standards of different countries. The requirement that calls for a ME > 150 kg/cm^2 and a CBR $> 8\%$ in the case of 100% Proctor standard is not achievable, in particular with cohesive ground. More realistic values for simple roads are ME > 100 kg/cm^2 and CBR $> 3\%$. These are minimum values, however, for construction to begin, that is, to consider installing and compacting the road body. There must be no mixing of the sand and gravel with the subsoil and, moreover, a suitable foundation must be available for the compacting. By way of improvement, the following measures should be foreseen:

- where the natural water content (w_{nat}) is approximately equal to the optimum water content by compression (w_{opt}), normal compacting is possible.
- where w_{nat} is much greater than w_{opt}, dewatering by airing or lime stabilization is possible.
- where organic soil or peat occurs, a pounded layer, a spur layer, fibrous matting, or stabilization of foreign material can be used in application of the "floating road" principle. (floating road: road structure has to be enough flexible to avoid fracture in case of unequal settlements caused by the underground)

The largest possible slope gradient and the lowest level of safety permissible in terms of stability under load should be preferred, in order to keep earthmoving to a minimum. In the event that the slope slips or is demolished as a result, repair during construction or when the snow next begins to melt in mountainous regions, will be more cost-effective than carrying out the bottoming with fairly level slopes and large cubatures. It would be advisable to model new slopes of earthworks on stable slope gradients in the immediate vicinity and in general to adapt the road to the surroundings. This generally means low slopes. In any case, fairly level slopes are necessary for the activities carried out in agricultural areas. The availability of a number of variations (steep slopes achieved with biotechnical measures, or retaining walls) should make it easier to reach consensus in the search for a cost-effective solution.

The safety of the slope should be an immediate priority (application of maximum safety level). When deciding on the method, attention should be paid to appropriateness to the landscape, good cover of the surface to prevent erosion, thorough rooting for consolidation, and good water runoff. The method and the choice of plants will depend on the purpose, the local conditions, the climate, the danger of erosion, the steepness, and so on. Biotechnical measures often are combined with civil engineering and landscape architecture techniques [26, 30].

Bear in mind that, even as early as the planning stage, few walls and only short ones are necessary for minor rural roads. With simple roads, the supporting structures are mainly deadweight constructions (wood boxes, wire ballast baskets, block bottoming). Wood boxes and wire ballast baskets, in particular, have to be renewed every 10 to 20 years. These can be combined with biotechnical measures. Guidelines for the dimensions of nonreinforced concrete walls are available in a number of countries. State-of-the-art supporting walls are required for greater heights. These are expensive, however, and their design loads are not suitable for minor rural roads.

3.4.4 Frost Problems

In areas where frost is common, with soils that are vulnerable to frost conditions and the associated water input problem, simple roads are particularly exposed to frost damage (ice formation). Frost-sensitive soils fall into two main categories:

1. silty soils, where frost causes lifting and thawing causes a loss of bearing capacity;
2. clay soils, where thawing results in a loss of bearing capacity.

Under normal conditions, a road with the proper dimensions will not suffer from frost. In gravelly or sandy soils, frost will not be a problem as long as the drainage is good. In the case of roads in Alpine regions that are not kept clear of snow, the snow acts as an insulator and no frost-related problems arise.

In the case of most soils, except extreme silt or very fine sand ($I_p < 12$, where I_p = plasticity index), building for bearing capacity will suffice (e.g., as per American Association of State Highway Officials (AASHO)).

3.4.5 Drainage

Drainage is also an important consideration when planning simple roads. If sufficient thought is not given to providing adequate drainage, then major damage may result. Large quantities of surface water can particularly be a problem with inclined terrain, so that it will pay to consider various ways to ensure flawless drainage. In this context, the first consideration should be to determine in which hydrological catchment area the road is situated. This depends on the adjacent area (upstream) and in particular its dimensions, inclination, and the ground cover. In mountain regions, the amount of melt water (from melting snow) is also important, as are the location and capacity of surface waterways in flat regions. Depending on the road environment, the following systems can be applied:

- In forests, open ditches are easy to manage and maintain. If the gradient exceeds 3% to 4%, or the water input is relatively large, fortification of the base is advisable.
- In agricultural areas, so that there is the least interference. With cultivation, rubble drains with pipes (usually plastic) should be used.

When it comes to drainage of the road surfaces, protection of the road body and the substructure from both surface and seepage water should be the foremost concern. Erosion of the cover layer is to be avoided. Aquaplaning should not be a concern, considering the low speeds normally involved on such roads.

To promote drainage, several types of carriage crossfalls, shown in Fig. 3.13, are proposed:

- Downstream carriageway crossfall. For ecological reasons (to avoid interference with natural circulation), an effort must be made to avoid concentrating the water drainage in a way that allows increasing high-water peaks in the drainage ditch; additional measures (retention basins) also should be avoided. A downstream road gradient of 1% to 3% is a good solution. It is also advisable to avoid long distances between discharges on the road surface because of the danger of erosion of the verges and of the downstream slope. Regular partitions should be made to allow drainage of the carriageway water. The subgrade surface also has to be slanted to ensure even thickness in the road body. A downstream carriageway crossfall of 3%,

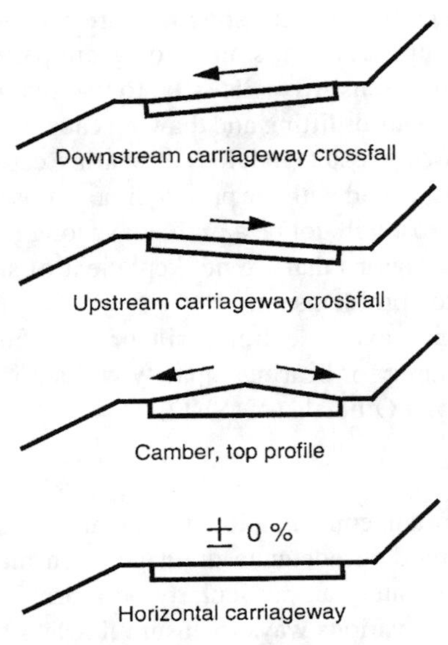

Figure 3.13. Types of carriageway crossfall.

however, can lead to a certain amount of anxiety for drivers. In the event of ice formation, there might be a danger for vehicles (the liability of the owners of the road will need to be clarified).

- Upstream carriageway crossfall. In addition to the variations already mentioned, upstream longitudinal drainage (weep drain) also must be provided, although it will increase the cost. In poor weather conditions (ice on the road, slushy snow), this will provide a certain additional safety, because vehicles tend to skid in the upstream direction.
- Camber, top profile. This method, common in earlier days, is still appropriate for dirt roads. The amount of water that falls on the road is channeled evenly to both sides of the road. Runoff grooves should be provided on the verge downstream, and longitudinal drainage (ditches) will be necessary upstream. The subgrade surface must be made horizontal, so that the greatest road body thickness occurs between lanes.
- Horizontal carriageway. This is a commonly employed method for simple covered roads. The water drainage occurs sporadically at the curves thanks to the camber of the road surface. This technique is not recommended for dirt roads.

The general rule to follow is that water must be evacuated from the road surface as quickly as possible. Cross drains should be provided every 20 m to 50 m, depending on the steepness of the road, the amount of precipitation, and the frequency of bad weather. These will affect the road comfort as well as increase the cost of road maintenance.

Longitudinal drainage ditches, with or without pipes, also are commonly used systems. Other possibilities, however, include asphalt reinforcement, an asphalted ditch, or prefabricated concrete forms. The latter are expensive and therefore not widely used for rural roads.

Intake constructions (shafts for road water) should be placed so as to ensure that the surface runoff water does, in fact, flow into them, and to minimize the danger of obstruction (leaves, twigs, branches, and road debris) as well as to make maintenance easy.

Particular attention also must be paid to stream crossings. If possible, bridges should be avoided as a solution because they represent a significant cost factor. Depending on the topography and other marginal considerations (the danger of flooding), there nevertheless may be a need for bridges, support viaducts, or other major engineering structures, which we do not discuss further at this point. Some possible simple solutions for stream crossings are as follows:

- Fords. This is an appropriate solution for ditch crossings involving large inflows of water on a periodic basis, avalanches, and landslips. A longitudinally cut pipe culvert, accessible from above so that blockages can be remedied simply and quickly, is recommended as a normal outlet.
- Plate outlets made of local concrete.
- Plate outlets made of prestressed and glued wooden slabs.
- Outlets of corrugated steel (corrugated pipe outlet).

It is important that both the in and out ends of the outlet be secured and that the road be protected (from erosion in flumes and on the road surface).

3.4.6 Road Body

Structure

The road body serves to transfer the load and to prevent penetration of water and thus erosion. It reduces the danger of frost and, depending on the type of surface finish, the development of dust. The course structure of simple roads has the following pattern (beginning with the bottom course, ending with the top, as shown in Fig. 3.14):

- Transition course. This serves to prevent interpenetration of the subsoil with the foundation or bearing bed (base course) in cases where there is poor subsoil bearing capacity. The various possibilities include lime stabilization, stabilized foreign matter, wool (fleece. [expensive]), beaten layers, branch layers (only small surfaces possible).
- Foundation course. In an effort to contain costs, local materials such as uncombined gravel and sand and crushed rock should be used whenever possible. The degree of consistency in bearing capacity, which is required in the construction of major roads, is not usually possible with minor rural roads.
- Bearing bed (base course). There is often no distinction made between a foundation course and a base course, particularly when the latter is an uncombined layer using

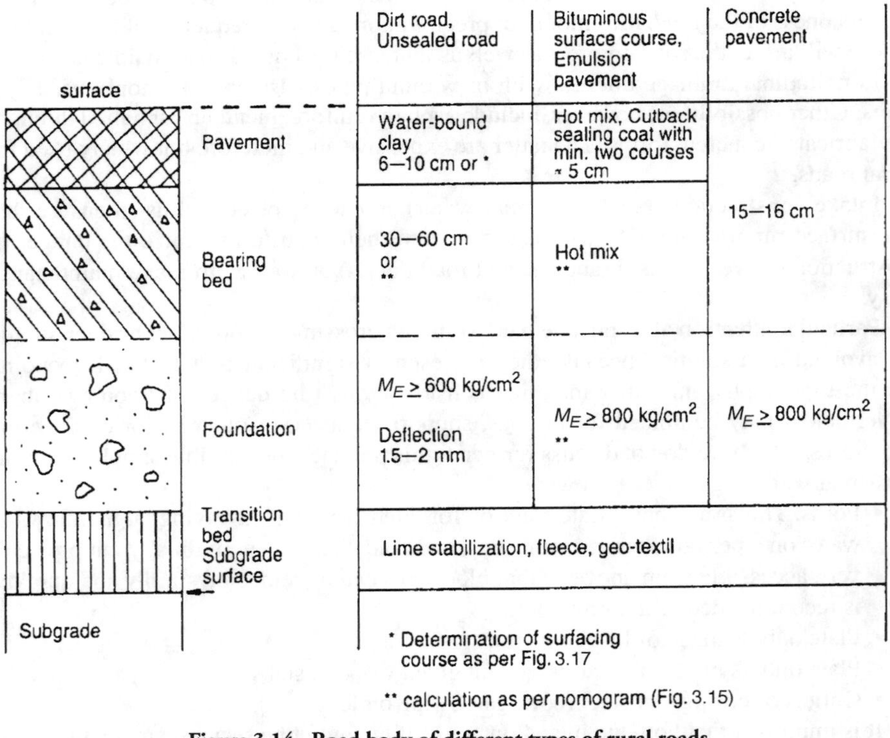

	Dirt road, Unsealed road	Bituminous surface course, Emulsion pavement	Concrete pavement
surface			
Pavement	Water-bound clay 6–10 cm or *	Hot mix, Cutback sealing coat with min. two courses ≈ 5 cm	15–16 cm
Bearing bed	30–60 cm or **	Hot mix **	
Foundation	$M_E \geq 600$ kg/cm^2 Deflection 1.5–2 mm	$M_E \geq 800$ kg/cm^2 **	$M_E \geq 800$ kg/cm^2
Transition bed	Lime stabilization, fleece, geo-textil		
Subgrade surface			
Subgrade			

* Determination of surfacing course as per Fig. 3.17

** calculation as per nomogram (Fig. 3.15)

Figure 3.14. Road body of different types of rural roads.

the same materials as for the foundation course. The term "base course" often is used for a layer above the subgrade, whereas pavement also means "bearing bed".

The following are other types of bearing beds:

- Stabilized bearing beds. Local and centralized mixing procedures for tar and cement are well known (mixed in place and mixed in plant). Cement-stabilized gravel paths also have proven their worth. The proportion of cement should be 50–70 kg/m^3 with an appropriate homogeneous grading curve. It is also important to use the prescribed water content and to mix thoroughly to ensure optimum durability. In any case, the meteorological conditions at the time of laying the bed, such as wind, air humidity, and sunshine, are also important. If meteorological conditions are bad, works must be stopped. Dry weather demands the moistening of the stabilized bearing bed. It is also possible that separation during transport or application will impair the quality. Otherwise, the result can be a compact surface that will be prone to cracking.
- Hot-mix bearing beds. The minimum thickness should be 6 cm at least, because anything less can lead quickly to damage, as indeed can too precise a thickness of the subgrade surface, which is technically possible.
- Concrete pavement. The application thickness together with the sliding moulding finish should be 15–16 cm. Concrete paving has shown itself to be exceptionally long-lasting and maintenance-free. If the subsoil has good bearing properties and the concrete slab can be laid directly, the construction costs can be quite low. On the other hand, if the concrete pavement is laid out on top of subsoil with poor bearing properties or insufficient homogeneity, uneven settling may occur when heavier loads move across it, reducing the road's practicability or even ending it altogether (should the slab be subject to subsidence).
- Pavement. This serves to seal the road and prevent water penetration. It is intended as a wearing course, i.e., to bear the direct brunt of abrasion wear caused by spike tyres, horseshoes, drip damage in forests, contamination by the passage of livestock.

As far as simple roads are concerned, the following types of pavement are commonly used:

- water-bound clay wearing course on unbound gravel and sand bearing bed;
- cutback pavements, as a flexible cover on unbound gravel, possibly with cement stabilization;
- sealing coat, requiring a minimum of two courses for an acceptable result.

3.4.7 Construction

Construction of the road body will depend on the subsoil (bearing characteristics, uniformity), environmental conditions (hydrological conditions, frost), traffic (frequency of standard single-axle weight), and construction materials (toughness, stability). Whereas proven standard values should govern the construction of dirt roads and rigid-type roads (concrete slabs), the following should conform more to a flexible construction style. In

Table 3.9. Type of road and traffic volume

Road Type	Frequency of Standard Single-Axle 8.2-t Load Vehicles
Connecting roads	100,000–150,000
Forestry roads	25,000–50,000
Feeder roads	
Agricultural roads	10,000–25,000
Forestry roads	10,000–25,000
Access roads	
Agricultural roads	10,000

Table 3.10. Determining the subsoil bearing capacity and the road body construction method in relation to subsoil conditions

Subsoil	Procedures for Determining Subsoil Bearing Capacity	Construction
Fine-grain soils, sand and gravel with many fine particles[a] CBR value <10% (scarcely or not at all practicable with trucks)	Determination of CBR value with the help of a Farnell hand penetrometer	Determination of the SN strength index by means of the construction formula or construction nomogram (Fig. 3.15)
Sand, gravel with many fine practicles[b] 10% <CBR value <20% (practicable with trucks)	Determination of deflection by means of Benkelman beam	Procedure similar to road body reinforcement: determination of SN with reinforcement diagram for CBR = 10%
Gravel with few fine particles[c] CBR value >20%	Determination with subsoil bearing capacity not necessary	Choice of a suitable surfacing course

[a] (USCS: CL, CH, ML, MH, OL, OH, SC–CL, SM–ML, GC–CL, GM–ML)
[b] (USCS: SW, SP, SM, SC, GC–CL, GM–ML)
[c] (USCS: GW, GP, GC, GM)
Source: [27].

this context, greatly simplified calculation principles should apply (traffic dimension of 100,000 to 200,000 standard single-axle loads, time dimension of 30 to 40 years).

The construction principles can be determined with the help of Table 3.9 (type of road and traffic volume), Table 3.10 (subsoil bearing capacity), and Table 3.11 (regional factors). This will help to determine the construction diagram of Fig. 3.15 (Nomogram) of the SN weighted strength index. The thickness and composition of the road body can be determined on the basis of the coefficient of bearing capacity for different courses and materials, shown in Table 3.12. An example in Fig. 3.16 illustrates this simple construction method.

In determining the covering course for dirt roads, a simple diagram can be of considerable assistance (Fig. 3.17). This permits an objective assessment of the situation when discussing a road's suitability in relation to the landscape.

Table 3.11. Regional factor (R) for simple roads

Type of road (conditions)	R
Roads requiring no snow clearance (forestry roads, miner rural roads)	1.0
Connecting roads <700 m above see level	1.3
Connecting roads:	
Favorable hydrological conditions	1.5
Unfavorable hydrological conditions	2.0

Table 3.12. Coefficient of bearing for the most commonly used building materials

		Course Thickness (cm)	
Construction Materials	a Value	Same Bearing Capacity	Minimum
AB 25/TA 25 (Asphalt pavement)	0.44	0.9	6
MHT B (Hot mix)	0.40	1.0	6
HMT A (Hot mix)	0.30	1.3	6
Bituminous stabilization	0.23	1.8	12
Stabilization with cement			
mix-in-place	0.20	2.0	15
mix-in plant	0.30	1.3	15
Stabilization with lime			
quick-acting	0.13	3.1	15
slow-setting	0.30	1.3	15
Crushed gravel	0.14	2.9	20
Round grave	0.11	3.6	20

Source: [24].

It should be borne in mind that the construction traffic will already involve a certain number of standard single-axle loads, that is, there is a preloading that somewhat reduces the theoretical mean life of the road. Logging, quarrying, and landfilling activities can likewise put great stress on minor rural road networks. All basic relevant information should be available at the planning stage. If at a later stage there is to be an extra burden, reinforcement of the road must be carefully planned. The following are useful general guidelines:

- Construction site transport $\approx$0.8 standard single-axle load per m^3.
- Timber transport $\approx$0.4 standard single-axle load per m^3.

3.4.8 Other Types of Road Construction

Roads with continuous pavement are much appreciated for such leisure activities as bicycling, jogging, or rollerskating, although avoided by other users such as ramblers. In vulnerable areas, objections often are raised to pavement for reasons that have to do with the landscape. Many spur roads are built in such areas, a variation that has proven its worth. The tracks for vehicles usually are made of concrete, with grass in between

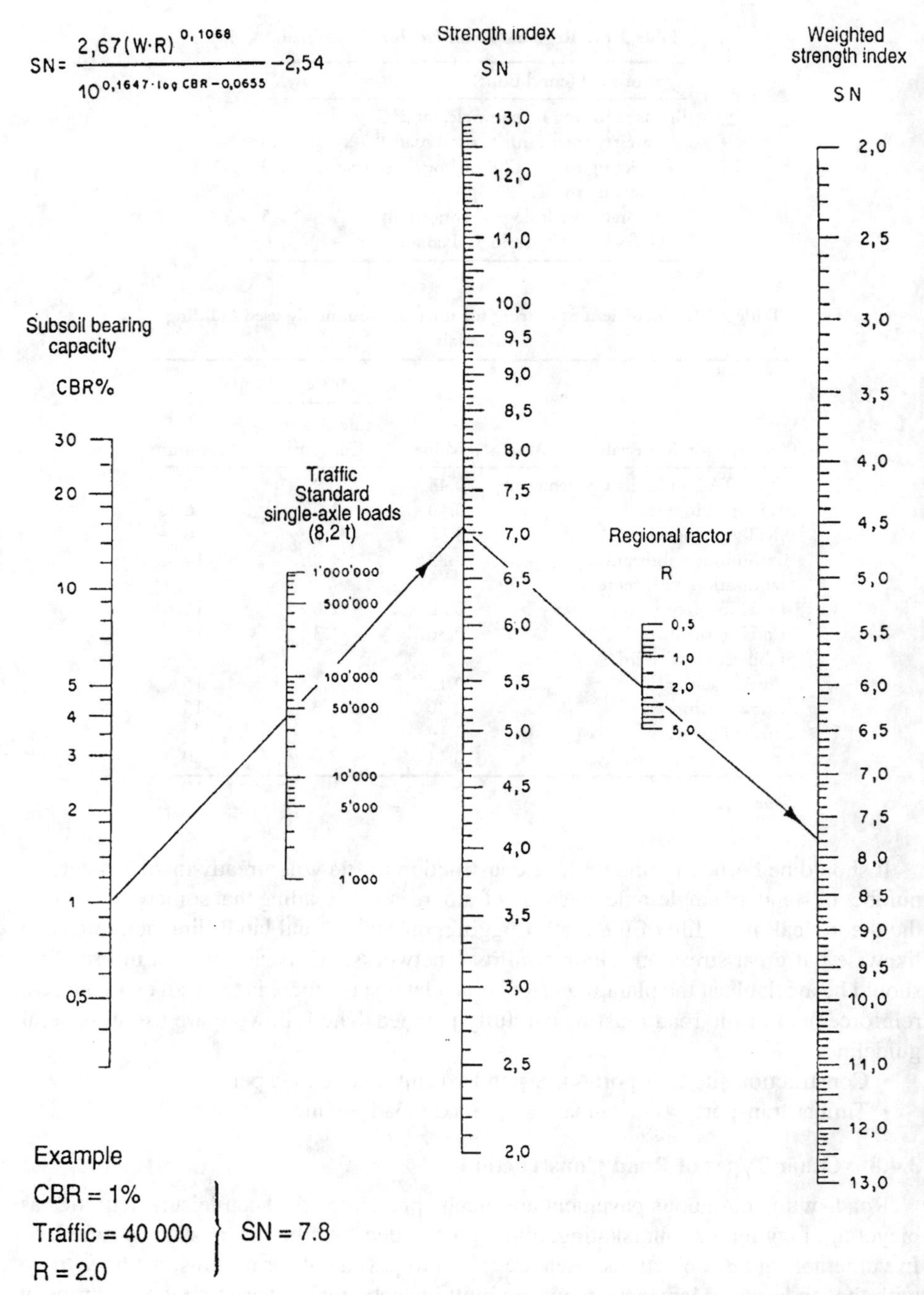

Figure 3.15. Nomogram for the construction of roads with low volume of traffic and flexible road body ($p = 1.5$). *Source:* [27].

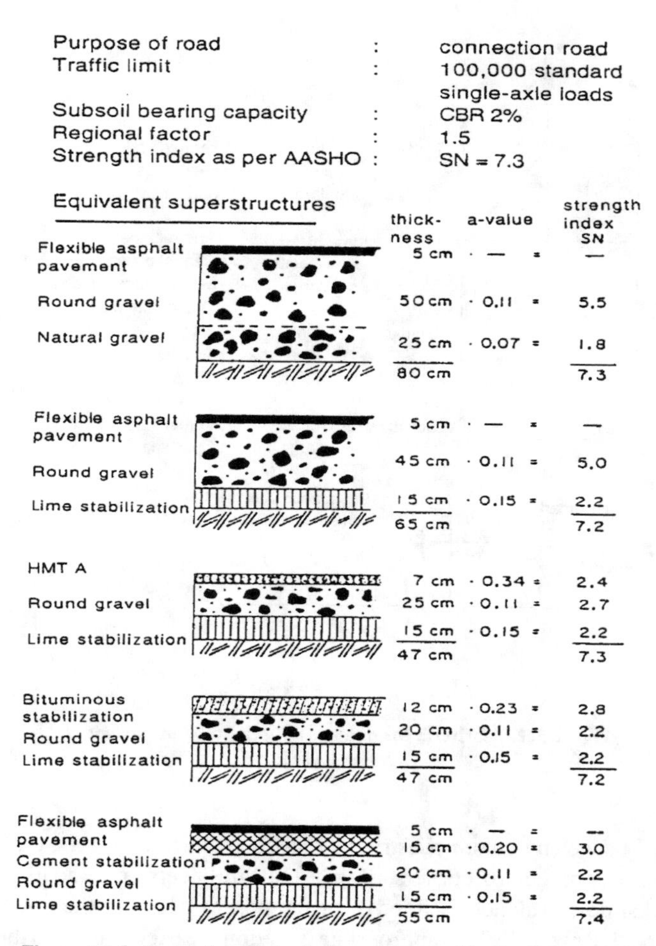

Figure 3.16. Example of calculation of the road body. ***Source:*** **[24].**

and on the verges (Fig. 3.18). This reduces the total extent of the sealed area, and the amount of surface water runoff, without having too much effect on the practicability. The visual impact is also better. In some countries, it has long been the custom to use hot-mix bearing beds for this application. For purely agricultural roads with steep gradients and involving a real danger of erosion, the tracks are made of permeable concrete sections (e.g., checker brick). Even with roads of this type, it is important to consider the question of drainage. It goes without saying that spur roads are suitable only for single-lane roads with little traffic. Performing ecological impact studies are important as well [28].

3.4.9 Road Safety

By virtue of their geometry, minor rural roads are suited for speeds of 30–40 kph. It must be clear, therefore, that they do not meet the same safety standards as high-performance roads. Even so, it may become necessary to take certain measures in the

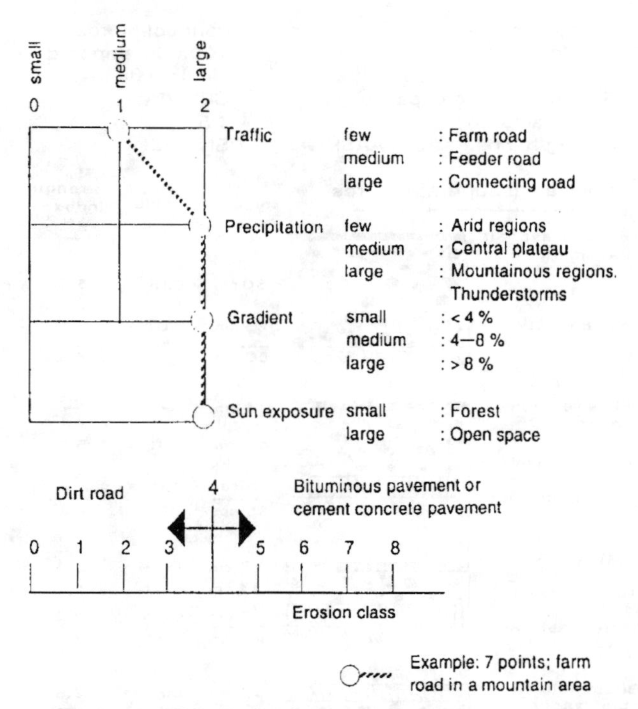

Figure 3.17. Criteria for determining the type of pavement.
***Source:* [24].**

event of mixed usage (by cars, pedestrians, bicyclists, in-line skaters, etc.), in relation to dangerous sections (sharp corners, slopes that are steep or ill-adapted to driving, danger of rockslides, avalanches, floods), or temporary peaks in the amount of traffic (weekends, special events). The appropriate solution in each case can be difficult to assess. The measures referred to include

- signs indicating use of the road by agricultural or forestry traffic, livestock, bicyclists, or pedestrians and alerting users to the danger of falling rocks, etc.;
- passing places within easy viewing range along one-track through roads, and along other simple roads at distances of 150–250 m (within earshot), light signal controls for one-track tunnels and long bridges;
- protective barriers along dangerous roads;
- Rockslide safety measures (netting, blasting at the time of construction, clearing away of rocks);
- avalanche and flood warning systems.

Galleries to protect against rockslides or avalanches are, in most cases, out of the question for cost reasons. To guard against such dangers, measures to combat the causes, such as afforestation or timbering in areas where cracks are a problem, may be more appropriate.

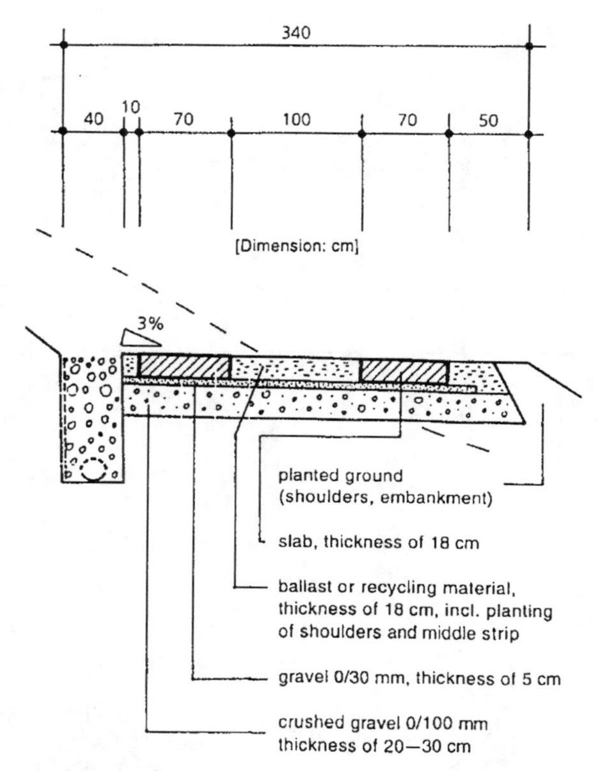

Figure 3.18. Standard cross section of a concrete spur road.

3.5 Maintenance of Rural Roads

J. Amsler

As already indicated in the preceding section, roads are built for a given life span, that is, a given number of standard axle weights. Once built, with the best possible quality, they are opened to traffic. As near-surface constructions they are particularly prone to wear and tear and to aging. Over time, the wear and tear from traffic, climate, timber harvesting, and agricultural and other activities reduces the original quality of the road. The state of the road, in particular of the track for vehicles, worsens because of this wear, and also as a result of damage and aging of the building materials. During the life cycle of the road, however, its quality must not be allowed to fall below a given level (see Fig. 3.19).

Roads built with public funds or by a local authority often are subject to legal minimum maintenance standards, designed to ensure safe conditions. This is achieved with the help of periodic road maintenance.

Simple, inexpensive solutions often are adopted for the building of minor rural roads. This inevitably results in slightly higher maintenance costs. If it is possible to increase the

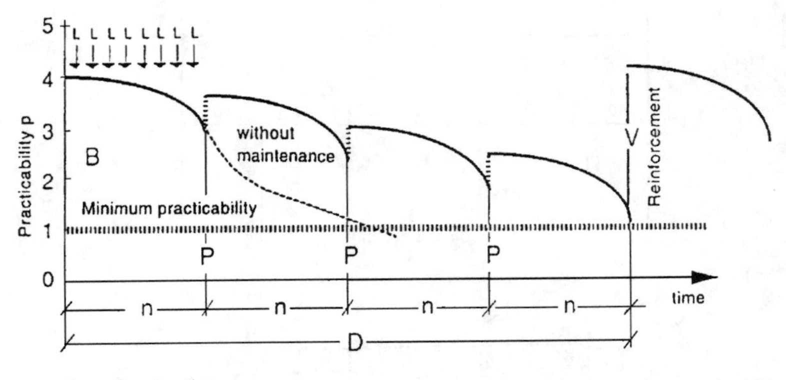

Figure 3.19. Life cycle of a road. ***Source:*** **[29].**

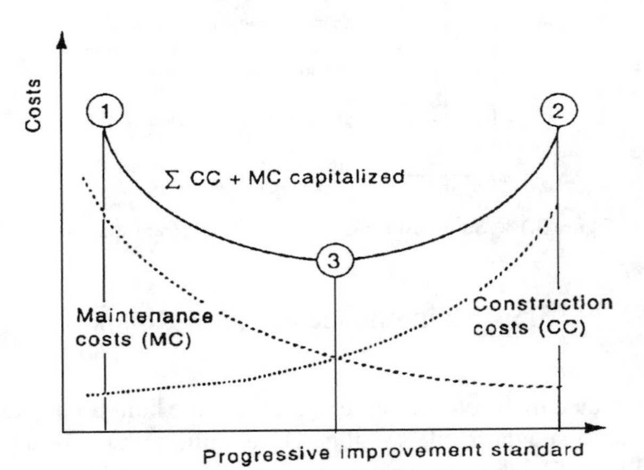

Figure 3.20. Construction and maintenance costs as a function of road standard (economic decision-making model). ***Source:*** **[29].**

service life of a road by proper maintenance, this will also reduce interest and depreciation costs. So, maintenance makes economic sense too (see Fig. 3.20).

The main objectives of road maintenance can be summed up as follows (key items in bold):

- **Structural maintenance**;
- **Upholding required traffic safety standards**;

- **Observing adequate standards of user comfort**;
- Ensuring permanent readiness for service;
- Preventing harm to the health of users or damage to vehicles through vibration or shocks;
- Keeping overall economic costs (construction and maintenance) to a minimum;
- Keeping environmental impacts (environment-friendly maintenance, economic use of material resources, recycling) to a minimum.

The networks of minor rural roads intended for agricultural and forestry use in different countries are often quite extensive. Total investment costs can amount to several billion ECU in the larger countries, and even in densely populated smaller countries. The associated maintenance costs can quickly add up to several hundred million U.S. dollars. In Switzerland, for example, the network covers a total of 60,000 km, for which the maintenance costs amount to about U.S. $70 million a year, to be paid for by the owners (individuals, cooperatives, local authorities). In such circumstances, it is worthwhile observing basic road maintenance planning principles, which will allow a clear overall view of the size, extent, and conditions of the road network. These are particularly useful when it comes to the technical and financial planning of the annual maintenance program. It is equally essential to make the road owners and political and public authorities aware of the importance of professional maintenance services, so that the necessary funds will be made available to protect the value of such a large investment, through proper maintenance. However, this will be possible only if a comprehensible model is available for the construction, maintenance, and refurbishment of simple roads. Although there are no spectacular ready-made recipes on the market, researchers have been looking into this matter for some time now, and so, there is at least a solid basis on which to work [29, 34].

The main thrust of maintenance on minor rural roads concerns the actual road surface, the essential requirement being maintaining practicability. On roads with little traffic, this involves such factors as the evenness of the surface, both lengthwise and crosswise; the extent of cracked, damaged, or patched spots; surface condition; and so on. The practicability is measured or estimated as a practicability value p, so that, depending on the significance, the burden of traffic, and the practicability requirements, review of the practicability may become necessary. The initial practicability of newly built or extended roads is gradually lost through wear and tear and damage, at first slowly, but if maintenance is neglected, with increasing speed until it falls below an acceptable level. Work then will be urgently needed to restore the road's initial practicability and user safety.

3.5.1 Different Concepts

Distinctions should be made between the four different concepts (see Table 3.13). The first two are the only ones that can be defined as maintenance in the strict sense of the term:

- Continuous maintenance. This is designed to ensure the road's continuing practicability and user safety. It requires constant monitoring, cleaning, the repair of damage as and when it occurs under the supervision of the surveyor (put in charge of road maintenance by the local authority or the cooperative). Maintenance must

Table 3.13. Outline of maintenance and repair work

	Continuous Maintenance	Periodic Maintenance	Reinforcement	Repair
Cause of damage	• Traffic • Weather • Forestry and agricultural activity	• Traffic • Weather: rain, snow	• Traffic, very heavy	• Natural events
Damage	• Pollution • Local damage (potholes, isolated cracks)	• Wear and tear • Erosion of surfacing course • Drainage works • Engineering works	• Destruction of road body • Major deformations	• Landslides • Destruction by landslips, avalanches, etc.
Measures	• Inspection • Cleaning • Repairs	• Renovation of surfacing course • Rebuilding of profile (dirt roads)	• Reinforcement of bearing course • Laying of pavement	• Repair
Objective	• Conservation: practicability, traffic safety	• Improvement: practicability, traffic safety • Conservation of material	• Repair: practicability, traffic safety	• Repair: practicability, traffic safety
Extent of work	• Localized • Small surfaces	• Large surfaces • Usually entire vehicle track	• Entire road surface	• Damaged areas
Execution of work	• Road maintenance crew • Responsible persons	• Construction company • Forestry service or local-authority maintenance crew	• Construction company	• Construction company
Time of execution	• Whenever required, at least once a year	• On a periodic basis, every 6–12 years	• At the end of the calculated life cycle, e.g., 40 years	• After the event
Financing	• Owners	• Owners	• Owners + subsidies	• Owners + subsidies

Source: [29].

be carried out before the onset of winter, after the last snows have melted, after storms, or after heavy use (harvesting or transport of timber), and in any case at least once a year. The damage that occurs in such conditions will not have been caused by any insufficient bearing capacity of the road. The structural integrity of the road and its various courses will not be in any danger.

- Periodic maintenance. This has the additional aim of upholding the substance and value of the construction as a whole, and improving the road's practicability, that is, increasing the practicability *p* value. Restoration of the surface profile, resurfacing or retouching of the surface course (wear course, pavement), drainage installations, and engineering work have to be done in a cycle of 8 to 12 years. As a result of wear and aging of the construction materials, the surface course may no longer be in a fit state to fulfill its purpose, putting the bearing bed at risk. In this case, a new surface course will have to be laid, to restore protection of the bearing bed. The need for periodic maintenance, that is, for refurbishing of the surface course, applies to the entire road surface. The length of the maintenance cycle will depend on several factors: the wear resistance, the nature and thickness of the surface course, and the intensity of traffic- and climate-related stress and strain. In most cases, these works are beyond the means of the surveyor in charge and will have to be taken on by small companies or road maintenance groups. It is up to the surveyor, however, to decide on what measures are needed, and when. Once the construction has lost its suitability for traffic, falling below the minimum requirements of practicability, its condition must be restored by means of reinforcement work designed specifically to bring the road back to the required standard for traffic use. This work of restoration and reinforcement, as well as the repair of basic damage, is usually beyond the financial means of the road owners. The appropriate public authorities therefore should be able and willing to provide any additional funds necessary.
- Repairs. This term covers all unforeseeable, exceptional damage caused by major storms, avalanches, landslides, or other events that cannot be met from the normal maintenance budget. The work often has to be organized in projects, and execution might be placed in the hands of a private-sector company, or a state-controlled consortium, often with the involvement of civil protection or military units.
- Improvements. This concept covers adaptation of the road to new and generally more demanding requirements. It calls for wide-ranging project planning and engineering, often requiring a full review of the existing construction.

3.5.2 Causes of Damage

Right from the start, a road can be subject to certain external influences that will cause extra wear and damage. These include the following:

- Conditions of use. Excessive driving speeds or vehicle weight, particularly in humid conditions or during thaw, as well as dense traffic, inevitably lead to excessive wear of the surface course and the formation of potholes, cracks, and similar evidence of damage; crushing and subsidence of the road body, verges, and drainage installations; and pollution. Mechanical clearance of snow from the road results in cracking of the surface course and damage to the verges, not to mention the harm caused by road salt. Timber harvesting causes abrasion wear and compaction, and agricultural use leads to pollution from dung and urine (livestock) and from soil material (farming).

- Climate. Precipitation and melting snow cause erosion, leaching, landslips, and drip damage. Frost and dew cause cracks, whereas strong sunshine, wind, and heat result in a loss of fine matter and of binding materials.
- Vegetation. The fall of leaves and pine needles pollutes the road surface, and vegetation grows from cracks in the road and from drainage installations. Trees and bushes reduce the profile and the visibility and cause drip damage.
- Civil engineering defects. Faulty or inappropriate drainage installations increase the risk of erosion, whereas insufficient bearing capacity increases the likelihood of subsidence and of damage by vehicles.
- Natural aging. All construction materials are subject to natural aging, which leads to thinning of the surface course, segregation, aging of the concrete, and the structural disintegration of the bearing and surface courses.

3.5.3 Planning and Execution of Work

So that the necessary organizational and financial planning arrangements can be taken care of in an optimum way, and to ensure the most rational execution and supervision of the maintenance work in hand, it is best to have all relevant data on the conditions of the existing road network readily available in a data bank. A file card is shown in Fig. 3.21. At the local-authority level, the data can be integrated into existing information systems such as LIS or GIS. For cooperatives that are not able to call on electronic data processing facilities, card file systems remain a practical solution.

Maintenance tasks can be subdivided into the following categories:

- Inspection. Routine visual inspections (on foot by difficult road conditions) are the best way to keep check on the condition of a road, and are the main task of the surveyor in charge. Such inspections are especially important during and after periods of snowfall and bad weather, to make sure the drainage systems are working properly. The aim is to discover damage and its causes at as early a stage as possible, in order to draw the necessary conclusions and take remedial action in good time. Note, however, that visual inspection alone is not always a sufficient basis on which to make a clear-cut decision as to the best way to deal with damage of a given type and extent, that is, whether periodic maintenance will be enough to guarantee the necessary practicability, or whether extensive reinforcement is required. Where there is doubt, measurements of the road's bearing capability can help to clarify matters. Well-grounded experimental values for this assessment can be obtained with the help of the Benkelman beam deflection measurements. These should reveal whether or not the bearing capacity values measured (see deflection in Fig. 3.22) are still in a permissible range.
- Cleaning. Road cleaning work, such as ensuring that water drainage systems are not blocked, cleaning shafts, removing harmful contaminants (especially after agricultural activities), and eliminating unwanted vegetation, should be carried out during tours of inspection.

File card: maintenance	
Owner:	
Person in charge	
Road description: Grosser Runs No. 21.11	
Year of construction:	1977
Cost of construction:	215.- CHF/m
Length of road (m):	2,800
Vehicle track width (m):	3.40
Longitudinal gradient (%):	3/11
Carriage way:	horizontal
Culverts:	—
Drainage:	—
Signaling:	No vehicles allowed (exceptions indicated)
Snow clearance:	Yes
Special features:	Access to gravel pit
Type of road:	Feeder
Traffic (std. single-axle load):	50,000
Road body	
Wearing course:	6 cm HMT Melio
Bearing course:	45 cm gravel 2 15 cm stabilization with lime
SN strength index:	~ 6
Subsoil/foundation	
Soil type (USCS):	Clay with great plasticity (CH)
Bearing capacity CBR:	1.6%
Road condition/measures	
Year / Date	Road condition / Practicability: good sufficient insufficient — Measures / Inspections

Figure 3.21. Example of file card. *Source:* [29].

- Simple repair work. For this work, the surveyor usually will require additional equipment and construction materials. Attention should be paid above all to the proper functioning of the drainage installations because unchecked water flows present the greatest danger of rapidly spreading destruction. Care also should be taken to remedy any cracks and potholes found in the road surfaces.
- Comprehensive repairs and work of renovation. This is also an essential part of

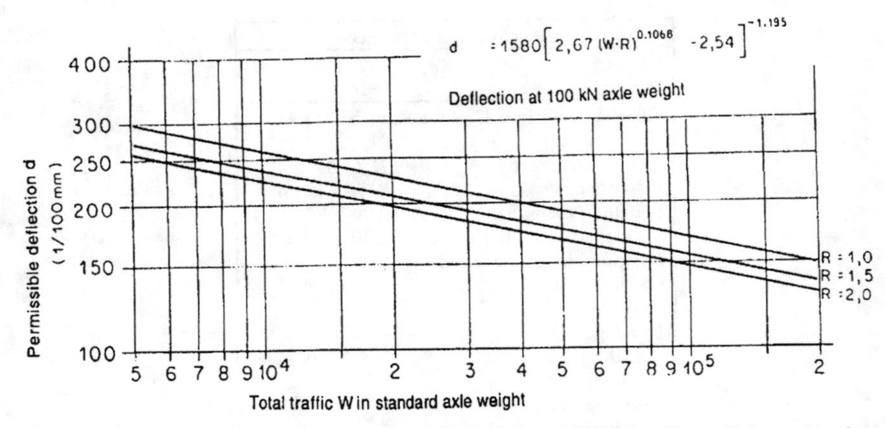

Figure 3.22. Permissible deflection in relation to the traffic factor *W* and the regional factor *R*. *Source:* [24].

periodic maintenance, calling for a certain amount of planning and for the help of specialists. Such repairs also require a considerable investment in both time and money.

In addition to these active measures, certain passive measures, which can also help to prevent or limit damage, often are overlooked. These include general or time-limited restrictions on traffic or on certain axle weights (e.g., during periods of frost or humidity). These should always be seen as companion measures, for use when the amount of traffic and the burden it causes begins to go out of control. Such cost-free measures help to protect against excessive stress, thereby lowering the amount that needs to be budgeted for maintenance, while increasing the service life of the roads.

General restrictions on road use also can be in the interest of nature and the environment, and indeed of the users. When minor rural roads built to make life easier for agricultural and forestry workers are overrun by weekend tourists, they are far less able to fulfill their original purpose. Nature and landscape protection groups tend to look on such developments with suspicion. General restrictions therefore should be discussed with those concerned including the authorities at the earliest possible project stage. Otherwise, they will be difficult to implement.

3.5.4 Financing

Tolls on public roads are not acceptable in most countries. However, there can be, legal grounds for collecting tolls on motorways: at the entrance to long tunnels and prior to passage across large and expensive civil engineering works. When tolls are to be considered for minor rural roads, the legal aspects should be clarified immediately. The appropriate location of the toll booth (as with a parking lot) and the provision of a barrier must be given careful consideration. Should a toll be unacceptable, however, it might be worthwhile to consider charging for parking space, and earmarking the funds gathered from road users in this way for the upkeep and renovation of the road.

Experience has shown that it is advisable to arrange for enforcement by the police. Private individuals, even if they have the necessary abilities, do not have the legal authority to force people to obey the rules.

References

1. Tolley, R. S., and B. J. Turton. 1995. *Transport Systems, Policy and Planning*: A *Geographical Approach*. Harlow, UK: Longman.
2. Organization for Economic Cooperation and Development. 1986. Economic Design of Low-Traffic Roads. Paris: OECD.
3. Jaarsma, C. F. 1994. Rural low-traffic roads (LTRs): The challenge for improvement of traffic safety for all road users. *Proceedings for the Dedicated Conference on Road and Vehicle Safety, 27th ISATA*. pp. 177–183. London: ISATA.
4. Boss, C., and U. Flury. 1988. Integral planning and design of minor rural roads in Switzerland. *Minor Rural Roads. Planning, Design and Evaluation*, eds. Michels, T., and C. F. Jaarsma, pp. 67–84. Wageningen: Pudoc.
5. European Commission, Directorate-General for Development. 1996. Towards Sustainable Transport Infrastructure. A Sectoral Approach in Practice. Transport Sector Guidelines. Brussels/Luxembourg: Office for Official Publications of the European Communities.
6. Jaarsma, C. F. 1991. Categories of rural roads in The Netherlands: Functions, traffic characteristics and impacts. Some problems and solutions. XIXth World Road Congress PIARC-publication 19.52E, pp. 101–103. Paris: PIARC.
7. Hawkins, M. R., and M. J. Hatt. 1988. Managing a rural road network. *Minor Rural Roads. Planning, Design and Evaluation*, eds. Michels, T., and C. F. Jaarsma, pp. 55–65. Wageningen: Pudoc.
8. Kollmer, K. 1988. Planning, design and realisation of minor rural road networks in German land consolidation projects. *Minor Rural Roads. Planning, Design and Evaluation*, eds. Michels, T., and C. F. Jaarsma, pp. 25–38. Wageningen: Pudoc.
9. Murty, A. V. S. R., P. K. Dhawan, and R. Kanagadurai. 1994. Rural road development in India. Paper presented at the Second CIGR/PIARC Workshop on Minor/Local Roads, Pitesti, Romania.
10. Wegman, F., 1997. The Concept of a Sustainable Safe Road Traffic System. A New Vision for Road Safety Policy in The Netherlands. Report D-97-3, Leidschendam: SWOV (Institute for Road Safety Research).
11. van Langevelde, F., and C. F. Jaarsma. 1997. Habitat fragmentation and infrastructure: The role of minor rural roads (MRRs) and their traversability. *Habitat Fragmentation and Infrastructure*, eds. Canters, K. J. et al., pp. 171–182. Ministry of transport, Road and Hydraulic Engineering Division: Delft.
12. Jaarsma, C. F., and F. van Langevelde. 1996. The motor vehicle and the environment: Balancing between accessibility and habitat fragmentation. *Proceedings for the Dedicated Conference on the Motorvehicle and the Environment—Demands of the Nineties and Beyond, 29th ISATA*, pp. 299–306. London: ISATA.

13. Jaarsma, C. F. 1997. Approaches for the planning of rural road networks according to sustainable land use planning. Landscape and urban planning 39(1): 47–54.
14. van Iwaarden, J. W. 1967. Bepaling van de voordeligste kavelafmetingen. (Calculation of the cheapest parcel sizes). *Cultuurtechnisch Tijdschr.* 7(2):64–74 (in Dutch).
15. van Lier, H. N. 1994. Land use planning in perspective of sustainability: An introduction. *Sustainable Land Use Planning. Proceedings of an International Workshop*, eds. van Lier, H. N., et al., pp. 1–11. Amsterdam: Elsevier.
16. Centre for Research and Contract Standardization in Civil and Traffic Engineering (CROW). 1989. Verkeersmaatregelen in Het Buitengebied (Traffic Measures in Rural Areas.) Mededeling 21, Ede: CROW (in Dutch.)
17. Macpherson, G. 1993. Highway and Transportation Engineering and Planning. Harlow, UK: Longman.
18. Koornstra, M. J., M. P. M. Mathijsse, and J. A. G. Mulder. 1992. Naar een Duurzaam Veilig Wegverkeer: Nationale Verkeersveiligheid voor de Jaren 1990–2010. (Towards a Sustainable Safe Road Traffic: National Traffic Safety for the Period 1990–2010). Leidschendam: SWOV (in Dutch).
19. Jaarsma, C. F., J. O. K. Luimstra, and T. J. de Wit. 1995. De Kortste Weg naar een Verkeersleefbaar Platteland. Onderzoek Ruraal Verblijfsgebied Ooststellingwerf. (The Shortest Path to a Liveable Rural Area. A traffic and Transportation Plan for a Residential Rural Area in Ooststellingwerf.) Nota vakgroep Ruimtelijke Planvorming 58 Wageningen: Agricultural University (in Dutch).
20. Jaarsma, C. F., and F. van Langevelde. 1997. Right-of-way management and habitat fragmentation: An integral approach with the spatial concept of the traffic calmed rural area. *Environmental Concerns in Right-of-Way Management*, eds. Williams et al., pp. 383–392. Amsterdam: Elsevier Science.
21. Jaarsma, C. F. 1996. Energy saving for traffic on minor rural roads. Opportunities within re-development Projects. *World Transport Research*, Vol. 3. *Transport Policy*, eds. Hensher, D. A., J. King, and T. Oum, pp. 347–360. Amsterdam: Elsevier Science.
22. Cullinane, S., K. Cullinane, and J. Fewings. 1996. Traffic management in Dartmoor National Park: Lessons to be learnt. *Traffic Eng. Control* 37(10): 572–573, 576.
23. Kuonen, V. 1983. *Forst- und Güterstrassen, Lehrbuch (Forest- and Minor Rural Roads).* Zurich: Federal Institute of Technology (in German).
24. Hirt, R., 1990. Einfache Strassen, Lehrschrift (Simple roads, Instruction Note). Zurich: Federal Institute of Technology (in German).
25. Billington, D. P. 1990. *Robert Maillart and the Art of Reinforced Concrete*. Zurich and Munich: Verlag für Architektur Artemis.
26. Schiechtl, H. M. 1973. *Sicherungsarbeiten im Landschaftsbau (Measures Against Landslide Caused by Earthworks in Rural Areas).* Munich: Verlag Callwey (in German).
27. Burlet, E., 1980. Dimensionering und Verstärkung van Strassen mit geringem Verkehr und flexiblem Oberbau (Design and Reinforcement of Low-Volume Roads with Flexible Pavement), Thesis ETH No. 6711. Zurich: Federal Institute of Technology (in German).

28. Thorens, P., et al. 1996. Etude des Potentialités Naturelles d'un Chemin avec Bandes de Roulement Bétonnées (Study of the Ecological Value of Spur Roads). Neuchâtel: Bureau d'étude des invertébrés, case postale, (in French).
29. Hirt, R., et al. 1992. Unterhalt von Wald- und Güterstrassen (Maintenance of Forest- and Minor Rural Roads). Zurich: Federal Institute of Technology (in German).
30. Schiechtl, H. M., and R. Stern. 1992. *Handbuch für Naturnahen Erdbau (Handbook of Biological Methods in Foundation Practice and Earthworks).* Vienna: Österreichischer Agrarverlag (in German).
31. CEMAGREF, 1981. Caractéristiques des Voies et Réseaux de Desserte (Characteristics of Minor Rural Roads and Road Networks). Technical Note No. 46. Parc de Tourvoie, Antony Cedex: Groupement d'Antony (in French).
32. Cain, C. 1981. Maximum grades for Log Trucks on Forest Roads. Eng. Field Notes, Vol. 13, No. 6. Washington D.C.: USDA Forest Service.
33. Glennon, J. C. 1979. Design and Traffic Control guidelines for Low-Volume Rural Roads. Washington D.C.: Transportation Research Board 214.
34. Paterson, W. D. O. 1987. *Road Deterioration and Maintenance Effects: Model for Planning and Management.* Highway Design and Maintenance Standards Series. Baltimore, MD: Johns Hopkins University Press.
35. Paterson, W. D. O. 1991. Deterioration and maintenance of unpaved roads: models of roughness and material loss. *Transport. Res. Rec.* 1291:143–156.
36. United Nations Food and Agriculture Organization (FAO). 1989. Watershed Management Field Manual. Road Design and Construction in Sensitive Watersheds. Conservation Guide 13/5. Rome: FAO.
37. Jordaan & Joubert Inc. 1993. Cost-Benefit Analysis of Rural Road Projects: Program CB-Roads. Pretoria: South African Roads Board, Department of Transport.
38. Pienaar, P. A. 1993. The Management of Tertiary Road Networks in Rural Areas, Ph.D. dissertation. University of Pretoria, South Africa.
39. McFarlane, H. W. et al. 1973. Use of Benkelman Beam on Forest Roads. Pulp and Paper Institute of Canada, Logging Research Rept. LRR/49. Department of Civil Engineering, University of New Brunswick, Canada.
40. Litzka, J., 1993. Low-volume rural roads in Austria: Guidelines to minimize environmental impacts. Presented at the 72nd Annual Meeting, Transportation Research Board, Washington D.C.

4 Land Reclamation and Conservation

4.1 Soil Reclamation

D. De Wrachien and G. Chisci

4.1.1 Assessment of the Problem

There is an increasing awareness that the world's resources are not boundless. In many countries, there is now a concern about husbanding and protecting the environmental resources upon which life depends. Among the most basic resources is the soil.

Reclamation, in its broad sense, means modifying the physical, chemical, and biological characteristics of the soil to restore its capacity for beneficial use while protecting the environment. The soil could have lost its suitability for cropping or other use by means of natural or anthropogenic causes. In any case, we have to deal with an ecosystem in which normal biological processes are at a standstill. These processes must be restored so that a normally functioning ecosystem of soil, plants, and animals is reactivated, such that the natural processes of nutrient release, plant growth, and nutrient cycling proceed at a natural rate. However, the soil is a complex biological system built up over long periods of time. If we are to restore it, we must understand how it functions biologically, physically, and chemically. It also involves understanding the requirements of plants and the other organisms that interact with soil processes.

Soil restoration can take a long time and generally depends on the actual conditions of the soil and on the processes employed in restoring it. Modern methods can hasten the process considerably. The key is the identification and correction of the problems that brought the soil into its degraded state. This identification process requires imagination and sensitivity as well as science.

Drainage of wetlands, amendment of saline soils, restoration of soil contaminated with urban waste materials, and the rehabilitation of floodplains and tidal forelands are all reclamation processes. In some cases they may not lead to complete restoration of the soil and landscape, although a self-sustaining ecosystem always must be created.

Figure 4.1 shows the steps involved in the planning of a thorough soil reclamation scheme [1].

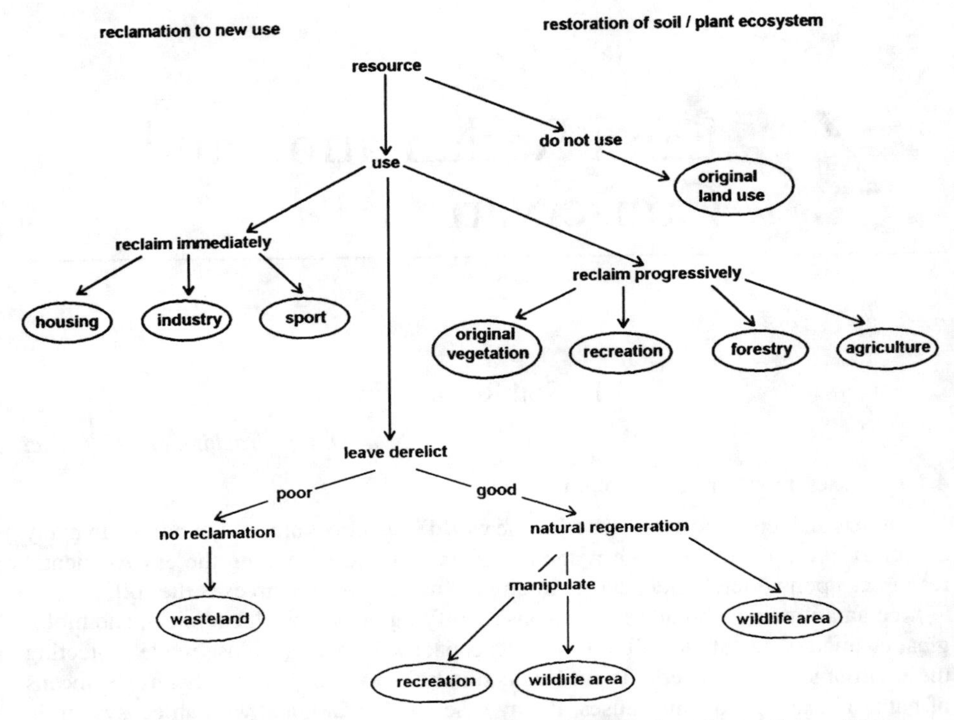

Figure 4.1. Steps involved in the planning of a soil reclamation scheme. ***Source:*** **[1].**

4.1.2 Wet Soils

Problems

In general terms, wet soils are defined as soils where saturation with water is the dominant factor determining the nature of soil development and the types of plant and animal communities living on the surface [2]. Wetness occurs whenever the water supply by rainfall, overland flow, and seepage from upslope exceeds the water loss by drainage, throughflow, and evapotranspiration. Poor drainage usually is associated with soils that have low permeability. Often, soils are shallow and/or overlie an impermeable barrier such as rock or very dense clay pan. The impermeable subsoil prevents the water from moving downward and the natural subsurface drainage system from functioning properly. Many of these areas have flat topography with depressions and, in some instances, the areas may lack adequate drainage outlets. Representative examples include the basin clay soils of fluvial deposits, pseudogley soils of central and eastern Europe, planosols of the semihumid or semiarid tropics, vertisols of the semiarid tropics, and glacial till soils.

Experiments that have been conducted to determine the effect of soil wetness on crop yield generally show a reduction in yield at shallow water-table depths (see Fig. 4.2). Frequency and duration of high water tables as well as the crop sensitivity to waterlogging

Table 4.1. Relative tolerance of plants

Tolerance Level	High Groundwater Levels	Waterlogging
High	Sugarcane Potatoes Broad beans	Plums Strawberries Several grasses
Medium	Sugar beets Wheat Barley Oats Peas Cotton	Citrus Bananas Apples Pears Blackberries
Sensitive	Maize	Peaches Cherries Raspberries Date palms Olives

Source: Modified from [4].

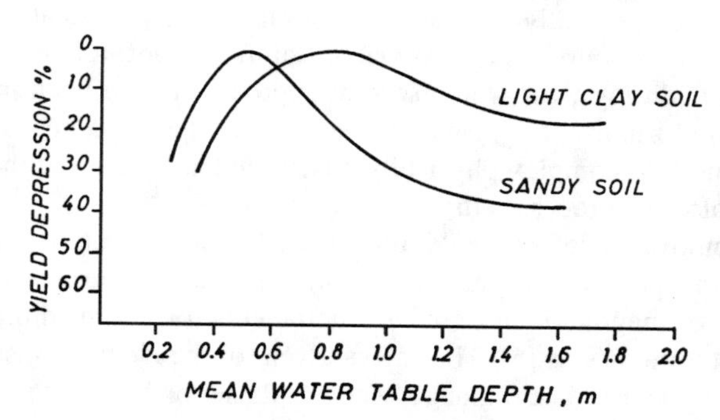

Figure 4.2. Yield depression as a function of mean water-table depth during the growing season for two types of soil. *Source:* [3].

also have a relevant impact on yield (see Chapter 5, Section 5.5). Table 4.1 gives some indications of the relative tolerance of some plants to waterlogging.

Wet soils are an obstacle to man's use of land resources for cropping and, in general terms, for economic development. However, wetlands play a major role as a natural resource, which may be superior to the benefits from agricultural resources. To bring these soils back into beneficial agricultural use, they must be reclaimed. To this end, it is necessary to eliminate the depressions, to provide sufficient slope for overland flow, and to construct channels to convey the water away from affected areas.

Table 4.2. Drainage coefficients for typical agricultural basins

Region	Crop	Water Source	P_5[a]	Slope of land (%)	Drainage Coefficient ($s^{-1} \cdot ha^{-1}$)
Netherlands[b]	Grasses, crops	Rain	31–44	0.05–0.20	1.0–1.3
Yugoslavia[c]	Grasses, crops	Rain	52–64	0.25–0.50	3.5
Sudan	Cotton	Furrow irrigation	64–90	0.05–0.10	4.0
Japan	Rice	Flooded	80–115	0.05–0.20	5.0
Tanzania	Sugarcane	Sprinkler irrigation	145–165	0.20	7.0

[a] 24- to 28-hr rainfall for a recurrence interval of 5 years.
[b] Subsurface drainage.
[c] Subsurface and surface drainage.
Source: From [6].

Reclamation Procedures

The reclamation practice, known as surface drainage, can be defined as the diversion or orderly removal of excess water from the land through ditches or by shaping of the ground surface. Sometimes, a subsurface system of drain pipes is needed in conjunction with the surface drainage measures (see Chapter 5, Section 5.6).

Design discharges generally are based on rainfall intensity because this is almost always the most critical source of excess water. Rainfall–runoff relationships are a classical problem in hydrology, for which several approaches have been proposed. These range from simple formulas that give only the peak discharge value, such as the rational formula method, to complex physically based models that allow the simulation of distributed hydrological processes [5].

The rate of removal is called the "drainage coefficient," which usually is expressed in terms of flow rate per unit area and varies with the size of the area. Table 4.2 gives an idea of the order of magnitude of drainage coefficients for flat, medium-size (10,000–25,000 ha) agricultural basins [6]. The values given for the Netherlands apply to basins with predominantly subsurface drainage, whereas those for Yugoslavia apply to a combination of subsurface and shallow drainage.

Surface drainage, when properly planned and designed, eliminates ponding, prevents prolonged saturation, and accelerates flow to an outlet without siltation or soil erosion.

Surface Drainage Systems

Surface drainage systems can be described and classified in several ways. However, there are basically three types of systems that are in common use today [7]: parallel, random, and cross slope. Under some circumstances encountered in the field, combinations of two or more of these systems may be required.

All these systems should

- fit the farming layout;
- remove water quickly from land to ditch, avoiding erosion and siltation;
- have adequate capacity to carry the flow;

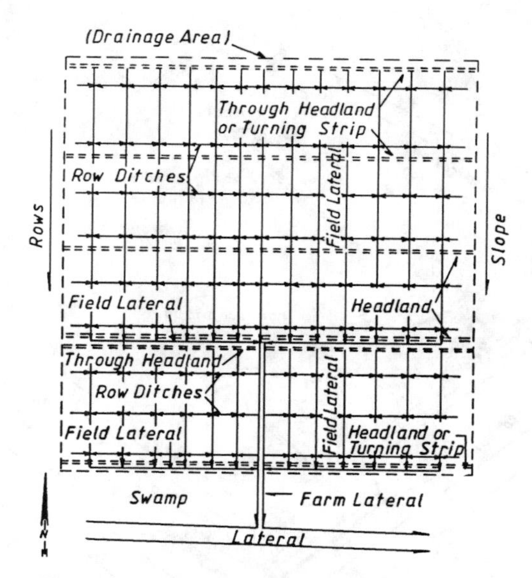

Figure 4.3. Layout of parallel system. ***Source:*** [7].

- be designed for construction and maintenance with appropriate equipment locally available.

Parallel System. This system is predominantly applicable to fairly regular, even-surfaced, flatland, which adapts well to a parallel alignment of field ditches, which least hinders farm operations. The field ditches are dug in a parallel, but not necessarily equidistant, pattern, as shown in Fig. 4.3. The orientation of field ditches normally depends on the direction of surface slope, location of diversions, cross-slope ditches, and access of the drained lands to farming equipment. The ditches are well incised and definitely aimed at achieving considerable profile drainage together with collecting excess surface water. Usually, the field ditches discharge into lateral ditches, which should be deeper than the field ditches to allow free outflow. The land surface should be shaped and smoothed to favor overland flow. Spacing of parallel field ditches depends mainly on soil and slope characteristics and on the water tolerance of the crops grown. Design specifications for this system are given in Table 4.3.

Table 4.3. Ditch specifications for water-table control

Dimension	Sandy Soil	Other Mineral Soils	Organic Soils, Peat, and Muck
Maximum spacing (m)	200	100	60
Minimum side slopes	1:1	$1\frac{1}{2}$:1	up to 1:1
Minimum bottom width (m)	1.2	0.3	0.3
Minimum depth (m)	1.2	0.8	0.9

Source: From [8].

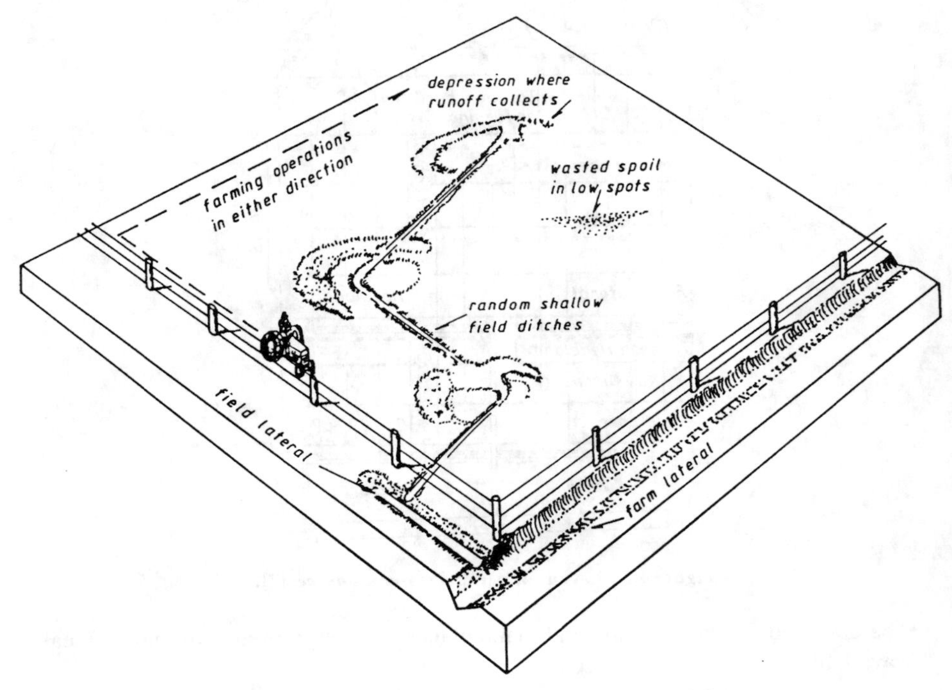

Figure 4.4. Layout of random system. ***Source:*** [7].

For water-table control during dry seasons, namely in cases of peat and acid sulfated soil, adjustable weirs or gates are placed at various points in the ditches to maintain the water surface at the required level. During wet seasons these controls may be removed, to let the system provide surface drainage.

Random System. This system, also called a depression ditch system, is applicable to fields in which a limited number of pronounced depressions exist, lending themselves to drainage by means of ditches. The depressions may be drained individually or linked. The layout of the field ditches has to be such that as many depressions as possible are intersected along a path through the lowest part of the field toward an available outlet (Fig. 4.4). The path should be chosen so as to provide the least interference with farming operations and a minimum of deep earth cuts. Therefore, only the main depressions should be provided with ditches; the small depressions should be filled in with, for example, the spoil from the excavated ditches. Where the overland flow rate toward the outlets is too slow, land shaping will be necessary to ensure complete surface water removal.

The design of the field ditches is similar to the design of waterways (see Section 4.4 of this chapter). Where farming operations cross the channel, the side slopes should be flat [8]: 8:1 or greater for depths of 0.3 m or less and 10:1 or greater for depths over 0.6 m. Minimum side slopes of 4:1 are allowed if the field is farmed parallel to the ditch. The depth is determined primarily by the topography of the area, outlet conditions,

Table 4.4. Limitations on flow velocity and on side slope in unlined drainage canals[a]

Soil Type	Max. Permissible Mean Flow Velocity (in m/s)	Max. Permissible Side Slope
Fine sand	0.15–0.30	1:2 to 3
Coarse sand	0.20–0.50	1:1$\frac{1}{2}$ to 3
Loam	0.30–0.60	1:1$\frac{1}{2}$ to 2
Heavy clay	0.60–0.80	1:1 to 2

[a] Highest velocities and steepest side slopes apply to well-vegetated canals.
Source: [6].

and capacity of the channel. The grade in the ditch should be such that the velocity does not cause erosion or sedimentation. Maximum allowable velocities for various soil conditions are given in Table 4.4.

Cross-Slope System. This system, also called a diversion system, is applicable on gently sloping land where sufficient runoff may occur but where water also may remain ponded on the land. The system consists of one or more diversions, terraces, or field ditches built across the slope, as shown in Fig. 4.5. As water flows downhill, it is intercepted and

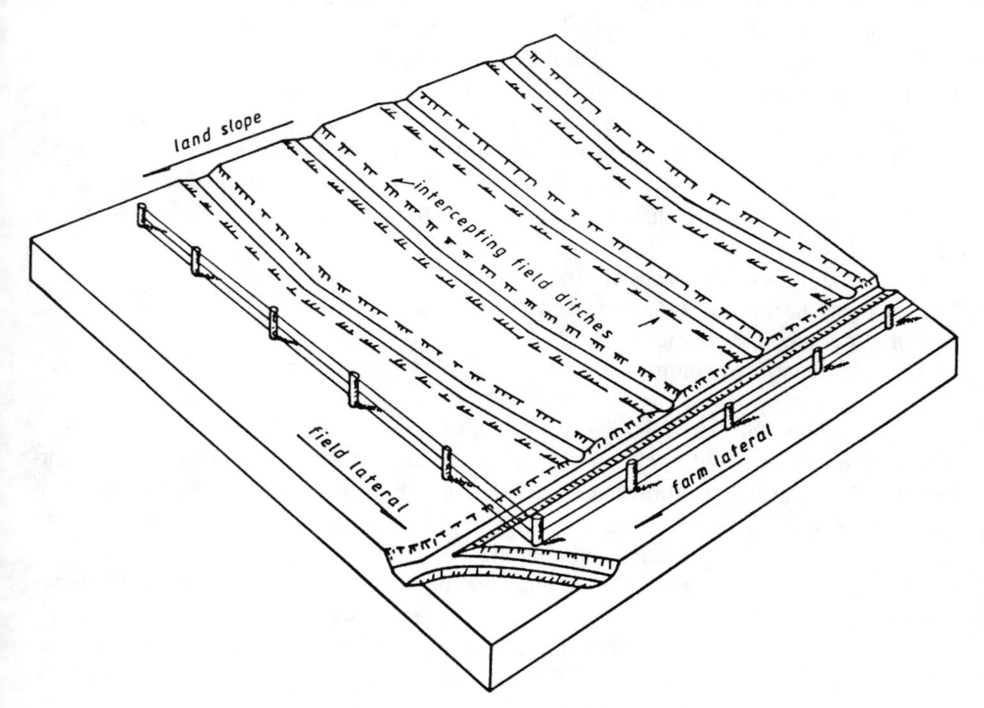

Figure 4.5. Layout of cross-slope system. ***Source:*** [7].

carried off. The cross-slope system is used mainly with the aim of

- draining gentle sloping land that may get wet because of slowly permeable soil,
- preventing the accumulation of water from upslope areas,
- avoiding the concentration of water in shallow pockets within the field.

Such ditches usually provide both surface drainage and erosion control. When designed specifically for the control of erosion, they are called terraces (see Section 4.4). Diversion ditches sometimes are used to divert runoff from low-lying areas, thus reducing the drainage problem.

Design and Construction Procedures. A general description of design and construction criteria is given here. For more detailed information the reader is referred elsewhere [9–14].

Design. Generally, the first stage of the design procedures involves identification of the study area, the definition of natural catchment basins and natural drainage lines, and the characterization of the areas requiring surface drainage. The latter step may involve field investigations, including

- soil and land-use surveys,
- monitoring of groundwater levels,
- soil and groundwater salinity assessment,
- characterization of rainfall intensities and establishing rainfall patterns,
- topographic surveys and mapping,
- identification of approximate locations for drains and outlets, and
- examination of aerial photographs and remote-sensing information.

The next stage consists in the definition of the hydraulic design criteria, which may include a range of options, accompanied by preliminary estimates of costs and benefits. A community involvement strategy also should be developed, paying attention to the possible involvement in drainage responsibilities of both public and private groups. Particular attention has to be paid to the perceived social and environmental impacts.

The final choice among different options leads to the selected design solution, which has to include

- detailed design for construction;
- evaluation of risks of flooding or drain failure;
- hydrogeological characteristics;
- assessment of environmental impacts relative to nutrient loads, hydrological impacts on waterways, lakes, and estuaries; and
- effects on wildlife and biodiversity.

The design of drains and ditches will take into account the area and slope of the catchment, soil permeability, storm characteristics, groundwater interactions, crop requirements, landholdings to be served, construction and maintenance costs and benefits, and maintenance and management requirements.

Construction. The selection of equipment and procedures for drain construction varies with the drainage depth and the land and soil characteristics.

Blade graders, scrapers, and heavier terracing machines are suitable for depth of cut up to 0.75 m. For cuts deeper than 0.75 m, bulldozers equipped with push or pull back bladers and carryall scrapers can be employed to fill the pothole area or other depression near the point of excavation.

Land Forming

Soils that require surface drainage may have depressions, which vary in size and shape. To prevent ponding and to keep the water moving at a uniform rate over the ground, it is necessary prior to the digging of the ditches to fill the depressions and to smooth out the high points in the field. This procedure is called land forming and consists in mechanically smoothing, grading, and bedding the soil surface. This allows better drainage and more efficient operation of farm equipment, cuts down the cost of ditch maintenance, and reduces ice crusting.

Shaping land to a smooth surface improves surface drainage. In general terms, land smoothing does not change the main contour of the land but it does eliminate minor differences in field elevation. Soil to be smoothed must have such a profile that small cuts can be made without hindering the soil itself and plant growth.

Land grading consists of shaping the land by cutting, filling, and smoothing until continuous surface grades are reached. The purpose is to prevent ponding and prolonged saturation of the soil. Land grading for drainage normally does not require shaping of the land into a plane surface with uniform slopes, but special care must be taken in filling depressions with soil from neighboring ridges and mounds. In areas with little or no slope, grades can be established to direct surface runoff into the ditches.

Bedding is the classical land-forming method for flat, heavy land in humid climates. It resembles a system of parallel field ditches with intervening lands shaped to a convex surface. The excess surface water is drained by lateral overland flow toward the field drains (Fig. 4.6). For the typical conditions under which bedding is used, flat beds generally should be no wider than 10 m to obtain good drainage. The beds are made by plowing, blading, or otherwise elevating the surface of flatland into a series of broad, low ridges separated by shallow, parallel furrows.

Optimal land-forming design aims at obtaining a properly graded surface of concave, convex, or plane shape from a natural ground surface, with the minimum volume of earthwork involved [15–19]. The slope of the graded surface may be different in the longitudinal and transversal directions and there may be restrictions on the slope in

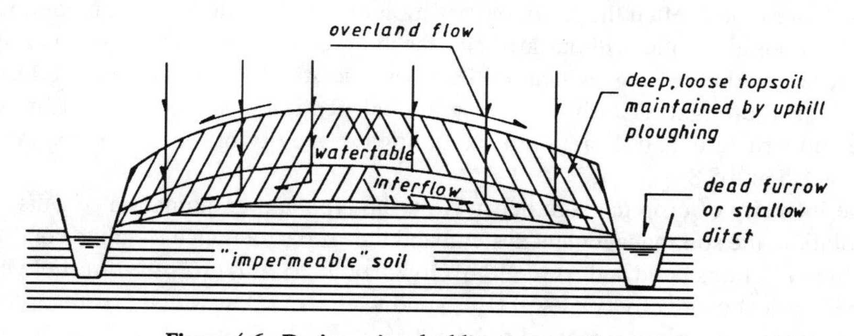

Figure 4.6. Drainage in a bedding system. ***Source:*** **[6].**

either or both directions. These factors make land-grading design a difficult problem to solve. Systematic procedures for land grading were introduced first by Givan [20] for rectangular fields and by Chugg [21] for irregularly shaped fields. Both authors applied least-squares methods to define a plane surface that fits the natural ground surface with minimum earthwork.

Hamad and Ali [22] proposed a method using the box-complex approach. Although this method is better than those based on the least-squares techniques, it lacks robustness and requires an initial feasible set of decision variables. Recently, Reddy [23] devised a new method for optimal land-grading design, based on a nonlinear genetic algorithm, suitable for both plane and curved surfaces while minimizing the total volume of earthwork. The application of this procedure shows that the optimal search is carried out simultaneously in several promising zones of the decision plane and a variety of design constraints can be considered, including curved surfaces. Pareto-optimal solutions, in addition to the best solution, are produced. Pareto-optimal solutions give an opportunity to the designer to use his/her judgment in selecting the most suitable surface based on practical considerations.

4.1.3 Salty Soils

Problems

Salts occur in the soil in one of the following three forms: salt ions dissolved in the soil water (the soil solution), cations adsorbed on the negatively charged surfaces of the soil particles (adsorption complex), and precipitated salts.

The principal salts are sodium, magnesium, and calcium chlorides; sodium and magnesium sulfates; sodium carbonates and bicarbonates; and nitrates and borates. These originate from rocks as they weather into soil. In this process the salts are carried downward (leached) with the percolating water. In the lower layers of the soil, they may either precipitate or continue to be transported in solution, eventually ending up in the sea. In certain cases, a high salt content in the soil may be related directly to the soil's parent material. This type of soil salinity is normally termed "primary" or "residual" salinity.

The most common cause of high soil salinity is salinization, that is, the accumulation of salts in the upper layers of the soil from some outside source. Salinization can be either a natural or an anthropogenic process.

The latter type is often the result of irrigation in regions with low precipitation, high rate of evaporation, and without artificial leaching or drainage. The result is a rather rapid accumulation of salts, which influence the growth of crops. The main effects are physiological drought, disturbance of the ion balance in the soil solution, degradation of the soil structure, and decrease of the biological activity of the soil (see Chapter 5, Sections 5.5 and 5.8).

The intensity of crop damage depends on the type and combination of salts in the soil solution, the soil management system, and the crop's tolerance to saline conditions. The United Nations Food and Agriculture Organization (FAO) gives the threshold values of the electrical conductivity of soil (EC_e) and water (EC_w) at which potential yield of various crops normally is expected to be 100%, 90%, 75%, and 50% of the maximum yield [24].

Reclamation Procedures

The term "salty" generally refers to a soil that contains sufficient salts to impair its productivity.

There are several methods of soil analysis to determine the salt content. Most of them, however, are insufficient for characterizing salinity status because the influence of the salts on the crop growth depends on many factors. These include the composition and properties of the different soil layers and horizons, seasonal variations of the capillary zone and the depth of groundwater, soil temperature and its seasonal fluctuations, and the stage of development of crops. Moreover, salinization is a dynamic process, changing with time and varying over short distances in the field and along the soil profile.

The system developed by the U.S. Salinity Laboratory is currently the most commonly used for the classification of salty soils [24]. The diagnostic parameters in this system follow:

- The electrical conductivity of the saturated soil extract, EC_e (dS/m), is a temperature-sensitive measurement and therefore is standardized at 25°C.
- The exchangeable sodium percentage (ESP) can be written as

$$\mathrm{ESP} = \frac{100(0.015 \text{ adj } R_{\mathrm{Na}})}{1 + 0.015 \text{ adj } R_{\mathrm{Na}}}, \tag{4.1}$$

where adj R_{Na} (adjusted sodium adsorption ratio) is defined as [24]

$$\text{adj } R_{\mathrm{Na}} = \frac{\mathrm{Na}}{\sqrt{\frac{\mathrm{Ca}_x + \mathrm{Mg}}{2}}} \tag{4.2}$$

in which Na and Mg are the concentrations (in me/L) of sodium and magnesium in the soil extract and Ca_x is a modified value of the calcium concentration, which takes into account the effects of carbon dioxide (CO_2), bicarbonate (HCO_3), and salinity of the soil water (EC_w) upon the calcium originally present. Values for Ca_x are given in Table 4.5.

With reference to the reclamation procedures, the salty soils have been separated into the following three groups [25]: saline, saline-sodic, and sodic.

Saline Soils

Saline soils are those for which the conductivity of the saturated soil extract $EC_e > 4$ dS/m at 25°C and ESP < 15. Ordinarily, the pH is less than 8.5.

The chemical characteristics of saline soils are the following:

- Sodium seldom comprises more than half of the soluble cations and hence is not adsorbed to any significant extent.
- The amounts of calcium and magnesium may vary considerably.
- Soluble and exchangeable potassium are ordinarily minor constituents, but sometimes they may appear as major constituents.
- The main anions are chloride, sulfate, and nitrate.

Because of the presence of excess salts and the absence of significant amounts of exchangeable sodium, saline soils generally are flocculated and therefore have permeabilities similar to those of nonsaline soils.

Table 4.5. Modified calcium concentration Ca_x (meq/L) as function of electrical conductivity of soil water and of ratio of bicarbonate to calcium

Ratio of Bicarbonate (meq/L^3) to Calcium (meq/L^3)	Electrical Conductivity of Applied Water, EC_w (dS/m)											
	0.1	0.2	0.3	0.5	0.7	1.0	1.5	2.0	3.0	4.0	6.0	8.0
0.05	13.20	13.61	13.92	14.40	14.79	15.26	15.91	16.43	17.28	17.97	19.07	19.94
0.10	8.31	8.57	8.77	9.07	9.31	9.62	10.02	10.35	10.89	11.32	12.01	12.56
0.15	6.34	6.54	6.69	6.92	7.11	7.34	7.65	7.90	8.31	8.64	9.17	9.58
0.20	5.24	5.40	5.52	5.71	5.87	6.06	6.31	6.52	6.86	7.13	7.57	7.91
0.25	4.51	4.65	4.76	4.92	5.06	5.22	5.44	5.62	5.91	6.15	6.52	6.82
0.30	4.00	4.12	4.21	4.36	4.48	4.62	4.82	4.98	5.24	5.44	5.77	6.04
0.35	3.61	3.72	3.80	3.94	4.04	4.17	4.35	4.49	4.72	4.91	5.21	5.45
0.40	3.30	3.40	3.48	3.60	3.70	3.82	3.98	4.11	4.32	4.49	4.77	4.98
0.45	3.05	3.14	3.22	3.33	3.42	3.53	3.68	3.80	4.00	4.15	4.41	4.61
0.50	2.84	2.93	3.00	3.10	3.19	3.29	3.43	3.54	3.72	3.87	4.11	4.30
0.75	2.17	2.24	2.29	2.37	2.43	2.51	2.62	2.70	2.84	2.95	3.14	3.28
1.00	1.79	1.85	1.89	1.96	2.01	2.09	2.16	2.23	2.35	2.44	2.59	2.71
1.25	1.54	1.59	1.63	1.68	1.73	1.78	1.86	1.92	2.02	2.10	2.23	2.33
1.50	1.37	1.41	1.44	1.49	1.53	1.58	1.65	1.70	1.79	1.86	1.97	2.07
1.75	1.23	1.27	1.30	1.35	1.38	1.43	1.49	1.54	1.62	1.68	1.78	1.86
2.00	1.13	1.16	1.19	1.23	1.26	1.31	1.36	1.40	1.48	1.54	1.63	1.70
2.25	1.04	1.08	1.10	1.14	1.17	1.21	1.26	1.30	1.37	1.42	1.51	1.58
2.50	0.97	1.00	1.02	1.06	1.09	1.12	1.17	1.21	1.27	1.32	1.40	1.47
3.00	0.85	0.89	0.91	0.94	0.96	1.00	1.04	1.07	1.13	1.17	1.24	1.30
3.50	0.78	0.80	0.82	0.85	0.87	0.90	0.94	0.97	1.02	1.06	1.12	1.17
4.00	0.71	0.73	0.75	0.78	0.80	0.82	0.86	0.88	0.93	0.97	1.03	1.07
4.50	0.66	0.68	0.69	0.72	0.74	0.76	0.79	0.82	0.86	0.90	0.95	0.99
5.00	0.61	0.63	0.65	0.67	0.69	0.71	0.74	0.76	0.80	0.83	0.88	0.93
7.00	0.49	0.50	0.52	0.53	0.55	0.57	0.59	0.61	0.64	0.67	0.71	0.74
10.00	0.39	0.40	0.41	0.42	0.43	0.45	0.47	0.48	0.51	0.53	0.56	0.58
20.00	0.24	0.25	0.26	0.26	0.27	0.28	0.29	0.30	0.32	0.33	0.35	0.37
30.00	0.18	0.19	0.20	0.20	0.21	0.21	0.22	0.23	0.24	0.25	0.27	0.28

Source: [24].

Reclamation techniques normally depend on suitable irrigation water, appropriate means of water application, and good drainage. Large quantities of water with low salt content are needed to leach salts from the soil profile. Surface basins and sprinkle irrigation are usually appropriate because these apply water on the entire soil surface. Salt-tolerant vegetation can help to reclaim saline soils, especially those with fine textures. Plant roots help to keep the soil permeable and the top growth prevents erosion. Good drainage is required to remove the leaching water fast enough and to keep groundwater levels from rising. Sometimes leaching is done intermittently both to save water and to give more time for drainage.

Reclamation requires that land be well managed afterward. The water table has to be kept low enough to keep from making the soil saline again. Suitable amounts of irrigation water have to be applied to let drainage remove the salt brought in by the irrigation water.

Saline-Sodic Soils

The term "saline-sodic" is applied to soils for which $EC_e > 4$ dS/m at 25°C and ESP >15. Normally, the pH is seldom higher than 8.5. These soils form as a result of the combined processes of salinization and sodification.

As the concentration of salts in the soil solution decreases, some of the exchangeable sodium hydrolyzes and forms sodium hydroxide. This salt may change to sodium carbonate on reaction with carbon dioxide absorbed from the atmosphere. Upon leaching, the soil may become strongly alkaline, and after the consequent dispersion of the fine particles, unfavorable for the entry and movement of water and for tillage. Although the return of the soluble salts may help particles to flocculate, the management of saline-sodic soils continues to be a problem until the excess salts' exchangeable sodium is removed from the root zone.

When gypsum is added to saline-sodic soils and such soils are leached, the excess salts are removed and calcium replaces the exchangeable sodium. The reclamation of saline-sodic soils requires amendments—normally gypsum, sulfur, or calcium chloride—to replace sodium. Amendments have to be applied prior to the leaching process.

The amendment requirement (AR) can be calculated from the following formula [4]:

$$\mathrm{AR} = \left(\frac{\mathrm{ESP_{in}} - \mathrm{ESP_{fn}}}{100}\right)\mathrm{CEC} \tag{4.3}$$

where CEC is the cation exchange capacity in meq/100 g of soil; $\mathrm{ESP_{in}}$ and $\mathrm{ESP_{fn}}$ are, respectively, the initial and final exchangeable sodium percentages (before and after the reclamation); and AR is the amendment requirement in meq/100 g of soil.

CEC usually is measured by saturating the soil complex with one type of cation (e.g., Na^+), then replacing this cation and measuring the displaced quantity in the soil water extract (standardized at pH $= 7.0$). If, for example, $\mathrm{ESP_{in}} = 30$, $\mathrm{ESP_{fn}} = 10$, and $\mathrm{CEC} = 24$, then $\mathrm{AR} = 4.8$ meq of amendment/100 g of soil.

When the amendment is gypsum [1 meq of gypsum/100 g of soil equals 860 ppm (parts per million) of gypsum], the amount of gypsum required for 1 ha to depth of 20 cm (roughly 3.10^6 kg) will be 12,384 kg.

This calculation is based on 100% replacement of sodium by calcium. Because of the presence in some saline-sodic soils of free soda, the actual efficiency is lower. Thus, it is recommended that the amount of applied gypsum be increased in accordance with equivalents of free sodium carbonate and bicarbonate.

Reclamation with gypsum—sometimes added to the irrigation water—is usually less expensive than with other amendments, but the amount required is larger and the process is slow. Sulfur also is cheap and widely used, but also has it own limitations. Soil bacteria must first oxidize the sulfur to sulfuric acid. The hydrogen ion of the acid then must react with the soil lime to release calcium, which, in turn, exchanges for the soil-adsorbed sodium. Only then is the sodium ready for leaching and the whole process may take months or years.

Sodic Soils

Sodic soils are those for which the percentage of exchangeable sodium is greater than 15 and the conductivity of the saturated soil extract is less than 4 dS/m at 25°C. The pH readings usually range between 8.5 and 10. These soils frequently occur in small irregular areas of arid and semiarid regions, which often are referred to as "slick spots."

The removal of excess salts due to leaching tends to increase the rate of hydrolysis of the exchangeable sodium and can lead to a rise in soil pH. Moreover, evaporation may deposit organic matter present in the soil solution on the ground surface causing the formation of a dark crust.

These soils are characterized by the presence of sodium carbonates, sodium hydrogen-carbonates, and other sodium salts in the soil solution. These salts are very harmful, particularly if the soil does not contain gypsum. Physical properties are affected by the presence of exchangeable sodium in the soil. As ESP increases, the soil tends to become more dispersed. The pH may increase, sometimes becoming as high as 10. At these high pH values and in the presence of carbonate ions, calcium and magnesium are precipitated, resulting in an increase in the concentration of sodium in the soil solution.

Sodic soils are difficult to reclaim because their permeability is too low for water to carry amendments to the soil colloids. Mechanical mixing and deep plowing may help because they get the amendment directly into the soil and open up some passages for water percolation.

Few plants will grow in sodic soils until some reclamation has been achieved. To this end, even weeds should be encouraged because their roots open channels and crevices that improve soil permeability.

In the initial stages of sodic soil reclamation, the application of salty water is much better than pure water because salts help to flocculate the soil colloids and increase the soil permeability, often by one or two orders of magnitude. In this regard, water containing calcium is especially helpful because this ion replaces exchangeable sodium.

4.1.4 Derelict and Degraded Soils

Problems

A soil that was once biologically productive and has suffered to such an extent from human activities that it is less capable or incapable of beneficial uses (even for leisure or recreation) is described as "derelict." Attempts at renewal of the degraded resource are sometimes given different names such as redemption, rehabilitation, and restoration.

Reclamation often is used as a blanket term to describe all of the activities that seek to upgrade damaged soil and to bring it back into beneficial use.

Soils may be degraded in a variety of ways and to very different degrees. Productive areas have been spoiled by acid precipitation, or have been covered by mining loads, smelter slags, or by the waste of chemical and manufacturing industries.

When a human activity adversely affects the land, it is generally the biologically active surface layer—soil—that is most easily destroyed or degraded. In addition to the impairment of biological activity, the handling and tipping of loads often can lead to compaction and degradation of soil structure, rendering the environment inhospitable for plant growth.

These problems must be overcome if a derelict soil is to be upgraded so as to appropriately support plant and animal communities. When this process begins, some of the features of a fertile soil will reappear in deeper layers. Plant growth will improve steadily and, in turn, will introduce residues that will improve the physical and chemical characteristics of the soil. The initiation of degraded resource renovation is the broad aim of derelict soil reclamation.

Reclamation Procedures

Methods of reclamation have to be tailored to the specific problems at hand. However, there are some principles that are common to all reclamation procedures. Figure 4.7 shows the steps and the considerations that are necessary in the design of a suitable reclamation scheme. We have to bring back to full life an ecosystem in which normal biological processes are at a standstill. The natural development of such an ecosystem normally takes a long time, and so, the key to improvement is upgrading all of its activities by the input of materials that the soil has in short supply. This can be done directly with nutrients and/or fertilizers, or indirectly with organic matter built up by the growth of plants.

Once the soil is charged adequately, it has the nutrients and the physical properties that allow plants and soil organisms to grow and develop. The cycling of materials can begin and a self-sustaining ecosystem results. The soil structure also improves with the input of organic matter and the concurrent increase in the activities of soil microorganisms.

The most common techniques are topsoil conservation, topsoil application, application of other materials, hydraulic seeding, and revegetation.

Topsoil Conservation

If a land is expected to be disturbed by anthropogenic activities, the fertile topsoil and sometimes the subsoil have to be removed, stored, and replaced later when the disturbance has finished. This technique is suitable mainly for progressive, fast-moving, strip-mining operations taking a relatively thin layer of material, such as coal, bauxite, or sand. Generally, if the procedure is carried out adequately, the topsoil loses very little of its original properties and, provided that care is taken to prevent consolidation, crops and other vegetation can be reestablished immediately [26].

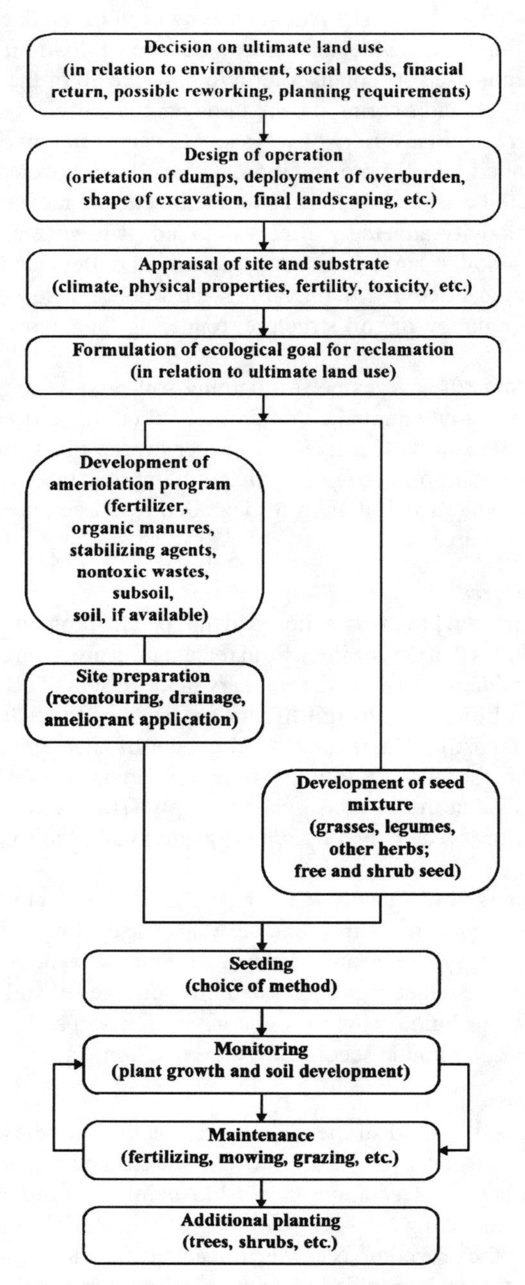

Figure 4.7. Steps involved in the development of a soil reclamation scheme. *Source:* Modified from [1].

In many sites, however, the depth of work may be considerable and/or the area for storage small. In these cases the topsoil has to be stripped at the outset, stored in large heaps, and replaced later. This procedure tends to cause a deterioration in the topsoil structure and some loss of fertility. Therefore, the topsoil has to be removed and replaced carefully, preferably during dry conditions, without being mixed with other materials if the natural processes are to be restored quickly.

Topsoil Application

Sometimes, it may be possible to remove the top layers of a good soil from the site and spread it over an another area where the soil has been degraded. This procedure may provide a fully developed and fertile soil of good structure and texture. Generally, the establishment and growth of plants when topsoil is used as a covering material are satisfactory, usually with fertilizer addition. However, there always will be a limit to the depth of topsoil that can be applied. The underlying layers will be part of the rooting zone of the plants, and their characteristics, therefore, cannot be disregarded totally [27]. If they are toxic or impermeable, the roots will be confined to the topsoil. Whereas impermeability of the underlying material can be ameliorated by ripping operations, problems of toxicity are more difficult to overcome. Moreover, some wastes, such as chemical refuse, are not only toxic in themselves but can release substances that will poison added soil above. In these cases, a deeper layer of soil will be necessary as a covering.

On the whole, topsoil application is recommended in small sites where high quality and rapid solutions are required and when it is necessary to restore the original ecosystem as faithfully as possible.

Application of Other Materials

This technique is based on the idea of using some waste products to deal with the problems caused by others [1]. As a matter of fact, there are many waste products that are good media for plant growth, such as sewage sludge, mushroom compost, farmyard manure and pig slurry, domestic refuse, fuel ash, and mining and chemical wastes. These materials, notwithstanding their different origins, are similar in the sense of being non-toxic and water-and nutrient-retaining.

Hydraulic Seeding

Hydraulic seeding or hydro-mulching is a technique which consists of spraying seeds and nutrients over the ground in the form of a slurry. Normally the seed is suspended in an aqueous solution that can be sprayed over distances of as much as fifty meters or more, from a machine that does not have to pass over the ground. When the seed reaches the ground surface it is carried into crevices by the liquid. To keep the solution properly mixed, starch derivatives or oil-based emulsions are added to increase the viscosity of the mixture. These substances also play an important role in improving the adhesion of the seeds to the soil when they hit the ground. Adequate nutrients—provided either by fertilizer or by a liquid manure—must also be added to prevent the young plants from failing to establish due to unfavorable soil conditions.

Normally, not more than 200 kg/ha of a compound fertilizer, containing about 30 kg/ha of nitrogen, phosphorus and potassium, can be added to the mixture containing

Table 4.6. Hydroseeding indicative application rates of materials suitable for the establishment of grass–legume cover in temperate climates

Material	Application Rate
Grass-seed mixture appropriate for situation	70 kg/ha
Wild white clover or other inoculated legumes	10 kg/ha
Mulch: wood fiber, chopped straw, or glass wool	1–2 t/ha
Stabilizer: alginate, PVA, or latex	Depends on stabilizer
Fertilizer: complete 15:15:15	200 kg/ha, followed by 300 kg/ha after 8 weeks
Dry organic manure	500 kg/ha
Lime	0–5,000 kg/ha

Source: [1].

the seed. Table 4.6 gives indicative application rates of some materials suitable for the establishment of grass/legume cover in temperate climates. This technique is quite expensive and not yet totally reliable for all situations. On the whole, it is cost effective mainly on steep, less accessible slopes or where there is a serious erosion problem requiring the use of a soil-binding material [28].

Re-vegetation

One of the most common components of all reclamation schemes is the establishment of vegetation. Soils drastically disturbed by anthropogenic activities are often difficult to revegetate because the soil layers have been compacted to the extent that plant roots are unable to penetrate them. Essential plant nutrients may not be present in an available form or in properly balanced proportions. Nitrogen is invariably deficient and phosphorus is generally in short supply. Moreover, the sites may be too dry or too wet. In spite of these problems, considerable success has been attained in vegetating degraded soils on mining dumps and air fields, along highways and pipelines, or around buildings. The first step is normally the establishment of a grass cover, usually containing legumes. Legumes, in fact, are of paramount importance in almost all grass mixtures because they provide and maintain an adequate supply of nitrogen and ensure the build up of an suitable amount of organic matter in the newly forming soil. The choice of the proper legumes depends on soil conditions and on climate. Generally, the most valuable legumes are those which are used in agriculture since they have high rates of nitrogen fixation. A list of these plants on a world-wide basis is given in Table 4.7.

An alternative to sowing grass and legumes is to establish trees or shrubs. These can be planted on ground where agriculture is impossible, such as steep-sided heaps and mining loads. However, they are less effective than grasses in stabilizing soils against erosion. To overcome this problem, because trees and shrubs are established at low density and a lot of surface is left unprotected, a grassy ground cover (grass–legume mixture or pure legume) may play an important role.

In any case, the establishment of either perennial vegetation or cultivated crops represents the final step in the development of any successful derelict soil reclamation procedure.

Table 4.7. Perennial legumes suitable for derelict soils

Legume Species	Soil Preference[a]	Climate Preference[b]
Amorpha fruticosa (indigo bush)	NC	W
Centrosema pubescens (centro)	AN	W
Coronilla varia (crown vetch)	AN	CW
Desmodium uncinatum (silver leaf desmodium)	AN	W
Lathyrus sylvestris (mat peavine)	NC	W
Lespedeza bicolor (lespedeza)	AN	W
Lespedeza cuneata (sericea lespedeza)	AN	W
Lespedeza japonica (Japanese lespedeza)	AN	W
Lotus corniculatus (birdsfoot trefoil)	NC	CW
Lupinus arboreus (tree lupin)	ANC	CW
Medicago sativa (lucerne [alfalfa])	NC	CW
Melilotus alba (sweet clover [white]	ANC	CW
Melilotus officinalis (sweet clover [yellow]	ANC	W
Phaseolus atropurpureus (siratro)	ANC	W
Stylosanthes humilis (Townsville stylo)	AN	W
Trifolium hybridum (alsike clover)	ANC	C
Trifolium pratense (red clover)	NC	C
Trifolium repens (white clover)	NC	CW
Ulex europaeus (gorse)	ANC	C

[a] A = acid, C = calcareous, N = neutral.
[b] C = cool, W = warm.
Source: [1].

4.1.5 Tidal Soils

Problems

Soil in coastal areas, along river margins, in estuaries and bays, and on land reclaimed from the open sea are subjected in varying degrees of overflow and restricted drainage caused by tidal water. The extent and frequency of overflow and drainage inpairment may vary widely, depending on the elevation and exposure of the sites to tidal water. The action of waves may cause erosion in some regions and the deposition of sand, silt, an clay in others. For centuries efforts have been made to rescue parts of these coastal areas from the sea and turn them into productive land. In this regard, perhaps the most famous project is the impoldering of the former Zuiderzee in the Netherlands, which began in the 1920s and continues to the present day.

Reclamation Procedures

Reclamation and drainage of tidal soils are closely related. Infact, during the initial stages of the process, they are largely synonymous.

In low-lying coastal soils, outflow from interior areas is blocked temporarily by high tides when it does not have free outlet to the sea. Thus, in these areas, draining water is ponded (stored) twice daily in drainage channels and adjacent low areas. This water may be fresh, salty, or a mixture of the two, depending on the hydraulic gradient between the sea and the ponded water. The basic purpose of reclamation in these areas is to prevent seawater from moving landward during high tide and to provide for the disposal of stored fresh water during periods of low tide.

Protection from overflow usually is achieved by enclosing such areas with dikes. Drainage may be obtained by establishing a system of internal drains with water discharged over the dikes by pumps, by gravity flow through gates, or by a combination of pumps and gates.

Pumps are necessary when storage for accumulating drainage water within ponding areas is not available; when flow through gates is prevented over long periods by tides, floods, or inadequate outlets into the sea; or when the construction and maintenance of foreshore channels are impractical.

The exact steps to be followed in the reclamation of a particular tidal area depend upon local conditions, but in all cases, when the area on the land side of the dike is wholly dried up, the initial overall appearance of the soils is that of soft mud. These soils, generally called "unripened soils" consist of loose alluvial deposits that have very little stability. Upon drainage, changes take place in the mud, which together are referred to as "ripening." Ripening precedes the common soil formation processes leading to full development of a soil profile [6].

Soil Ripening Processes

Soil ripening of fresh subaqueous deposits starts with the loss of excess water from the mud, by evaporation and drainage. The water table falls and the soil above the phreatic surface becomes exposed to capillary forces that pull the soil particles into a closer packing arrangement. This induces a reduction of the soil volume, subsidence, and, in the end, the structural development of the soil. Together, these processes comprise the physical ripening of the soil.

The ripening process normally begins at the surface and slowly extends to the deeper soil layers. At the same time, chemical and biological changes occur in the soil (respectively referred to as "chemical" and "biological" ripening). Chemical ripening consists of oxidation processes and adjustments in the cation composition of the adsorption complex. Biological ripening involves the development of aerobic microbial life within the soil.

During the first steps of the ripening process, the hydraulic conductivity of the soil is usually very low and only artificial drainage keeps the surface clear of ponded water. To this end, the most suitable system is that of parallel field ditches, with spacing of the order of 10 m or more. The field ditches should be deepened gradually, in parallel with the progress of the ripening. Starting with depths of 30–40 cm, a depth of 60–70 cm normally is reached over a period of about five years.

By the time the ripening front has reached well into the subsoil (e.g., to about 60- to 70-cm depth) pipe drains also may be installed. In this way, during the rain period, the excess rainwater flows vertically through the cracks and then flows laterally, over the impermeable unripened layers, to the trenches and then down to the pipes.

The spacing to be used depends very much on the changes of the values of the hydraulic conductivity during the development of the ripening process. When a good number of ripening cracks are established, a spacing of up to 30–50 m may be used. On the whole, drainage can only remove the soil water above the field capacity. The removal of the pore water held below field capacity depends mostly on evaporative drying. This process is enhanced if there is a vegetative cover with a deep and extensive root system. Reclaimed land in northern Europe (the Netherlands, England, and parts of Germany) often is sown with reed just before emergence mainly for this reason. Reeds are followed in turn by the first crops. A grass ley, with legumes, may be sown and used for grazing sheep or cattle for some 10 years. During this time, soluble salts are leached from the surface layers to lower levels. Calcium gradually replaces sodium as the dominant ion in the cation exchange complex, organic matter accumulates, and soil structure is improved greatly. Sometimes, ripening may be considerably retarded by seepage of brackish water or seawater into the land. In such areas the process can be accelerated by intercepting the seepage in flow.

Coastal Sands and Dunes

Coastal sands represent a valuable resource not only for their own sake as wild open country with an interesting vegetation, but also as a support area for beaches and other resort facilities. Recently, they have been subjected to an enormous pressure from recreation and even mining activities.

Sand dunes generally are made up almost entirely of silica particles and usually are lacking in water, organic matter, and nutrients. Moreover, they are unstable and shift under the influence of wind and waves.

When dune areas are being reclaimed, the first step is to ensure protection from sea erosion, by wave screens or groines. At the same time, the original vegetation must be reestablished in places where it has been removed or destroyed [29, 30]. Sea lyme grass is valuable for building foredunes because it can tolerate a high degree of salinity. A number of other species that are adapted to the unfavorable coastal environment also can be planted or sown. These include sea couch, red fescue, sea oats, and beach bean. After planting, the sand surface often requires stabilization. Bituminous stabilizers are suitable for this purpose and can be very effective if applied properly. To this end, cover crops of large-seeded annuals such as cereal rye and sorghum can be very valuable if sown at a low density along with fertilizer dressings. Experience from many different parts of the world shows that, in the first year, about 100 kg/ha of nitrogen and 25 kg/ha of phosphorus (given in two or three applications) are imperative. Where it is not essential to reestablish the native vegetation, attention should be focused on shrubs, mainly legumes, that grow rapidly and can act as slow releasers of nutrients for subsequent tree planting.

References

1. Bradshaw, A. D., and M. J. Chadwich. 1980. *The Restoration of Land. The Ecology and Reclamation of Derelict and Degraded Land.* Berkeley, CA.: University of California Press.
2. Cowardin, L. M., V. Carter, F. C. Golet, and E. T. La Roe, 1979. Classification of Wetlands and Deepwater Habitats in the United States, Rep. FWS/OBS-79/31. Washington DC: U.S. Fish and Wildlife Service.

3. Skaggs, R. W. 1985. Drainage. *Water and Water Policy in the World Food Supply*. ed. Jordan, W. R. College Station: Texas A&M University Press.
4. Food and Agriculture Organization and United Nations Educational, Scientific and Cultural Organization. 1973. *Irrigation, Drainage, and Salinity*. London: Camelot Press.
5. Maidment D. R. (ed.). 1992. *Handbook of Hydrology*. New York: McGraw-Hill.
6. Smedema, L. K., and D. W. Rycroft. 1983. *Land Drainage. Planning and Design of Agricultural Drainage Systems*. London: London Batsford Academic and Educational.
7. U.S. Department of Agriculture. 1973. *Drainage of Agricultural Land*. Washington, DC: Water Information Center.
8. Schwab, G. O., R. K. Frevert, T. W. Edminster, and K. K. Barnes. 1981. *Soil and Water Conservation Engineering*. New York: Wiley.
9. Food and Agriculture Organization. 1980. Drainage design factors. Paper No. 38. Rome: Food and Agricultural Organization.
10. De Wrachien, D. 1991. Short guidelines on the hydraulic design of drainage canals. *J. Irrigat. Drain. Eng*. 1:33–39.
11. International Commission on Irrigation and Drainage - Australian Committee. 1991. Guideline on the Construction of Surface Drainage Systems. Riverland: (Australia) Engineering and Water Supply Department.
12. Easa, S. M. 1992. Probabilistic design of open drainage channels. *J. Irrigat. Drain. Eng*. 94(5): 949–956.
13. Barua G., and K. N. Tiwari. 1996. Ditch drainage theories for homogeneous anisotropic soils. *J. Irrigat. Drain. Eng*. 122(5):276–285.
14. Barua G., and K. N. Tiwari. 1996. Theories of ditch drainage in layered anisotropic soil. *J. Irrigat. Drain. Eng*. 122(6):321–330.
15. U.S. Department of Agriculture. 1970. Land leveling. *SCS National Engineering Handbook*. (USDA) Washington, DC.
16. Scaloppi E. J., and L. S. Willardoson, 1986. Practical land grading based upon least squares. *J. Irrigat. Drain. Eng*. 112(2):98–109.
17. De Wrachien, D. 1991. Lay-out of surface drainage systems. *J. Irrigat. Agric*. n.3: 168–173.
18. Schorn, H., and D. H. Gray. 1995. Landforming, grading, and slope evolution. *J. Geotech. Eng*. 121(6):729–734.
19. Gray, D. H., and R. B. Sotir. 1996. *Biotechnical and Soil Bioengineering. Slope Stabilization: A Practical Guide for Erosion Control*. New York: Wiley.
20. Givan, C. E. 1940. Land grading calculations. *Agric. Eng*. 21(1):11–12.
21. Chugg, G. E. 1947. Calculation for land gradation. *Agric. Eng*. 28(10):461.
22. Hamad, S. N., and A. M. Ali. 1990. Land grading design by using nonlinear programming. *J. Irrigat. Drain. Eng*. 116(2):219–226.
23. Reddy, S. L. 1996. Optimal land grading based on genetic algorithms. *J. Irrigat. Drain. Eng*. 122(4):183–188.
24. Ayers, R. S., and D. W. Westcot. 1985. Water quality for agriculture, irrigation and drainage. Paper No. 29, Rev. 1. Rome: Food and Agriculture Organization.

25. Troeh, F. R., J. A. Hobbs, and R. L. Donahue. 1980. Soil and Water Conservation for Productivity and Environmental Protection. London: Prentice-Hall.
26. Knabe, W. 1965. Observations on worldwide efforts to reclaim industrial waste land. *Ecology and Industrial Society*. ed. Goodman, G. T., R. W. Edwards, and J. M. Lambert, pp. 263–296. Oxford: Blackwell.
27. Cooke, G. W. 1967. *The Control of Soil Fertility*. London: Crosby, Lockwood.
28. Sheldon, J. C., and A. D. Bradshaw. 1978. The development of a hydraulic seeding technique for unstable sand slopes. I. Effects of fertilisers, mulches and stabilisers. *J. Appl. Ecol.* 14:905–918.
29. Coaldrake, J. E. 1973. Conservation problems of coastal sand and open-cast mining. *Nature Conservation in the Pacific*, ed. Costin, A. B., and R. H. Groves, pp. 299–314, Canberra, Australia.
30. Zak, J. M. 1965. Sand dune erosion control at Provincetown, Massachusetts. *J. Soil Water Conserv.* 20:188–189.

4.2 Soil Regeneration and Improvement

G. Chisci and D. De Wrachien

4.2.1 Soil Degradation

For a soil to provide the most desirable hydraulic and mechanical properties as an ideal medium for crop production, it is necessary that all of the different hierarchical orders of its solid-phase components be well developed and stable against the action of water and external mechanical stresses. According to Dexter [1], to accommodate almost all of the different soil aspects affecting a soil's capability to provide an ideal medium for crop production, it is convenient to widen the definition of soil structure to the "spatial heterogeneity of the different components or properties of the soil." Such a definition accommodates many aspects of the soil that manifest themselves at different scales and their variable distribution in time and space (soil anisotrophy).

Synthetically, Dexter considers soil structure as the spatial variability of its components and lists a number of soil features with different dimentions (Fig. 4.8). It follows that structure, in the broad sense outlined above, is the most important feature of a soil in relation to plant growth and crop production. Consequently, soil degradation can be envisaged as the consequence of any process or human action negatively influencing soil structure. Conversely, soil regeneration is the result of any natural process and/or human action capable of recovering soil structure from a deteriorated condition and possibly aimed at an improvement of soil structure by ameliorative measures.

The main factors involved in the physical degradation of a soil, in the short and medium term, are

- compaction,
- microstructure degradation, and
- crusting.

Such processes (alone and/or in combination) depend on many factors that are discussed later and are responsible not only for reducing the suitability of a soil for plant growth and

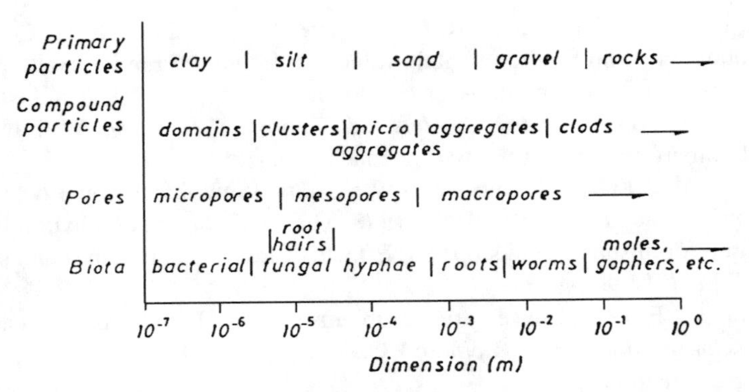

Figure 4.8. Spatial variability of soil components and soil features with different dimensions. *Source:* [1].

crop production, but also for impairing the soil system in its capabilities as a depurator and water regulator.

Compaction

Compaction can be defined as the response of a soil to external forces and implies a decrease in its volume (or an increase in its density). The extent of the compaction depends on both the soil and the forces applied. Several soil parameters can be used to characterize soil compaction, such as bulk density, void ratio, specific volume, or bulk weight volume [2]. The compactability of the soil (i.e., its response to compaction forces) depends on soil type, moisture content, and initial state of compaction.

The resistence of each soil to compaction (soil strength) decreases rapidly when soil moisture content increases. Soil strength generally increases with depth so that the strength of the subsoil is generally higher than that of the topsoil, which may be loose and soft under certain operative conditions and therefore more compactable. A moderate recompaction of a plowed soil may be favorable to plant growth, but intensive traffic frequently causes excessive compaction. The tilled layer often undergoes an annual cycle of loosening by tillage and recompaction by natural processes and machinery traffic [3–6].

The extent of recompaction by machinery traffic depends on many factors. It tends to increase with the soil moisture content, the wheel load, the inflation pressure of tires, the number of machinery passes, the wheel slip, and the velocity of the vehicle [6–8].

As previously stated, soil compaction for a specific site can be characterized by the bulk density or by the porosity of the soil. However, it is extremely difficult to identify all of the effects of soil compaction on crop growth and yield. So far, the experimental measurement of crop response in field experiments is probably the only way to assess the integrated effect of soil compaction on crop yield. Generally speaking, yields decrease in overly dense soils, probably as a consequence of poor aeration and mechanical impedence to root growth. In overly loose soils, the unsaturated hydraulic conductivity, the volumetric water content, and the concentration of nutrients are often lower. These conditions may restrict water and nutrient transport to roots.

Other soil properties besides bulk density or directly related quantities also have been used for characterizing soil compaction. Among these, penetration resistence is the most common. Provided that soil water content is uniform, penetrometer resistance may be a more sensitive parameter than bulk density [9]. Nevertheless, this parameter has the disadvantage of having high variability and is unsuitable for use in stony soils.

A more universal method for measuring soil compaction was devised by Eriksson et al. [10] and Hakansson [2], by normalizing the bulk density values so that they are comparable in their effect on crop production. The maximum bulk density obtained in the laboratory using a long-term, confined, uniaxial compression test and a static pressure of 200 kPa was established as a reference point; the bulk density of the same soil in the field, expressed as a percentage of the reference maximum bulk density, was adopted as the degree of compaction. This parameter has been shown to have biological significance in experiments carried out by Hakansson [2], which indicated that maximum crop yield can be attained at the same degree of compactness irrespectively of soil type.

Scientists, technicians, and farmers are much more concerned about soil compaction today than in the past, because of the increasing mechanization of crop production in most parts of the world. This trend has enhanced the risks of soil physical degradation when wheels pass over soil used as growing medium for crops [10–12]. Nowadays, farming almost invariably involves the passage of wheeled and/or tracked vehicles for primary and secondary cultivation, sowing, spraying, and harvesting operations. However, soils that have been cultivated for a long time often will not conserve the strength to support the most modern vehicles without considerable compression and rutting. Continuously tilled soils that are less suited to crop growth often need corrective treatments between crops, such as more intensive or deeper primary tillage operations, thus increasing the costs of production. In addition, such operations are rarely completely effective. Compaction is becoming a very serious soil degradation problem where heavy wheeled vehicles are used in crop production, with the possible exception of arid nonirrigated areas. As a result, it has become increasingly necessary to relate the compaction produced by agricultural vehicles to the complete system of soil management.

Crop rotation also may be important in relation to compaction under vehicles. Garwood et al. [13] found that a free-draining sandy-loam soil was considerably more compact following 20 years of arable cropping then after a similar period of grass pasture (Fig. 4.9).

Wheeled vehicles are becoming larger and more powerful. McKibben [14] reported that the average mass of tractors increased from 2.7 t in 1948 to 4.5 t in 1968. Since then, the average power has increased at a rate of about 5% to 7% per annum. Vehicles used for transporting and spreading lime and slurry now have a mass that sometimes exceds 20 t [15]. Furthermore, the development of combine machinery has increased the risk of soil compaction in the actual crop management systems.

Traditionally arable soils pass through an annual cycle of loosening and compaction. Loosening generally is obtained by primary tillage operations or by subsoiling, whereas compaction may occur at several stages of the crop cycle, for instance at seed-bed preparation or at harvesting. In relation to seed-bed preparation, the soil density at the time of sowing is often as high as it is prior to plowing [3]. The mechanism of compaction that occurs during seed-bed preparation is related to the passage of both the implement

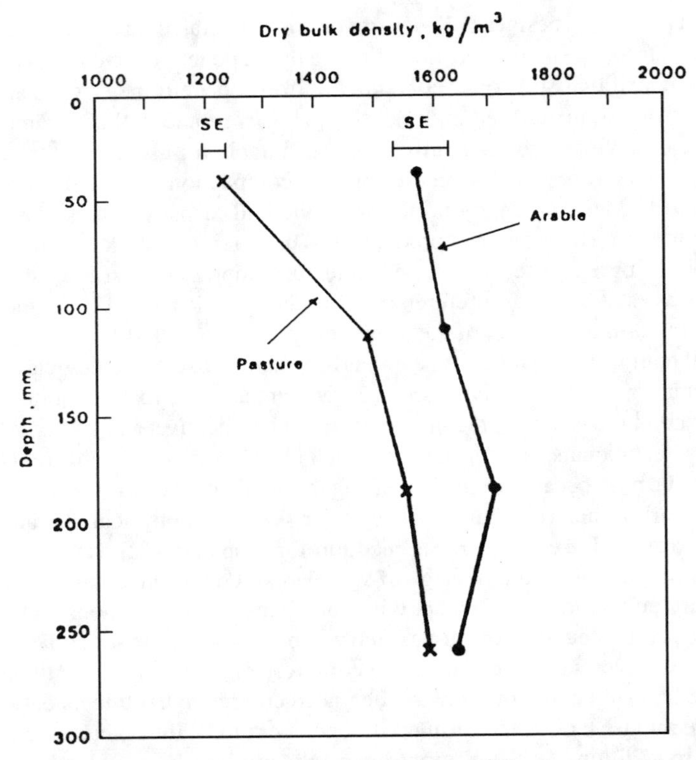

Figure 4.9. Soil compaction resulting from two different crop rotation systems. ***Source:*** **[13].**

itself and the tractor tires. The ability of a soil to support such traffic without structural damage beyond the limits for good crop growth has been used as a definition of the "trafficability" of an agricultural soil [16].

Subsoil compaction represents a very serious problem in that it is much more permanent and recovery from it is more difficult, and its negative influence on soil hydrology, microbial activity, and root development may be very consistent, especially for deep-rooting crops.

It has been demonstrated that subsoil compaction under a tractor tire is related to the bulk density of the subsoil at the moment of the passage and to the width of the tire [17]. In practical situations, however, the above postulates will be dependent on the relative compactability of the surface soil and of the subsoil.

Under continuous zero-tillage systems, the cycle of compaction and loosening, typical for arable soils subject to annual tillage operations, does not occur. Similarly, in continuous direct drilling systems, there is not evidence of a progressive increase in bulk density or of a variation in relation to soil type [18]. Soils subject to continuous direct drilling for more than two years are likely to become compacted but may acquire sufficient strength to carry normal agricultural traffic without any further compaction.

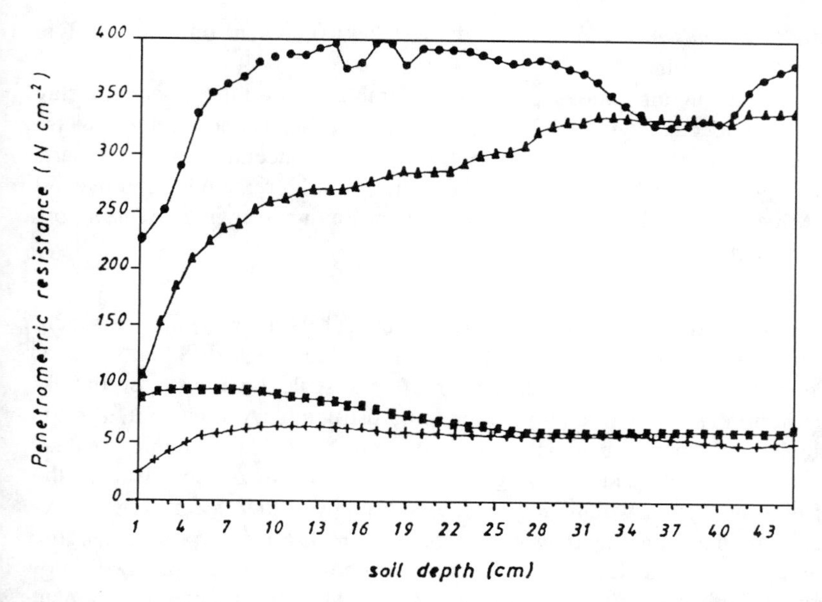

Figure 4.10. Soil compaction as a result of different management systems. ***Source:*** **[19].**

However, if during harvesting or other mechanical operations the soil is moist, rutting may occur and this may be particularly deleterious if the subsequent crop is still to be planted by direct drilling. Moreover, the decrease of both macropores and permeability below machinery wheel tracks produces much less favorable conditions for crop establishment than wheel-free areas (Fig. 4.10) [19].

Many soil microorganisms, both beneficial and pathogenic to crops, are known to be sensitive to change in aeration, pore-size distribution and soil-water status, which may result from the passage of vehicle wheels. For instance, a reduction in the number of soybean nodule bacteria has been observed in compacted soils [20].

Tractor wheeling on silty-loam and clay soils was found to decrease the number of soil fauna (*Collembula*, *Acari*, and earthworms) due to direct physical damage to the fauna more that to unfavorable soil physical conditions.

Microstructure Degradation

Quirk [22], in reporting a quotation of Bradfield [23], emphasized that soil "granulation is flocculation plus" with the intent to stress that soil aggregation depends not only on flocculation of clay particles but on many natural and unnatural factors.

Mineral Soil Microstructure

When the stability of the aggregates of different sizes is considered, it can be said that such stability is primarily dependent on the flocculation of the clay particles by electrolites of the soil solution developing a mineral structure, which, in turn, depends, among other things, on the cations adsorbed on the surfaces of clays and on the electrolyte concentration in the soil solution.

Adsorbed iron and calcium cations are highly desirable because they make the clay particles form stable flocculated particles, whereas sodium cation is undesirable because it make clay particles repel and disperse. A high electrolyte concentration also imparts stability to soil aggregates but only as long as the concentration remain high in the soil solution [24]. Mineral soil colloids (clay, oxides), in contributing to aggregate formation, are relatively static in time.

Organic Binding Agents

Organic binding agents have been grouped in transient, temporary, and persistent categories according to their endurance in the soil [25]. Moreover, different kinds of organic binding agents tend to locate themself in different scale dimentions, eventually justifying the differentiation between macro- and microstructure. At this point, a soil is structurally stable on the macroscale only if it is structurally stable on the microscale.

On the microscale, roots and microorganisms within the rhizosphere lead to the production of high-molecular-weight polysaccharides, among other decomposition by-products. Such components act as transient binding agents lasting for periods ranging from a few months to a year. Such compounds are responsible for the stabilization of aggregates greater than about 2 μm [25]. Recent reviews have shown a general agreement in considering polysaccharides as the main factor of aggregate stability in cultivated soils [26–28].

Roots and fungal hyphae also act as temporary binding agents at a larger scale, especially mycorrhizae. Their permanency in soil ranges from months to a few years and they are generally associated with stable aggregates larger than 250 μm. They are also quite sensitive to changes of soil environmental conditions prompted by tillage operation.

Persistent binding agents are represented mainly by aromatic humic material in association with amorphous iron, aluminium oxides, and aluminosilicates. Such agents bind the smallest microaggregates of sizes from about 0.2 to 250 μm [29, 30]. Within these dimensions, chemical bonds and electrochemical forces (mainly London–Van der Walls) control the type and number of bonds formed [31]. For instance, in Ultisols and Oxisols, polycations of iron and aluminium are basic elements in the following microstructural units [29]:

[clay]–[polyvalent cation]–[humified organic matter].

It follows that the removal of these polycations leads to an high degree of aggregate breakdown [32, 33]. The amount of clay particles in the soil relative to soil particles of greater size regulates the frequency of bonding between the two.

In agricultural soils under intensive cultivation, inorganic colloids are more important as elements of structural stability in comparison to undisturbed soils where, apparently, humified organic matter becomes the main agent of soil structural stability [25, 34].

Mechanical Structure

The main unnatural process affecting soil structure, particularly soil macroporosity, is represented by tillage practices. According to Dexter [1, 24, 45], given enough energy input, a soil with a bad physical condition (nonaggregated, massive, hard, anaerobic) can be temporarily turned into a soil with an apparently near-perfect structure (seed-bed of 1- to 5-mm diameter aggregates overlying a loosened, well-drained subsoil) by mechanical manipulation. However, a mechanically produced structure may be consistently unstable. Lacking the intrinsic stability of microstructure, the soil may collapse when wet and/or subject to mechanical actions (rainfall, wind, animal trampling, machinery traffic) and become as bad, if not worse, than it was before the tillage operations.

As a consequence of the tillage operations, a soil that has been sheared and molded by tillage implements and/or beneath vehicle wheels is weaker than an undisturbed soil at the same density and water content. Hence, a soil that has been disturbed mechanically is more susceptible to erosion and compaction than an undisturbed soil. However, if a soil is left after disturbance at constant water content and constant density, then its aggregate water stability and strength will be regained gradually over time. Such recovery is partly due to the rearrangement of the clay particles to new positions of lower free energy and partly to the reformation of cementing bonds between soil particles [52–54]. The rearrangement of clay particles at constant water content leads to a change in the size of the micropore distribution and hence to a change in matrix water potential.

Crusting

Over the past few decades, it has been well established that a variety of soils may develop a layer of reduced permeability at the surface when exposed to rainfall, after tillage practices have been applied and no cover has been provided. This process is referred to as surface crusting [59, 60]. The process has been extensively studied for well-aggregated soils [61–67] and, to a lesser extent also for poorly aggregated, coarse-textured soils [68–71].

Falling raindrops are the most effective agent of both soil consolidation and soil dispersion. Their action is understood easily by considering the momentum of a single raindrop falling on the soil surface. Such momentum is partly reflected and partly transferred, each component depending on the slope of the soil. The transfer of momentum to soil particles has two effects. First, it provides a consolidating force compacting the soil and, second, it produces a disruptive force as the water rapidly disperses away from the point of impact and falls again to the surface.

The consolidation effect is seen in the formation of a surface crust, usually only a few millimeters thick, which results from clogging of the pores by soil compaction. This is associated with the dispersal of fine particles from soil aggregates and/or clods, which then fill in the pores between soil aggregates [72]. The most recent studies have pointed out that crusts have a dense surface skin or seal about 0.1-mm thick formed of well-oriented clay particles. Beneath this skin, there is a layer 1- to 3-mm thick, where the larger pore spaces are filled by the finer detached and washed-in material. Whereas Hillel [73] has tried to explain crusting as the consequence of collapsing of the saturated soil aggregates, Farres [74] has demonstrated that raindrop impact is the most important agent of crust development.

A distinction often is made between a crust and a seal. In fact, sealing refers especially to the reorganization of the surface soil layer during a rainstorm, whereas crusting refers to the hardening of the surface seals when the soil dries out. Over a number of rainstorms, a structural crust initially develops in situ following the destruction of aggregates and the

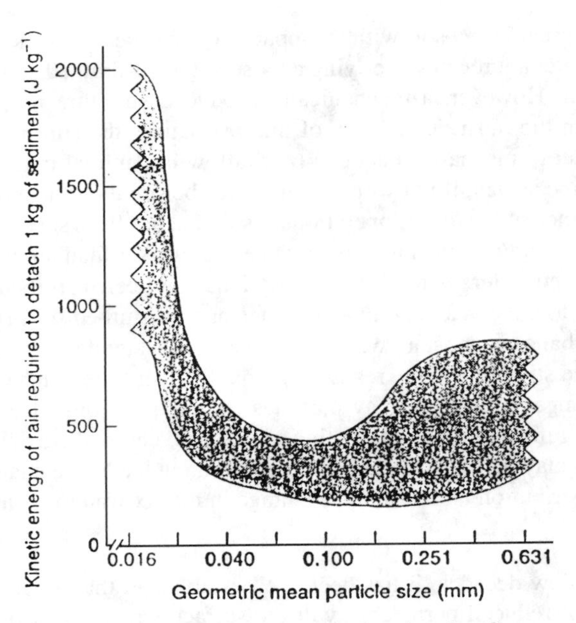

Figure 4.11. Kinetic energy required to detach sediment.
Source: [77].

clogging of pores by finer material. This is followed by the formation of a depositional crust due to the deposition of the finer laminated sediments transported by the overland flow and deposited in soil surface depressions where puddles form during storms [75]. Generally speaking, larger areas of depositional crusts may form downslope where the gradient generally decreases and the amount of sediment input is increased by the deposition of sediment transported by rills.

The most important consequence of the formation of a surface crust is the reduction of the infiltration capacity and, consequently, an increase in surface runoff. In loamy soils, a reduction of the infiltration capacity from about 45 mm/h on an uncrusted soil to about 6 mm/h on the same soil with a structural crust has been observed [76].

In general terms, the intensity of the crusting process decreases with an increase in the clay and organic-matter content. It follows that loam and sandy-loam soils are the most vulnerable to crust formation.

Studies carried out by Poesen [77] on the kinetic energy required to detach 1 kg of sediment by raindrop impact have shown that a minimum energy is required for particles of 0.125 mm, and particles ranging from 0.063 to 0.250 mm are most vulnerable to detachment (Fig. 4.11). Coarser particles are more resistant to detachment because of their weight, whereas the finer clay particles are resistant to detachment because the raindrop energy has to overcome the adhesive and/or chemical-bonding forces linking the minerals of the clay particles [78]. This means that the soil with high percentage of particles within the most vulnerable range (i.e., silty-loams, loams, fine sands, and sandy-loams) will be the most susceptible to detachment and crusting.

The actual response of a soil to a given rainfall depends upon its moisture content and structural state on the one side and on the intensity of the rainfall on the other. Le Bissonnais [79] describes three possible responses:

1. When the soil is dry and the rainfall intensity is high, the soil aggregates dissolve quickly by slaking due to the breakdown of bonds by air compression ahead of the wetting front. The infiltration capacity decreases rapidly as a consequence of the sealing of pores.
2. When the surface aggregates initially are wetted partially or the rainfall intensity is low, then microcracking occurs, and the larger aggregates break into smaller ones. Surface roughness thus decreases and the infiltration capacity remains high because the pore spaces within the microaggregates are not sealed.
3. When the aggregates are initially saturated, the infiltration capacity depends on the saturated hydraulic conductivity of the soil, and a large quantity of rain is required to seal the surface.

As well as inducing a decrease in infiltration and an increase in runoff, soil crusting and sealing also represent an impedence to seedling emergence as a consequence of the dense surface skin or seal of oriented clay particles, which occurs when the surface dries up. Moreover, the orientation of soil particles produces a continuity in the direction of macropores, favoring an increase in surface evaporation and, consequently, a greater loss of water from the rhizosphere.

4.2.2 Soil Regeneration

Soil Compaction

Regeneration from compaction due to machinery use in agricultural exploitation of the land may be achieved in four primary ways:

1. by reducing the number of passes of conventional machines,
2. by reducing the vehicle mass and/or the contact pressure of the wheel system,
3. by confining the traffic to permanent or temporary wheel tracks (controlled traffic),
4. by modifying the cultivation system, so as to reduce the overall use of machinery.

A diagrammatic representation of the main options in relation to the type of vehicles is shown in Fig. 4.12 [21].

Traffic reduction also may be achieved by combining several operations in one pass, such as cultivation and spraying, or certain types of harvesting operations using combines and common-sense attitudes in machinery management.

Finally, the introduction of a consistent reduction of wheel contact pressure or the adoption of a controlled traffic system may require the use of special vehicles or wheel systems, which are still under development.

Mineral Structure

The state and stability of soil structure under different management systems are strictly connected with several natural processes. The most important are the hydrology and the biological activity of the soil [24, 35].

An important distinction can be made between natural and human-induced processes in relation to the amelioration of soil physical conditions dependent on soil structure.

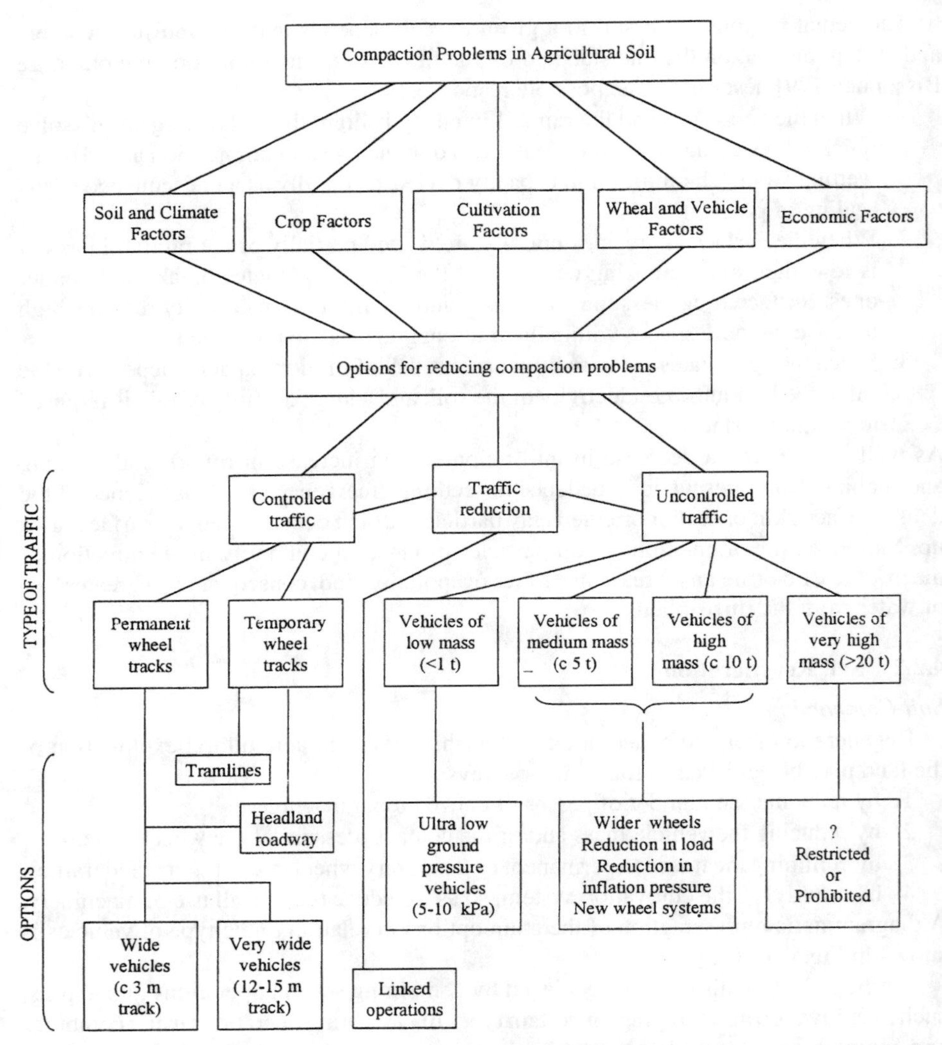

Figure 4.12. Main options for regenerating soil compaction due to machinery use. *Source:* [21].

Human actions such as tillage practices, crop rotation, crop fertilization, herbicide and pesticide applications, and use of soil conditioners can affect soil processes. On the other hand, soil structural features are directly responsible of soil physical conditions influencing plant growth and crop production as well as soil physical degradation due to compaction, erosion, and soil crusting and sealing.

In more detail, the status and stability of soil structure influence soil hydrology and aeration through the pore-size distribution in the soil mass. It follows that soil structural characteristics also can be assessed by an evaluation of pore-size distribution of the soil [22].

Porosity assumes different features when a volume of soil is considered. Microporosity is connected to pore-size dimensions ranging from 10^{-7} to 10^{-4} m and can be practically separated from macroporosity, which is associated with pore-size dimentions greater than 10^{-4} m. Such a division appears to be functionally useful because microporosity is the soil domain where capillary forces prevail whereas macroporosity is the domain where gravitational forces prevail in terms of movement of water within the soil.

The other important physical parameter strictly connected with soil structural conditions is soil cohesion, which plays an important role in relation to plant germination and seedling sprouting and to soil erodibility.

Considering natural processes of structural recovery, cycles of soil wetting and drying are responsible for macropore dynamics. According to Dexter [24], when a soil of medium to high clay content dries, it shrinks and vertical desiccation cracks form. If the drying is rapid, the cracks will be closely spaced and narrow. If the process is slow, the cracks will be spaced more widely. Cracks form important pathways for rapid water infiltration, aeration, and deep penetration of roots through the soil layers, which may otherwise exert mechanical impedence. When these vertical (primary) cracks become wider than about 4 mm, significant convection of air currents may occur within them so that drying occurs on their faces, producing secondary cracks at right angles to the primary cleavage direction. Occasionally, tertiary cracks also can form on the surface of secondary cracks in the same way.

When the soil becomes wet, it swells and the desiccation cracks close. The rate at which cracks close varies widely. In some soils it happens almost immediately after wetting whereas in others it may take days or even weeks before some cracks finally close. When a soil is wetted rapidly, for example during an intense rainstorm, the combined effect of differential swelling and pressure buildup in entrapped air can cause mechanical failure of aggregates, sometime to the point of complete slaking into microaggregates and/or primary particles $<250\ \mu m$. Such a phenomenon was described many years ago for clay soils and designated as hydromolecular aggregate disruption [36–38]. With slower wetting, complete slaking may not occur. Instead, there may be only partial slaking or mellowing in which arrays of microcracks form throughout the soil mass, which retains its coherence and shape [39]. Soil that has been wetted rapidly appears to offer less resistence to root penetration [40]. Slaking of a compacted layer caused by rapid wetting seems to make the soil more penetrable to growing roots.

Cycles of freezing and thawing also influence soil structural dynamics. When soil water content is less than 20%, ice crystals that form during freezing are often smaller than pores and therefore cause little disruption of particle-to-particle bonds. However, when water contents are higher and freezing takes place slowly, some of the crystals become larger than the pores and exert stresses that break many of the bonds between soil particles.

In fall, as the soil surface cools, moisture moves to the surface layers from both the atmosphere and the underlaying warmer soil layers. Consequently, the water content of the soil near the surface, at the moment when it starts to freeze in winter, is often higher than 20%, even in arid climates. Major disruption of the soil aggregates, therefore, may take place.

Plant roots have an important role in soil structural dynamics, directly affecting the macrostructure by the formation of biopores. The mean soil density remains constant while the volume of roots is accommodated by the loss of pore space of the sorrounding soil. Thus, the soil around the roots can be compacted to some extent for a distance of the order of one root diameter beyond the surface of the root.

When roots decompose, which usually occurs within about one year for the nonlignified tissues of annual crops, a biopore remains. The amount of roots and, hence of biopores, can be quite impressive. For the soil profile of wheat crops in South Australia, a development of 15 km of roots was estimated per square meter of soil surface.

Roots play a double role in aggregate formation and soil stabilization. A well-known example is the increase in aggregation of soil particles observed under grass-based pastures [41]. In such systems, an improvement of aggregation occurs as a consequence of the high density of the grass roots that bind the aggregates. In addition, soil structural stabilization is enhanced as a consequence of the organic-matter turnover in a well- balanced aerobic/anaerobic soil environment. This is the case of the grass/legumes perennial swards, which favor the production of humified organic compounds by microbial activity, capable of binding soil particles at the microscale level [42, 43].

Besides grass roots, fungal hyphae also may play an important role in binding larger aggregates. Moreover, exudates from roots are important in the stabilization of microaggregates ($<250\ \mu m$), whereas other exudates also are produced by bacteria and fungi involved in the decomposition of root material.

Biopores formed by one crop often can provide channels for deep rooting of a subsequent crop [44]. There is also evidence that roots grow preferentially toward biopores under conditions of poor aeration [45].

Soil fauna is another important natural factor in structural amelioration. Earthworms can have a profound effect on soil structure both via their production of burrows (macrostructure) and via the casts that they excrete (microstructure). There are numerous species of earthworms, which behave in different ways. Worm number is greatest in clayey soils of humid regions and it is virtually zero in sandy soils in arid regions. At the Waite Institute in Australia, 250 to 750 worms per square meter were observed in old pastures, compared with only about 20 worms per square meter in a wheat–fallow rotation [46]. Tillage exerts a rather drastic effect on earthworm number, reducing population density to about 15% of its value in nontilled soils [47]. Worms can move around in the soil either by pushing soil aside or by ingesting the soil, having the overall effect of making tunnels into the soil mass. McKenzie and Dexter [48, 49] have reported that earthworms can exert a mean maximum axial pressure of 73 kPa and a mean radial pressure of 230 kPa. Ingested soil is molded in the guts of earthworms at pressures of about 260 Pa and cast, when excreted, at much lower bulk density (around 1.15 t/m^3) in comparison withe the soil in which the worms live (1.5–1.6 t/m^3). Earthworm burrows provide pathways of reduced mechanical impedence to root penetration, so that extensive rooting in earthworm tunnels often is observed [50, 51].

Mechanical Structure

Tillage normally is used to modify the structural organization of the soil by mechanically loosening the soil surface. This loosening results in a reduction of bulk density and an associated increase in the porosity and hydraulic conductivity. Water retention is enhanced and root proliferation is encouraged to exploit available soil water and nutrients. Although beneficial, the physical soil properties associated with these freshly tilled surfaces are only temporary because rainfall and other mechanical actions tend to modify them. The rate at which the permeability, porosity, and surface structural change is a function of the soil texture, plant cover, previous management, and rainfall characteristics [55]. Amelioration of soil microstructure also can be achieved unnaturally by the application of soil structural conditioning substances.

Besides the addition of organic matter to the soil by crop-residue incorporation and organic manuring (force yard manure, organic wastes, sludges), which are converted to cementing humic substances by soil microorganisms, the main soil stuctural conditioners include the following:

- organic by-products,
- polyvalent salts,
- various synthetic polymers.

Polyvalent salts, such as gypsum, bring about flocculation of the clay particles, whereas the organic by-products and synthetic polymers act as cementing substances, binding the soil particles into more or less stable aggregates. Trivalent iron compounds may be included in the polyvalent salt group. Flotal and Glotal act as flocculating agents in slightly acid soil environments. They pass from sol to gel, changing pH and level of dehydration and can become stable cementing agents of soil particles [56].

Temporary stability can be achieved by using either oil- or rubber-based stabilizers or conditioners, or polyfunctional polymers, which develop chemical bonds with the mineral constituents of the soil. Although often too expensive for general agricultural use, they are helpful at special sites such as sand dunes, road cuttings, embankments, and stream banks, to provide a short period of stability prior to the establishment of a plant cover.

Soil structural conditioners fall under the following groups:

- those that render the soil hydrophobic and therefore decrease infiltration and increase runoff; and
- those that make the soil hydrophilic, favoring infiltration and, consequently, decreasing runoff.

For instance, hydrophilic conditioners based on bitumen, asphalt, and latex emulsion, can be used temporarily for controlling erosion and for soil sealing in relation to water harvesting. Other applications include the increase of infiltration rate and/or the prevention of crusting. However, the critical factor is that the size of aggregates that are produced may be so small that the infiltration rate remains low. A more effective improvement of the infiltration rate can be achieved with hydrophilic conditioners. In this case, the soil aggregates should be at least 2 mm and, ideally, larger than 5 mm [57].

Considerable interest has been shown in combining polyacrylamide conditioners with polysaccharides, which are biodegradable and swell in presence of water, thereby increasing the water-holding capacity of the soil.

Polyurea polymers contain a mixture of hydrophobic propylene oxides in a proportion determined according to the degree of hydrophilicity or hydrophobicity required. They have been used successfully, especially in dune stabilization [58].

Numerous other materials have been tested in soil stabilization and structural conditioning. Most are too expensive for generalized agricultural use, except with high-value crops such as vegetables and flowers.

Organic products, such as slurries and/or sludges, when free of inorganic and/or synthetic organic pollutants, could be disposed beneficially for soil improvement and environmental cleansing. Unfortunately, they are unpleasant to handle, create odors, and in the absence of special soil incorporation equipment, can be applied only in small quantities.

Other products obtained by processing different kind of wastes (including selected urban wastes) using either wet microbial processing or incineration, also can be used conveniently to give temporary stability to soil structure of road embankments and/or construction sites in urban areas.

Soil Crusting

Soil crusting can be prevented by several direct and indirect operations, which account for the mechanics of soil crusting and sealing due to rainfall impact, splash-erosion watering of clods, and, eventually, trampling by animals, and machinery compaction of tilled soil. For example, soil crusting and sealing do not occur or are greatly reduced in a soil with a dense plant cover. In arable, freshly tilled soils, especially when in seedbed condition, all factors favoring soil erodibility exist and make the soil susceptible to crusting and sealing. It follows that any natural and/or unnatural factor that reduces soil erodibility is also suitable for crust control.

The soil structural stability is the most important intrinsic condition that can decrease the risk of crust formation. Clay-particle flocculation and cementation by inorganic and organic compounds will consistently restrain the formation of a dense crust under intense rainfall.

Organic fertilization, crop-residue management, rotation including perennial forage crops or rotational pasture, and soil structural conditioning are also suitable measures for reducing the risk of soil crusting and sealing.

Another important direct measure for the control of soil crusting, besides cover crops, is mulching (see Section 4.4).

Direct intervention by cultivation practices (arrowing, disking, chiseling) can be used extensively to break surface crusts, restore soil infiltration capacity, reduce soil evaporation, and enhance the storage of moisture in the rhizosphere, mainly in semiarid rainfed agriculture.

References

1. Dexter, A. R. 1988. Advances in characterization of soil structure. *Soil Tillage Res.* 11:199–238.
2. Hakansson, I. 1988. A method for characterizing the state of compactness of an arable soil. *Impact of Water and External Forces on Soil Structure*, Catena Suppl. 11, eds. Drescher, J., R. Horn, and M. De Boodt. Cremlingen: Catena Verlag.

3. Kuipers, H., and C. Van Ouwekerk. 1963. Total pore-space estimation in freshly ploughed soil. *Neth. J. Agric. Sci.* 11:45–53.
4. Allmaras, R. R., R. E. Burwell, W. E. Larson, R. F. Holt, and W. W. Nelson. 1966. Total Porosity and Random Roughness of the Interrow Zone as Influenced by Tillage. USDA, Agricultural Research Service Conservation Research Rept. No. 7. Washington, DC: U.S. Department of Agriculture.
5. Andersson, S., and I. Hakansson. 1966. Structure dynamics in the topsoil. A field study. *Grundfoerbacttring* 19:191–228.
6. Carter, M. R. 1986. Temporal variation of soil physical properties following moldboard plowing and direct drilling of a sandy loam soil. *Soil Tillage Res.* 8:355.
7. Raghavan, G. V. S., E. McKyes, and E. Chasse. 1977. Effect of wheel slip on soil compaction. *J. Agric. Eng. Res.* 22:79–83.
8. Raghavan, G. V. S., E. Mac Kyes, E. Stenshorn, A. Gray, and B. Beaulieu. 1977. Vehicle compaction patterns in clay soils. *Trans. ASAE* 20:218–220.
9. Voorhees, W. B., C. G. Senst, and W. W. Nelson. 1978. Compaction and soil structure modification by wheel traffic in the northern Corn-Belt. *Soil Sci. Soc. Am. J.* 42:344–349.
10. Eriksson, J., I. Hakansson, and B. Danfors. 1974. The Effect of Soil Compaction on Soil Structure and Crop Yield. Rept. 354. Uppsala: Swedish Institute of Agricultural Engineering.
11. Soane, B. D. 1970. The effect of traffic and implements on soil compaction. *J. Proc. Inst. Agric. Eng.* 25:115–126.
12. Barnes, K. K., W. M. Carleton, H. M. Taylor, R. I. Throckmorton, and G. E. Vandenberg (eds.). 1971. *Compaction of Agricultural Soils*. St. Joseph, MI: American Society of Agricultural Engineers.
13. Garwood, E. A., K. C. Tyson, and C. R. Clement. 1977. A Comparison of Yield and Soil Conditions During 20 Years of Grazed Grass and Arable Cropping. Tech. Rept. 21. Hurley, U.K.: Grassland Research Institute.
14. McKibben, E. G. 1971. Introduction. *Compaction of Agricultural Soils*. eds. Barnes, K. K., W. M. Carlton, J. M. Taylor, R. I. Throckmorton, and G. E. eds. Vanderberg, pp. 3–6. St. Joseph, MI: American Society of Agricultural Engineers.
15. Hakansson, I. 1979. Experiments with Soil Compaction at High Axle Load. Soil Investigation 1-2 Years After the Experimental Compaction. Division of Soil Management Rept. 57, 15 pp. Uppsala: Swedish University of Agricultural Science.
16. Paul, C. L., and J. De Vries. 1979. Effect of soil water status and strength on trafficability. *Can. J. Soil Sci.* 59:313–324.
17. Fekete, A. 1972. Studies of the soil compacting effect of tyres. *Mezogazd. Gepes. Tamul.* 19:16–26.
18. Cannell, R. Q., F. B. Ellis, D. G. Christian, J. P. Graham, and G. T. Douglas. 1980. Growth and yield of winter cereals after direct drilling, shallow cultivation and ploughing on non-calcareus clay soils. *J. Agric. Sci. Camb.* 94:345–359.
19. Bazzoffi, P., and G. Chisci. 1986. Effetto del passaggio di macchinari agricoli e di differenti pratiche agronomiche su alcune caratteristiche fisiche di un suolo limoso-

argilloso del Mugello (Toscana). *Ann. 1st. Studio e Difesa del Suolo di Firenze*. XVII:41–56.
20. Voorhees, W. B., V. A. Carlson, and C. G. Senst. 1976. Soybean nodulation as affected by wheel traffic. *Agron. J.* 68:976–979.
21. Soane, B. D., J. W. Dickson, and P. S. Blackwell. 1979. Some options for reducing compaction under wheels on loose soils. Proc. 8th Conf. Int. Soil & Tillage Res. Organ., Stuttgart: Report University of Hohenhein, Vol. 2:347–352.
22. Quirk, J. P. 1978. Some physico-chemical aspects of soil structural stability—A review. *Modification of Soil Structure*. eds. Emerson, W. W., R. D. Bond, and A. R. Dexter, pp. 3–16. London: Wiley.
23. Bradfield, R. 1936. Am. *Soil Surv. Assoc. Bull.* XVII:31–32.
24. Dexter, A. R. 1991. Amelioration of soil by natural processes. *Soil Tillage Res.* 20:87–100.
25. Tisdall, J. M., and J. M. Oades. 1982. Organic matter and water stable aggregates in soils. *J. Soil Sci.* 33:141–163.
26. Lynch, J. M. 1984. Interaction between biological processes, cultivation and soil structure. *Plant Soil* 76:307–318.
27. Oades, J. M. 1984. Soil organic matter and structural stability: Mechanisms and implications for management. *Plant Soil* 76:319–317.
28. Lynch, J. M., and E. Bragg. 1985. Microorganisms and soil aggregate stability. *Adv. Soil Sci.* Vol. 2:133–171.
29. Edwards, A. P., and J. M. Bremner. 1967. Microaggregates in soils. *J. Soil Sci.* 18:64–73.
30. Cambier, P. R., and R. Proust. 1981. Etude des association des costituants d'un materiaux ferralitique. *Agronomie* 1:713–722.
31. Harris, R. F., G. Chester, and O. N. Allen. 1966. Dynamics of soil aggregation. *Adv. Agron.* 18:107–160.
32. Hamblin, A. P., and D. J. Greenland. 1977. Effects of organic constituents and complexed organic ions on aggregate stability of some East Anglian soils. *J. Soil Sci.* 28:410–416.
33. Reid, J. B., M. J. Goss, and P. D. Robertson. 1982. Relationship between the decrease in soil stability affected by the growth of maize roots and changes in organically bound iron and aluminum. *J. Soil Sci.* 33:397–410.
34. Tisdall, J. M., and J. M. Oades. 1979. Stabilization of soil aggregates by the root systems of ryegrass. *Aust. J. Soil Res.* 17:429–441.
35. Heinonen, R. 1986. Alleviation of soil compaction by natural forces and cultural practices. *Land Clearing and Development in the Tropics*. eds. Lal, R., P. A. Sanchez, and R. W. Cummings, pp. 285–297. Rotterdam: A. A. Balkema.
36. Passerini, G. 1934. La degradazione delle argille plioceniche. *La Bonifica Nelle Colline Argillose*. Rome: Ministero dell'Agricoltura e delle Foreste.
37. Passerini, G. 1941. Studi di meccanica pedologica. Dinamismo strutturale del Suolo per Azioni Idromolecolari. Firenze: Barbera Tipografia.
38. Giudici, P. 1954. Evaluation des facteurs déterminant la désintegration hidro-moleculaire des aggregates terraux. *V Congrès Int. de la Science du Sol. Leopoldville:ISSS*, Vol. II, pp. 219–228.

39. Stengel, P. 1988. Cracks formation during swelling: effects on soil structure regeneration after compaction. *Proc. 11th Confer. Inter. Soil & Tillage Res. Organ.* Hedinburg: Int. Soil & Tillage Res. Organ. pp. 147–152.
40. Whiteley, G. M., W. H. Utomo, and A. R. Dexter. 1981. A comparison of penetrometer pressures and the pressures exerted by roots. *Plant Soil* 61:351–365.
41. Low, A. J. 1955. Improvement in the structural state of soil under ley. *J. Soil Sci.* 6:179–199.
42. Haussmann, G., and G. Chisci. 1958. Prime osservazioni sull'evoluzione della fertilità nelle cotiche monofite di graminacee pratensi. *Ann. Sper. Agrar.* XII(5):1483–1514.
43. Chisci, G. 1980. Prati e pascoli e fertilità del suolo. *Ital. Agric.*, 117(4):162–192.
44. Elkins, C. B., R. L. Haaland, and C. S. Hoveland. 1977. Grass roots as a tool for penetrating soil hardpans and increasing crop yield. *Proc. 34th Southern Pasture and Forage Crop Improvement Conference*, Auburn, AL: Auburn Univ. pp. 21–26.
45. Dexter, A. R. 1987. Mechanics of root growth. *Plant Soil* 98:303–312.
46. Barley, K, P. 1970. The influence of earthworms on soil fertility. I. Earthworm population found in agricultural land near Adelaide. *Aust. J. Agric. Res.* 10: 171–178.
47. Low, A. J. 1972. The effect of cultivation on the structure and other physical characteristics of grassland and arable soils (1945–1970). *J. Soil Sci.* 23:363–380.
48. McKenzie, B. M., and A. R. Dexter. 1987. Physical properties of casts of the earthworm *Aporrectodea rosea. Biol. Fertil. Soils* 5:152–157.
49. McKenzie, B. M., and A. R. Dexter. 1988. Axial pressure generated by the earthworm *Aporrectodea rosea. Biol. Fertil. Soils* 5:323–327.
50. Ehlers, W., U. Kopke, F. Hesse, and W. Bohm. 1983. Penetration resistence and root growth of oats in tilled and untilled loess soil. *Soil Tillage Res.* 3:261–275.
51. Wang, J., J. D. Esketh, and J. T. Wolley. 1986. Preexisting channels and soybean rooting patterns. *Soil Sci.* 141:432–437.
52. Utomo, W. H., and A. R. Dexter. 1981. Age-hardening of agricultural top-soil. *J. Soil Sci.* 32:335–350.
53. Kemper, W. D., and R. C. Rosenau. 1984. Soil cohesionas affected by time and water content. *Soil Sci. Soc. Am. J.* 49:979–983.
54. Molore, M. B., I. C. Grieve, and E. R. Page. 1985. Thixotropic changes in the stability of molded soil aggregates. *Soil Sci. Soc. Am. J.* 49:979–983.
55. Mwendera, E. J., and J. Feyen. 1994. Effects of tillage and rainfall on soil surface roughness and properties. *Soil Technol.* 7:93–103.
56. Chisci, G., G. Lorenzi, and L. Piccolo. 1978. Effect of a ferric conditioner on clay soil. *Modification of Soil Structure*. eds. Emerson, W. W., R. D. Bound, and A. R. Dexter, pp. 309–314. London: Wiley.
57. Pla, I. 1977. Aggregate size and erosion control on sloping land treated with hydrophobic bitumen emulsion. *Soil Conservation and Management in the Humid Tropics*. eds. Greenland, D. J., and R. Lal, pp. 109–115. London: Wiley.
58. De Kesel, M., and D. DeWleeschauwer. 1981. Sand dune fixation in Tunisia by means of polyurea polyalkilene oxide (Uresol). *Tropical Agriculture Hydrology*. eds. Lal, R., and E. W. Russell, pp. 273–281. Chichester: Wiley.

59. Sumner, M. E., and B. A. Stewart. 1992. *Soil Crusting: Chemical and Physical Processes.* Vol. 20, *Adv. Soil Sci*: Chelsea, MI: Lewis.
60. Morgan, R. P. C. 1995. *Soil Erosion and Conservation*, 2nd ed. Longman.
61. Ferries, P. J. 1985. Feedback relationship between aggregate stability, rainsplash erosion and soil crusting. *Assessment of Soil Surface Crusting and Sealing*. eds. Collebaut, F., D. Gabriels, and M. De Boodt, pp. 82–90. Ghent: State University of Ghent.
62. Le Bissonnais, Y., A. Bruand, and M. Jamagne. 1989. Laboratory experimental study of soil crusting: Relation between aggregate break-down mechanisms and crust structure. *Catena* 16:377–392.
63. Bresson, L. M., and J. Boiffin. 1990. Morphological characterization of soil crust development stages on an experimental field. *Geoderma* 47:301–325.
64. Bresson, L. M., and L. Cadot. 1992. Illuviation and structural crust formation on loamy temperate soils. *Soil Sci. Soc. Am. J.* 56:1565–1570.
65. Tanaka, U., Y. Yokio, and K. Kiuma. 1992. Morphological characteristics of soil surface crusts formed under simulated rainfall. *Soil Sci. Plant Nutr.* 38:655–664.
66. Shaimberg, L., G. H. Levy, P. Rengasamy, and H. Frenkel. 1982. Aggregate stability and seal formation as affected by drops inpact energy and soil amendments. *Soil Sci.* 154:113–119.
67. Slattery, M. C., and R. B. Bryan. 1994. Surface seal development under simulated rainfall on an actively eroded surface. *Catena* 22:17–34.
68. Hoogmoed, W. B., and L. Stroosmijder. 1984. Crust formation on sandy soils in the Sahel. Rainfall and infiltration. *Soil Tillage Res.* 4:5–23.
69. Collins, J. F., G. W. Smillie, and S. M. Hussain. 1986. Laboratory studies of crust development in Irish and Iraqui soils. III. Micromorphological observations of artificially formed crusts. *Soil Tillage Res.* 6:337–350.
70. Kooistra, M. J., and W. Siderius. 1986. Micromorphological aspects of crust formation in a Savanna climate under rainfed subsistence agriculture. *Assessment of Soil Surface Sealing and Crusting.* eds. Callebaut, F., D. Gabriels, and M. De Boodt, Ghent, Belgium: University of Ghent.
71. Valentin, C., and L. M. Bresson. 1992. Morphology, genesis and classification of surface crusts in loamy and sandy soils. *Geoderma* 55:225–245.
72. Poesen, J. 1992. Mechanisms of overland-flow generation and sediment production on loamy and sandy soils with and without rock fragments. *Overland Flow Hydraulics and Erosion Mechanics.* eds. Parsons, A. J., and A. D. Abrahms, pp. 275–305. London: UCL Press.
73. Hillel, D. 1960. Crust formation in loessial soils. Trans. VI Int. Congress Soil Science, Madison WI: ISSS, pp. 330–337.
74. Farres, P. 1978. The role of time and aggregate size in the crusting process. *Earth Surface Processes*, 3:243–254.
75. Boiffin, J. 1985. Stage and time-dependency of soil crusting in situ. *Assessment of Soil Surface Crusting and Sealing*. eds. Callebaut, F., D. Gabriels, and M. De Boodt, pp. 91–98. Ghent, Belgium: State University of Ghent.
76. Boiffin, J., and G. Monnier 1985. Infiltration rate as affected by soil surface crusting caused by rainfall. Assessment of Soil Surface Crusting and Sealing. eds. Callebaut,

F., D. Gabriels, and M. DeBoodt, pp. 210–217, Ghent, Belgium: State University of Ghent.
77. Poesen, J. 1985. An improved splash transport model. *Z. Geomorphol.* 29:193–212.
78. Yariv, S. 1976. Comment on the mechanism of soil detachment by rainfall. *Geoderma* 15:393–399.
79. Le Bissonnais, Y. 1990. Experimental study and modelling of soil surface crusting processes. *Catena Suppl.* 17:87–93.

4.3 Assessment of Soil Erosion

V. Bagarello and V. Ferro

4.3.1 Types of Erosion and Its Assessment

Soil is a vital resource for the production of renewable resources for the necessities of human life, such as food and fiber. Soils, however, essentially are nonrenewable resources [1].

According to Golubev [2], the area of cultivated land in the world is 14.3 million km^2. In cultivated areas, drastic changes in vegetation have occurred and instead of dense natural vegetation cover, bare soil often is exposed for most of the year with sparse crop vegetation existing for a few months. These changes in vegetation cover are the main reason for the increase of soil erosion on cropland as compared to that on natural landscapes. Results of computations by Golubev [2] show that soil erosion in the world is 5.5 times more than during the preagricultural period. According to Brown [3], the world is currently losing 23 billion tonnes of soil from cropland in excess of new soil formation each year [4]; therefore, accelerated soil erosion is a serious problem to consider for the development of a sustainable agriculture. Other environmental problems caused by severe soil erosion are reservoir sedimentation, which results in a lowering of the available surface water resources, and nonpoint-source pollution due to sediment transport phenomena.

On a global scale, even if the mean annual sediment yield estimate is based on the available suspended sediment transport measurements, Walling and Webb [5] gave a reliable assessment of the global pattern of water erosion. This assessment established that the semiarid and semihumid areas of the world (China, India, western United States, and Mediterranean lands) are the most vulnerable to soil erosion.

Soil erosion losses are often due to a few severe storms with high rainfall intensity and/or high rainfall depth [6], or to high wind velocity values. Figure 4.13 shows that, on a given site, with an invariable land-use and crop management, the long-term average soil loss is dominated by a few and relatively rare events.

Soil erosion is generally a normal aspect of landscape development in which soil particles are removed by wind or water. In some parts of the world, other processes of denudation such as soil mass movement can dominate.

Wind erosion is the process of detachment and transport caused by fluid (air) action on the soil surface [7]. The process removes the finer particles and the organic matter from the top soil. Redeposition of the soil particles can bury soil and vegetation. The process

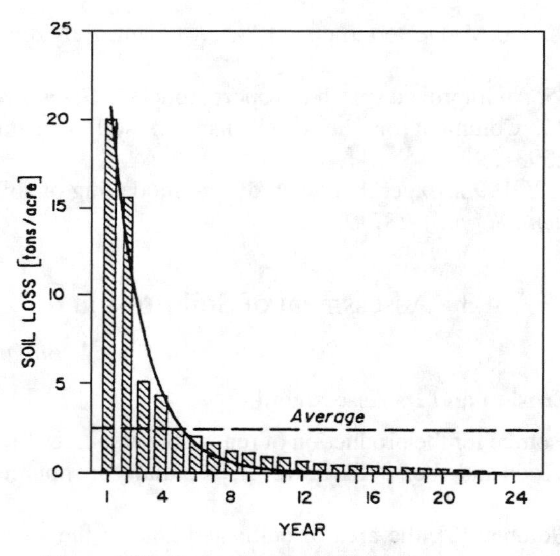

Figure 4.13. Ordered annual soil erosion amounts measured, for a 24-year period, at Kingdom City, MO. ***Source:*** **[6].**

operates in a variety of natural environments that lack a protective cover of vegetation. Such areas include the Great Plains of North America, the fringes of arid Africa, India, Australia, and the steppes of Russia, Mongolia, and China.

Soil particles are carried by wind into suspension, by saltation, and by surface creep, depending on their size [1]. Soil particles and small aggregates (<0.05 mm in diameter, ϕ) are kept in suspension by air turbulence unless the wind velocity is drastically reduced. Intermediate-size grains ($0.05 < \phi < 0.5$ mm) move in a series of short leaps, jumping into the air and bouncing back on the soil surface. Soil particles larger than 0.5 mm are not lifted. However, grains that are $0.5 < \phi < 1$ mm are bumped along the surface by jumping particles.

The wind erosion phenomenon is controlled by soil susceptibility to particle detachment and by the detachment and transport capacity of wind. Factors affecting soil susceptibility to wind erosion are dry aggregate size distribution, mechanical stability of soil structure, surface ridges, rainfall, length of the exposed area, and vegetative cover.

Grains larger than 1 mm in diameter are non-erodible whereas particles that are $0.5 < \phi < 1$ mm are only eroded by high wind velocities. Soil particles able to move into suspension are highly erodible. Obviously, soil properties such as texture, organic matter, and exchangeable cations, which promote aggregate stability, reduce wind erosion susceptibility. Surface ridges, which increase soil surface roughness, reduce wind velocity near the ground and promote trapping of the eroded particles. Rainfall moistens the soil surface, which transitorily reduces wind erosion. However, rainfall also can promote wind erosion by breaking soil aggregates and smoothing soil surface. Because wind transport capacity at a specific shear velocity $u*$ [m/s] can be considered constant,

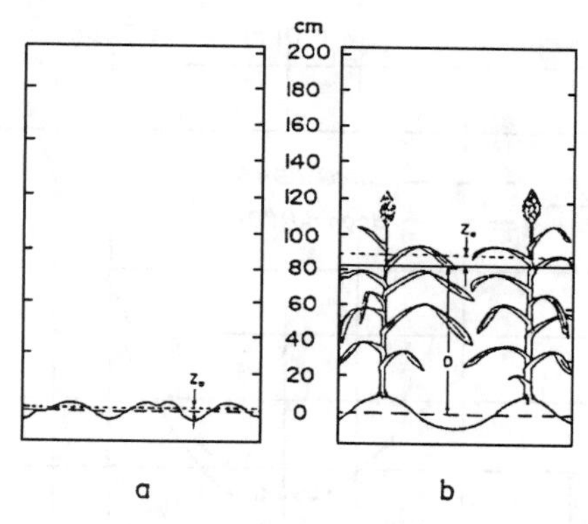

Figure 4.14. Displacement of the zero velocity plane due to the vegetation cover.
Source: **[1].**

the distance the wind must travel to reach its load capacity depends on soil erodibility. Vegetative cover is the most effective way to reduce wind erosion because plant cover determines a displacement $D + Z_0$ (m) of the zero plane, in which Z_0 is the effective roughness height, that is, the plane at which wind velocity is zero (Fig. 4.14). Plant protection is affected by the amount of cover and time of year in which it is provided, plant geometry, and row orientation. Crop residues left on the soil surface act usefully to reduce wind erosion.

The erosive power of wind is controlled by shear velocity, $u* = (\tau/e)^{1/2}$, in which τ is the surface shear stress (kg/m^2) and e is air density (kg s$^2/m^4$); $u*$ is related to the velocity profile and to the drag exerted by wind on the soil surface. For highly turbulent air flow (for shear Reynolds number $Re^* = u* Z_0/\nu > 90$, ν being the kinematic viscosity of the fluid), shear velocity is related to the local mean wind velocity u_z (m/s) at height z (m) by the logarithmic velocity profile:

$$\frac{u_z}{u*} = \frac{1}{k} \ln\left(\frac{z - D}{Z_0}\right) \tag{4.4}$$

in which k = von Karman's constant approximately equal to 0.4. Both the detachment (D_c) and the transport (T_c) capacities of wind depend on $u*$. In particular, D_c (kg $\cdot$ s^{-1} m^{-2}) depends on the square of the shear velocity and the size of the erodible particles; T_c (kg $\cdot$ s^{-1}m^{-2}) is essentially proportional to the third power of the shear velocity [1].

For each soil and surface condition, a threshold shear velocity $u*_c$, that is, a minimum wind velocity starting soil particle movement, can be defined. Bagnold [8] and Chepil [9] showed that the critical shear velocity varies with the particle size (Fig. 4.15). In particular, finer particles are characterized by $u*_c$ values decreasing for increasing grain size. In fact, the cohesiveness forces are most effective for small soil particles, which also are protected by the surrounding coarser particles. For grain sizes greater than 0.1 mm, $u*_c$ increases with the particle diameter because of the increase of the grain weight.

The soil instability process, called "mass movement," usually is neglected in soil erosion studies because this process generally involves high volumes and deep layers of soil.

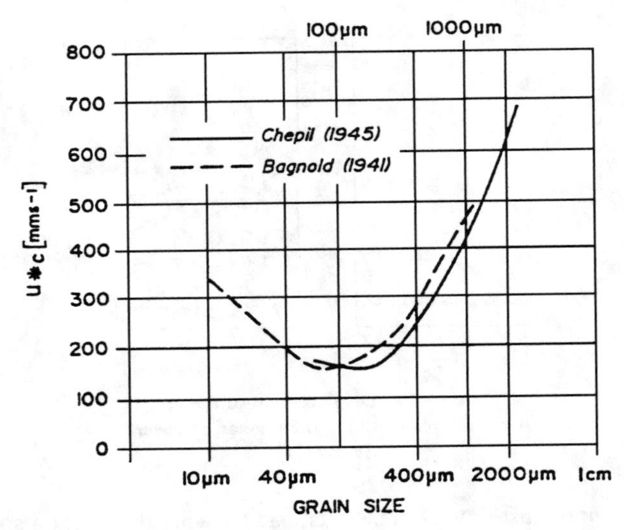

Figure 4.15. Relationship between critical shear velocity of wind and particle size. *Source:* [29].

The instability mechanism depends on the breaking and mass transport processes (breakdown, sliding, rolling, or mixed mechanism). Mass movement occurs as creep, slides, rock falls, debris flow, and mudflow, depending on the ratio between the solid and liquid components of the moving mass. In other words, the different types of mass movement can be considered as part of a continuum of solid transport phenomena ranging from slides, in which the solid/liquid ratio is high, to mudflow having a low solid/liquid ratio.

In 1947, Ellison [10, 11] defined soil erosion as "a process of detachment and transportation of soil materials by erosive agents." For water soil erosion, these agents are rainfall and runoff. Ellison's definition can be extended to take into account deposition processes occurring when the energy of the transporting agent is no longer available to transport soil particles. The intensity of the erosion process depends on the quantity of soil supplied by detachment and the capacity of the erosive agents to transport it. When the agent has the capacity to transport more soil than is supplied by detachment, the erosion is detachment limited. When more soil is supplied than can be transported, the process is transport limited.

According to a classic scheme of the erosive process, the following four phases are distinguished: rainsplash, sheet, rill, and gully erosion.

The impelling force, caused by the raindrops hitting the soil surface, determines soil particle detachment and transport (splash erosion). Waterdrop impact forces depend on the number of hitting drops per unit area and time, drop size distribution, and drop fall velocity. Both rainfall kinetic energy and momentum, which are the most used erosivity parameters, can be calculated using this basic information. The drop size distribution of rainfall is represented briefly by the median (d_{50}) drop diameter. Figure 4.16 shows that median drop size increases with rainfall intensity I up to 100 mm/h; at higher intensities,

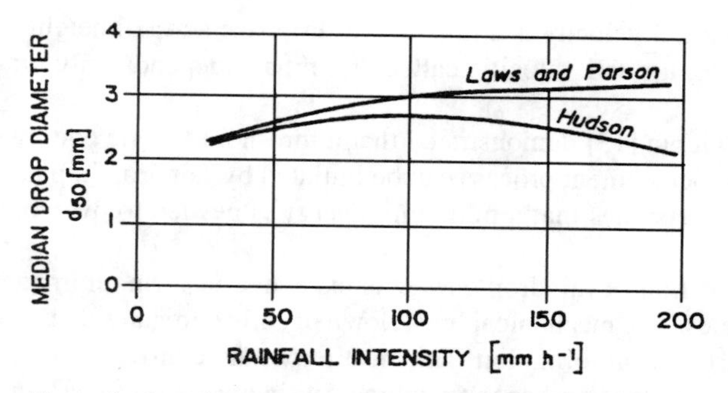

Figure 4.16. Relationship between median drop diameter and rainfall intensity. ***Source:*** **[29].**

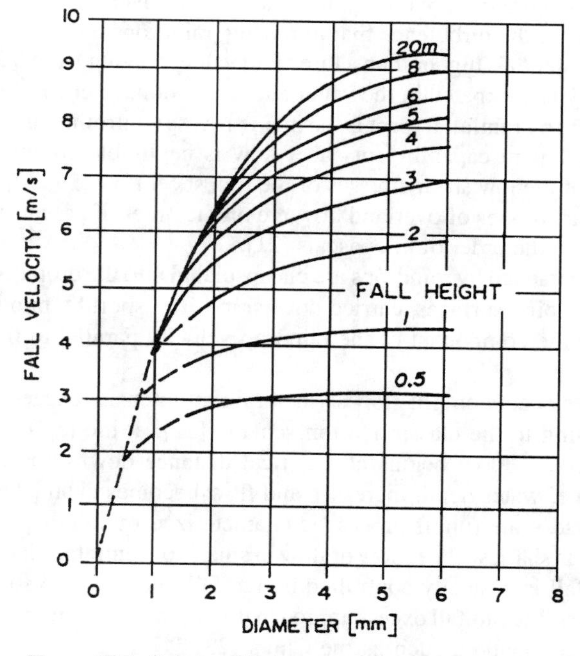

Figure 4.17. Relationship between fall velocity and drop diameter for a given fall height. ***Source:*** **[14].**

d_{50} remains essentially constant [12] or decreases [13] because of drop size instability, typical of tropical rainfall. In fact, at higher intensities, drop diameter d is unstable and drop breaks due to both turbulence phenomena and the weight action not counteracted by surface tension. For these reasons, natural drops have a maximum size equal to 6–7 mm. Drop fall velocity is also strongly dependent on drop size (Fig. 4.17) because the drag force of the waterdrops is contrasted by the gravity force. Figure 4.17 also demonstrates that fall velocity is a function of drop size and fall height; for fall heights greater than 10 m, the fall velocity, called "terminal," depends only on drop diameter [14].

Sharma and Gupta [15] demonstrated that a threshold kinetic energy or momentum exists before the detachment process can be initiated by raindrop impact. The threshold erosivity concept assumes that a minimum energy is needed to overcome the inherent soil strength.

The largest portion of raindrop energy is expended to form an impact crater and to move soil particles. The mechanical breakdown of soil aggregates due to drop impacting can induce a surface seal formation. The most important consequence of seal formation is a reduction of infiltration capacity, which, by increasing surface runoff, can cause an increase in soil erosion. Splash detachment is higher in soils that are not highly susceptible to surface sealing.

Drop impact is more effective if a thin water layer covers the soil surface. This is believed to be due to the turbulence that impacting raindrops induce in the water layer. However, if water depth is higher than a threshold value, ranging from $0.2d$ to d [16–18], the rainfall energy is dissipated in the water and does not have erosive effects.

Soil detachment by rainfall impact is the main process controlling interrill soil erosion because the detachment capability of sheet flow is negligible compared with that of rainfall because of the low shear stresses of the thin sheet flow [19, 20]. In fact, for soil surfaces, the shear stresses of overland flow are on the order of pascals whereas the soil shear strength is on the order of kilopascals [21].

Soil particles detached by raindrops are encapsulated into the droplets generated after impact and, for sloping surfaces, carried downslope. Transport by rain is generally low and is caused by the component of the raindrop velocity parallel to the surface of the slope.

Rainfall excess occurs on hill slopes when rainfall intensity exceeds soil infiltration capacity. According to the classic Horton scheme [22], at the top of the hill slope, a flowless zone occurs. Flow begins at a critical distance downslope from the divide. Farther downslope, water depth increases and flow becomes channeled and breaks up into rills. The field-scale runoff process is characterized by rainfall excess dominated runoff occurring as shallow sheet flow or flow in small concentrated channels. The runoff response to rainfall is basically controlled by rainfall intensity and soil properties. For modeling purposes, the rainfall excess approach uses a time-intensity rainfall distribution and an infiltration equation, such as the Chu's [23, 24], to compute a rainfall excess distribution over the field. For field-scale applications, some form of the De Saint Venant shallow water equations has been used recently to route the rainfall excess over the flow surface [25].

Moss [26] showed that overland-flow sediment transport is a combined action of raindrops and flow: the raindrop impact induces the flow to transport particles locally increasing its turbulence. In other words, without rainfall, the flow would be incapable of transportation. Flow transport processes associated with raindrop action are called rain-induced flow transport (RIFT) [27, 28]. RIFT acts for shallow flow depth (less than $3d$) impacted by medium to large-size raindrops [21].

The hill-slope flow rarely is distributed evenly on the soil surface. More commonly, it appears as a mass of anastomosing water courses with no pronounced channels [29]. Rills are ephemeral features, that is, small and intermittent water channels that do not interfere with conventional tillage operations. Once obliterated, rills will not reform at the same site [30]. Merritt [31] identified four subsequent stages of rill development: sheet flow, flow lines (starting of flow concentration), microchannels without headcuts, and microchannels with headcuts (channeled flow).

Compared with rill erosion, interrill erosion contributes a very small proportion to the sediment transported downward [20].

Rill initiation usually is described by the threshold value of a variable crucial for this erosion mechanism. Generally, a hydraulic variable, such as discharge [32], Froude number [33], shear velocity [34], or unit stream power [35], is used to describe the ability of the erosive agent to start rilling. Other authors suggest the consideration of the soil susceptibility to rilling. For example, Savat [36] explained rilling by defining a critical Froude number F_c depending on the median soil grain size. A more complex approach, proposed by Boon and Savat [37], includes both F_c and the sediment concentration in rill flow.

For a recently tilled field, rill initiation also can be induced by piping. When the topsoil is saturated, in some isolated location an unequal settlement of the surface layer can take place. This phenomenon may be due to the large pores among clods. Runoff from the upper area flows into the crevices, resulting from the unequal settlement, and creates pipes just above the undisturbed, more compact subsoil [35]. The main factors controlling piping are soil properties, such as porosity and soil erodibility.

Since rill discharge significantly affects the ability of the rill to detach and transport sediment particles, the knowledge of the number of rills that may form per unit of cross-sectional area and the variation in flow rate between individual rills is necessary. Gilley et al. [38] studied partitioning flow between rills and determined the relative frequency of flow rates among rills on a given plot. Figure 4.18 shows the distribution of the relative discharge, equal to the discharge for each rill divided by the maximum discharge on the plot among rills. The figure shows that differences in discharge existed between individual rills and that 30% of the rills had discharge equal to the maximum value. In

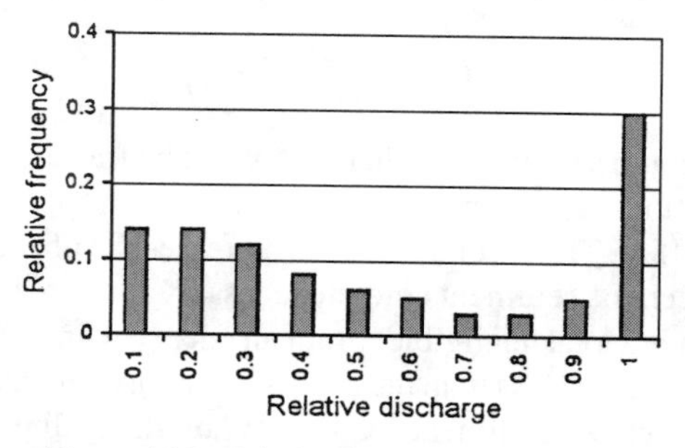

Figure 4.18. Relative frequency of measured relative discharge. *Source:* [38].

addition to discharge, identification of other rill hydraulic variables, such as rill width, hydraulic roughness coefficient, and flow velocity, also may be important [39, 40].

Rill erosion is the detachment and transport of soil particles by concentrated flow. Soil particle detachment by flow depends on the rill erodibility, the hydraulic characteristics of rill flow, and the actual flow sediment load. The simplest approach is considering rill detachment as due only to the scouring processes of the wetted perimeter. The maximum rill detachment, called detachment capacity, D_c ($kg \cdot s^{-1}m^{-2}$), occurs when a clear flow moves on an erodible soil. The soil is characterized by a rill erodibility parameter, K_r, lumping the effects of different factors such as grain size distribution, rock fragment cover [41], and soil structure and its stability. For a given soil, the detachment capacity D_c depends on the excess of flow energy content as related to a threshold value. The most widely applied equation to estimate D_c is the modified Du Boys sediment transport equation [42],

$$D_c = K_r(\tau - \tau_c)^a, \tag{4.5}$$

in which τ is the bed shear stress (Pa), τ_c is its threshold value, and a is a constant quasi-equal to 1 [43, 44]. Other approaches assume as flow variable the discharge or the stream power [45]. A more detailed approach for estimating D_c needs to take into account scouring, headcutting [46], side-wall sloughing, and slaking [47]. Kohl [48] found that headcutting accounted for up to 60% of total rill erosion on some of the soils considered during Water Erosion Prediction Project development. Flow stream power is used as an indicator of detachment due to headcutting [47]. Side-wall sloughing could be a major erosion component in freshly tilled soils with low cohesion and high capillary pressures that have a rill caused by scour or headcut erosion [49]. Slaking affects D_c only for soils with high clay content, low organic matter, and low antecedent water content [50] or soils with a weak structural stability [51]. Establishing the influence of the above-mentioned factors on D_c needs experimental evaluation of soil erodibility parameters corresponding to each process. Soil structure mostly affects the values of these erodibility parameters [47].

Since sediment generally is carried by runoff water, the actual detachment rate D_r ($kg \cdot s^{-1}m^{-2}$) is less than D_c. According to Foster and Meyer [52], the detachment capacity has to be reduced by a feedback factor f_c that depends on the ratio between the actual sediment transport G ($kg \cdot s^{-1}m^{-1}$) and the transport capacity T_c ($kg \cdot s^{-1}m^{-1}$):

$$D_r = D_c f_c = D_c\left(1 - \frac{G}{T_c}\right). \tag{4.6}$$

T_c expresses the maximum sediment discharge that can be transported by a rill flow with given hydraulic conditions. T_c generally is assumed to be proportional to the 1.5 power of the bed shear stress [53]. The feedback factor expresses the physical circumstance that the rill flow has to detach the sediment amount necessary to make the difference between T_c and the actual sediment load negligible. From an energy point of view, the flow energy available for rill detachment is less than the total flow energy because a quota is expended to carry the actual suspended sediment load G. When the sediment transport capacity is exceeded by the sediment load, deposition occurs. For small channels, such as rills, the deposition rate is assumed to be proportional to the difference between the actual sediment load and the transport capacity. Foster [20] assumed that the proportionality constant was directly proportional to the settling velocity and inversely proportional to the discharge.

Gullies are relatively permanent, steep-sided channels in which ephemeral flows occur during rainstorms [29] and cannot be eliminated by usual tillage operations. Gullies are usually deep channels with a narrow cross section. In the first stage, the gully cross section is V-shaped. As the gully develops, its cross section can be modeled by scouring and side-sliding phenomena for assuming a triangular, trapezoidal, and U-shape. The gully channel is characterized by an overfall at the gully head, advancing upstream [54].

The initiation and growth of gullies are dependent on a flow concentration sufficient to form a definable channel. Schumm [55] suggested that the channel length is dependent on the contributing drainage area.

According to Mitchell and Bubenzer [54], gullies are formed when rills combine and develop to the extent that they cannot be eliminated by tillage operations. Morgan [29] established that gully initiation is a more complex process. Small depressions on the hill slope, for example due to a break in vegetation cover (Fig. 4.19), determine flow concentration inducing localized erosion processes. In particular, the erosion is concentrated at the head of the depression where a near-vertical scarp develops. Water falling from the upstream hill slope into the depression determines scouring at the base of the headcut leading to collapse and retreat of the scarp upslope. Flow concentration induces gully floor incision and the development of a stable channel by the scouring action of a running channel flow.

Gully development is not always due to surface erosion processes. In fact, concentrated runoff occurring as subsurface pipe flow can determine erosion processes giving rise to the development of a subsurface tunnel network. Heavy rain can induce subsidence of the ground surface, so exposing the pipe network as gullies.

Haigh [56] described for a desert gully system a complex mechanism of gully enlargement (Fig. 4.20a) caused by both scouring surface processes and tunnel erosion. According to this scheme, an increase of gully cross-sectional area is due to a parallel retreat of gully walls, an aggradation of gully bed, and an enlargement of soil pipe by collapse (Fig. 4.20b). Pipe breaks the gully bed, creating a narrow, vertical sided slot (Fig. 4.20c) inducing a parallel retreat of the former soil pipe and aggradation of the channel bed (Fig. 4.20d).

4.3.2 Measurement of Soil Losses from Erosion

The characteristics of the experimental site are determined by the experimental objectives and the type of data to be obtained. Generally, plots are used to study physical phenomena affecting soil detachment and transport whereas complex hill slopes or small watersheds are gauged for examining deposition processes and therefore sediment yield.

In an experimental station for soil erosion studies, bounded runoff plots of known width, slope length, slope steepness, and soil type are monitored. Plot size depends on the physical process under investigation. Early basic data used for the development of the Universal Soil Loss Equation (USLE) [57] were often obtained from plots 40.5 m^2 in size, having a standard length of 22.1 m [54].

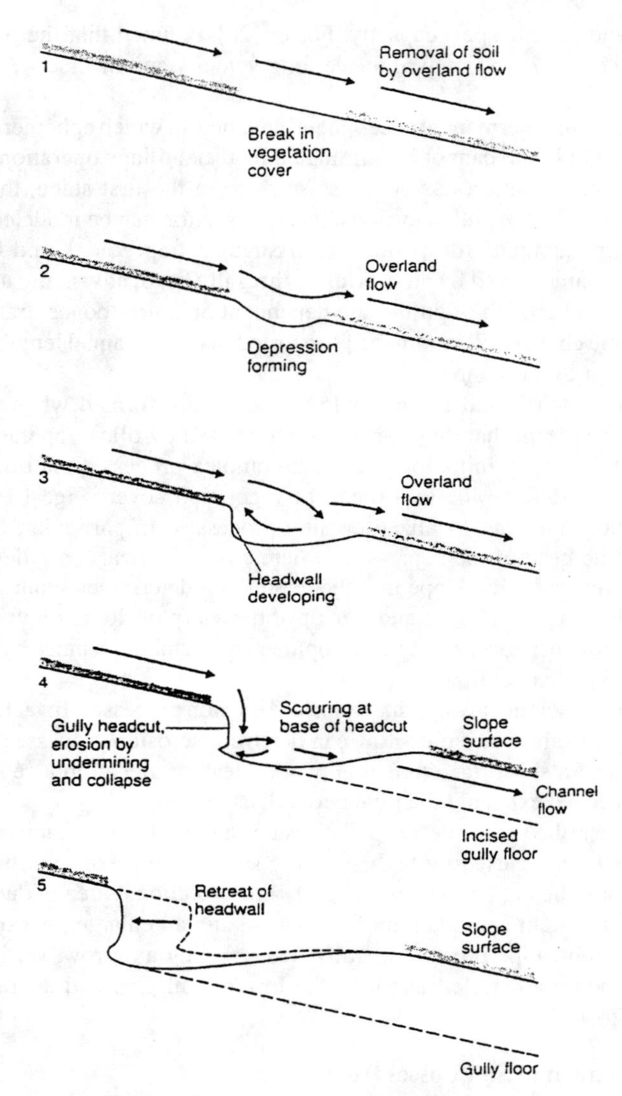

Figure 4.19. Stages of the surface development of a gully on a hill slope. ***Source:*** **[29].**

Studies on sheet erosion due to overland flow need small plot width (2-3 m) and length (<10 m). Studies on rill erosion require plot lengths greater than 6-13 m [29]. Sites should be selected to represent the range of uniform slopes prevailing in the farming area under consideration. Slopes ranging from convex to concave are used to evaluate the influence of the slope shape on soil loss. Because the plot area is known, soil loss can be expressed as kg/m^2 per unit time, which assumes a uniform soil loss over the plot.

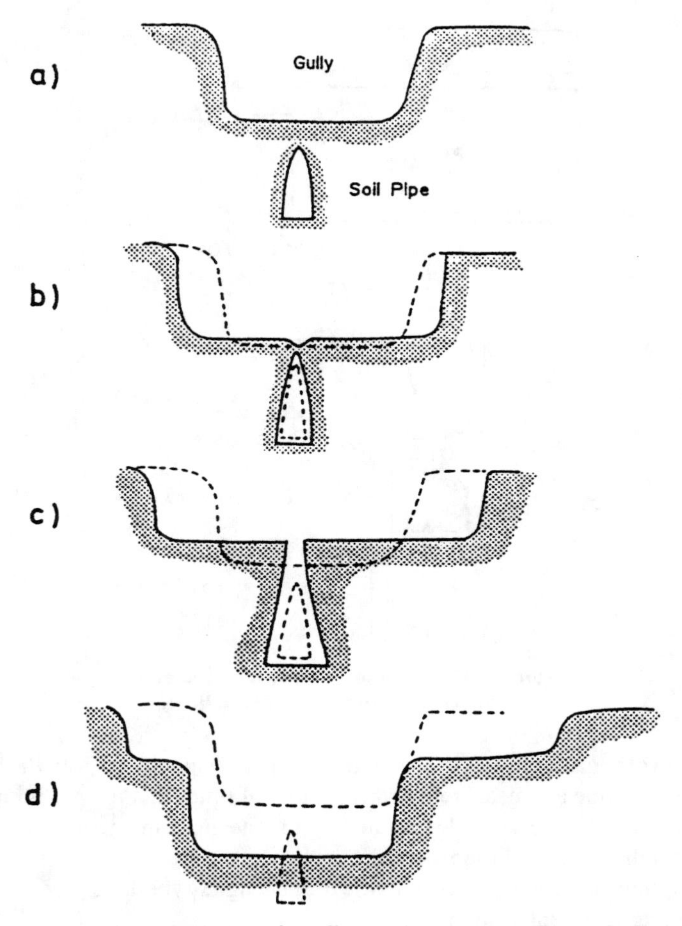

Figure 4.20. Evolution of a gully in a desert system. ***Source:*** **[56].**

The equipment needs for a runoff plot are boundaries around the plot to define the measurement area, collecting equipment to catch plot runoff, a conveyance system (H-flume or pipe) to carry total runoff to a sampling unit, and a storage system (Fig. 4.21). The sampling device may consist of a Coshocton wheel [58] (Fig. 4.22) which has a water wheel, slightly inclined from the vertical, and a sampling head with a narrow opening (slot). With each revolution of the wheel, the slot cuts across the jet from the flume and collects a given portion of the runoff, which is transported to a storage tank.

A recently developed sampler is the Fagna-type hydrological unit [59] (Fig. 4.23). Runoff, cleaned of the coarser materials by passing it through a sedimentation tank, falls on a revolving pot supported by two U-shaped forks. When the pot is full, it turns completely upside down. A few cubic centimeters of the outgoing jet are intercepted by a sampling hole and conveyed to a small tank, below the hole level. When the pot is empty,

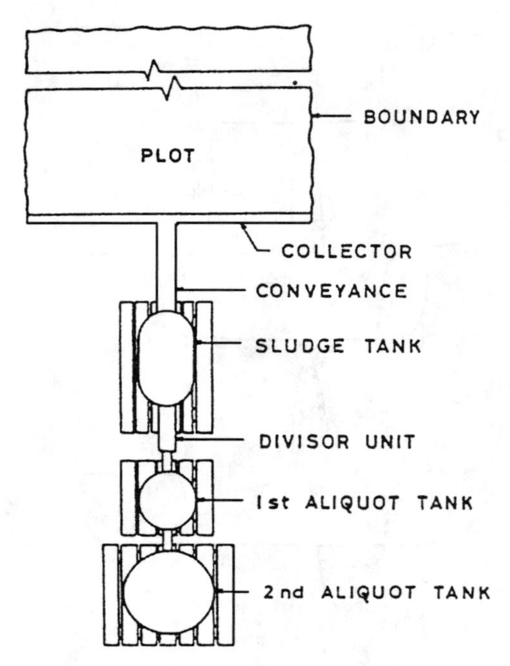

Figure 4.21. Layout of an experimental plot for soil erosion studies. ***Source:*** **[29].**

two coaxial pivots enable the pot mouth to return in the up position. Both the hydrograph and the runoff volume are measured by the number of times that the pot is emptied. The weight of suspended material is determined by the mean sediment concentration in the small tank and the measured runoff volume.

Both sampling devices (Coshocton wheel and Fagna-type unit) need field tests to calibrate the measurement system.

A simple method for measuring the sediment concentration is to store all runoff or to divide it using a sequence of tanks. In each tank the mean concentration C_m (g/L) is estimated from the concentration profile obtained by collecting samples of given volume at different depths. If C_m is calculated by integrating the concentration profile measured along the axial vertical of a tank wall, Bagarello and Ferro [60] illustrated that the actual concentration C (g/L) is linked to C_m according to the following relationship:

$$C = bC_m, \tag{4.7}$$

where the coefficient b depends on the eroded soil type, the sampled suspension volume from each location, the water depth h, the number of sampling locations, and the mixing time before sampling. Figure 4.24 shows, for three soil types (clay C, loam L, sand S), the relationship between the coefficient b and water depth.

Investigations of sediment yield are carried out at hill-slope or small-watershed scale. To measure discharge and suspended sediment concentration, a channel of known cross

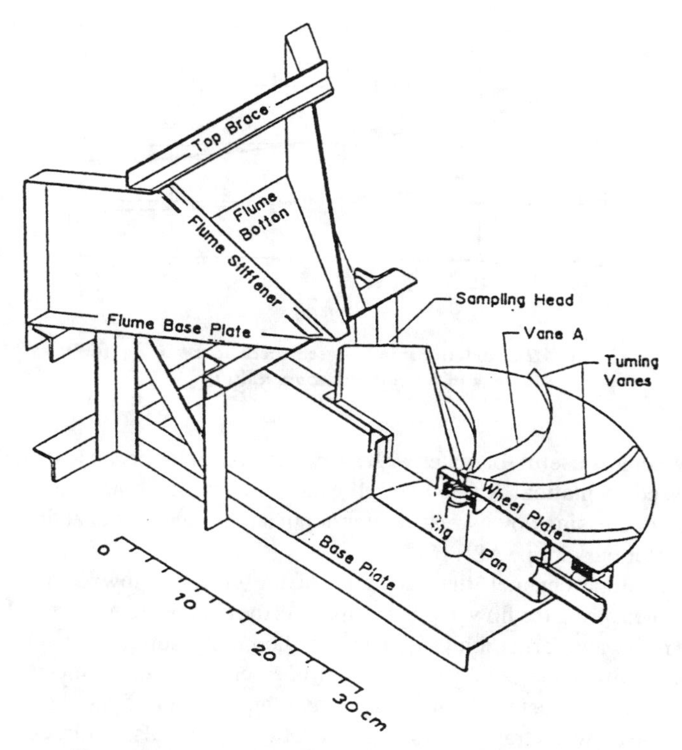

Figure 4.22. Coshocton-type wheel sampler. ***Source:*** **[58].**

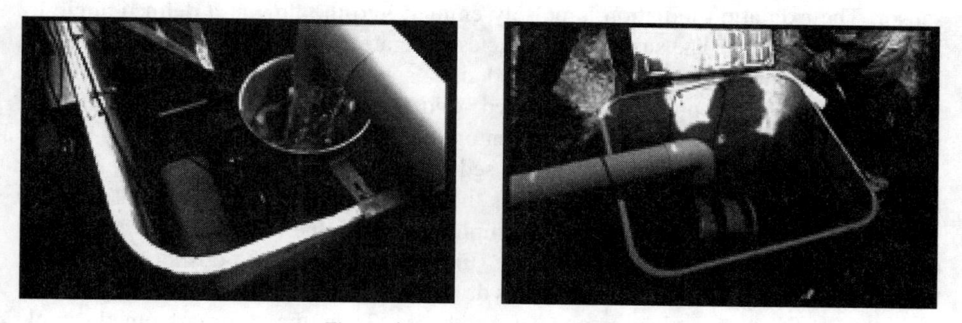

Figure 4.23. Fagna-type sampler.

section and slope usually is constructed. For channels of low slope, discharge measurement can be carried out using a Venturi meter channel [61] or by a free overfall [62]. The Venturi meter channel is based on the principle of critical control section in which the relationship between the critical depth h_c (m) and the discharge Q (m^3/s) is definitive, independent of the channel roughness and of other uncontrollable circumstances. The

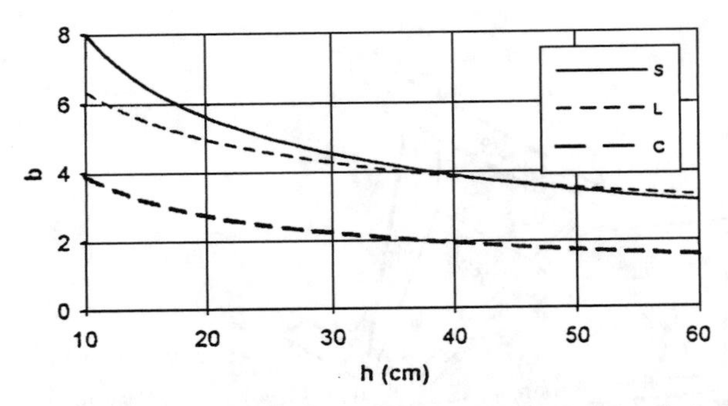

Figure 4.24. Relationship between coefficient *b* of Eq. (4.7) and water depth. *Source*: [60].

Venturi flowmeter is useful for water with suspended particles because it does not induce soil-particle sedimentation. A free overfall establishes in the mild channel a subcritical flow having a decreasing downstream water depth and a water depth at the overfall equal to h_c (backwater curve type M2) [61].

For a steep-slope channel, the discharge of the uniform flow is calculated by the Chezy's law, measuring the flow depth. In this case, the channel length has to be calculated to avoid the roll-wave formation [63]. The flow-depth measurement can be obtained by a nonintrusive, ultrasonic-level probe. The probe sends an ultrasonic pulse that, after partial reflection, is detected by the same sensor and converted back into an electrical signal. The time between transmission and reception of the pulse is directly proportional to the distance between the probe and the water surface.

The suspended sediment concentration can be measured by an infrared turbidity sensor. The excitation radiation is pulsely emitted into the flow at a defined angle by infrared transmitters. The suspended particles generate a scattered light that is received by scattered light receivers. The measured signals are processed to produce an output signal that is proportional to the suspended solid concentration. By coupling this local measurement to the discharge, the sedimentograph is obtained.

A global measurement of suspended sediment yield can be obtained by a water flow sample collector. Figure 4.25 shows the operating scheme of a typical portable sample collector. At the start of every sampling process, the built-in diaphragm pump pneumatically shuts off the dosing device (Fig. 4.25a) and via the dosing glass, blows the sample intake hose free. The sample is drawn in until the conductivity probe located at the top of the dosing glass responds (Fig. 4.25b). Then, the preprogrammed sample volume V_p is withheld and the surplus is returned to the source (Fig. 4.25c). The hose valve opens (Fig. 4.25d) and the sample under low pressure is drained off into the sampling bottle. The sampling procedure is controlled by a microprocessor that allows a sampling event proportional to time or to quantity.

Problems of representativeness arise when plot data are extrapolated to the hill slope and watershed scales. In fact, measurements from a bounded area provide little information on the local variability of erosion rates and the redistribution of soil within a field.

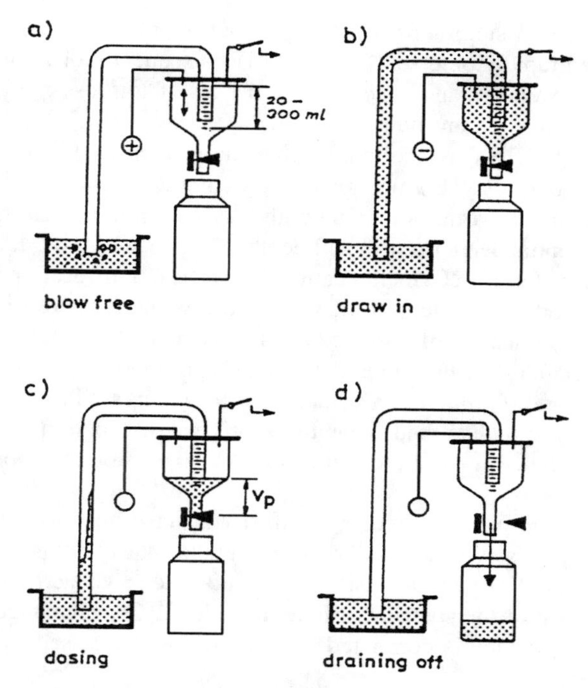

Figure 4.25. Operating scheme of a typical portable sample collector.

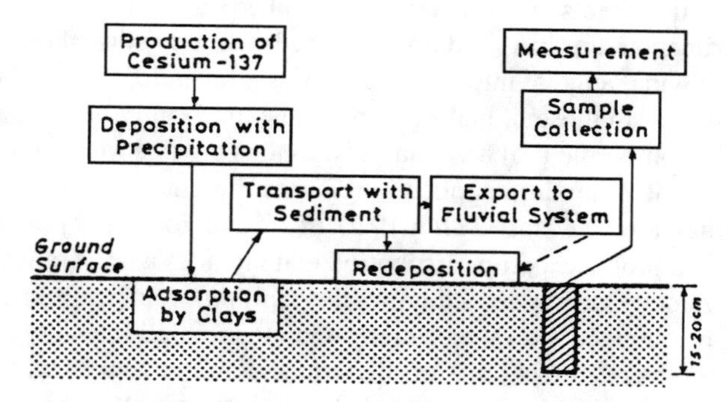

Figure 4.26. Schematic representation of the basis of the ^{137}Cs technique. *Source:* [65].

Tracer techniques afford an alternative to the use of plots and a means of overcoming the problems of measurement representativeness and spatial variability. The most widely used tracer, which possesses the greatest potential in soil erosion and sediment yield studies, is the radionuclide caesium-137, which has a half-life of 30 years. It is a fission reaction product of atmospheric thermonuclear weapons test and has been released in significant quantities into the stratosphere. The ^{137}Cs technique (Fig. 4.26) is based on the following key assumptions: (1) local fallout distribution is uniform; (2) ^{137}Cs fallout is adsorbed rapidly onto soil particles; (3) subsequent redistribution of ^{137}Cs is due to sediment movement; and (4) estimates of rates of soil loss can be derived from measurements of soil ^{137}Cs inventories.

As shown in Fig. 4.26, ^{137}Cs fallout reaches the land surface by precipitation. It is strongly adsorbed onto clay and organic particles and is essentially nonexchangeable [64]. Adsorption on soil particles is rapid with distribution in undisturbed soil profiles showing an exponential decrease with soil depth. More than 75% of the total inventory usually occurs in the top 15–20 cm, indicating that vertical translocation is minimal. The measurement of the ^{137}Cs content of the soil in each point of interest is easy to obtain because it involves collection of a single core sample and a laboratory measurement of ^{137}Cs activity (mBq/cm^2) using standard gamma-spectrometry equipment. If little or no soil loss has occurred, the total ^{137}Cs loading will remain similar to that at the local reference site Cs_{rif} (an undisturbed site with the original distribution of fallout activity). The watershed reference site is a little area in which no erosion or deposition processes occur.

Where soil erosion has occurred, ^{137}Cs also will have been lost, leading to a reduced loading. Conversely, where soil deposition takes place, an increase in ^{137}Cs activity will be found. The geostatistical analysis of local ^{137}Cs activity measurements allows establishment of the spatial distribution of the ^{137}Cs percentage residuals $Y = (^{137}Cs - Cs_{rif})/Cs_{rif}$ which is correlated with the mean annual sediment yield spatial distribution [4, 65].

In many laboratory and plot studies, a rainfall simulator is used to reproduce a storm of known kinetic energy and drop-size distribution. Rainfall simulators are ideal tools for distinguishing the effects of single factors on soil erosion and for carrying out a large number of experimental runs in a short time. Desired characteristics of rainfall simulators [66, 67] include a wide range of intensities, drop sizes, fall velocities, and impact energy characteristics similar to those of natural rainstorms; uniformity of spray pattern and of drop-size distribution on the plot area; and continuous application of rain or minimum time between simulated raindrop applications if intermittent.

In rainfall simulators, developed after 1930, the use of nozzles is practically the only method available to produce a drop distribution that includes a large range of drop sizes. Drop size and velocity at impact, equal to the *terminal* velocity, similar to those of natural rain can be obtained; however, a high simulated rainfall intensity also is obtained because water is delivered under pressure. To reduce rainfall intensity without modifying drop sizes and fall velocities of the simulated rain, intermittent spray is used in many nozzle rainfall simulators. The main disadvantage of this last solution is that the average intensity is made up of high-intensity periods and zero-intensity periods.

Recently, Leone and Pica [68–70] demonstrated the need to reproduce a simulated rainfall with both the same *momentum* per unit time and surface area p_s [N/m^2] and the same *kinetic energy* per unit time and surface area P_s [W/m^2] as those of natural rainfall. According to those authors, the rainfall momentum is responsible for soil aggregate detachment due to the drop impact on the soil surface, whereas rainfall kinetic energy is responsible for rainfall detachability and transportability.

By reanalyzing the experimental data of Laws [14] and Gunn and Kinzer [71], Leone and Pica obtained the following relationship between the terminal velocity V_D (m/s) and the raindrop diameter d (cm):

$$V_D = 49d \exp(-2d) \tag{4.8}$$

By using Eq. (4.8) and describing the raindrop size distribution with a Gamma probabilty distribution (as suggested by meteorological data collected by Ulbrich [72]), Leone and Pica [68] deduced the following relationships for estimating natural rainfall momentum p_n (N/m^2) and kinetic energy per unit time and surface area P_n (W/m^2):

$$p_n = 1.25 \times 10^{-3} I^{1.09} \tag{4.9}$$

$$P_n = 3.16 \times 10^{-3} I^{1.17}, \tag{4.10}$$

where I is the natural rainfall intensity (mm/h).

Leone and Pica suggested that the same momentum and kinetic energy of the natural storm event can be reproduced using test duration and rain intensity in rainfall simulation [69]. The portable rainfall simulator proposed by these authors, which is a modified version of the Guelph II simulator [73], uses a Fulljet nozzle (Spraying System Co.) for which the following relationships for estimating momentum p_s (N/m^2) and kinetic energy per unit time and surface area P_s (W/m^2) are available:

$$p_s = 0.491 \times 10^{-3} I_s^{1.21} \tag{4.11}$$

$$P_s = 0.448 \times 10^{-3} I_s^{1.42} \tag{4.12}$$

in which I_s (mm/h) is the intensity of the simulated rainfall.

By equating natural and simulated rainfall momentum [Eqs. (4.9) and (4.11)], the following equation for calculating the simulated rainfall intensity is obtained:

$$I_s = 2.16 I^{0.9}. \tag{4.13}$$

The ratio between the duration of the simulated rainfall t_s (h) and the natural rainfall t_n (h) is obtained by equating the kinetic energy per unit surface:

$$\frac{t_s}{t_n} = \frac{P_n}{P_s} = \frac{3.16 \times 10^{-3} I^{1.17}}{0.448 \times 10^{-3} I_s^{1.42}} = \frac{3.16 I^{1.17}}{0.448(2.16 I^{0.9})^{1.42}} = \frac{2.363}{I^{0.108}}. \tag{4.14}$$

For example, a natural rainfall intensity equal to 100 mm/h has to be simulated with respect to momentum per unit time and area, using a simulated rainfall intensity I_s equal to 136 mm/h. With respect to kinetic energy per unit time and area, the same storm has to be simulated using a rainfall duration 1.44 times the natural rainfall duration.

Different rainfall simulators have been proposed. For example, Meyer and Harmon [66] developed an intermittent simulator for small fields or laboratory plots (1 × 1 m^2). Moore et al. [74] proposed the Kentucky rainfall simulator for use in field studies, which also may be used for relatively large areas. Recently, a small rainfall simulator, which does not allow reproduction of natural rainfall properties, was proposed by Kamphorst [75] for soil erodibility studies.

4.3.3 Estimation of Soil Losses from Erosion

For evaluating the present extent of soil erosion in a given area, two methods can be applied: a) soil erosion mapping by surveys; and b) erosion modelling [76].

Mapping from Surveys

The surveys are based on informed opinion, on the analysis of the loss of the original profile, and on maps in which the soil erosion features found in the landscape are plotted.

Surveys involving informed opinion have been developed by using published research on the area of interest. For example, informed opinion was used in the Global Assessment of Soil Degradation (GLASOD) for producing a 1:1,000,000 world map showing the extent and severity of various types of soil degradation including erosion by water, wind, and mass movement [76].

According to Morgan [76], one of the best examples of surveys involving informed opinion is the soil erosion map of western Europe developed by De Ploey [77], using the published research of many European scientists. In particular, the water soil erosion map shows the areas with arable land use having a mean annual soil loss greater than 10 t/ha.

Surveys of soil profiles are used for classifying areas using the percentage of eroded soil profile. For example, in the soil erosion map of Hungary, which is used for soil conservation policies, the following three categories of erosion are used: weakly eroded soils (less than 30% of the original profile has been lost), moderately eroded soils (30%–70%), and strongly eroded soils (>70%) [76].

Soil erosion mapping involves plotting the location of the areas of sheet, rill, and gully erosion; different types of mass movement using data from field surveys; and aerial photograph interpretation. If the available information is extremely detailed and difficult to interpret, the map can be simplified by categorizing the information in erosion severity classes. The main limit of a survey-based map is that it shows the situation for only one time period and gives no information about the rate of erosion [76]. The erosion survey maps become dynamic if the mapping work is repeated at regular time intervals using aerial photography or field survey data corresponding to different dates.

Soil erosion modeling is used for assessing the erosion risk of a given area. The *potential risk* can be evaluated by climate, soils, and morphological information. Land cover, land use, and management information also are needed for establishing the *actual risk*. To this aim, two main approaches can be used, one based on factorial scoring, which gives a qualitative assessment, and the other based on modeling, which gives a quantitative result.

Factorial scoring, applicable at different scales, allows the studied area to be divided into subareas for which the different factors affecting soil erosion have a numerical scale, for example, from 1 (low risk) to 5 (high risk). The factorial scoring system is simple and allows the consideration of all erosion processes and of qualitative influencing parameters like human activity [76]. Factorial scoring is based on both numerical and descriptive information. The numerical information is based on either *at site* measurement (soil erodibility and rainfall erosivity) or *areal* information (slope steepness and length, land use, and vegetation cover). The areal information can be obtained by topographic maps or digital terrain models (slope and length) and by analyzing remote-sensing images (land use and vegetation cover) [78, 79].

The descriptive information involves evaluating areas affected by different evolving phases of the erosion phenomenon. This last information is achieved by direct surveys or aerial photogrammetry.

All of the described data, having different sources, can be organized into different informatic layers of a geographical information system (GIS). The GIS database allows for the management of information and for the production of thematic maps, by overlaying the different layers.

Preliminarily, the studied area has to be divided into homogeneous areas of a given character (land use, slope, or aspect). Using a GIS, the subdivision of the area can be carried out automatically by a square mesh (grid format), which is useful for employing matrix calculus. In each square cell, a score is attributed to each factor that is considered. Overlaying allows for the automatic calculation of the total score of each cell and for its classification according to preestablished soil erosion classes (very low, low, moderate, or high). A map based on factorial scoring is not useful for evaluating the rate of soil erosion.

Use of Models

Erosion modeling is the most widely applied method for producing a soil erosion map and for estimating future soil loss using present information. The main problems of using a mathematical model, especially if it is empirically based, are the knowledge of the input data and the need for calibration in the area studied.

For estimating soil erosion, physically based and empirical models are available. Physically based models simulate different hydrological (infiltration and runoff) and erosive (soil detachment, transport, and deposition) components of the erosive process even if empirically derived relationships are often used. Although the physically based models represent the future of soil erosion prediction, at present they are useful only for research purposes or in highly controlled experimental areas.

The most widely applied soil erosion model, which represents the best compromise between applicability in terms of input data and reliability of soil loss estimate [80], is the empirically derived USLE by Wischmeier and Smith [57], which has been revised recently [81]. Note that most of the available soil erosion models put some elementary factors of the USLE into their basic equations to estimate the soil loss or to simulate the subprocesses contributing to soil erosion.

Field or Plot Scale

The USLE was developed to predict *average annual* soil loss A ($\mathrm{t \cdot ha^{-1} \cdot year^{-1}}$) and is based on more than 10,000 plot years of runoff/erosion measurements:

$$A = R\ K\ L\ S\ C\ P. \tag{4.15}$$

R is the rainfall factor (MJ · mm · ha^{-1} · h^{-1} · year^{-1}), K is the soil erodibility factor (t · ha · h · ha^{-1} · MJ^{-1} · mm^{-1}), L is the slope length factor, S is the slope steepness factor, C is the cropping management factor, and P is the erosion control practice factor. L, S, C, and P factors are dimensionless. C and P generally range from 0 to 1 even if C can be as high as 1.5 for a finely tilled, ridged surface that produces much runoff and leaves the soil highly susceptible to rill erosion [81].

The storm erosion index of each event, R_e, is calculated as

$$R_e = I_{30} \sum_{j=1}^{n} (0.119 + 0.0873 \log I_j) h_j \tag{4.16}$$

in which I_{30} is the maximum 30-min rainfall intensity for the storm (mm/h), h_j and I_j are the rainfall depth (mm) and intensity (mm/h) for the jth storm increment, and n is the number of storm increments. The R factor is calculated by adding the R_e values of all erosive events occurring in a period of N years ($N = 25$–30 years) and dividing by N. For many regions of the world, isoerodent maps and simplified criteria for estimating R are available [82–85].

The K factor has to be estimated by the nomograph of Wischmeier et al. [86], which uses five parameters for estimating the inherent soil erodibility: percentage silt (0.002–0.05 mm) plus very fine sand (0.05–0.10 mm), F; percentage sand (0.10–2 mm) G; organic matter content OM (%); a structural index SI; and a permeability index PI. The nomograph uses a particle size parameter $M = F(F+G)$ that explains 85% of the variation in K. The full nomograph expression for the calculation of K (t · ha · h · ha^{-1} · MJ^{-1} · mm^{-1}) is given by Rosewell and Edwards in [87] as

$$\begin{aligned} K = {} & 2.77 \times 10^{-7} M^{1.14} (12 - \mathrm{OM}) \\ & + 4.28 \times 10^{-3} (\mathrm{SI} - 2) + 3.29 \times 10^{-3} (\mathrm{PI} - 3). \end{aligned} \tag{4.17}$$

In the revised equation (RUSLE) [88], a new estimate procedure for K factor takes into account seasonal variability of soil erodibility due to freezing and thawing, soil moisture variation, and soil consolidation. The proposed procedure is complex and needs the evaluation of the annual minimum and maximum value of the soil erodibility factor and the knowledge of the seasonal distribution of rainfall erosivity. The strongly empirical and geographical dependent nature of the procedure needs further verification for different zones. At present, it seems that the difference between the new and the old K evaluation can be more than 20% [89].

The slope length factor L is defined as

$$L = \left(\frac{\lambda}{22.1} \right)^m, \tag{4.18}$$

where λ is the slope length (m) and the m exponent depends on the slope steepness s ($m = 0.5$ for $s \geq 0.05$). In RUSLE, the following expression for m is used:

$$m = \frac{f}{1 + f} \tag{4.19}$$

where f, which is the ratio between rill and interrill erosion, can be evaluated by the following relationship [90]:

$$f = \frac{\sin\beta}{0.0896} \frac{1}{(3\sin^{0.8}\beta + 0.56)} \tag{4.20}$$

in which β is the slope angle. The values of f obtained from Eq. (4.20) have to be multiplied by 0.5 if the ratio of rill to interrill erosion is low or by 2 if this ratio is high.

Soil loss is much more sensitive to changes in slope steepness than to changes in slope length. In USLE, the S factor is evaluated by the following relationship:

$$S = \frac{0.43 + 0.30s + 0.043s^2}{6.613} \tag{4.21}$$

in which s is slope steepness expressed as a percentage. The RUSLE has a more nearly linear slope steepness relationship than the USLE [90]:

$$S = 10.8\sin\beta + 0.03 \quad \text{if } \tan\beta < 0.09, \tag{4.22a}$$

$$S = 16.8\sin\beta - 0.5 \quad \text{if } \tan\beta \geq 0.09. \tag{4.22b}$$

On steep slopes, computed soil loss is reduced almost by half with RUSLE [81].

The cropping management factor represents the ratio of soil loss from a specific cropping or cover condition to the soil loss from a tilled, continuous-fallow condition for the same soil and slope and for the same rainfall. The procedure for calculating the C factor requires a knowledge of crop rotation, crop-stage periods, soil loss ratio (SLR) for each crop period, and temporal distribution of rainfall erosivity. SLR values for the most crop rotations in the United States have been given by Wischmeier and Smith [57]. To compute C, SLRs are weighted according to the distribution of erosivity during a year. The main limitations of this approach are the unavailability of SLRs for all crop covers (e.g., fruit trees) or vegetable crop rotations and the difficulties of transfering the SLR values calculated for American conditions to other regions.

To estimate the C factor for forest conditions, the procedure proposed by Dissmeyer and Foster [91] can be used. This is based on an evaluation of nine subfactors: amount of bare soil, canopy, soil reconsolidation, high organic content, fine roots, residual binding effect, onsite storage, steps, and contour tillage.

In the RUSLE, SLR values are computed using five subfactors expressing the prior land use, the canopy, the surface cover, the surface roughness, and the soil moisture status. The subfactor approach permits application of SLRs where values are not available from previously published experimental analyses [92].

The P factor mainly represents how surface conditions affect flow paths and flow hydraulics. The erosion control practices usually included in this factor are contouring, contour strip-cropping, and terracing [57]. In the RUSLE, an update of the evaluation procedure for the P factor has been proposed [81].

Soil loss tolerance T is defined as the maximum rate of soil erosion that permits a high level of productivity to be sustained. The soil loss tolerance for a specific soil is useful for establishing soil conservation planning. The most common T value is equal to

$10\ \mathrm{t \cdot ha^{-1} \cdot year^{-1}}$. If the USLE or the RUSLE is used for estimating soil loss, it results in

$$\frac{T}{RKS} = LCP \tag{4.23}$$

Equation (4.23) shows that slope length, land use, and erosion control practices are the three factors that can be practically modified for obtaining a given T value.

Watershed Scale

The prediction of sediment yield (i.e., the quantity of transferred sediments) in a given time interval, from eroding sources through the channel network to a watershed outlet, can be carried out using either an erosion model (USLE, RUSLE) with a mathematical operator expressing the sediment transport efficiency of the hill slopes and the channel network or a sediment yield model (MUSLE, WEPP) [93–95].

A physically based model is theoretically preferable, but its parameters, which are often numerous, may not be easy to measure or estimate. In addition, the measurement scale may not be at the same level as the scale of the watershed discretization for applying the model. Moreover, watershed soil erosion estimates obtained from physically based models are affected by the uncertainties in the equations used to simulate the detachment, transport, and deposition processes in morphologically complex areas. For this reason, a parametric approach, such as USLE and RUSLE, is a more attractive method even if the parameters have little or no physical meaning because they lump together both the effects of several different processes and inaccuracy [96].

Two different strategies can be followed for applying a soil erosion model at the watershed scale: (1) modifying the calculating procedure of the model factors in order to transform a watershed into an equivalent plot [94, 97]; or (2) dividing the watershed into morphological areas, that is, into areas for which all elementary factors of the selected erosion model can be evaluated [98, 99].

The first procedure allows the evaluation of total watershed soil loss, which then has to be coupled with an estimate of the sediment delivery ratio (SDR_w) for the whole watershed to predict total sediment yield. The second procedure allows the calculation of soil loss in each morphological area, which has to be coupled with a disaggregated criterion for estimating sediment delivery processes to obtain sediment yield spatial distribution.

SDR_w (% or dimensionless) generally decreases with increasing watershed size, indexed by area or stream length; the American Society of Civil Engineers (ASCE) [100] has suggested the use of the following power function (Fig. 4.27) [101]:

$$\mathrm{SDR}_w = kS_w^n \tag{4.24}$$

in which k and n are numerical constants and S_w (km^2) is the watershed area.

Spatial disaggregation of sediment delivery processes requires that the watershed be discretized into morphological units that have an irregular shape if they are bounded in such way as to define areas with a constant steepness and a clearly defined length

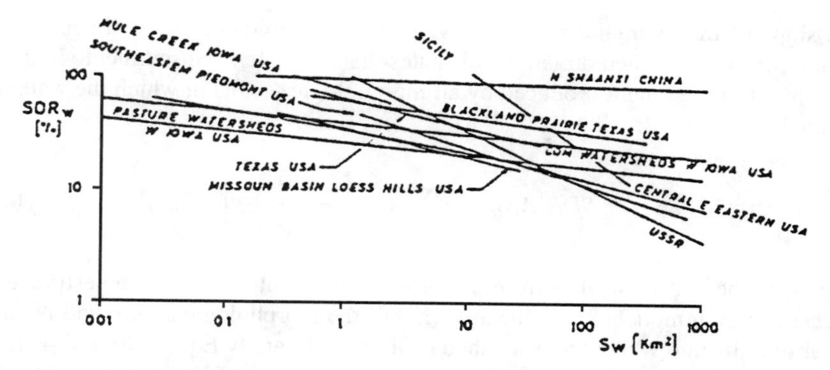

Figure 4.27. Comparison of different relationships between SDR$_w$ and watershed area S_w. *Source:* [101].

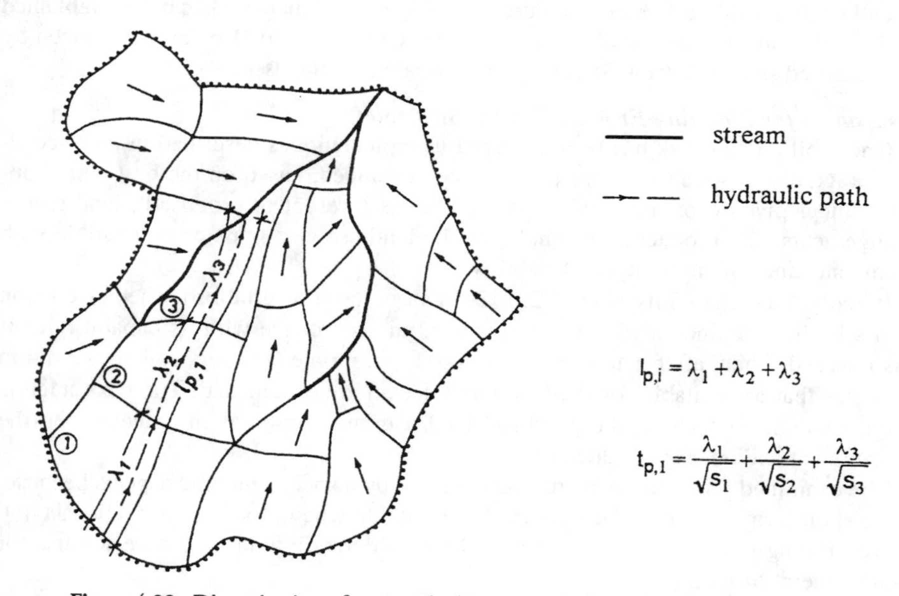

Figure 4.28. Discretization of a watershed into morphological units. *Source:* [103].

(Fig. 4.28). Discretization can be simplified by superimposing a regular grid with square or triangular cells onto the watershed area.

For each morphological unit, the sediment delivery ratio SDR$_i$ (dimensionless) has to be evaluated according to the following equation [102, 103]:

$$\mathrm{SDR}_i = \exp\left[-\varphi \frac{l_{p,i}}{\sqrt{s_{p,i}}}\right] \tag{4.25}$$

in which φ is a coefficient that is constant for a given watershed, $l_{p,i}$ (m) is the length of the hydraulic path from the morphological unit to the nearest stream reach, and $s_{p,i}$ (m/m) is the slope of the hydraulic path. For evaluating φ coefficient, the sediment balance equation for the watershed outlet, which states that the watershed sediment yield Y_s (t) is the sum of the sediment produced by all morphological units in which the watershed is divided, can be used:

$$Y_s = \sum_{i=1}^{N_u} SDR_i A_i S_{u,i} = \sum_{i=1}^{N_u} \exp\left[-\varphi \frac{l_{p,i}}{\sqrt{s_{p,i}}}\right] A_i S_{u,i}, \tag{4.26}$$

where A_i is the soil loss (t/ha) from a morphological unit that has to be estimated by a selected erosion model, $S_{u,i}$ is the area (ha) of the morphological unit, and N_u is the number of units into which the watershed is divided. To apply Eq. (4.26), a soil erosion model at the plot scale has to be selected and measurements of Y_s are necessary.

Using soil-loss data from plots or sediment yield estimates from a selected model, a potential relationship between radionuclide loss Y and sediment yield can be established [104, 105]. Thus, the spatial distribution of the sediment yield also can be deduced by the measured spatial distribution of ^{137}Cs residual percentages Y.

Assessment for Planning Erosion Control Strategies

Once soil erosion risk has been assessed using techniques described in the preceding paragraphs, a sound land-use plan can be developed, based on the best-suited options, under given economic and social conditions (present or proposed), land tenure arrangements, and production technology. The land use also must be compatible with the maintenance of environmental stability.

Initially, land capability (Fig. 4.29) [29] can be used to establish whether erosion occurs in the examined area when soil is used in accordance with its capability or if it is misused. Then, after determining the most appropriate land use, soil conservation strategies that are suitable for the land use selected are established. The next stage is to quantify the impacts of the proposed land use and conservation strategies on the environment and on crop productivity.

The examined problems also are characterized by a socioeconomic aspects because, in the short term, land use is economically profitable whereas soil conservation is not. However, long-term forecasting also must be considered in order to ensure soil use for future generations.

Figure 4.29 shows the general sequence of stages for soil conservation planning described above. This sequence can be specified for a particular land use (crop rotation, pasture, forest, rangeland, and urban areas).

4.3.4 Conclusions

Conserving soil for future uses has to be linked to the development of sustainable agriculture.

The phenomenological aspect of the soil erosion process is well understood and many recent developments (RIFT mechanism, overland flow hydraulics, rill flow velocity and rill shear stress studies, and gully initiation criteria) have increased the body of the knowledge.

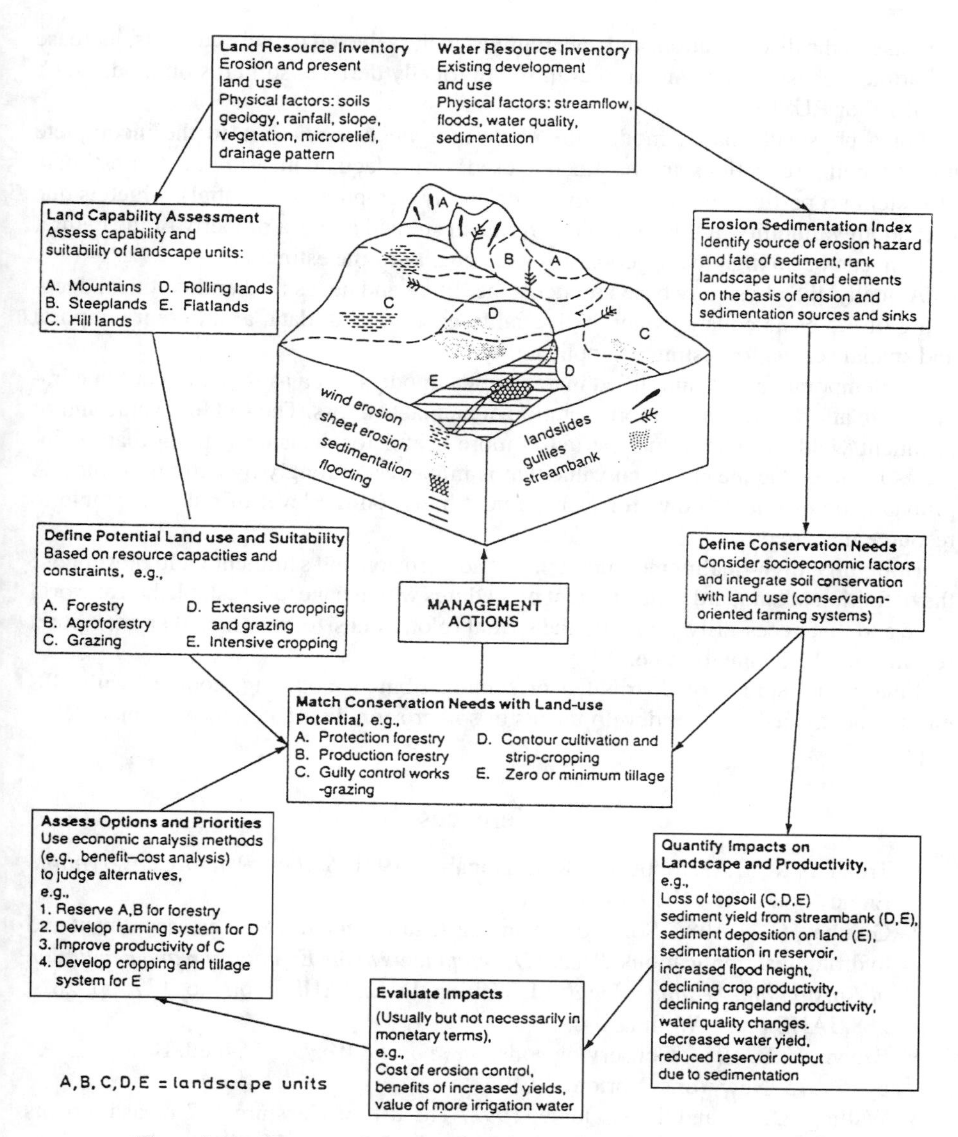

Figure 4.29. Sequence of stages in planning soil conservation. ***Source:*** **[29].**

The erosion phenomenon is very complex and the numerous hydrological subprocesses (infiltration, runoff, sediment transport) make very difficult an accurate physically based model.

The existing difficulties of the physically based models, such as the numerous input parameters, differences between the scale of measurements of the input parameters and the scale of the discretization, and uncertainties in the selected model equations, increase the attractiveness of a parametric, even if empirically derived, soil erosion model such as USLE or RUSLE.

For a physically based model, the *knowledge uncertainty*, due to the incomplete understanding (equations and parameters used) or inadequate measurement or estimate of system properties, and the *stochastic variability* (temporal and spatial), which is due to random variability of the natural environment studied and is a property of the natural system, can be so high that it jeopardizes the quality of the estimate.

Actually, the physically based model is the future and needs further theoretical deepening of the basic equations and high-quality experimental data, at different temporal and spatial scales, for testing its applicability.

Contemporaneously, simplified mathematical models, such as RUSLE, could be improved to allow a stochastic representation of annual soil loss. The soil loss value and/or sediment yield of given return period is more useful for decision making than a single estimate of the mean annual value that is influenced strongly by extreme values. A probabilistic estimate allows for management based on the level of risk acceptable to resource managers.

Hopefully, high-quality measurements of soil erosion and sediment yield presuppose the definition of standard measurement procedures which have to be established at world scale and that necessarily vary with the spatial (plot, field size, watershed) and temporal (event, monthly, annual) scale.

Finally, the spread of distributed or semidistributed modeling, coupled with GIS employment, needs further developments of soil erosion distributed measurement techniques.

References

1. Troeh, F. R., J. A. Hobbs, and R. L. Donahue. 1991. *Soil and Water Conservation*. Englewood Cliffs, NJ: Prentice-Hall.
2. Golubev, G. N. 1982. Soil erosion and agriculture in the world: an assessment and hydrological implications. *Recent Developments in the Explanation and Prediction of Erosion and Sediment Yield*. ed. Walling, D. E., IAHS Publ. No. 137, pp. 261–268, IAHS press, Wellingford.
3. Brown, L. R. 1984. Conserving soils. *State of the World, 1984*. ed. Brown, L. R., pp. 53–75. New York: Norton.
4. Walling, D. E., and T. A. Quine, 1992. The use of Caesium-137 measurements in soil erosion surveys. *Erosion and Sediment Transport Monitoring Programmes in River Basins. Proceeding of the Oslo Symposium*, IAHS Publ. No. 210, pp. 143–152, IAHS press, Wellingford.
5. Walling, D. E., and W. Webb. 1983. Patterns of sediment yield. *Background to Palaeohydrology*. ed. Gregory, K. J., pp. 69–100. Chichester: Wiley.
6. Larson, W. E., M. J. Lindstrom, and T. E. Schumacher. 1997. The role of severe storms in soil erosion: A problem needing consideration. *J. Soil Water Conserv.* 52(2): 90–95.

7. Chepil, W. S. 1957. Dust bowl: Causes and effects. *J. Soil Water Conserv.* 12:108–111.
8. Bagnold, R. A. 1941. *The Physics of Blown Sand and Desert Dunes.* London: Chapman and Hall.
9. Chepil, W. S. 1945. Dynamics of wind erosion. III. Transport capacity of the wind. *Soil Sci.* 61:331–340.
10. Ellison, W. D. 1947. Soil erosion studies—Part I. *Agric. Eng.* 28(4):145–146.
11. Ellison, W. D. 1947. Soil erosion studies—Part II. *Agric. Eng.* 28(5):197–201.
12. Laws, J. O., and D. A. Parsons. 1943. The relationship of raindrop size to intensity. *Trans. Am. Geophys. Union* 24:452–460.
13. Hudson, N. W. 1963. Raindrop size distribution in high intensity storms. *Rhod. J. Agric. Res.* 1:6–11.
14. Laws, J. O. 1941. Measurements of the fall-velocity of waterdrops and raindrops. *Trans. Am. Geophys. Union* 22:709–721.
15. Sharma, P. P., and S. C. Gupta. 1989. Sand detachment by single raindrops of varying kinetic energy and momentum. *Soil Sci. Soc. Am. J.* 45:1031–1034.
16. Palmer, R. S. 1965. Waterdrop impact forces. *Trans. ASAE* 8:69–70, 72.
17. Mutchler, C. K., and R. A. Young. 1975. Soil detachment by raindrops. *Proceedings of the Sediment Yield Workshop*, Rept. ARS-S-40, pp. 113–117. Oxford MS: US. Department of Agriculture.
18. Torri, D., and M. Sfalanga. 1984. Some problems on soil erosion modelling. *Proceedings of the Workshop on Prediction of Agricultural Non-point Source Pollution: Model Selection and Application*. pp. B1–B10, Venezia: University of Padua.
19. Young, R. A., and J. L. Wiersma. 1973. The role of rainfall impact on soil detachment and transport. *Water Resour. Res.* 9:1629–1636.
20. Foster, G. R. 1982. Modeling the erosion process. *Hydrologic Modeling of Small Watershed*. ed. Haan, C. T., ASAE Monograph No. 5, pp. 297–379. St. Joseph, MI: American Society of Agricultural Engineers.
21. Sharma, P. P. 1996. Interrill erosion. *Soil Erosion, Conservation, and Rehabilitation*, ed. M. Agassi, pp. 125–152. New York: Marcel Dekker.
22. Horton, R. E. 1945. Erosional development of streams and their drainage basins: A hydrophysical approach to quantitative morphology. *Bull. Geol. Soc. Am.* 56:275–370.
23. Chu, S. T. 1978. Infiltration during an unsteady rain. *Water Resour. Res.* 14(3):461–466.
24. Agnese, C., and V. Bagarello. 1997. Describing rate variability of storm events for infiltration prediction. *Trans. ASAE* 40(1):61–70.
25. Stone, J., K. G. Renard, and L. J. Lane. 1996. Runoff estimation on agricultural fields. *Soil Erosion, Conservation, and Rehabilitation*. ed. Agassi M., pp. 203–238. New York: Marcel Dekker.
26. Moss, A. J. 1988. Effects of flow-velocity variation of rain-driven transportation and the role of rain impact in the movement of solids. *Aust. J. Soil Res.* 26:443–450.
27. Kinnell, P. I. A. 1990. The mechanics of raindrop induced flow transport. *Aus. J. Soil Res.* 28:497–516.

28. Kinnell, P. I. A. 1991. The effect of flow depth on sediment transport induced by rain-drops impacting shallow flows. *Trans. ASAE* 34:161–168.
29. Morgan, R. P. C. 1986. *Soil Erosion and Conservation*. London: Longman.
30. Grissinger, E. H. 1996. Rill and gullies erosion. *Soil Erosion, Conservation, and Rehabilitation*. ed. Agassi, M., pp. 153–167. New York: Marcel Dekker.
31. Merritt, E. 1984. The identification of four stages during microrill development. *Earth Surf. Process. Landforms* 9:493–496.
32. Meyer, L. D. 1975. Effect of rate and canopy on rill erosion. *Trans. ASAE* 18:905–911.
33. Savat, J., and J. De Ploey. 1982. Sheet wash and rill development by surface flow. *Badland Geomorphology and Piping*. eds. Bryan, R. B., and A. Yair, pp. 113–126. Norwich: Geobooks.
34. Rauws, G. 1987. The initiation of rills on plane beds of non-cohesive sediments. *Rill Erosion Catena Suppl.* 8, ed. Bryan , R. B., pp. 107– 118.
35. Cai, Q., and G. Wang. 1993. Rill initiation and erosion on hillslopes in the hilly loess region, western Shanxi Province. *Int. J. Sediment Res.* 8(3):21–32.
36. Savat, J. 1979. Laboratory experiments on erosion and deposition of loess by laminar sheet flow and turbulent rill flow. *Colloque sur l'Érosion Agricole des Sols en Milieu Tempéré Non-Mediterranéen,* ed. Vogt, H., and T. Vogt, pp. 139–143. Strasbourg: University Louis Pasteur.
37. Boon, W., and J. Savat. 1981. A nomogram for the prediction of rill erosion. *Soil Conservation: Problems and Prospects,* ed. Morgan, R. P. C., pp. 303–319. London: Wiley.
38. Gilley, J. E., E. R. Kottwitz, and J. R. Simanton. 1990. Hydraulic characteristics of rills. *Trans. ASAE* 33(6):1900–1906.
39. Foster, G. R., L. F. Hugghins, and L. D. Meyer. 1984. A laboratory study of rill hydraulics: I. Velocity relationships. *Trans. ASAE* 21:790–796.
40. Foster, G. R., L. F. Hugghins, and L. D. Meyer. 1984. A laboratory study of rill hydraulics: II. Shear stress relationships. *Trans. ASAE* 21:797–804.
41. Poesen, J. 1987. Transport of rock fragments by rill flow: A field study. *Catena Supplement 8: Rill Erosion*. ed. Bryan, R. B., pp. 35–54.
42. Meyer, L. D. 1964. Mechanics of soil erosion by rainfall and runoff as influenced by slope length, slope steepness and particle size. Ph.D. Thesis, W. Lafayette, IN: Purdue University.
43. Foster, G. R., W. R., Osterkamp, L. J. Lane, and D. W. Hunt. 1982. An erosion equation derived from basic erosion principles. *Trans. ASAE* 19(4):678– 682.
44. Knisel, W. G. 1980. CREAMS: A Field-Scale Model for Chemicals, Runoff, and Erosion from Agricultural Management Systems. Conservation Research Report No. 26. Washington, DC: U.S. Department of Agriculture Science and Education Administration.
45. Rose, C. W., J. R., Williams, G. C. Sander, and D. A. Barry, 1983. A mathematical model of soil erosion and deposition processes: I. Theory for a plane land element. *Soil Sci. Soc. Am. J.*, 47: 991–995.
46. Meyer, L. D., G. R. Foster, and S. Nikolov. 1975. Effect of flow rate and canopy on rill erosion. *Trans. ASAE* 18(6):905–911.

47. Elliot, W. J., and J. M. Laflen. 1993. A process-based rill erosion model. *Trans. ASAE* 36(1):65–72.
48. Kohl, K. D. 1988. Mechanics of rill headcutting. Ph.D. Dissertation Ames, IA: Iowa State University.
49. Brown, L. C., G. R. Foster, and D. B. Beasley. 1989. Rill erosion as affected by incorporated crop residue and seasonal consolidation. *Trans. ASAE* 32(6):1967–1978.
50. Young, R. A., and C. A. Onstad. 1982. Erosion characteristics of three northwestern soils. *Trans. ASAE* 19(1):367–371.
51. Kemper, W. D., and R. C. Rosenau, 1984. Soil cohesion as affected by time and water content. *Soil Sci. Soc. Am. J.* 48(5):1001–1006.
52. Foster, G. R., and L. D. Meyer. 1972. A closed-form soil erosion equation for upland areas. *Sedimentation, Symposium to Honor Prof. H.A. Einstein*, ed. Shen, H. W., pp. 12.1–12.19. Fort Collins: Colorado State University.
53. Finkner, S. C., M. A., Nearing, G. R. Foster, and J. E. Gilley. 1989. A simplified equation for modeling sediment transport capacity. *Trans. ASAE* 32:1545–1550.
54. Mitchell, J. K., and G. D. Bubenzer. 1980. Soil loss estimation. *Soil Erosion.* eds. Kirkby M. J., and R. P. C., Morgan. pp. 17–62. Chichester: Wiley.
55. Schumm, S. A. 1956. Evolution of drainage systems and slopes in badlands in Perth Amboy, New Jersey. Office of Naval Research Tech. Rept. 8, Columbia University.
56. Haigh, M. J. 1990. Evolution of an anthropogenic desert gully system. *Erosion, Transport and Deposition Processes, Proceedings of the Jerusalem Workshop*, IAHS Publ. No. 189, pp. 65–77, IAHS press, Wellingford.
57. Wischmeier, W. H., and D. D. Smith. 1978. *Predicting Rainfall Erosion Losses. A Guide to Conservation Planning*, Agriculture Handbook No. 537. Washington, D.C.: U.S. Department of Agriculture.
58. Carter, C. E., and D. A. Parson. 1967. Field tests on the Coshocton-type wheel runoff sampler, *Trans. ASAE* 10(1):133–135.
59. Bazzoffi, P. 1993. Fagna-type hydrological unit for runoff measurements and sampling in experimental plot trials. *Soil Technol.* 6:251–259.
60. Bagarello, V., and V. Ferro. 1998. Calibrating storage tanks for plot soil erosion measurements. *Earth Surf. Process. Landforms*. 23.
61. Chow, V. T. 1959. *Open-Channel Hydraulics*, New York: McGraw Hill.
62. Ferro, V. 1992. Flow measurement with rectangular free overfall, *Proc. ASCE* 118 (IR6):956–964.
63. Montuori, C. 1984. Sviluppi recenti nello studio delle correnti supercritiche. *Proceedings of Workshop "Idraulica del Territorio Montano" in Honor of Prof. A. Ghetti*, pp. 205–256. Istituto di Idraulica "G. Poleni" dell'Università di Padova. Padova: Cortine.
64. Ritchie, J. C., and J. R. McHenry. 1990. Application of radioactive fallout cesium-137 for measuring soil erosion and sediment accumulation rates and patterns: A review. *J. Environ. Qual.* 19:215–233.
65. Walling, D. E., and T. A. Quine. 1991. Use of ^{137}Cs measurements to investigate soil erosion on arable fields in the UK: Potential applications and limitations, *J. Soil Sci.* 42:147–165.

66. Meyer, L. D., and W. C. Harmon. 1979. Multiple-intensity rainfall simulator for erosion research on row sideslopes. *Trans. ASAE* 22(1):100–103.
67. Torri, D., and C. Zanchi. 1991. I simulatori di pioggia: Caratteristiche ed utilizzazioni. *La Gestione delle Aree Collinari Argillose e Sabbiose*, Edizioni delle Autonomie SRL, pp. 121–127 (in Italian).
68. Leone, A., and M. Pica. 1993. Caratteristiche dinamiche e simulazione delle piogge, Parte prima: Fondamenti teorici. *Riv. Ing. Agrar.* 3:167–175 (in Italian).
69. Leone, A., and M. Pica. 1993. Caratteristiche dinamiche e simulazione delle piogge, Parte seconda: Procedure di prova e proposta di un simulatore. *Riv. Ing. Agrar.* 3:176–183 (in Italian).
70. Pica, M. 1994. Dynamic characteristics of rainfall. *Idrotecnica* 2:59–67.
71. Gunn, R., and G. D. Kinzer. 1948. The terminal velocity of fall for water droplets in stagnant air. *J. Meteorol.* 6:243–248.
72. Ulbrich, C. W. 1983. Natural variations in the analytical form of the raindrop size distribution. *J. Climatol. Appl. Meteorol.* 22:1764–1775.
73. Tossell, R. W., G. J. Wall, R. P. Rudra, W. T. Dickinson, and P. H. Groenevelt. 1990. The Guelph rainfall simulator II: Part 2, A comparison of natural and simulated rainfall characteristics, *Can. Agric. Eng.* 32:215–223.
74. Moore, I. D., M. C. Hirschi, and B. J. Barfield. 1983. Kentucky rainfall simulator. *Trans. ASAE* 26(4):1085–1089.
75. Kamphorst, A. 1987. A small rainfall simulator for the determination of soil erodibility. *Neth. J. Agric. Sci.* 35:407–415.
76. Morgan, R. P. C. 1993. Soil erosion assessment. ed. Morgan, R. P. C. *Proceedings of the Workshop on "Soil Erosion in Semi-arid Mediterranean areas,"* pp. 3–17. Taormina: C.N.R.-Progetto Finalizzato RAISA, ESSC and CSEI.
77. De Ploey, J. 1989. Soil erosion map of western Europe. Cremlingen–Destedt: Catena Verlag.
78. Ventura, S. J., N. R. Chrisman, K. Connors, R. F. Gurda, and R. W. Martin. 1988. A land information system for soil erosion control planning. *J. Soil Water Conserv.* 43(3):230–233.
79. Bocco, G., and C. R. Valenzuela. 1993. Integrating satellite-remote sensing and geographic information systems technologies in gully erosion research. *Remote Sensing Rev.* 7:233–240.
80. Risse, L. M., M. A. Nearing, A. D. Nicks, and J. M. Laflen. 1993. Error assessment in the Universal Soil Loss Equation, *Soil Sci. Soc. Am. J.* 57:825–833.
81. Renard, K. G., G. R. Foster, G. A., Weesies, and J. P. Porter. 1991. RUSLE—Revised Universal Soil Loss Equation. *J. Soil Water Conserv*. 46(1):30–33.
82. Bagarello, V., and F. D'Asaro. 1994. Estimating single storm erosion index. *Trans. ASAE* 37(3): 785–791.
83. Aronica, G., and V. Ferro. 1997. Rainfall erosivity over Calabrian region. *J. Hydrolog. Sci.* 42(1):35–48.
84. Banasich, K., and D. Gorski. 1994. Rainfall erosivity for south-east Poland. *Conserving Soil Resources. European Perspectives*. ed. Rickson, R. J. Lectures in Soil Erosion Control, pp. 201–207. Silsoe College, Cranfield University.

85. Bergsma, E. 1980. Provisional rain-erosivity map of the Netherlands. *Assessment of Erosion*, eds. De Boodt, M., and D. Gabriels. Chichester, UK: Wiley.
86. Wischmeier, W. H., C. D. Johnson, and B. V. Cross. 1971. A soil erodibility nomograph for farmland and construction sites, *J. Soil Water Conserv.* 29:189–193.
87. Loch, R. J., and C. J. Rosewell. 1992. Laboratory methods for measurement of soil erodibilities (*K* factor) for the Universal Soil Loss Equation. *Aust. J. Soil Res.* 30:233–248.
88. Young, R. A., M. J. M. Romkens, and D. K. McCool. 1990. Temporal variations in soil erodibility. *Catena Suppl.* 17:41–53.
89. Renard, K. G., G. R. Foster, D. C. Yoder, and D. K. McCool. 1994. RUSLE revisited: Status, questions, answers, and the future. *J. Soil Water Conserv.* 49(3):213–220.
90. Moore, I. D., and J. P. Wilson. 1992. Length-slope factors for the Revised Universal Soil Loss Equation: Simplified method of estimation. *J. Soil Water Conserv.* 47(5):423–428.
91. Dissmeyer, G. E., and G. R. Foster. 1981. Estimating the cover-management factor (C) in the Universal Soil Loss Equation for forest conditions, *J. Soil Water Conserv.* 36:235–240.
92. Renard, K. G., and V. A. Ferreira. 1993. RUSLE model description and database sensitivity. *J. Environ. Qual.* 22(3):458–466.
93. Williams, J. R. 1975. Sediment yield prediction with universal equation using runoff energy factor. Present and prospective technology for predicting sediment yields and source. *Proceedings of Sediment-Yield Workshop.* Oxford, MS: U.S. Department of Agriculture Sedimentation Lab, 244–252.
94. Williams, J. R., and H. D. Berndt. 1972. Sediment yield computed with universal equation. *Proc. ASCE* 98 (HY2):2087–2098.
95. U.S. Department of Agriculture. 1995. Water Erosion Prediction Project (WEPP). NSERL Report No. 10. West Lafayette, IN: National Soil Erosion Research Laboratory.
96. Richards, K. 1993. Sediment delivery and drainage network. *Channel Network Hydrology.* eds. Beven, K. and M. J. Kirkby, pp. 221–254. Wiley.
97. Bagarello, V., G. Baiamonte, V. Ferro, and G. Giordano. 1993. Evaluating the topographic factors for watershed soil erosion studies. ed. Morgan, R. P. C. *Proceedings of Workshop on "Soil Erosion in Semi-arid Mediterranean Areas,"* pp. 19–35. Taormina: C.N.R.-Progetto Finalizzato RAISA, ESSC and CSEI.
98. Julien, P. Y., and M. Frenette. 1986. Scale effects in predicting soil erosion. *Proceedings of "Drainage Basins Sediment Delivery,"* IAHS Publ. No. 159, pp. 253–259. IAHS Press, Wellingford.
99. Julien, P. Y., and M. Frenette. 1987. Macroscale analysis of upland erosion. *J. Hydrolog. Sci.* 32(3):347–358.
100. American Society of Civil Engineers. 1975. *Sedimentation Engineering.* ASCE Manuals and Reports on Engineering Practice, 54.
101. Walling, D. E. 1983. The sediment delivery problem. *J. Hydrol.* 65:209–237.
102. Ferro, V., and M. Minacapilli. 1995. Sediment delivery processes at basin scale. *J. Hydrolog. Sci.* 40(6):703–717.

103. Ferro, V. 1997. Further remarks on a distributed approach to sediment delivery. *J. Hydrolog. Sci.* 42(5):633–647.
104. Campbell, B. L., R. J. Loughran, J. L. Elliott, and D. J. Shelly. 1986. Mapping Drainage Basin Sediment Sources Using Caesium-137. IAHS Publ. No. 159, pp. 437–446.
105. Ritchie, J. C., J. A. Spraberry, and J. R. McHenry. 1974. Estimating soil erosion from redistribution of fallout 137-Cs, *Soil Sci. Soc. Am. Proc.* 38:137–139.

4.4 Soil Conservation: Erosion Control

D. De Wrachien and G. Chisci

In most cases a combination of measures is needed to reduce the effects of the processes that cause erosion and to stimulate land use so that the soil is kept permanently productive. However, it must be emphasized that none of these methods has universal application. In selecting from possible measures, the following issues must be considered [1]:

- Any measure must be suitable for the intended land use and cropping systems.
- The objectives must be relate to rainfall and soil. In high-rainfall areas, a common goal is to lead unavoidable runoff safely off the land using drains and ditches. In semiarid regions the objective is that of slowing down the runoff to nonscouring velocities to encourage infiltration or deposition of silt.
- The inputs, especially of labor, must be affordable and the benefits must be sufficient to justify the inputs.

Anyway, the final choice, among possible conservation procedures, should depend on the social and economic conditions of those involved.

There are so many different measures used in erosion control that some form of grouping is needed to describe them. Mechanical methods encompass all techniques that involve earthmoving, such as digging drains, building banks, and leveling sloping land. Anything else, nowadays, is lumped under agrobiological measures. This is appropriate for large mechanized farms where machines are used to do the earthmoving and this is followed up with improved farming methods. But the division does not suit the concept of erosion control through better land husbandry by means of mechanical protection. Moreover, it becomes artificial when it deals with progressive terracing using grass strips or live hedges. Anyway, for want of better terms, the above-mentioned approach is kept is this section. Measures special for wind erosion control are descibed in Section 4.4.3 of this chapter.

4.4.1 Mechanical Methods

Table 4.8 summarizes the main mechanical practices used for soil, water, and crop management, normally in conjunction with agronomic measures. These practices can be subdivided into the following mechanical methods:

- contouring,
- terraces,
- waterways,
- stabilizing structures.

Table 4.8. Mechanical methods for soil, water, and crop management

Main Objective	Function	Measure
Soil management	To modify soil slope	Bench terraces
	To slowly reduce soil slope	Progressive terracing
	To contain erosion with low inputs	Ladder terraces, Trash lines
	To contain erosion with minimal earthmoving on steep slopes	Step terraces, Hillside ditches, Intermittent terraces
Water management	To multiply effective rainfall	Conservation bench terraces
	To catch and hold all runoff	Absorption ridges
	To absorb some runoff with emergency overflow	Contour furrows, Contour bound
	To control unavoidable runoff	Graded channel terraces
	To control reduced runoff	Ridging, Tied ridging
	To reduce the velocity of runoff and promote infiltration	Strip cropping, Grass strips, Permeable barriers
Crop management	To provide level areas on steep slopes, or to ease cultivation according to whether by hand, ox, or machine	Step terraces, Hillside ditches, Orchard terraces, Platforms
	To ease harvesting according to whether the crop is heavy damageable, harvested regularly or seasonally	Footpaths and farm tracks associated with orchard terraces or hillside ditches
	Drainage for crops that suffer from wet feet	Ridges on 2% grade, Up- and downslope beds, Small open drains up to 15%

Source: [1].

The reader is referred to the literature [2–10] for more detailed information on design, construction, and maintenance criteria.

Contouring

This technique involves plowing, tilling, planting, and cultivating across the slope of the land so that elevations along rows are as near constant as is practical. In conjunction with contouring, protected waterways must be used in all drains or drainageways, where gullies tend to start. Contouring, or contour farming, does best on fields that slope uniformly in one or both directions. Its effectiveness varies with the length and steepness of the slope and with storm intensity. This measure is inadequate as the sole conservation procedure for lengths greater than 180 m at 1° steepness [11]. The allowable length declines with increasing steepness to 30 m at 5.5° and 20 m at 8.5°. Moreover, it provides almost complete protection against erosion from storms of low to moderate intensity, but little or no protection against occasional heavy storms that cause extensive

Table 4.9. Slope-length limits for contouring

Land Slope (%)	Maximum Slope Length (m)
1–2	330
3–5	125
6–8	65
9–12	40
13–16	25
17–20	20
20–25	15

Source: [12].

overtopping and breaking of the contour lines. Contouring is most efficient in reducing erosion on gentle slopes. On steeper slopes, erosion becomes progressively more severe. Sometimes more erosion actually occurs in gullies developed in contoured areas than in the rills between crop rows on the noncontoured land. Ridge height is an additional factor affecting the effectiveness of contouring. Data from field studies indicate that contouring is more effective with high ridges, whereas limited heights may greatly reduce its effectiveness, especially under severe storms. Table 4.9 gives the slope-length data for successful contouring in the case of moderate ridge height and storm intensity [12]. The slope length varies with soil properties (length can be greater on more permeable soils), with type of crop grown (longer for more protective crops, such as small grains), and with the area's rainfall characteristics (longer with less intense storms). Experience with no-till and other reduced-tillage systems that leave the soil surface well protected with crop residue has shown that field lengths far in excess of those given in Table 4.9 can be used safely [4, 6, 8, 10].

Wherever possible, the upper and lower field boundaries should be changed to follow the contour. This reduces the number of short rows in the field and makes it more convenient to farm. Natural drainage ways should be prepared to handle the water that will be guided onto them by tillage marks and crop rows. Any water flowing onto the field from higher land should be intercepted by a diversion built along the upper field boundary to catch the foreign water and carry it away from the field. To meet this requirement, a new waterway has to be installed or an existing one widened and reshaped to accommodate the diverted water. On silty and fine sandy soils, erosion may be greatly reduced by storing water on the surface rather than letting it run off. Limited increases in storage capacity can be obtained by forming ridges at regular intervals according to the slope steepness. The ridge-furrow system is usually stable for gentle slopes of up to about 7° and for soils that have a relatively stable structure [13]. Contour ridging is generally ineffective on its own as a soil conservation measure on slopes steeper than 4.5°. Filed data collected from trial plots in Venezuela showed that contour ridges reduced soil losses under cabbage and cauliflower to 9.9 t/ha compared with 15 t/ha without ridges [14].

Contouring has no direct effect in controlling wind erosion unless ridges formed by tillage increase surface roughness. Thus, where the risk for wind erosion is much greater

Table 4.10. Types and functions of terraces

Type	Function
Diversion terraces	Used to intercept overland flow on a hillside and channel it across a slope to a suitable outlet.
Magnum	Formed by taking soil from both sides of an embankment.
Nichols	Formed by taking soil from the upslope sides of an embankment only.
Broad-based	Bank and channel occupy width of 15 m.
Narrow-based	Bank and channel occupy width of 3–4 m.
Retention terrace	Level terraces used where water must be conserved by storage on the hillside.
Bench terraces	Alternating series of shelves and risers used to cultivate steep slopes.

Source: [11].

than that for water erosion, ridges formed by tillage should be at right angles to the prevailing wind direction without regard to the land slope.

Terraces

Terraces are earth embankments designed and constructed to intercept runoff and convey it to a protected outlet. They reduce soil erosion by shortening slope length, slowing the velocity of the running water and improving infiltration. Terraces can be classified into the following main types:

- diversion,
- retention,
- bench,
- hillside ditch.

The characteristics and functions are given in Table 4.10 and Fig. 4.30.

Diversion terraces are designed primarily to intercept runoff and divert it to a suitable outlet. Broad-based diversion terraces (Nichols and Magnum), with the embankment and channel occupying a width of about 15 m, are wide enough to be cultivated if bank slope does not exceed 14°. Narrow-based terraces do not exceed 4 m in width. Diversion terraces are suited mostly for moderate ground slopes (<7°) that allow a sufficient spacing and reasonable expense of construction. Modest slopes (<4.5°) and permeable soils are recommended also for the use of retention terraces, which are aimed at the conservation of water by storing it on the hillsides. They are generally level and the design storm has about a 10-year return period. Bench terraces are suited for cultivation of steep slopes (up to 30°). They consist of an alternation of shelves and risers. The latter are protected from erosion by vegetation cover or by concrete or stone facing. For several thousand years, bench terraces have been widely adopted over the world, particularly in Europe, Australia, and Asia. A special type of bench terrace is the orchard terrace. Orchard terraces are narrow bench terraces built on very steep slopes, from 25° to 30°, and their spaces are determined by the planting distance of the fruit or food trees. Because of steepness, the spaces between should be kept under permanent grass.

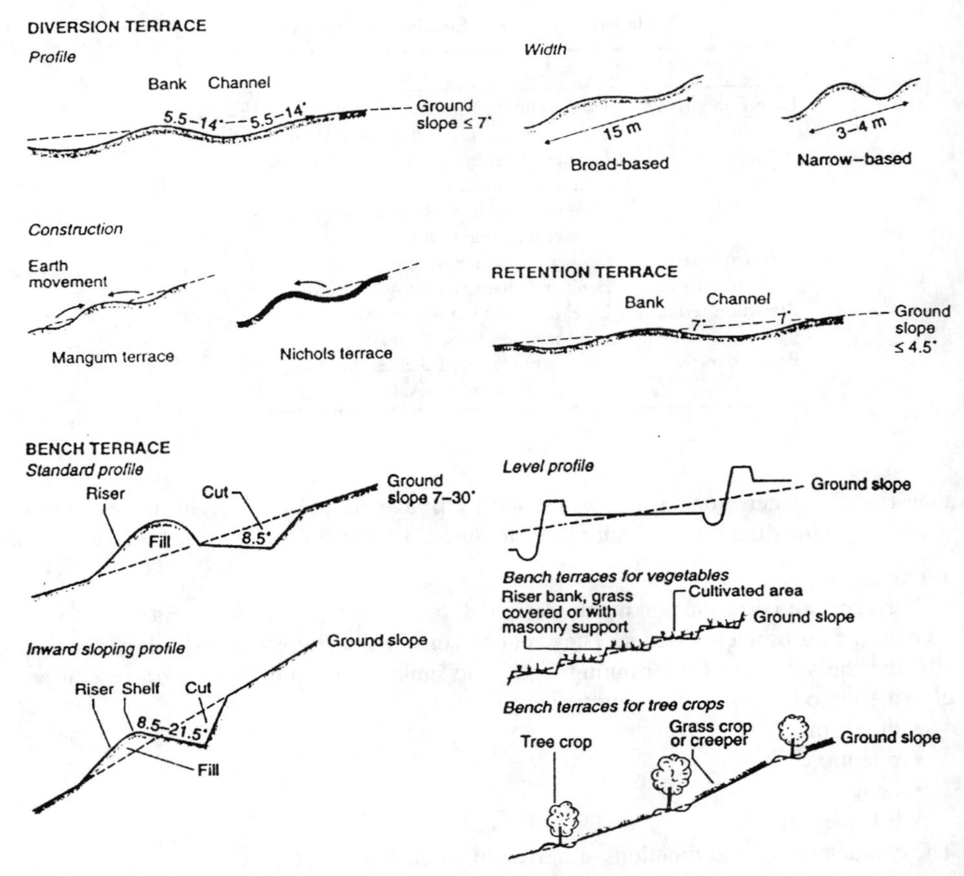

Figure 4.30. Types and main characteristics of terraces. ***Source:*** **[11].**

Another practice used on steep slopes is hillside ditches. The hillside ditch is a discontinuous kind of narrow, reverse-sloped terrace built across the land in order to break long slopes into a number of short slopes so that the runoff will be safely intercepted and drained before causing erosion. The cross section of this kind of narrow bench is more convenient for maintenance than the conventional type of ditch. They also can be used simultaneously as roads. The distance between two ditches is determined by the degree or percentage of the corresponding slope. The cultivable strip between two ditches should be supplemented with agronomic conservation measures.

As technology has advanced, terrace design (spacing, length, location of outlets) has been scientifically adapted to the hydrologic and erosion control needs of the treated areas. Channel cross sections have been modified to become more nearly compatible with modern mechanization. Table 4.11 gives some criteria that can help in decision making about slopes and dimensions for terraces [1]. Many formulas also have been

Table 4.11. Design criteria for terraces

Criterion	Dimension
Maximum length	
Normal	250 m (sandy soils) to 400 m (clay soils)
Absolute	400 m (sandy soils) to 400 m (clay soils)
Maximum grade	
First 100 m	1:1000
Second 100 m	1:500
Third 100 m	1:330
Fourth 100 m	1:250
Constant grade	1:250
Ground slopes	
Diversion terrace	Usable on slopes up to 7°; on steeper slopes the cost of construction is too great and the spacing too close to allow mechanized farming
Retention terraces	Recommended only on slopes up to 4.5°
Bench terraces	Recommended on slopes of 7° to 30°

Source: [1].

developed for determining the so-called vertical interval (VI), that is, the difference in height between two consecutive terraces [11]. Several empirical formulas have been calibrated on a national basis [11, 12]. To know what formula is most suitable in a given situation, one should find out which design criteria are successfully used in an similar situation (if possible in the same area). The most important design parameters are the width of the terrace and the height and the slope of the riser [1, 15]. From a theoretical point of view, terraces can be built on any slope with deep soil but, in practice, on very steep slopes the riser becomes too high and therefore difficult to maintain, and the terrace becomes too narrow. Practical limits are about 20° for terraces built by machine, and 25° to 30° when handmade. The width of the bench depends on three factors:

- the cost,
- what is desirable for ease of cultivation,
- what is practical for construction.

For a given slope, increasing the width of the terrace increases the amount of both earthwork and fill. Narrow terraces are therefore cheaper than wide terraces. For terraces cultivated by hand or walking tractor, bench width of 2 to 5 m is feasible. For animal draft or four-wheel tractor cultivation, a minimum width of 3 or 4 m is desirable. It is not practical to excavate down into the subsoil or bedrock, because of both the increased cost of excavation and the lower fertility of the exposed subsoil on the inner edge of the terrace. This may be avoided by building a terrace and then putting the topsoil removed from the next terrace on it [1].

A variety of equipment is available for terrace construction. Terracing machines include the bulldozer, pan or rotary scraper, motor patrol, and elevating grader. Smaller equipment, such as moldboard and disk plows are suitable for slopes of less than 8%.

The rate of construction is affected chiefly by the following factors:

- equipment,
- soil moisture,
- crops and crop residues,
- degree of regularity of land slope,
- soil tilth,
- gullies and other obstructions,
- terrace length,
- terrace cross section,
- experience and skill of the operator.

Proper maintenance plays an important role in the terrace conservation practices.

The procedures must not be expensive; normal farming operations usually will suffice. The work should be watched more carefully during the first years after construction because tillage tends to increase the base width of the terrace. Therefore, tillage practices such as stubble-mulch operations, disking and harrowing, and planting have to be performed parallel to the plow furrows.

On the whole, the main features required for terrace conservation are [1]:

- A gentle slope of 0.5%–1.5% is most desirable.
- A deep soil is required, both to provide sufficient soil moisture storage and to lessen the effect of cutting down during the construction process. Good permeability is required so that the contained runoff can be absorbed quickly.
- Smooth slopes are an advantage where mechanized farming can be made more convenient by constructing all of the terraces parallel to each other and of equal width.
- Precise leveling of the terrace is important to ensure uniform infiltration.
- If there is a risk that runoff from the catchment area will be greater than can be absorbed and stored on the terrace, there must be outlets at the end of the terrace.
- Typical ratios of catchment area to terrace area are from 1:1 to 2:1.

Waterways

Water that overflows from terraces and contoured fields or requires draining must be conveyed in a non-erosive manner for overall soil conservation. The runoff must be discharged into either natural streams, watercourses, or artificial excavated waterways. Their purpose is to convey runoff to a suitable outlet. Normally, their dimensions must provide sufficient capacity to cope with peak runoff rates with a 10-year return period. There are three types of waterways (Fig. 4.31):

- diversion channels,
- terrace channels,
- artificial watercourses, (grass) waterways.

The characteristics and aims are described in Table 4.12. More detailed information on design criteria can be found in the literature [16–22]. Waterways are surface drainage devices. Therefore, for design criteria, reference should also be made to Chapter 5, Section 5.6, "Drainage."

Artificial watercourses may need special attention, depending on the topographic conditions and the equipment available. The major types of waterways are

Table 4.12. Types and functions of waterways

Type	Function
Diversion ditches	Placed upslope of areas where protection is required to intercept water from top of hillside; built across slope at a slight grade so as to convey the intercepted runoff to a suitable outlet.
Terrace channels	Placed upslope of terrace bank to collect runoff from interterraced area; built across slope at a slight grade so as to convey the runoff to a suitable outlet.
Grass waterways	Used as the outlet for diversions and terrace channels; run downslope, at grade of the sloping surface; empty into river system or other outlet; located in natural depressions on hillside.

Source: [11].

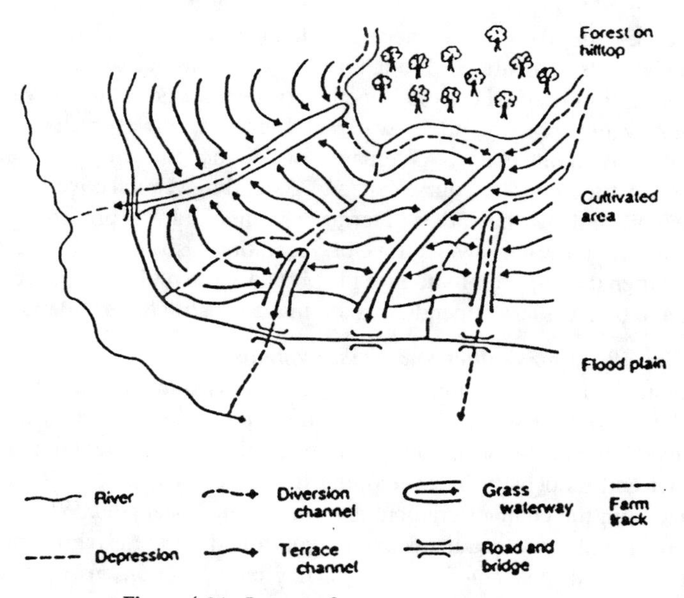

Figure 4.31. Layout of waterways. ***Source:*** **[11].**

- Grass waterway, a parabolic-shaped channel planted with low and rhizome-type grasses. The channel should be shaped as uniform cross section and as consistent in gradient as possible. It is the most inexpensive type of waterway on gentle slopes and its maintenance is easy.
- Grass waterway with drop structures. On moderately steep slopes or in a discontinuous type of channel, small drop structures and check dams can be used in conjunction with grass to take care of the steep sections.

- Ballasted waterway. On moderately steep slopes where large quantities of head-size stones are available, ballasting the parabolic channel with stones keyed in the ground can provide good protection.
- Prefabricated concrete waterway. On very steep slopes and where it rains so frequently that normal construction is hampered, these prefabricated structures, parabolic or V-notch type, can be used readily to protect the center part of a waterway and leave two sides protected by grass. They also can be used in these channels where there are constant small flows due to seepage or groundwater.

Maintenance procedures prevent the failure of waterways due to insufficient capacity, excessive water velocity, or inadequate vegetation cover. The solution to the first two of these problems is largely a matter of design. The condition of the vegetation is affected by both the construction criteria and the subsequent management procedures. With reference to this issue, grass waterways should be mowed and raked several times a season to stimulate new growth and control weeds. Annual application of manure and fertilizer maintains a dense sod, any break of which should be repaired. When land adjacent to the waterway is being plowed, the ends of furrows abutting against the vegetated strip should be staggered to prevent flow concentration down the edges of the water course. Sometimes, underground outlets are necessary to take runoff from low points in terrace or diversion channels and carry it through a pipeline to a place of safe discharge. This procedure normally is used where the slope is too steep for non-erosive water flow.

Good conservation practice at the watershed level is, however, the most effective means of waterway maintenance. As a matter of fact, accumulated sediments smother vegetation and restrict the watercourse capacity. Extending vegetal cover well up the side slopes of the work and into the outlets of terrace channels helps to prevent sediment from being deposited in the waterway. Control of vegetation also reduces the accumulation of sediment. Often the opportunity to provide protective cover by natural revegetation is overlooked, and unnecessary expenditures are made for structures and planting.

Stabilizing Structures for Gullies and Embankments

The mechanics of gully erosion can be reduced to two main processes [24]: downcutting and headcutting. Downcutting of the gully bottom leads to gully deepening and widening. Headcutting extends the channel into ungullied headwater areas and increases the stream net and its density by developing tributaries. Thus, effective gully control must stabilize both the channel gradient and the channel headcuts. Where an effective vegetation cover will grow, gradients can be controlled by the establishment of plants without supplemental mechanical measures. If growing conditions do not permit the direct establishment of an effective vegetation cover, engineering measures will be required at the critical locations where the erosion processes take place. The main objective of these structures is to stabilize the gully gradient. Once the gully gradient is stabilized, vegetation can establish and gully banks can stabilize. Because of sediment accumulation in the gully above the structures, the storage capacity of the channels increase, channel gradients decrease, and thus peak flows decrease.

In practice, the control of gully erosion is difficult and expensive. Therefore, the decision whether control structures are justified depends on other objectives, such as the prevention a storage dam downstream from being silted up by sedimentation.

The structures consist in small dams and weirs built across gullies to trap sediment and thereby reduce channel depth and slope. These works can provide temporary or permanent stability and are normally used in association with agronomic treatment of the surrounding land where grasses, trees, and shrubs are well established.

The spacing of the dams can be obtained from the following formula [24]:

$$\mathrm{DS} = \frac{\mathrm{HE}}{K \tan\theta \cos\theta} \tag{4.27}$$

where DS = dam spacing (m); HE = dam height, measured from the crest of the spillway to the gully floor (m); and K = constant coefficient that depends on local conditions. The equation is based on the assumption that the gradient of the sediment deposits is $(1 - K)\tan\theta$, with θ = slope angle (degree). In Colorado, for example, the estimated values for K are 0.3 for $\tan\theta \leq 0.2$ and 0.5 for $\tan\theta > 0.2$.

The height not only influences structural spacing but also volume of sediment deposits. The structures are provided with spillways to deal with overtopping during storms. The spillway or notch must be big enough to pass the whole of the flood; otherwise, the water will pass over the whole width of the structure and erode the vulnerable banks on either side of the structure. At the foot of the structure, where the flood water falling over the structure strikes the ground with a destructive force, an adequate protection must be provided to prevent scour and undercutting of the structure.

The danger of scouring and tunneling around the check dam can be minimized by a key in such way that lateral seepage around the end of the dam is prevented. Therefore the route of seepage must be considerably lengthened. The dam stability will be greatly increased by keying the dam into the sides and floor of the gully by digging a trench, usually 0.6 m deep and 6 m wide. Where excessive instability is demonstrated by large amounts of loose materials on the lower part of the channel side slopes or by large cracks and fissures in the bank walls, the depth of the trench should be increased to 1.2 or 1.8 m [24]. The trench is filled with loose rock in such way that no large voids will remain in the key. If available, use a mixture of pebbles with 80% smaller than 14 cm. When log dams are used, the bottom layer should be sunk below ground surface or long posts driven deep into firm soil.

Based on the cost of construction, the type of dams best suited for different depths of the gully are listed in Table 4.13. The dividing line between temporary and permanent structures is quite arbitrary. Many works could have a variable life depending on how they are constructed and maintained, and what pressure of use they had to withstand. As a general rule, low cost and simplicity are often principal objectives. The materials and

Table 4.13. Types of dams for gully stabilization

Dam Type	Recommended Height (m)	Gully Depth (m)
Loose rock	≤ 0.45	≤ 1.2
Single fence	$0.45 \div 0.75$	$1.2 \div 1.5$
Double fence	$0.75 \div 1.70$	$1.5 \div 2.1$

Source: Modified from [11].

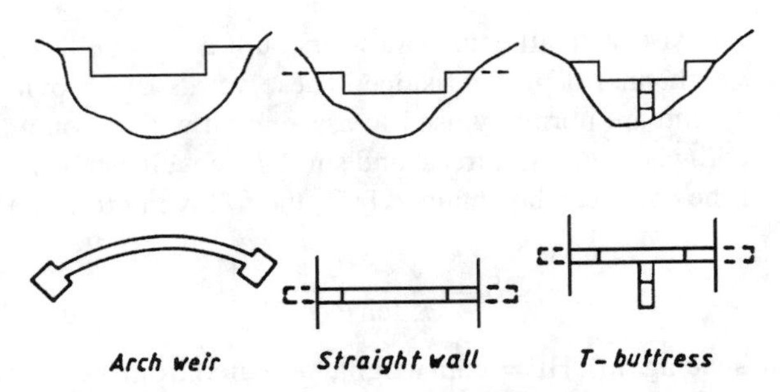

Figure 4.32. Types of small stabilizing structures for gully erosion control. ***Source:*** **[1].**

the design must be chosen ad hoc to suit the individual conditions of each site. Some simple designs are shown in Fig. 4.32.

The shape that gives the best strength/weight ratio is the arch weir. However, in the situation of a rock that runs across the bed of the gully, a straight gravity wall also is indicated. More detailed information on structure designs and design criteria can be found in the literature [1, 23–31].

Stabilization structures are also used for erosion control on steep slopes. A special case is represented by soil bioengineering measures where stabilization is achieved primarily by imbedding and arranging in the ground plants and plant parts [32]. For example, driving live willow stakes into the soil along the contour to a depth of 0.2 m may guarantee stability for short time periods, to allow growing of a denser vegetation cover. Stakes should be placed approximately 1 m apart and form rows with a distance from 10 to 20 m, according to the slope steepness.

Stabilization structures as a component of biotechnical slope protection may be necessary in areas of potential mass movement. A structure placed at the foot of a slope helps to stabilize the slope against mass movement and protects the toe and face against scour and erosion. Selection of a suitable retaining structure entails a variety of choices. Several basic types are available, each with its particular advantages, requirements, and limitations. Selection of a suitable retaining wall will depend upon such considerations as site constraints, availability of materials, appearance of the wall, ease of construction, opportunities for incorporation of vegetation into structure, and cost [28]. Figure 4.33 shows the two basic types: the toe wall and the toe-bench structure. More detailed information on structure designs and design criteria can be found in the literature [1, 28].

4.4.2 Agrobiological Measures

Agroecosystem management is one of the most important factors controlling soil erosion at field, farm, and watershed scales.

In natural vegetation systems, such as forest, savannah, and prairie, soil erosion depends on the interaction between climatic parameters (rainfall, wind, humidity, solar radiation, and temperature) and vegetation growth, provided that other conditions such as geomorphology and soil type are constant.

(a)

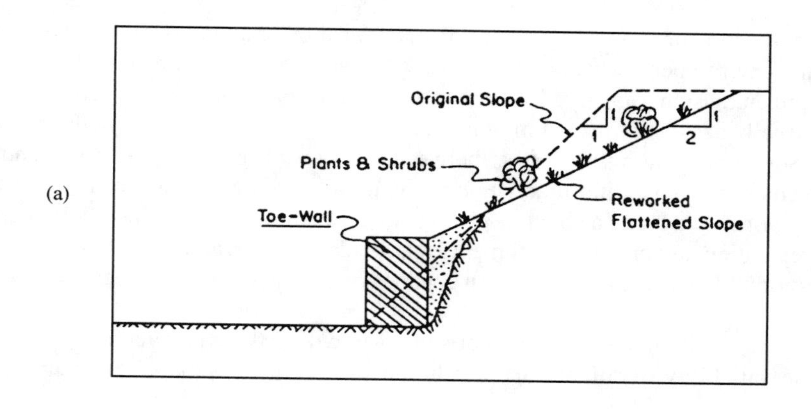

(b)

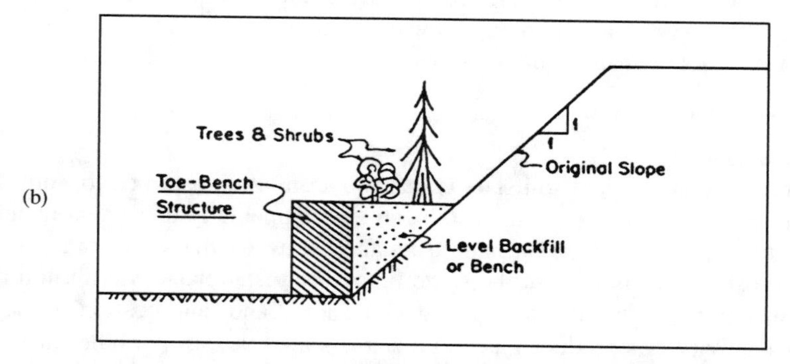

Figure 4.33. (a) Low toe wall, which permits slope flattening and establishment of vegetation on slope above. Encroachment on land use or right of way at foot of slope is minimized.
(b) Toe-bench structure, which buttresses base of slope and creates level bench on which to establish vegetative screen. Bench also catches debris coming off the slope above. *Source:* [28].

In agricultural systems, on the other hand, vegetation cover and soil conditions can assume different (management-derived) characteristics with the seasons of the year, depending on the

- crop rotation,
- choice of annual and/or perennial crops,
- crop stage from sowing to harvest,
- cropping system (monoculture, intercropping, or mixed cropping),
- planned spatial geometry of the plants on the soil,
- soil type and crop management technology.

Seeking the maximum economic productivity of any crop, it is fundamental to avoid competition between plants of the same cultivated species (intraspecific competition), and between cultivated plants and weeds (interspecific competition).

Plant architecture represents a specific manifestation of an interaction between genotype and environment, whereas the geometry of plant distribution in a field and the soil management system employed are a consequence of crop cultivation technology. The latter usually has the ultimate aim of maximizing economic productivity with respect to costs. Sometimes there is a conflict between soil protection and optimal soil and crop management for maximum production. Sowing of annual crops, for example, is traditionally done on a bare seedbed of plowed soil, which means that the whole period of soil preparation and crop establishment is particularly vulnerable to soil erosion.

In agricultural systems, soil erosion depends on the interaction between the following factors:

- the variability in the protection of soil cover during the crop cycle,
- the variability of soil surface conditions due to tillage practices during the crop cycle,
- The residual effects of soil and crop management from preceding crops.

Specific soil conservation technology for water erosion control, therefore, concerns both soil and vegetation management measures.

Soil Management

Reconsolidation

A naturally consolidated soil state is that found in an undisturbed topsoil of forest, savannah, prairie, or steppe. In such soils, an environmentally dependent natural plant community is established and a natural balance between the soil-forming processes and soil erosion is achieved. At the same time, a semistationary equilibrium between organic-matter production by plants, and its turnover and humification in the soil by biological activity, has produced a certain stable soil structure. All other factors being equal, natural vegetation cover and the structural conditions of the topsoil maintain soil erosion at a certain stable level, which allows soil development and plant dynamics that are characteristic for the site in question.

Reconsolidation, on the other hand, is the reduction of soil disturbance caused by cultivation practices. Generally speaking, the reconsolidation process is characterized by compaction and rearrangement of the soil particles, after a period during which the soil is loosened by plowing or other tillage practices, into aggregates. Reconsolidation processes are characterized by the type of vegetation involved and the length of time from initiation to new plowing (sod turning).

A grass sod protects the soil and reduces the effects of raindrop impact and scouring by overland flow. Furthermore, the improvement of soil structure and infiltration will decrease runoff generation with the overall effect of a very consistent abatement of soil erosion.

In natural reconsolidation, the vegetation cover depends on the types of weed seeds that the soil contains. This cover affords variable protection for soil erosion, related to the prevailing environmental conditions and to the proceeding cultivation history. In general, soil protection will be scarce in the first year of reconsolidation, especially on marginally arable land. In fact, natural vegetation recovery on abandoned fields often is represented by weeds of modest growth that offer sparse cover to the soil, especially

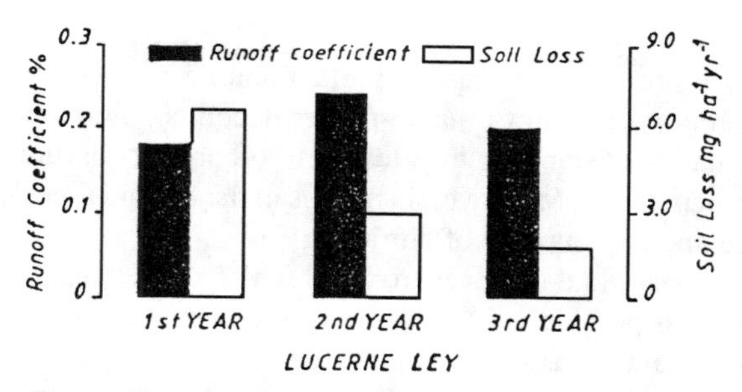

Figure 4.34. Runoff and soil loss responses as related to lucerne ley in the first, second, and third years.
Source: **[33].**

in extreme climatic conditions. Where climatic and soil characteristics allow the quick development of a good natural vegetation cover, soil protection against erosion will be much better and will reach a stationary equilibrium after a few years.

The reconsolidation process may be aided greatly by sowing a mixture of good grasses that are well adapted to the local environment. The use of an artificial mixture of well-adapted perennial forage grasses and legumes in land reconsolidation is defined as "steered reconsolidation."

Reconsolidation may represent a permanent change in the state of the soil or simply a period of repose for the soil between successive phases of arable cultivation, as in agricultural rotations with perennial forage crops. Although perennial forage crops, such as lucerne and clover leys, can provide considerable soil protection in the years after the first one (Fig. 4.34) [33], they are somewhat less protective than close permanent grassland.

Reconsolidation by forage crops used in rotational agriculture is sometimes called "rotational reconsolidation." This term covers systems ranging from the shifting cultivation of the tropics to the fallow agriculture of the semiarid zones and to the dynamic rotations of arable crops and perennial forage species in temperate areas.

In agricultural systems based on trees, a "grass-sodding" type of steered reconsolidation is being adopted increasingly as a very promising solution to erosion control or for ameliorating soil carrying capacity for machinery traffic. Where competition problems for water and nutrient uptake arise between cultivated trees and grasses, grass-sod management that follows the contours has proved to be a valuable measure for erosion control. Grass-sod management in tree cultivation can be done in several ways:

- by grazing the phytomass using different species of animals,
- by cutting the phytomass for hay production or for covering the soil with mulch,
- by using herbicides for killing and/or drying the aboveground vegetation.

In grass-sod management, it is particularly important to avoid the formation of up- and downslope tractor tracks because such tracks may become preferential pathways for concentrated runoff and erosion.

Organic Matter

Organic matter represents the most important source of energy for soil biological activity. The different products of organic-matter decomposition are very important in promoting the formation of stable aggregates from soil primary particles, by flocculation and cementation phenomena. Moreover, increases in the colloidal organic fraction assist in the slow release and plant uptake of nutrients.

Organic-matter release and turnover from perennial herbaceous plants in the topsoil during reconsolidation provide a basic source of soil humus, thereby playing a very important role for the amelioration of soil structure.

Organic matter can be added to the soil in different forms as green manure, straw, stubble, industrial residues, sludges, or animal manure. This may have already undergone some form of fermentation, as in the case of fermented urban garbage and farmyard manure (FYM).

The addition of raw organic material that contributes to the soil in building up the humus content is expressed in the isohumic factor, as shown in Table 4.14 for some raw organic materials [34].

When green manures (often leguminous crops) are added to soil, they may be subject to a high degree of fermentation and rapid mineralization. Low amounts of humic substances (low isohumic factor) probably will be produced in these cases. The possible increase in soil structural stability will be of short duration and the main effect of the addition of such material is to increase the medium-term supply of nutrients in the soil.

Straw, on the other hand, decomposes more slowly because it is richer in cellulose and lignin. The rate of decomposition depends, to a large extent, on the availability of soil nitrogen sources for the nutrition of cellulolytic microorganisms. The C/N ratio dynamics of organic matter in the soil are mediated primarily by microorganisms. The higher the C/N ratio, the more stable the humic acids binding the soil particles in structural crumbs. Low C/N ratios may favor microorganisms producing humic acids that have little or no effect on soil structural stability. The isohumic factor of this material is high and the buildup of humic acids increases soil structural stability to a higher degree than that achieved with green manure.

Table 4.14. Examples of isohumic factors for several organic materials

Organic Material	Isohumic Factor
Plant foliage	0.20
Green manure	0.25
Cereal straw	0.30
Roots of crops	0.35
FYM	0.50
Deciduous tree litter	0.60
Coniferous tree litter	0.65
Peat moss	0.85

Source: [34].

The influence of the organic matter on soil structure is enhanced by the presence of base minerals in the soil. Such minerals, in fact, favor the flocculation of colloidal clay and humus to form compounds that comprise soil aggregates.

Tillage Practices

Tillage is a fundamental soil management technique in agricultural systems to provide a suitable seedbed for plant germination and growth, helping to control weeds and assisting the infiltration of fertilizers and pesticides into the soil.

Tillage loosens compacted soil, regenerates porosity, and increases permeability, thereby generally improving soil physical conditions for plant growth, especially where intrinsic soil physical characteristics are particularly adverse. For instance, the formation of a temporary mechanical structure in a compacted clay soil is a prerequisite condition for the cultivation of annual crops. There are, in fact, a number of different techniques for tilling the soil. All of them, however, represent some degree of soil disturbance which, although necessary for agricultural exploitation of the land, often produce an increase in potential erodibility of the soil.

The most common system of preparing the soil for cultivation is by plowing the soil to loosen it and to turn over weeds, previous crop residuals, organic matter, and remaining pesticides and chemical fertilizers in the plow layer. Subsequent seedbed preparation consists of arrowing the soil with disk or dent cultivators, to eliminate any weed seedlings that may have sprouted in the meantime and to produce a mechanical structure suitable for sowing the crop.

The conventional system of tilling the soil, recognized in principle since the origins of agriculture, has been subjected to many improvements over time in terms of the implements used, the mode of execution, and the kind of power employed to operate the implements. The average plow-layer depth ranges from 100 mm to 1,000 mm. Deep plow layers are common in clay soils where plowing also has the function of improving infiltration.

Different implements may be used to plow the soil. Examples include the moldboard plow, the disk, the chisel, and the rotovator. Sometimes, the work is done using a combination of implements as, for instance, in double-layer plowing. This allows subsoiling and an inversion of the surface layer (to imbed fertilizers, weeds, and crop residues) to be achieved simultaneously.

Plowing generally produces a more or less rough cloddy surface, depending on soil characteristics and moisture content. A clod height of 400–600 mm may be produced in clay soils plowed to a depth of 400–500 mm in dry tilth. In loamy soils, on the other hand, a clod height of 120–160 mm is observed more commonly.

Secondary cultivation consistently reduces the roughness of the soil surface in the same way as raindrop impact and wind action. Seedbed formation through secondary cultivation, however, generates a finer mechanical aggregate distribution in sandy-loamy soils than in clay soils.

Rougher freshly plowed soils are less susceptible to erosion. Cogo [35], in fact, has demonstrated that soil loss by water erosion will decrease with increasing roughness.

In conclusion, it is generally accepted that, all other factors being equal, a traditional seedbed preparation is the most susceptible condition for soil erosion. For this reason, tilled bare fallow is taken as the standard condition with which other cover and management situations are compared [36].

In some circumstances, tillage practices such as plowing and seedbed preparation may be less susceptible to soil erosion when practiced on the contour; see also "Contouring" in Section 4.4.1 of this chapter.

Conservation Tillage

Many studies and experiments have been carried out recently to evaluate the effects of innovative tillage systems in reducing soil erosion while maintaining adequate soil conditions for optimal plant growth and crop yield.

Different kinds of conservation tillage practices have been devised, each one more or less soil specific but also dependent on how well weeds, pests, and diseases are controlled in the cultivation system under consideration (Table 4.15).

In all cases, the basic principle of conservation tillage is to reduce tillage operations, thereby avoiding soil disturbance as much as possible and maintaining the most efficient cover of vegetation and residues throughout the crop cycle in order to reduce soil erosion but provide appropriate crop germination.

Table 4.15. Tillage practices used for soil conservation

Practice	Description
Conventional	Standard practice of plowing with disc or moldboard plow, one or more disc harrowings, a spike-tooth harrowing, and surface planting.
No tillage	Soil undisturbed prior to planting, which takes place in a narrow, 2.5- to 7.5-cm-wide seedbed; crop residue covers of 50%–100% retained on surface; weed control by herbicides.
Strip tillage	Soil undisturbed prior to planting, which is done in narrow strips using rotary tiller or in-row chisel, plow-plant, wheeltrack planting, or listing; intervening areas of soil untilled; weed control by herbicides and cultivation.
Mulch tillage	Soil surface disturbed by tillage prior to planting using chisels, field cultivators, discs, or sweeps; at least 30% residue cover left on surface as a protective mulch; weed control by herbicides and cultivation.
Reduced or minimum tillage	Any other tillage practice that retains at least 30% residue cover.

Source: [11].

Generally, the better-drained and well-structured coarse- and medium-textured soils, with average organic-matter content respond better to reduced tillage. On the other hand, reduced tillage operations are unsuccessful on poorly drained soils unless the crop is planted on a ridge.

All conservation tillage practices are characterized by the amount of crop residues left on the surface of the soil at the time of greatest erosion risk. In any case, it is difficult to isolate the role of tillage in controlling erosion from that of the residues left on the surface or incorporated into the topsoil as mulch. The tillage system also may be integrated with special cover crops, represented by leguminous or grass species, which mainly are used to provide a cover to the soil and/or to protect the seedling of the main crop against adverse climatic conditions during the first stages of growth.

Crop and Vegetation Management

Agronomic measures for soil conservation are based on the protective effect of plant cover as one of the most effective tools for controlling soil erosion.

However, considering the different densities and characteristics of plants in both natural and agricultural ecosystems, differences in their ability to control erosion can be considerable. Annual field crops, for instance, are the least effective in protecting the soil and cause the most serious erosion problems. This is due to the high percentage of bare ground exposed under these systems, especially in the early stages of crop growth and during the preparation of the seedbed.

Crop Rotations

Annual field crops must be combined with protection-effective crops, such as multiannual or perennial forage crops. The frequency with which row crops are grown depends upon the severity of erosion. A high rate of soil loss under a row crop is acceptable when it is counteracted by low rates under the other crops so that, averaged over the total rotation period, the annual erosion rate remains low.

Row crops represent a planting system that currently is used because it facilitates mechanized farming operations. Row crops are planted in wide or narrow rows, depending on plant architecture and the economic characteristics of production. Soil erosion problems are associated mainly with wide-row crops such as maize, sugar beets potatoes, sunflower, sorghum, and soybeans because of the large amount of bare soil left per unit area and because they are tilled and subjected to clean weeding during the crop production cycle. Crops such as wheat, oats, and barley are more protection effective in relation to soil erosion.

All crop types are particularly subject to erosion risk during the period from sowing to complete canopy development. The extent of the risk depends on the seasonal distribution of the erosive rainfall events in relation to the absence of protective cover during the first stages of crop growth. For instance, in Mediterranean areas, erosive rainfall events are concentrated during the fall, which corresponds to the period when the seedbed usually is prepared for sowing winter grains. Another case is that of the corn belt in the United States, where the erosive summer rainfall events also coincide with bare seedbeds or fields characterized by the first stages of maize or soybean growth.

Suitable crops for use in rotation are legumes and grasses. These provide good ground cover, help to maintain or even improve the organic status of the soil, thereby contributing to soil fertility, and enable the development of a more stable aggregate structure. In temperate climates, this process is slow but the effects last for several years of cropping. Hudson [1] found that, under tropical conditions, this process is fast; however, the effects are often sufficiently long-lasting to reduce erosion and increase yield during the first year of row-crop cultivation, and they rarely extend into the second year. Therefore, two continuous years of planting with a row crop should be avoided.

Cover Crops

Cover crops are grown as a conservation measure as annual crops or as perennial stands of natural species or sown perennial grasses, legumes, and/or mixtures. They are grown either during the off-season period or as ground protection under trees. Leguminous crops, grown as winter catch crops, often are plowed into the soil as green manure. In other situations the crop is harvested as forage or used for seed production.

Different climates, soils, and agricultural situations demand different species for annual cover crops. Typical winter cover crops, where winter vegetation is possible, include rye, oats, Italian ryegrass, hairy vetch, crimson clover, sweet clover, lupins, and winter peas. In milder climates, common vetch, field beans, Egyptian clover, and alsike clover also are commonly used. In warmer climates, Crotalaria, Lespedeza, and Dolichos can be utilized conveniently. Finally, in tropical climates, *Pueraria phaseoloides*, *Calopogonium mucunoides*, and *Centrosema pubescens* are better adapted.

Perennial cover crops under cultivated trees may be a very good measure to consistently reduce erosion. In tropical areas, under oil palms, a reduction of erosion from $20\ t \cdot ha^{-1} \cdot year^{-1}$ on bare soil to $0.05\ t \cdot ha^{-1} \cdot year^{-1}$ with a cover crop has been reported [37]. In the Apennine area of Italy, soil loss from a vineyard was reduced from $95\ t \cdot ha^{-1} \cdot year^{-1}$ to $0.05\ t \cdot ha^{-1} \cdot year^{-1}$ by a perennial cover crop [38].

One major disadvantage of cover crops, especially perennial grass sod in orchards, is that, although they can increase the carrying capacity of the soil for machinery traffic, they may compete for available moisture and nutrients with the main crop.

Strip Cropping

In some situations, strip-crop cultivation, which follows the contours (Fig. 4.35), may increase storage and reduce competition for available soil water and nutrients and/or reduce soil erosion because the grass-sod strips act as traps for runoff and sediments [11, 39]. Sediment deposition behind the grass-sod strip produces a gradual buildup of soil over time and leads to the formation of bench-type terraces, reducing the overall gradient of the slope.

Strip cropping is especially suited for well-drained soils. In clay soils, the reduction in runoff velocity, combined with a low infiltration rate, may produce ponding and waterlogging.

Grass-sod contour strips are not required on slopes below 5%. On slopes of about 9%, the grass strip retards runoff by increasing the infiltration rate. The control of erosion by deposition of sediment on the grass strips may be effective on slopes between 9% and 15%. Grass stripping may be insufficient as a single measure of erosion control on

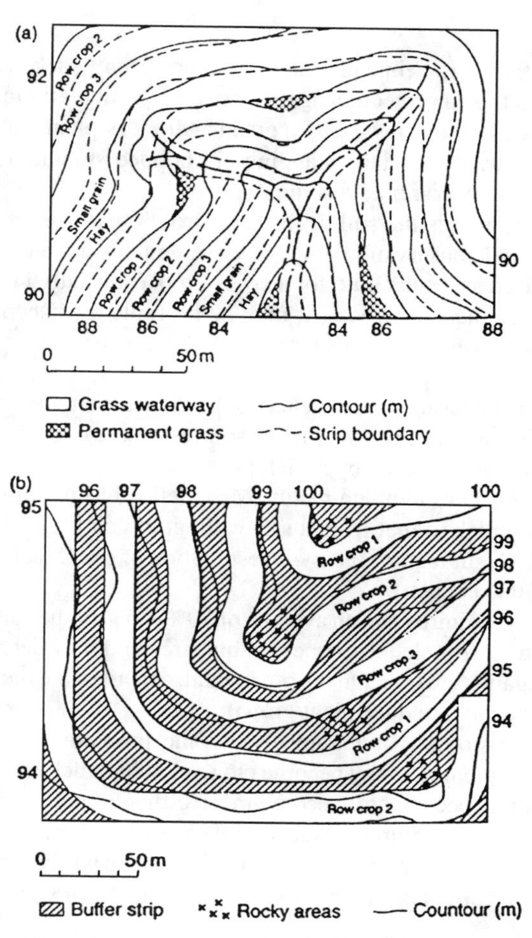

Figure 4.35. Contour strip-cropping design for (a) a 5-year crop rotation and (b) the use of buffer strips. The contour lines are in meters above an arbitrary datum. Note the use of a grass waterway to evacuate excess runoff from the strip-cropped area and the inclusion of rocky areas that cannot be farmed within the buffer strips. *Source:* [39].

slopes above about 15% [40]. However, these limits for slopes largely vary with soil characteristics: texture, structure, and stability of aggregates.

Plants utilized for buffer-strip formation must be well adapted to the specific location in which they are used. Normally, they are perennial grasses and/or legumes that have deep-rooted systems. The species must be perennial, quick to establish, and able to survive flooding and/or drought periods [11].

Mulching

Mulching is another agrotechnical practice that has been widely subjected to experimentation in relation to soil protection against erosion, particularly in the United States. It consists of covering the soil surface with green and/or dry plant material such as grass or crop residues (straw, maize stalk, standing or chopped stubble, residual leaves and stems) after harvesting by combine machinery.

The mulch cover protects the soil against raindrop impact and reduces the velocity of runoff. From the soil conservation point of view, mulch simulates the effects of plant cover and may be used as a substitute for a cover crop, especially in dry areas.

When the slope gradient increases, it may be a problem to keep the mulch spread uniformly on the soil surface, given its tendency to be transported by wind and overland flow.

There is consistent experimental evidence supporting the very important role of mulch in decreasing soil erosion. In fact, the rate of soil loss decreases exponentially with the percentage of soil surface covered by mulch [41–44].

This can be expressed as a mulch factor (MF), defined as the ratio of soil loss from a soil covered with mulch to that from a soil without mulch [42]. The MF is similar to the residue cover (RC) subfactor used in the estimation of the C factor in the USLE (see also Section 4.3 of this chapter).

Optimal protection requires that an average of 70%–75% of the soil surface is covered by mulch. A reduction of mulch cover below this threshold may decrease the protective value of the cover against erosion. On the other hand, considerable increases in the mulch cover above this value may reduce plant growth.

The control of erosion by mulching gives some problems in relation to farming operations. Tillage implements may become clogged with mulch residue. Weed and pest control, which require special care, become more difficult, and sowing and/or planting under a mulch residue is sometimes unsuccessful and can be done only using specifically designed machinery for sod-seeding under mulch. However, where problems of mulching in relation to crop performance can be overcome, it remains a very useful technique in reducing soil erosion.

Small rock fragments on the soil surface may offer very good protection against erosion. Special attention should be paid to the fact that, although organic mulching decomposes reasonably quickly on the soil surface or when embedded in the soil, rock fragments become a permanent soil feature. When the amount of rock material is above a certain threshold (about 20%), the volume of soil available to roots and the nutrients available to plants are reduced.

Although mulching with small rock fragments is of limited value in arable land, making tillage and seedbed preparation more difficult, it is a traditional soil cover in long-duration tree crops in semiarid areas. The rock cover on the soil surface does not interfere significantly with root development in the soil beneath the rock layer. Instead, it has several favorable effects, not only protecting the soil against erosion, but also enhancing water collection from dew and water concentration in the tree root zone, besides keeping weeds under control.

Rotational Grazing and Rangelands

Considering briefly other kinds of land exploitation (for instance pastures and/or forests), the management system of the vegetation cover is also very important in relation to erosion risks.

Rotational grazing (e.g., moving livestock from one plot to another in turn) is, generally speaking, the best system because it allows the grass to recover and thereby maintains a continuous protective cover against erosion.

Grazing must be carefully managed, by adjusting the stocking density to the available biomass of the pasture. In fact, whereas overgrazing leads to a deterioration of the vegetation cover and topsoil structure and to an increase in erosion risk, undergrazing results in the loss of nutritious grasses and legumes in favor of thorny weeds and shrubs of low palatability.

Pasture burning is one of the most commonly employed methods for the removal of undesirable species and for favoring regrowth of palatable species. Burning modifies the density, stature, and composition of bush species within a plant community but kills only a few species outright. Most brush species sprout vigorously again after fire to the detriment of grasses. Grasses, however, fill the gaps between the brushes and reach their maximum development within a year after burning. In any case, the increase of bare ground due to burning enhances the erosion process in the period before the vegetation cover is reestablished. Moreover, it has been demonstrated that, in the long run, continuous cycles of uncontrolled burning and natural vegetation recovery cause progressive deterioration of the composition of the vegetation and topsoil.

Controlled burning on a rotational basis may be more profitably used to remove undesirable species, especially in temperate areas. In the Mediterranean area, on the other hand, uncontrolled pasture burning is always dangerous, not only in relation to pasture deterioration and soil erosion, but also in relation to its potential to burn forests and settlements.

The use of mechanical brush clearing for pasture renewal is much safer than burning in the semiarid and Mediterranean areas. Such a technique has, in fact, given interesting results in the renewal of pastures in Sardinia (Italy). Leaving the chopped plants on the surface of the pasture as a mulch after mechanical brush clearing is also a very effective system of soil protection against erosion.

Forests

In forest management for the exploitation of timber resources, the principle of rotation also can be conveniently applied. The commercial exploitation by patch cutting on a rotational basis is safer than the complete logging, or clear-cutting, of a forest area. The elimination of forest undergrowth is especially dangerous in relation to the acceleration of soil erosion.

Erosion rates are generally higher in the period immediately after logging but decrease consistently in subsequent years with the regrowth of natural vegetation.

Generally, reforestation is a good protection measure against erosion but it takes an average of 7–12 years or more to become effective. However, reforestation plays a detrimental role in soil protection related to the extent of soil disturbance that sometimes occurs during the planting of new trees. Planting on undisturbed land is the most protection-effective practice because of the quick regrowth of natural grass and bush vegetation. However, the planted tree growth may be reduced in such conditions in comparison to the growth of trees planted on tilled soil, because of the competition with natural grasses and bushes.

Forest fires represent an additional hazard for increased soil erosion, not only in terms of the elimination of the vegetation cover, but also in terms of the increase in the hydrophobicity of the soil that occurs (depending on burning temperature). The latter effect produces higher runoff generation and increases soil erodibility [45].

Agroforestry is a system of land management based on the alternation of arable crops with reforested areas and/or with pasture and rangeland. Agroforestry may assume different geometries but generally has a consistent effect on protection of soil from erosion. Rows, bands, or plots of trees producing wood for fire or quality wood for construction alternate with arable fields for food production and/or with pasture or rangeland.

The system is especially important in tropical countries but it is also part of set-aside policies in temperate areas for the production of fruits and valuable timber.

Alternation of reforested areas and pastureland also is suggested for optimal animal husbandry on rangeland in semiarid areas, which allows rotational grazing of the pastureland during the winter season and the grazing of the woodland during the summer. The implementation of forage shrubs (e.g., *Atriplex halymus*) in inland Sicily (Italy) was a measure that resulted on both a higher production of forage available for the grazing of sheep and a very efficient measure for soil protection against erosion.

4.4.3 Measures for Wind Erosion Control

Wind erosion occurs whenever conditions are favorable for detachment and transportation of soil material by wind. Five factors influence the intensity of soil erosion:

1. soil erodibility,
2. surface roughness,
3. climatic conditions (wind velocity and humidity),
4. length of exposed surface,
5. vegetative cover.

Little can be done to change the climate of an area but it is usually possible to alter one or more of the other factors to reduce erosion. Regardless of the type of procedure, the aims or principles for success in control of soil drifting are the same, in all areas where wind erosion occurs. These are

- to reduce wind velocity near the ground level below the threshold velocity that will initiate soil movement,
- to remove the abrasive material from the wind stream,
- to reduce the erodibility of the soil.

Any practice that accomplishes one or more of these objectives, therefore, will reduce the severity of wind erosion. In this section, the most common measures are described. More detailed information on wind erosion mechanics and methods for wind erosion control can be found in the literature [1, 11, 39, 46–58].

Cultivated Crops

Vegetation is generally the most effective means of wind erosion control. Cultivated crops reduce wind velocity and hold soil against the tractive force of the wind. Woody plants, such as shrubs and trees, can be planted to reduce wind velocities over large areas. Whenever possible, cultural practices should be used before blowing starts, because wind erosion is easier to prevent than to arrest.

In general, close-growing crops are more efficient for erosion control than are intertilled crops. The crop effectiveness depends on the following factors:

- kind of crop,
- stage of growth,
- density of cover,
- row direction,
- width of rows,
- climatic conditions.

Pasture or meadow tend to accumulate soil from neighboring cultivated fields that is deposited by sedimentation. Field studies have shown that close-growing crops in the Texas Panhandle gained soil, but intertilled crops lost soil by wind erosion [39].

The best practice is to seed the crop normal to prevailing winds. A good crop rotation that maintains soil structure and conserves moisture is highly advisable. Crops adapted to soil and climate conditions and providing protection against blowing are also recommended. Stubble mulch farming and cover crops between intertilled crops in humid regions help control blowing until the plants become established. In dry regions, crops with low moisture requirements planted on summer fallow land may reduce the intensity of wind erosion. Vegetation should have the ability to grow on open sandy soils, the firmness against the prevailing winds and long life. It should also

- supply a dense cover during critical seasons,
- provide an obstruction to the wind that is as uniform as possible,
- reduce the surface wind velocity,
- form an abundance of crop residue.

Windbreaks and Shelterbelts

Windbreaks can be defined as any type of barrier for protection from wind. Commonly, they are associated with mechanical or vegetative barriers for buildings, gardens, orchards, and feedlots. Living windbreaks are known as shelterbelts. These works consist of shrubs and trees that form barriers longer than windbreaks. In addition to reducing wind speed, shelterbelts result in lower evapotranspiration, higher soil temperatures in winter and lower in summer, and higher soil moisture; in many cases, these effects can lead to increases in crop yield. A shelterbelt is designed so that it rises abruptly on the windward side and provides both a barrier and a filter to wind movement. Such structures can reduce wind speed to less than half of that in the open, as show in Fig. 4.36 [39].

A complete belt can vary from a single line of trees, to two or three tree rows and up to three shrub rows, one of which normally is placed on the windward side. Belt widths may vary from about 9 m for a two-row tree belt, to about 3 m for a single-row hedge belt. Shelterbelts should be moderately dense from ground level to tree tops if they are

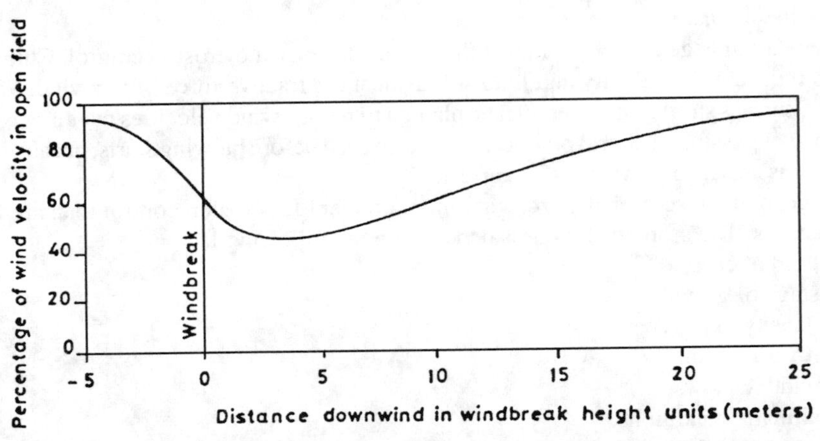

Figure 4.36. Wind velocity distribution (40 cm above ground level) in the vicinity of a windbreak as a percentage of what it would be without the structure. ***Source:*** **[39].**

to be effective in filtering the wind and lifting it from the surface. The correct density is equivalent to a porosity of 40% to 50%. Less-dense barriers do not reduce wind velocity sufficiently. Field investigations have shown that, with the increase of the density, there is a greater reduction in wind speed initially. However, because the velocity increases more rapidly with distance downwind than in the case for more porous barriers, these works are effective only for very short distances [11]. Moreover, taking into account that the wind velocity at the ends of the belts is as much as 20% greater than velocities in the open, long shelterbelts are more effective than short ones [31]. Belt lengths should be a minimum of 12 times the belt height (H), which depends on the plant species selected. To allow for deviations in wind directions, a longer length is desirable and a length of 24 times the height is generally recommended [11]. In establishing the direction of shelterbelts, records of wind direction and velocity, particularly during vulnerable seasons, should be considered, and the barriers should be oriented at right angles to the prevailing direction of winds. Where erosive winds come from several directions, grid or herringbone layouts may be necessary. The requirement is to provide maximum protection, averaged over all wind directions and all wind velocities, above the threshold level.

Wind-tunnel tests have shown that the effectiveness of these works depends on porous break. Hagen [59] determined the relative wind speed and erosion for 20% and 40% porous windbreaks. According to Hagen's experiments, the velocity and erosion were reduced from $6H$ windward to $24H$ leeward with the 20% porous windbreak and from $6H$ windward to $32H$ leeward with the 40% porous break. Erosion in the protected area was less than 50% of that in the open from $2H$ windward to $16H$ leeward, and from $2H$ windward to $17H$ leeward, respectively.

The plant species selected should be rapid growing and resistant to wind, light, and frost. The branches should be pliable and the roots should provide a firm anchorage to the soil. In any case, preference should be given to local rather than imported species. These works need to be protected in their early years against livestock and spray drift.

At intervals of every three or four years, mechanical cutting is required to maintain the sharp rise from the ground on the windward side.

References

1. Hudson, N. W. 1995. *Soil Conservation*, London: Batsford.
2. Food and Agriculture Organization of the United Nations. 1987. Soil and Water Conservation in Semi-Arid Areas. Soils Bulletin 57. Rome: FAO.
3. William, L. S., and R. J. Walter. 1988. Controlled erosion terraces in Venezuela. eds. Moldenhaner, W. C., and N. W. Hudson. *Conservation Farming on Steep Lands*. Ankeny, IA:
4. Elwell, H. A., and A. J. Norton 1988. No-till tied-ridging: A recommended sustained crop production system. ed. Institute of Agricultural Engineering. Harare, Zimbabwe.
5. Shaxson, T. E., N. W. Hudson, D. W. Sanders, E. Roose, and W. C. Moldenhauer. 1989. Land Husbandry: A Framework for Soil and Water Conservation. Ankeny, IA:
6. Iowa State University. 1990. *Conservation Tillage*. Iowa Cooperative Extension Service, Iowa State University, Waterloo.
7. World Bank. 1990. Vetiver Grass: The Hedge Against Erosion. Washington DC:
8. Brown, L. C., L. D. Norton, and R. C. Reeder 1991. Erosion and yield benefits from ridge-tillage systems. *Proceedings of 12th International Conference of Soil Tillage Research Association*.
9. Pellek, R. 1992. Contour hedgerow and other soil conservation interventions for the hilly terrain of Haiti. *Erosion, Conservation, and Small Scale Farming*, Vol 2, eds. Tato, K., and H. Hurni, pp. 313–320. Geographica Bernensia, Bangkok.
10. Vogel, H. 1992. Effects of conservation tillage on sheet erosion from sandy soils at two experimental sites in Zimbabwe. *Appl. Geogr.* 12: 29–42.
11. Morgan, R. P. C. 1995. *Soil Erosion and Conservation*. Burnt Mill: Longman House.
12. Agricultural Research Service. 1991. Predicting Soil Erosion by Water - A Guide to Conservation Planning with the Revised Universal Soil Loss Equation (RUSLE). Washington DC: U.S. Department of Agriculture.
13. Lal, R. 1981. Conservation tillage. Soil Erosion in the Tropics: Principles and Management, Chichester: Wiley.
14. Rodriguez, O. S., N. Fernandez de la Paz. 1992. Conservation practices for horticulture production in the mountainous region of Venezuela. *Erosion, Conservation and Small-Scale Farming*. eds. Hurni, H., and K. Toto, pp. 393–406. Geographica Bernensia, Bangkok.
15. Sheng, T. C. 1981. The need for soil conservation structures for steep cultivated slopes in the humid tropics. *Tropical Agricultural Hydrology*. eds. Lal, R. and E. W. Russel. Chichester: Wiley.
16. Griessel, O., and R. P. Beasly 1971. Design Criteria for Underground Terrace Outlet, Science and Tecnology Guide 1525. University of Missouri Extension Division.

17. Ree, W. O. 1976. Effect of seepage flow on reed canarygrass and its ability to protect waterways. City: Agricultural Research Service, U.S. Department of Agriculture. Washington D.C.
18. Food and Agriculture Organization of the United Nations. 1979. Watershed Development with Special Reference to Soil and Water Conservation, Soil Bulletin No. 40. Rome: FAO.
19. Gwinn, W. R., and W. O. Ree 1979. Maintenance Effects on the Hydraulics Properties of Vegetation-Lined Channel. ASAE Paper 79-2063. St. Joseph, MI, USA.
20. Food and Agriculture Organization of the United Nations. 1988. Slope Treatment Measures and Practices. Conservation Guide 13/3. Rome: FAO.
21. Temple, D. N. 1991. Changes in vegetal flow retardance during long-duration flows. *Trans. ASAE* 34:1769–1774. St. Joseph, MI, USA.
22. Temple, D. N., and D. Alspach. 1992. Failure and recovery of a grass-lined channel. *Trans. ASAE* 35:171–173. St. Joseph, MI, USA.
23. Heede, B. H., and J. G. Mufich. 1973. Funtional relationship and a computer program for structural gully control. *J. Environ. Manage.* 1:321–344.
24. Heede, B. H. 1976. Gully development and control: The status of our knowledge. Paper RM-169, Fort Collins: U.S. Department of Agriculture.
25. Heede, B. H. 1977. Gully control structures and systems. *Guidelines for Watershed Management: FAO Conservation Guide*, Vol. 1 Rome: FAO.
26. Imeson, A. 1980. Gully types and gully prediction. *KNAG Geol.Tij.*, XIV(5): 430–441.
27. Stocking, M. A. 1980. Examination of the factors controlling gully growth. Assessment of Erosion. eds: de Boodt, M., D. Gabriels, pp. 505–521. Chichester: Wiley.
28. Gray, D. H., and A. T. Leiser. 1982. *Biotechnical Slope Protection and Erosion Control*. New York: Van Nostrand Reinhold.
29. Food and Agriculture Organization of the United Nations. 1987. Gully Control Conservation Guide 13/2. Rome: FAO.
30. Heede, B. H. 1987. Opportunities and limits of erosion control in stream and gully systems. *Proceedings of the 18th* Annual Conference IECA, pp. 205–209. Pinola, California.
31. Schwab, G. O., W. Elliot, D. Fangmeier, and R. K. Frevert. 1993. *Soil and Water Conservation Engineering*. New York: Wiley.
32. Gray, D. H., and R. B. Sotir. 1996. *Biotechnical and Soil Bioengineering Slope Stabilization*. New York: Wiley.
33. Chisci, G., and V. Boschi. 1988. Runoff and erosion control with hill farming in the sub-coastal Apennines climate. *Soil Tillage Res.* 12:105–120.
34. Kolenbrander, G. J. 1974. Efficency of organic manure in increasing soil organic matter content. *Transactions of the 10th International Congress of Soil Science*, Vol. 2. pp. 129–36.
35. Cogo, N. P., W. C. Moldenhauer, and G. R. Foster, 1984. Soil loss reduction from conservation tillage practices. *Soil Sci. Soc. Am. J.* 48:368–373.
36. Wischmeier, D. H., and D. D. Smith. 1978. Predicting Rainfall Erosion Losses. A Guide to *Conservation Planning*, Agricultural Handbook No. 537. Washington, DC: U.S. Department of Agriculture.

37. Lim, K. H. 1988. A study of soil erosion control under mature oil palm in Malaysia. *Land Conservation for Future Generations*. ed. S. Rimwanich, pp. 783–795. Bangkok: Department of Land Development.
38. Chisci, G., and P. Bazzoffi. 1995. Interventi agrobiologici. In: *Frutticoltura di Collina. Limitazione dell'Erosione e dell'Inquinamento*. "Agricoltura," 10, pp. 41–45. ed. Agriculture Office, Rome.
39. Troeh, F. R., J. A. Hobbs, and R. L. Donahue 1991. *Soil and Water Conservation for Productivity and Environmental Protection*, pp. 1–718. London: Prentice-Hall.
40. Food and Agriculture Organization of the United Nations. 1965. *Soil Erosion by Water. Some Measures for Its Control on Cultivated Lands*. Rome: FAO.
41. Wischmeier, D. W. 1973. Conservation tillage to control water erosion. *Proceedings* of the National Conservation Tillage Conference, pp. 133–141. Ankeny, IA: Soil Conservation Society of America.
42. Laflen, J. M., and T. S. Colvin. 1981. Effect of crop residues on soil loss from continuous row-cropping. *Trans. ASAE* 24:605–609. St. Joseph, MI, USA.
43. Norton, L. D., N. P. Cogo, and W. C. Moldenhauer. 1985. Effectiveness of mulch in controlling erosion. *Soil Erosion and Conservation*, eds. El-Swaify, S. A., and W. C. Moldenhauer, pp. 598–606. Ankeny, IA: Conservation Society of America.
44. Foster, G. R., C. B. Johnson, and W. C. Moldenhauer. 1982. Hydraulics of failure of ananhored corn-stalk and wheat straw mulches for erosion control. *Trans. ASAE* 24:1253–1263. St. Joseph, MI, USA.
45. Giovannini, G., S. Lucchesi, and M. Giachetti. 1988. Effects of heating on some physical and chemical parameters related to soil aggregation and erodibility. *Soil Sci.* 146:255–261.
46. Chepil, W. S., and N. P. Woodruff. 1963. The physics of soil erosion and its control. *Adv. Agric.* 15:211–302.
47. Stoeckeler, J. H. 1965. The design of shelterbelts in relation to crop and yield improvement. *World Crops* 3–8, Grampian Press. Denver, CO, USA.
48. Marshall, J. K. 1967. The effect of shelter on the productivity of grasslands and field crops. *Field Crops Abstr.* 20:1–14.
49. Savage, R. P., and W. W. Woodhouse. 1968. Creation and stabilization of coastal barrier dunes. *Proceedings of Coastal Engineering Conference*, London, Wiley, pp. 671–700.
50. Skidmore, E. L., and N. P. Woodruff. 1968. Wind erosion forces in the United States and their use in predicting soil loss. Agricultural Research Service Handbook 346. Washington, DC: U.S. Department of Agriculture.
51. Borsy, Z. 1972. Studies on wind erosion in the windblown sand areas of Hungary. *Acta Geogr. Debrecina* 10:123–132.
52. Skidmore, E. L., and L. J. Hagen. 1977. Reducing wind erosion with barriers. *Trans. ASAE* 20:911–915. St. Joseph, MI, USA.
53. Wilson, S. J., and R. U. Cooke. 1980. Wind erosion. *Soil Erosion.* eds. Kirkby, M. J., and R. P. Morgan. Chichester: Wiley.
54. Li, J. H. 1984. Benefits draws from network of shelterbelts in farmland of Pengwa Production Brigade, Youyu Country. *Soil Water Conserv. China* 22:24–25.

55. Marshall, C. J., and A. R. de Fegely. 1987. Erosion control for tree plantation establishment. *J. Soil Conserv. NSW* 43:32–35.
56. Rosewell, C. J., and K. Edwards. 1988. SOILOSS: A Program to Assist in the Selection of Management Practices to Reduce Erosion. Technical Manual 11. Soil Conservation Service of New South Wales, Sydney, Australia.
57. Skidmore, E. L., and J. R. Williams. 1991. Modified EPIC wind erosion model. *Modeling Plant and Soil Systems*. Agronomy, Monograph No. 31. Wisconsin University Press, Madison.
58. McTainsh, G. H., C. W. Rose, G. E. Okwach, and R. G. Palis. 1992. Water and wind erosion: Similarities and differences. *Erosion, Conservation and Small-Scale Farming*. eds. Hurni, H., and K. Tato. Geographica Bernensia, Bangkok.
59. Hagen, L. J. 1976. Windbreak design for optimum wind erosion control. *Proceedings of the Symposium on Shelterbelts on the Great Plains* ed. Am. Ass. Civil Eng., Denver, CO, USA.

5 Irrigation and Drainage

5.1 Crop Water Requirements

L. S. Pereira and R. G. Allen

5.1.1 Main Concepts

The utilization of water for crop production has been recognized as essential since agricultural practices began to be developed. Irrigation was practiced in old civilizations, mainly in the Orient and the Mediterranean basin.

A few centuries ago, agricultural production changed, mainly because new crops began to be cultivated and agricultural production became progressively oriented toward the market. Farmers soon recognized that yields could be increased when water was applied in larger amounts. The concept of crop water requirements (CWR) probably did not exist when irrigation was practiced for subsistence purposes but surely developed as evidence began to accumulate on the dependency of yields on the water applications. These concepts became important when large engineering works began to be built to provide for water supply to newly irrigated areas and it became necessary to estimate the water volumes to be supplied.

The empirical concept of CWR became progressively better described through the combination of scientific progress in physics and physiology. The following definitions are utilized in describing crop water requirements [1, 2]:

- Evaporation is the physical process by which a liquid is transferred to the gaseous state.
- Transpiration is the evaporation process of liquid water within a plant through the stomata and plant surfaces into the air.
- Evapotranspiration (ET) is the combined process by which water is transferred from a vegetative cover and soil into the atmosphere through both transpiration from plants and evaporation from the soil, from dew, and from intercepted water on the plant surfaces.

The concept of CWR is still diverse according to the discipline that approaches the question. However, an agreement on this concept is essential because it is in the basis of irrigation planning, irrigation scheduling, and water delivery scheduling, as well as water resources planning and management. A well-accepted concept was introduced by Doorenbos and Pruitt [3], who defined CWR as the depth of water (mm) needed to meet the water loss through evapotranspiration from a disease-free crop, growing in large fields under nonrestricting soil conditions including soil water and fertility, and achieving full production potential under the given growing environment.

This concept accommodates all processes affecting the water use by a crop but excludes the influences of local advection, water stress, poor soil and poor fertility management, or inappropriate farming conditions. Thus, a complementary concept is used when actual crop and field conditions are considered. This is the consumptive crop water use (CWU) defined as the depth of water (mm) utilized by a crop through ET and cultivated under given farming conditions in a given growing environment. Therefore, when optimal cropping conditions are met, CWU = CWR. Both CWR and CWU apply to irrigated and rainfed crops. However, the first corresponds to potential yield production whereas the second relates to the real crop condition.

The concept of crop evapotranspiration (ET_c) is intimately connected with CWR and CWU. It is defined as the rate of ET (mm day^{-1}) of a given crop as influenced by its growth stages, environmental conditions, and crop management. To determine the crop ET, the reference evapotranspiration (ET_0) is used and it refers to a reference crop cultivated in reference conditions such that its rate of ET (mm day^{-1}) reflects the climatic conditions characterizing the local environment. The transfer from ET_0 to ET_c is done by adopting the crop coefficients (K_c), which represent the ratio between the rates of ET of the cultivated crop and of the reference crop, that is, $K_c = ET_c/ET_0$.

For irrigated crops, another main concept is the irrigation water requirement (IWR), defined as the net depth of water (mm) that is required to be applied to a crop to fully satisfy its specific CWR. The IWR is the fraction of CWR not satisfied by rainfall, soil water storage, and groundwater contribution.

When irrigated crops are grown under suboptimal conditions, particularly deficit irrigation—that is, they do not achieve full production potential—the net depth of water to be applied then is termed deficit irrigation water requirement (DIWR), which corresponds to the adoption of a given allowable deficit in the water supply to the crop.

Depending on the irrigation supply conditions, irrigation method, equipment utilized, irrigation scheduling, and the ability of the irrigator, the amounts of water to be applied have to compensate for all system losses and lack of efficiency. These include losses through percolation, seepage or leakage in the conveyance and distribution systems, management spills of excess water in canal systems, and runoff and deep percolation in the fields. Therefore, the gross irrigation water requirement (GIWR) corresponds to the gross depth of water (mm) to be applied to a crop to fully satisfy its CWR. When it is necessary to add a leaching fraction to ensure appropriate leaching of salts from the soil profile, this depth of water is also included in GIWR.

5.1.2 Evapotranspiration: The Penman-Monteith Equation

Background

Evaporation of water requires relatively large amounts of energy, either in the form of radiant energy or sensible heat. Therefore, the evapotranspiration process is governed by energy exchange at the vegetation surface and is limited by the amount of energy available. It is therefore possible to predict the rate of ET given a net balance of energy fluxes.

The primary energy components which supply (or diminish) energy at a vegetation surface are net radiation from the atmosphere (R_n), sensible heat to the air or the equilibrium boundary layer (H), sensible heat to or from the soil (G) and evaporation or evapotranspiration expressed as a flux density of latent heat (λET). Other fluxes or sinks are present, such as energy requirements for photosynthesis, but these are quite small relative to R_n, λET, H, and G. All terms are expressed in units of energy per horizontal area per unit of time. λET is the latent heat required to vaporize one unit of water [MJ Kg^{-1}]. Therefore the energy balance can be written in terms of these four components as

$$\lambda ET = R_n - H - G \tag{5.1}$$

The terms on the right side of this equation can be computed from measured or estimated climatic and vegetation factors. The climatic factors include short wave and long wave radiant fluxes from and into the atmosphere (R_n), effects of horizontal air movement (wind speed) and air and surface temperatures on H, and soil heat fluxes (G). Vegetation factors include the resistance to diffusion of vapour from within plant leaves and stems and the resistance to diffusion of vapour from near the vegetation or soil surface upward into the atmosphere.

For general prediction purposes, the complex turbulent structures within and above vegetation canopies and the effects of partitioning of net radiation and energy within the canopies can be described in terms of simple exchange coefficients or their inverses, resistances. Generally this is accomplished using the linear "big leaf" model of Monteith [4, 5] and Rijtema [6] where two resistances, canopy and aerodynamic, operate in series between leaf interiors and some reference height above the vegetation (Fig. 5.1). Canopy,

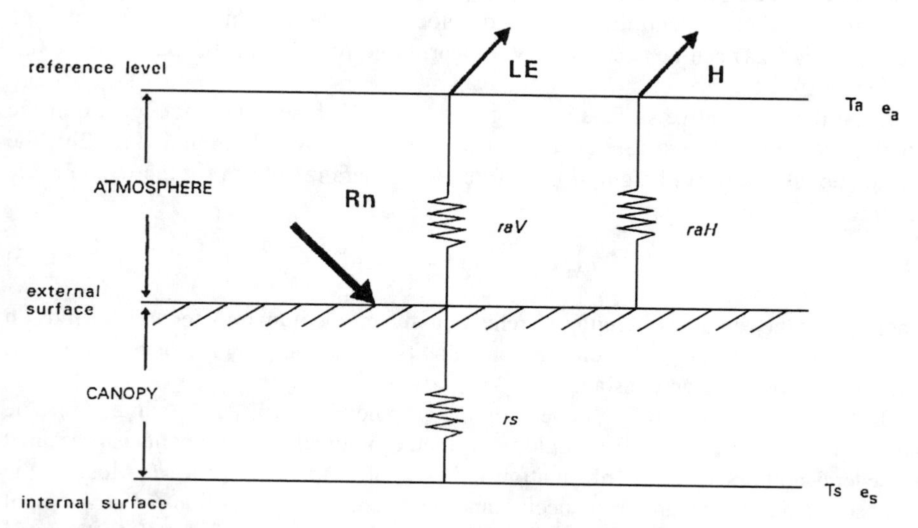

Figure 5.1. Schematic representation of the "big leaf" model with representations of the aerodynamic resistance for heat and vapor $r_{ah} = r_{av} = r_a$ and the surface resistance r_s. ***Source:*** **Adapted from [7].**

or bulk surface resistance (r_s) can be computed from the resistance of vapour flow through individual stomata openings (r_l) and total leaf area of the vegetation.

The aerodynamic resistance (r_a) describes the resistance to the random, turbulent transfer of vapour from the vegetation upward to the reference (weather measurement) height and the corresponding vertical transfer of sensible heat away from or toward the vegetation.

The energy balance equation can be arranged in terms of parameters R_n and G and parameters within the H and λET components. If one assumes that eddy-diffusion transfer factors for λET and H are the same and that differences between transfer factors for momentum and those for heat can be quantified through a simple ratio, then the Penman-Monteith (PM) form of the combination equation [4] results:

$$\lambda ET = \frac{\Delta(R_n - G) + \rho c_p(e_s - e_a)/r_a}{\Delta + \gamma(1 + r_s/r_a)} \tag{5.2}$$

where $(e_s - e_a)$ represents the vapor pressure deficit (VPD) of air at the reference (weather measurement) height (kPa), ρ represents mean air density (kg m^{-3}), c_p represents specific heat of air at a constant pressure (MJ kg^{-1} °C^{-1}), Δ represents the slope of the saturation vapor pressure–temperature relationship at mean air temperature (kPa °C^{-1}), γ is the psychometric constant (kPa °C^{-1}), r_s is the bulk surface resistance (s m^{-1}), and r_a is the aerodynamic resistance (s m^{-1}).

Crop Parameters in the Penman-Monteith Equation

The fluxes of sensible and latent heat between the evaporative surface and the atmosphere depend upon the gradients of temperature and vapor from the surface to the atmosphere. However, these processes are also governed by the momentum transfer, which affects the turbulent transfer coefficients.

Assuming thermal neutrality, the wind velocity profile $u(z)$ (m s^{-1}) above a plane surface (Fig. 5.2a) can be described by a logarithmic function of the elevation z_m (m) above that surface.

When the evaporative surface is a vegetated one, the height and architecture of the vegetation bring the zero reference level to a plane above the ground (Fig. 2b), the zero plane displacement height, d (m). Then, the generalized form for the wind profile becomes

$$u(z) = \frac{u_*}{k} \ln\left(\frac{z_m - d}{z_{om}}\right) \tag{5.3}$$

where z_{om} is the roughness length (m) relative to the momentum transfer and depends on the nature of the surface, u_* is the friction velocity due to the eddy momentum transfer, and k is the von Kármán constant.

Both parameters, d and z_{om}, depend on the crop height, h (m), and architecture. The determination of z_{om} and d from field micrometeorological measurements is presented by several authors [9, 10]. Information exists relating d and z_{om} to h. Most of the proposed relationships are crop specific and represent unique functions of crop height [10]. However, more general functions also consider the leaf area index (LAI) [11, 12] or a plant area index [13].

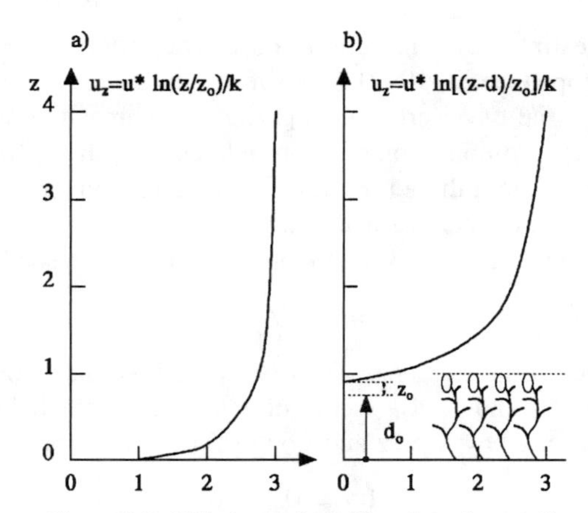

Figure 5.2. Wind-speed profiles above bare soil (a) and above a cereal-crop canopy (b) when wind speed at 4-m elevation is the same for both conditons. *Source:* [8].

From the preceding assumptions, the transfer of heat and vapor from the evaporative surface into the air in the turbulent layer above a canopy is determined by the aerodynamic resistance r_a between the surface and the reference level (at height z_m) above the canopy; that is,

$$r_a = \frac{\ln\left(\frac{z_m - d}{z_{om}}\right)\ln\left(\frac{z_h - d}{z_{oh}}\right)}{k^2 u_z}, \qquad (5.4)$$

where r_a is the aerodynamic resistance (s m^{-1}), z_m is the height of the measurements of wind velocity (m), z_h is the height of air temperature and humidity measurements (m), d is the zero plane displacement height (m), z_{om} is the roughness length relative to momentum transfer (m), z_{oh} is the roughness length relative to heat and vapor transfer (m), u_z is the wind velocity at height z_m (m s^{-1}), and k is the von Kármán constant (= 0.41).

Equation (5.4) assumes that the evaporative surface represented by the "big leaf" is inside the canopy. However, exchanges in the top layer of the canopy, between heights $d + z_{om}$ and h, are important as a source of vapor fluxes. Adopting the height $d + z_{om}$ as the level of the evaporative surface would lead to overestimation of r_a and underestimation of r_s [7]. Thus, in the alternative, r_a can be computed from the top of the canopy [14]:

$$r_a = \frac{\ln\left(\frac{z_m - d}{z_{om}}\right)\ln\left(\frac{z_h - d}{h - d}\right)}{k^2 u_z}. \qquad (5.5)$$

The application of Eqs. (5.4) and (5.5) for short time periods (hourly or less) using aerodynamic approaches requires the inclusion of corrections for stability [10, 12, 15–18]. These corrections are not considered when weather factors are measured at only one (reference) level, that is, under the assumption of neutral conditions stated earlier, and are less important under well-watered conditions.

The height z_{oh} is estimated as a fraction of z_{om}, commonly $z_{oh} = 0.1\, z_{om}$, for short and fully covering crops (see [2, 19–21]). However, the factor 0.2 is prefered by some authors for tall and partial cover crops (see [17, 22]) whereas the ratio 1:7 is assumed by others (see [23–26]). Computation of the roughness length z_{oh} for heat and vapor is not a problem in itself because there is relatively little effect on the ET calculations from selecting a z_{oh}/z_{om} ratio between 0.1 and 0.2.

The surface resistance for full-cover canopies is often expressed [4] as

$$r_s = r_l/\mathrm{LAI_{eff}}, \tag{5.6}$$

where r_s is the surface resistance ($\mathrm{s\,m^{-1}}$), r_l is the bulk stomatal resistance of a well-illuminated leaf ($\mathrm{s\,m^{-1}}$), and $\mathrm{LAI_{eff}}$ is the effective leaf area index ( ). A common formulation is that assumed by Szeicz and Long [27] and reviewed by Allen *et al.* [21],

$$\mathrm{LAI_{eff}} = 0.5\,\mathrm{LAI}, \tag{5.7}$$

which takes into consideration that only the upper half of a dense canopy is actively contributing to the surface heat and vapor transfer. Other formulations have been proposed, namely by Ben-Mehrez *et al.* [28],

$$\mathrm{LAI_{eff}} = \mathrm{LAI}/(0.3\,\mathrm{LAI} + 1.2). \tag{5.8}$$

This equation provides for a somewhat higher effective LAI for the initial crop stages (small LAI) and somewhat smaller values when the crop develops and LAI increases.

Applicability of the Penman-Monteith Equation

The PM equation (5.2) can be utilized for the direct calculation of crop ET because the surface and aerodynamic resistances are crop specific. However, despite research, data are available only for a limited number of crops. Typical values for r_l and r_s are given in reviews by Garratt [29] and Allen *et al.* [10]. For z_{om} and d and information on predicting LAI, see Allen *et al.* [10].

The computation of r_a with Eqs. (5.4) and (5.5) for partial cover crops presents some problems relative to the predictive calculation of the roughness length z_{om} and the zero plane displacement height d. For partial cover crops, it is necessary to treat the soil and the crop separately, thus considering two surface resistances [30, 31]. Shuttleworth and Wallace [32] developed a specific model for partial cover crops, which considers a more complex but rational separation of fluxes from the soil surface up to the canopy boundary layer. Good results have been obtained with that model [33]. The applicability of the "big leaf" approach to partial cover crops is questionable.

There is very little information on the early crop growth stages, when LAI is small. For the partial cover crops, approaches are still insufficient. Information is also scarce for fruit crops. For conditions of incomplete ground cover, other problems include the roughness of the surface and the variable albedo of the surface, which is now a combination of vegetation and soil, thus affecting the calculation of net radiation R_n.

Other difficulties concern the prediction of the surface resistance r_s. There is relatively little information on bulk stomatal resistances r_l characterizing a given crop; r_l usually increases as a crop matures and begins to ripen. The use of prediction equations is difficult because of differences among varieties and in crop management. Information on stomatal conductance or stomatal resistances available in the literature is often oriented to physiological or ecophysiological studies and is not easy to use predictively with Eq. (5.6). There are also difficulties in measuring or estimating the LAI and plant height h over time [10].

It is well known that resistances r_l and r_s are influenced by climate and water availability to the crop. Influences change from one crop to another and different varieties can be affected differently. Surface resistance r_s increases when the crop is water stressed or when soil water availability limits crop ET, when the VPD increases, and when r_a is higher. On the contrary, r_s decreases when energy available at the surface increases. However, all factors act together. The dependency of surface resistances on the climate is described by several authors [4, 11, 34–38]. It can be demonstrated [39] that r_s varies according to

$$r_s = r_a\left(\frac{\Delta}{\gamma}\beta - 1\right) + (1+\beta)\frac{\rho c_p \mathrm{VPD}}{\gamma(R_n - G)}, \tag{5.9}$$

where all parameters are defined as for the PM equation (5.2) and β is the Bowen ratio, $\beta = H/\lambda ET$. In this equation, β plays the role of a water stress indicator. This equation illustrates that weather variables interact and that their influences are interdependent. This brings into consideration the difficulties in appropriately selecting the r_s to be utilized predictively, unless appropriate simplifications are assumed.

Despite these difficulties, many recent studies have advocated using the PM formulation for estimating ET. McNaughton and Jarvis [40], Sharma [41], Hatfield and Fuchs [42], and Burman and Pochop [43] considered the PM equation to be the most acceptable form of the combination equation for computing crop and reference ET. The PM approach is commonly utilized in field studies of crop evapotranspiration (e.g., [7, 44–48]). Several crop–water models use the PM equation predictively, as do environmental models (e.g., [49–51]). This equation also is utilized widely in hydrology, particularly to predict ET from forests [52].

The difficulties referred to earlier create challenges in applying the PM equation or other more complex multilayer resistance equations to directly estimate ET from complex crop canopies. The algorithms and equations needed to describe changes in resistances and net radiation require more than a few simple equations. Current work at research locations around the globe is focusing on improving our ability to directly apply the PM equation or multilayer ET models to specific agricultural crops. However, the research community is probably some 10 to 15 years away from producing one-step procedures that are consistent, predictable, and reliable. When these one-step procedures do become available, they will likely be in the form of relatively complex computer models. Meanwhile, at present, full advantage in irrigation practice can be taken of the PM equation only to compute the reference ET and to determine the crop ET through use of the crop coefficients.

5.1.3 Description of the Reference Evapotranspiration

Introductory Aspects

The concept of reference ET (ET_0) was introduced [53, 54] to avoid ambiguities that existed with the definition of potential ET. ET_0 refers to the ET rates from vegetation over which weather measurements are made. Adopting this concept and grass as the reference crop, Pruitt and Doorenbos [55] were able to develop and validate methods to predict ET_0 on the basis of actual field measurements in several locations.

Potential evaporation (E_p) is the evaporation from a surface when all surface–atmospheric interfaces are wet (saturated) so that there is no restriction on the rate of evaporation from the surface, except for atmospheric conditions and energy available at the surface [1, 2]. Similarly, potential ET (ET_p) is the rate at which water would be removed from a wet (saturated) complex of vegetated surface and soil so that, without restriction, ET_p depends on atmospheric conditions and energy available at the evaporative surface but is influenced by the geometric characteristics of the vegetated surface, such as aerodynamic resistance, vegetative structure, and density. An updated discussion on the above concepts in reference to actual and crop ET is given by Pereira *et al.* [14].

Numerous researchers (e.g., [56–58]) have compared measured ET of grass crops with estimates of ET computed from meteorological measurements and have contributed to the identification of sources of inaccuracy of standardized methods and of improvements that could contribute to the consistency of estimation. It has been possible to test and validate improvements to the reference methodology because it was simple, clear, and based on solid, knowledgeably made measurements. Meanwhile, research has provided abundant information that has permited appropriate revision and updating of the reference ET concept as well as calculation procedures [2, 59, 60].

The reference ET takes into consideration the effects of climate on CWR, but refers to ET from vegetation over which weather measurements are made and provides for a consistent set of crop coefficients to be used to determine ET for other crops. Relating ET_0 to a specific crop has the advantage of representing the biological and physical processes involved in the energy balance at the cropped surface. Adoption of ET_0 has made it easier to select consistent crop coefficients and to make reliable crop ET estimates in new areas. The use of the crop coefficient–ET_0 approach has been enormously successful in obviating the need to calibrate a separate ET equation for each crop and stage of growth. It also has provided a working model [14] that can be used until more sophisticated methods become available for direct estimation of actual crop ET. Therefore, it is important to retain the use of ET_0 and to promote its use in making routine estimates of crop ET.

Adopting a living reference grass crop as defined by Doorenbos and Pruitt [55] provides for a height ranging between 0.08 and 0.15 m, and thus a large variation in bulk surface resistance r_s (100 to 50 s m^{-1}). Therefore, the average r_s and the associated aerodynamic resistances may vary appreciably with time between clippings and locations, depending on the structural characteristics and regrowth rates of the grass variety and management schedules. Difficulties with a living grass reference result from the fact that the grass variety and morphology can significantly affect the rate of ET_0, mainly during peak periods of water consumption. Maintenance of a living reference crop for use in computing crop coefficients for other crops is difficult.

The definition and concept of ET_0 must be one-dimensional with respect to evaporation and energy exchange processes. This means that all fluxes within the energy balance (net radiation, sensible heat, soil heat, and latent heat) must be uniformly vertical along the horizontal surface so that the reference surface completely represents the one-dimensional ET processes of large plantings. The requirement that the measurement of ET_0 be one-dimensional often is violated in lysimeter studies, in which the crop in the lysimeter or measurement area is taller or shorter than that outside the area or extends beyond the horizontal dimensions of the lysimeter. Various problems and effects of the maintenance of the necessary environmental, site, and equipment conditions have been discussed in the literature [57, 61–66]. In the discussion by Allen *et al.* [59], errors of 20% and up to 30% in measuring ET are reported.

Boundary-layer measurements of ET also can be beset with operational problems. These systems, which include eddy correlation and Bowen ratio systems, require long fetch in the upwind direction. In addition, boundary-layer measurement-equipment components are delicate and require special maintenance. These requirements often limit eddy correlation and Bowen ratio systems to research studies. A number of researchers have analyzed the requirements for and measurements errors in ET boundary-layer measuring systems [47, 56, 67–69].

The measurement of ET by performing a continuous soil water balance from data collected through the monitoring of soil water has never received the preference of researchers mainly because of difficulties in instrumentation and spatial variability. An analysis of the performance of neutron probe readings in estimating ET is given by Carrijo and Cuenca [70].

In summary, despite progress in measuring ET, it is preferable for irrigation practice, to compute ET_0 from weather data, and hence to use a climatic reference ET definition, and to adopt a reference grass crop with constant height.

Definition of Reference Evapotranspiration

The primary purpose for developing ET_0 equations is to estimate a reference ET by which a crop coefficient is multiplied to obtain an estimate of crop ET. In many instances, the ET_0 equation has represented a hypothetical living crop reference in order to provide for a complete record of ET_0 during the development of crop coefficients and during calibration or analysis of other ET equations. Wright and Jensen [53] and Wright [71] utilized a hypothetical alfalfa ET_0; Jensen [54] and Jensen *et al.* [2] used an equation to represent "lysimeter ET"; George *et al.* [72], Pruitt and Swann [56], and Feddes [73] used an ET_0 equation to represent a hypothetical grass ET_0; and Martin and Gilley [74] also assumed a hypothetical grass reference crop.

Allen *et al.* [21] and Jensen *et al.* [2] supported the use of the PM equation to represent both alfalfa and grass ET_0. The PM method was ranked high among all methods evaluated in the American Society of Civil Engineers' (ASCE) study [2], both as a grass and as an alfalfa reference. Based on results of studies reported here, the Food and Agriculture Organization (FAO) Expert Consultation on Revision of FAO Methodologies for Crop Water Requirements [75] recommended the PM method as the primary ET_0 method for defining grass ET_0 and for determining crop coefficients. A hypothetical grass reference crop with constant height and surface resistance was selected to avoid problems with local calibration that would require demanding and expensive studies.

The surface resistance [Eq. (5.6)] of a grass crop can be well estimated because it is commonly accepted that the average minimum daytime value of stomatal resistance for a single leaf is approximated as $r_l = 100\ \mathrm{s\,m^{-1}}$; the LAI effective for transpiration can be appropriately calculated using $\mathrm{LAI_{eff}} = 0.5\ \mathrm{LAI}$; and the LAI for a clipped grass with height not larger than 0.15 m can be approximated as $\mathrm{LAI} = 24\ h$. The above approximations produce the following relationship for the surface resistance:

$$r_s = \frac{100}{0.5\,\mathrm{LAI}} = \frac{200}{24\,h}. \tag{5.10}$$

If a constant $h = 0.12$ m is assumed hypothetically [75], Eq. (5.10) yields a constant LAI = 2.88, which produces the constant value $r_s = 70\ \mathrm{s\,m^{-1}}$. These characteristics, with the additional standard crop reflectivity factor (albedo), fully characterize a hypothetical grass reference crop as discussed by Allen *et al.* [59].

The same assumption is adopted in the Soil Conservation Service (SCS) Handbook [74], where the clipped grass reference crop is considered with constant height $h = 0.125$ m. This results in also adopting a constant $r_s = 66\ \mathrm{s\,m^{-1}}$.

The selection of constant height and bulk surface resistance parameters is a compromise and may not represent reality in all climatic regimes. However, this selection provides consistent ET_0 values in all regions and climates and among research locations. The PM equation is a close, simple representation of the physical and physiological factors governing the ET process. By using the PM definition for ET_0, one may calculate crop coefficients at research sites where ET_0 is not measured simultaneously with measured ET from other crops. In addition, various crops may exhibit variations in r_s with climate that are different from those exhibited by grass. In the $K_c\ ET_0$ approach, the variations in r_s and r_a relative to the reference crop are accounted for within the crop coefficient, which serves as an aggregation of the physical and physiological differences between crops (see Pereira *et al.* [14]). The aggregation of various differences and the development of crop coefficients should be simplified by using the constant r_s in the ET_0 definition [76, 77].

Therefore, the reference ET (ET_0) is defined as the rate of ET from a hypothetical reference crop with an assumed crop height of 0.12 m, a fixed surface resistance of $70\ \mathrm{s\,m^{-1}}$, and an albedo of 0.23, closely resembling the ET from an extensive surface of green grass of uniform height, actively growing, completely shading the ground, and having adequate water.

The Penman-Monteith ET_0 Equation

The grass reference crop in the above definition can be fully described using the PM approach. The grass reference crop is probably, together with alfalfa, the best-studied crop regarding its aerodynamic and bulk stomatal characteristics [2, 21, 59, 60].

The aerodynamic resistance r_a (Eq. 5.4) can be calculated easily because the zero plane displacement height can be estimated from the crop height h, $d = 0.67h$; the roughness height for momentum transfer can be given by $z_{om} = 0.123h$; and the roughness height for heat and vapor transfer can be approximated by $z_{oh} = 0.1z_{om}$.

From the constant crop height $h = 0.12$ m, the following standardized grass reference parameters result: zero plane displacement height $d = 0.08$ m; roughness height for momentum transfer $z_{om} = 0.015$ m; roughness height for heat and vapor transfer $z_{oh} = 0.0015$ m.

Assuming a standardized height for wind-speed, temperature, and humidity measurements at 2.0 m ($z_m = z_h = 2.0$ m), and introducing the above parameters into Eq. (5.4), the following aerodynamic resistance results:

$$r_a = 208/U_2, \tag{5.11}$$

where r_a is the aerodynamic resistance (s m^{-1}), and U_2 is the wind-speed measurement at 2-m height (m s^{-1}).

When wind speed is measured at a height higher than 2.0 m, assuming the logarithmic profile of the wind (Eq. 5.3) the wind speed at 2.0 m (U_2) can be obtained from the wind speed at height z_m (U_z) by

$$U_2 = U_z \frac{\ln\left(\frac{2-d}{z_{om}}\right)}{\ln\left(\frac{z_m-d}{z_{om}}\right)}. \tag{5.12}$$

For the standard heights $d = 0.08$ m and $z_{om} = 0.015$ m, the following simplified equation results:

$$U_2 = U_z \frac{4.87}{\ln(67.8 z_m - 5.42)}. \tag{5.13}$$

When r_a from Eq. (5.11) is combined with $r_s = 70$ s m^{-1}, the FAO-PM ET_0 equation for 24-h periods becomes [59, 60, 75]:

$$ET_0 = \frac{0.408\Delta(R_n - G) + \gamma \frac{900}{T+273} U_2 (e_s - e_a)}{\Delta + \gamma(1 + 0.34 U_2)}, \tag{5.14}$$

where ET_0 is the reference ET (mm day^{-1}); R_n is the net radiation at the crop surface (MJ m^{-2} day^{-1}); G is the soil heat flux density (MJ m^{-2} d^{-1}); T is the average temperature at 2-m height (°C); U_2 is the wind speed measured at 2-m height (m s^{-1}); $(e_s - e_a)$ is the vapor pressure deficit for measurements at 2-m height (kPa); Δ is the slope vapor pressure curve (kPa °C^{-1}); γ is the psychometric constant (kPa °C^{-1}); 900 is the coefficient for the reference crop (kJ^{-1} kg K day^{-1}) resulting from conversion of units and substitution of variables ρ, c_p, and r_a; 0.34 is the wind coefficient for the reference crop (kJ^{-1} kg K) resulting from the ratio $r_s/r_a (70/208 = 0.34)$; and 0.408 is the value for $1/\lambda$ with $\lambda = 2.45$ MJ kg^{-1}. Details on the derivation of this equation are given by Allen *et al.* [60].

To ensure the integrity of computations, the weather measurements for this equation must be taken above an extensive surface of green grass, with shading of the ground and no shortage of water.

The form of Eq. (5.14) is not very different from other Penman expressions except for the addition of T in the numerator and U_2 in both the numerator and the denominator.

A form of the FAO-PM equation for calculating hourly ET_0 is analyzed by Allen *et al.* [60].

The FAO-PM equation should not require local calibration or use of a localized wind function if wind speed is measured at a height of 2 m or is adjusted to this height [Eq. (5.13)].

No weather-based ET equation can be expected to predict ET perfectly under every climatic situation because of simplifications in formulation and errors in data measurement. Precision instruments under excellent environmental and biological management conditions probably will show that the FAO-PM equation (5.14) deviates at times from true measurements of grass ET_0. However, it has been the opinion of the FAO Expert Consultation [75] that the hypothetical reference definition of the FAO-PM equation should be used as the definition for grass ET_0 when deriving crop coefficients. This recommendation is based on the important need to standardize the ET_0 concept and its use.

5.1.4 Calculation of the Reference Evapotranspiration

Weather Data Requirements and Availability

The FAO-PM method requires that the following data be available for the daily (or monthly) calculations:

- temperature, that is, maximum and minimum daily air temperature (T_{max} and T_{min});
- humidity, that is, dewpoint temperature (T_{dew}), maximum and minimum relative humidity (RH_{max} and RH_{min}), or dry- and wet-bulb temperature (T_{dry} and T_{wet});
- wind speed, daily average (U_z or U_2);
- radiation, that is, net radiation R_n, shortwave incoming solar radiation (R_s), or bright sunshine hours per day (n).

An optimal situation occurs when all four types of these observations are made in a weather station (see "Daily ET calculations with FAO-PM Method," below). Many weather stations have only limited instrumentation and do not provide data for all types of required meteorological variables. This calls for the use of appropriate estimation methods (see "Estimation of ET_0 with Limited Weather Data," below) or for the application of other ET equations (see "Estimation of ET_0 with Other Equations," below). ET_0 also can be estimated from pan evaporation data (see "Pan-Evaporation Method," below). Unfortunately, it is not enough that observations be made, but also that integrity of data be assessed. Often, data sets are incomplete or do not comply with homogeneity requirements. Although accuracy of measurements is important, errors resulting from inadequacy of the site may be more important than normal errors of observations.

The quality of reference ET depends upon the quality of the weather data utilized. It therefore is advisable that the quality of weather data be assessed before these data are utilized in ET_0 calculations. Data integrity screening should be a part of every agricultural weather network to ensure that high-quality and representative data are being collected and provided to users. Procedures presented by Allen [78] concern the calculation of clear sky envelopes for solar radiation, computational validation of net radiation measurements, and expected trends and relationships between air vapor content and air temperature.

Because the ET_0 definition and concept apply to a grass crop that is adequately watered and actively transpiring, it is important to follow these conditions in selecting and screening weather data. Any deviations in weather measurements from those expected within an agricultural situation will violate equilibrium boundary-layer profiles and energy exchange conditions representative of a grass reference condition.

ET in an arid, nonirrigated environment is low during periods of low rainfall. Air temperatures T and VPDs in these environments are higher relative to an irrigated environment because of a greater conversion of net radiation into sensible heat. Use of data (particularly humidity and temperature) from nonirrigated weather stations may introduce a bias into ET equations, generally causing overestimation [79]. Differences in mean daily air temperatures as high as 4-5°C on a monthly basis between arid and irrigated stations located less than 10 km apart caused an overestimation of 16% in seasonally computed ET_0. Procedures for analysis and correction of weather data from nonreference sites are presented by Allen [78] and Allen *et al.* [10].

Daily ET_0 Calculations with FAO-PM Method

Several parameters characterizing the state of humidity in the atmosphere and the vapor transfer from the canopy into the air, and the energy available at the surface are required in the ET calculations with the FAO-PM. Details on these parameters are given by Allen *et al.* [60]. Calculations of ET_0 that are made using monthly average weather data are very similar to calculations made using daily average weather data on daily calculation time steps and summed over each day of the month [60]. The standardized daily ET_0 calculation procedures are described in the following subsections:

Slope of Vapor Pressure Curve (Δ)

The relationship between air temperature and saturation vapor pressure is characterized by the slope of the vapor pressure curve Δ (kPa °C^{-1}) as follows:

$$\Delta = \frac{2504 \exp\left[\frac{17.27T}{T+237.3}\right]}{(T+237.3)^2}, \tag{5.15}$$

where T is the mean daily air temperature (°C)

Psychrometric Constant (γ)

The saturation vapor pressure at wet-bulb temperature is related to the actual air conditions by

$$\gamma = 0.00163(P/\lambda), \tag{5.16}$$

where γ is the psychrometric constant (kPa °C^{-1}), P is the atmospheric pressure (kPa), and λ is the latent heat of vaporization = 2.45 MJ kg^{-1}.

A simplified equation derived from the ideal-gas law can be assumed for the atmospheric pressure as a function of elevation z (m):

$$P = 101.3\left(\frac{293 - 0.0065z}{293}\right)^{5.26}. \tag{5.17}$$

Saturation Vapor Pressure (e_s)

The saturation vapor pressure is given by

$$e^0(T) = 0.611 \exp\left(\frac{17.27T}{T + 237.3}\right), \tag{5.18}$$

where $e^0(T)$ is the saturation vapor pressure function (kPa) and T is the air temperature (°C). For 24-h time periods, e_s should be computed for the maximum and minimum daily temperature ($T_{\max}$ and $T_{\min}$):

$$e_s = \frac{e^0(T_{\max}) + e^0(T_{\min})}{2}. \tag{5.19}$$

Computations using the mean daily temperature should be avoided because the temperature–saturation vapor pressure relationship is not linear, and this would cause underprediction of e_s. An estimation of $T_{\max}$ and $T_{\min}$ from T_{mean} is given in Eqs. (5.46) and (5.47).

Actual Vapor Pressure (e_a)

The actual vapor pressure e_a is the saturation vapor pressure at dewpoint temperature (T_{dew}). The following approximations [60] are given, in a decreasing order of accuracy, for daily e_a computations.

Dewpoint Temperature. When available, it provides the best estimate:

$$e_a = e^0(T_{\text{dew}}) = 0.611 \exp\left(\frac{17.27T_{\text{dew}}}{T_{\text{dew}} + 237.3}\right), \tag{5.20}$$

where e_a is the actual vapor pressure (kPa) and T_{dew} is the dewpoint temperature (°C).

Psychrometer Measurements. Using measurements with dry- and wet-bulb thermometers e_a can be approximated by use of of the following equation:

$$e_a = e^0(T_{\text{wet}}) - \gamma_{\text{asp}}(T_{\text{dry}} - T_{\text{wet}})P, \tag{5.21}$$

where γ_{asp} is 0.00066 for Assmann aspiration at 5 m s^{-1}, 0.0008 for natural ventilation at 1 m s^{-1}, and 0.0012 for indoor ventilation of 0 m s^{-1}(°C^{-1}); T_{dry} is the dry-bulb temperature (°C); T_{wet} is the wet-bulb temperature (°C); P is the atmospheric pressure (kPa), and $e^0(T_{\text{wet}})$ is the saturation vapor pressure at wet-bulb temperature (kPa).

Hygrometer (or Psychrometer) Measurements of Relative Humidity (RH). When two RH measurements are available daily at early morning and at early afternoon, the following equation can be used:

$$e_a = \frac{1}{2}e^0(T_{\min})\frac{RH_{\max}}{100} + \frac{1}{2}e^0(T_{\max})\frac{RH_{\min}}{100}, \tag{5.22}$$

where $e^0(T_{\min})$ and $e^0(T_{\max})$ are the saturation vapor pressure (kPa) computed at $T_{\min}$ and $T_{\max}$, respectively, and $RH_{\max}$ and $RH_{\min}$ are, respectively, the maximum and minimum daily RH (%).

When only mean daily relative humidity (RH_{mean}) data are available, average daily vapor pressure e_a can be computed as

$$e_a = \frac{RH_{\text{mean}}}{\frac{50}{e^0(T_{\min})} + \frac{50}{e^0(T_{\max})}}. \tag{5.23}$$

Calculation of RH_{mean} by averaging hourly readings and calculation of e_a from T_{mean} are discouraged because computations may be biased.

Vapor Pressure Deficit (VPD)

The VPD (kPa) is computed as

$$\text{VPD} = e_s - e_a, \tag{5.24}$$

where e_s is the saturation vapor pressure (kPa) and e_a is the actual vapor pressure (kPa).

Net Radiation (R_n)

The net radiant energy R_n available at the evaporating surfaces is a fraction of the extraterrestrial radiation R_a and is calculated from the radiation balance.

Part of R_a is lost by absorption or reflection when passing through the atmosphere. The total incoming shortwave solar radiation R_s is termed global radiation. A fraction of the global radiation is reflected back to the atmosphere, depending on the reflectivity characteristics of the surface, that is, the albedo α. A part of that which is absorbed by the surface is reradiated back as long wave radiation R_{lu}. A fraction of R_{lu} is returned as incoming long wave radiation from the atmosphere, R_{ld}. The radiation balance therefore can be written as

$$R_n = R_s{\downarrow} - \alpha R_s{\uparrow} - R_{lu}{\uparrow} + R_{ld}{\downarrow}, \tag{5.25}$$

where the arrow indicates when the radiation component is incoming (↓) or outgoing (↑).

Extraterrestrial Radiation (R_a). This radiation at the top of the atmosphere can be computed as a function of the latitude of the site and the day of the year as

$$R_a = 37.6\, d_r(\omega_s \sin\varphi \sin\delta + \cos\varphi \cos\delta \sin\omega_s), \tag{5.26}$$

where R_a is the daily total extraterrestrial radiation (MJ m^{-2} day^{-1}); 37.6 is a coefficient relative to the time duration of calculation and the solar constant (MJ m^{-2} day^{-1}), d_r is the relative distance from Earth to the Sun ( ); δ is the solar declination (rad); φ is the latitude (rad) and is negative for the Southern Hemisphere); and ω_s is the sunset hour angle (rad).

The sunset hour angle is a function of the latitude and the solar declination:

$$\omega_s = \arccos(-\tan\varphi \tan\delta). \tag{5.27}$$

The relative distance from Earth to the Sun and the solar declination can be estimated as functions of the day in the year:

$$d_r = 1 + 0.033 \cos(0.0172J), \tag{5.28}$$

$$\delta = 0.409 \sin(0.0172J - 1.39), \tag{5.29}$$

where J is the number of the day in the year ( ), with January 1st = 1. For daily values, J can be determined by

$$J = \text{integer}[275(M/9) - 30 + D] - 2, \tag{5.30}$$

provided that, if $M < 3$, then $J = J+2$ and in a leap year when $M > 2$, then $J = J+1$, where M is the month number (1–12), starting in January, and D is the day in the month.

For *monthly values*, J can be determined for the middle of the month by

$$J = \text{integer}(30.5M - 14.6). \tag{5.31}$$

Global Radiation R_s. When not measured, R_s can be estimated from sunshine hours and R_a as

$$R_s = [a_s + b_s(n/N)]R_a, \tag{5.32}$$

where R_s is the incoming solar radiation (MJ m^{-2} day^{-1}), a_s is the fraction of extraterrestrial radiation R_a on overcast days, b_s is the proportionality factor, n is the bright sunshine hours per day (h), and N is the total length of the day (h), $a_s = 0.25$ and $b_s = 0.50$ for average climate conditions.

The daylight hours N are calculated from the sunset hour angle as

$$N = (24/\pi)\omega_s = 7.64\,\omega_s. \tag{5.33}$$

Net Shortwave Radiation R_{ns}. Representing the balance between incoming and reflected solar radiation, R_{ns} (MJ m^{-2} day^{-1}) is given by

$$R_{ns} = (1 - \alpha)R_s, \tag{5.34}$$

where α is the albedo or canopy reflection coefficient ( ); $\alpha = 0.23$ for the grass reference crop.

Net Long-wave Radiation (R_{nl}). Defined as the balance between the downcoming long-wave radiation from the atmosphere ($R_{ld}\downarrow$) and the outgoing long-wave radiation emitted by the vegetation and the soil ($R_{lu}\uparrow$), R_{nl} (MJ m^{-2} day^{-1}) is computed as

$$R_{nl} = -f\varepsilon'\sigma\frac{T_{K_x}^4 + T_{K_n}^4}{2}, \tag{5.35}$$

where f is the cloudiness factor ( ), ε' is the net emissivity of the surface ( ), σ is the Stefan-Boltzmann constant = 4.90×10^{-9} MJ m^{-2} K^{-4} day^{-1}, and T_{K_x} and T_{K_n} are, respectively, the maximum and minimum daily air temperatures (K).

The cloudiness factor f represents the ratio between actual net long-wave radiation and the net long-wave radiation for a clear-sky day and is computed as

$$f = a_c\frac{R_s}{R_{so}} + b_c, \tag{5.36}$$

where R_{so} is the shortwave solar radiation for a clear-sky day (MJ m^{-2} day^{-1}), $a_c \approx 1.35$, and $b_c \approx -0.35$ for average climate conditions, with $a_c + b_c \approx 1.0$.

R_{so} for daily periods can be estimated as

$$R_{so} = (0.75 + 2 \times 10^{-5} z)\, R_a, \tag{5.37}$$

where $0.75 = a_s + b_s$ Eq. (5.36), z is the station elevation (m), and R_a is the extraterrestrial radiation (MJ m^{-2} day^{-1}). This equation is valid for station elevations less than 6,000 m that have low air turbidity. Alternative equations for locations with high air turbidity and variable humidity are given by Allen [78].

The net emissivity of the surface, ε', represents the difference between the emissivity by the vegetation and the soil and the effective emissivity of the atmosphere and is computed as

$$\varepsilon' = 0.34 - 0.14\sqrt{e_a}, \tag{5.38}$$

where e_a is the actual vapor pressure (kPa). The coefficients $a_1 = 0.34$ and $b_1 = -0.14$ are recommended for average atmospheric conditions.

Net Radiation R_n. This is computed as the algebraic sum of the net shortwave radiation R_{ns} and the net long-wave radiation R_{nl}:

$$R_n = R_{ns} + R_{nl}. \tag{5.39}$$

Soil Heat Flux Density G

For daily calculations, G (MJ m^{-2} day^{-1}) can be estimated using a simplified approach to the heat balance of the soil profile:

$$G = 0.1(T_i - T_{i-1}), \tag{5.40}$$

where T_i is the mean daily air temperature (°C), T_{i-1} is the mean air temperature (°C) of the preceding three-day period, and 0.1 is an empirical conversion factor (MJ m^{-2} day^{-1} °C^{-1}). Because the magnitude of daily soil heat flux beneath densely planted grass is relatively small, it can be neglected (i.e., $G = 0$).

For *monthly* time-step calculations, G can be computed from the mean monthly temperatures of the current and preceding months as

$$G = 0.14(T_{\text{month}\,i} - T_{\text{month}\,i-1}) \tag{5.41}$$

Hourly ET_0 Calculations with the FAO-PM Method

It is recommended that ET_0 for hourly or shorter time periods be computed using the FAO-PM equation for hourly calculations (hourly FAO-PM):

$$ET_0 = \frac{0.408\Delta(R_n - G) + \gamma \frac{37}{T+273} U_2(e_s - e_a)}{\Delta + \gamma(1 + 0.34\,U_2)}, \tag{5.42}$$

where the units for ET_0 are mm h^{-1}; R_n and G have units of MJ m^{-2} h^{-1}; U_2 is hourly wind speed at 2 m in m s^{-1}; and T (°C), e_s (kPa), and e_a (kPa) are measured at 2.0 m and are relative to the hour period. The only distinction between the daily Eq. (5.14) and this one is that the 900 coefficient has become 37. Calculation procedures are given by Allen *et al.* [60].

Calculation of ET_0 on hourly or shorter time steps is generally better than using 24-h calculation time steps in areas where substantial changes in wind speed, dewpoint, or cloudiness occur during the day.

The hourly FAO-PM also assumed that bulk surface resistance r_s remains constant throughout a day at 70 s m^{-1}. As noted by many authors [4, 34, 35, 37], r_s has been observed to change with solar radiation, VPD, leaf temperature, and soil moisture deficit. However, relationships determined at various sites and with various data sets do not provide more consistent estimates than a constant r_s. This is discussed by Allen *et al.* [10, 60]. Thus, assuming a hypothetical grass reference with $r_s = 70$ s m^{-1} for all hourly periods is realistic, especially because adjustments for atmospheric stability are also absent from r_a in the FAO-PM equations.

Estimation of ET_0 with Limited Weather Data

For both daily and monthly calculations, data for some weather variables frequently are missing. This requires appropriate procedures to estimate such meteorological variables and then use these estimates in the FAO-PM calculations presented in the section entitled "Daily ET Calculations with FAO-PM Method," above.

Case When Radiation Data Are Missing

Several alternative methodologies can be utilized to estimate shortwave incoming radiation R_s. These have been evaluated using several data sets, namely those for a large variety of climates in the world included in CLIMWAT [80] for a total of 918 weather stations in 20 countries [81].

Use of R_s/R_a Data from a Nearby Weather Station. This method relies on the fact that, for the same month, very often for the same day, the variables commanding incoming solar radiation R_s in a given region are similar. This assumption implies that a nearby station should be at a distance not greater than 300 km; the physiography of the region should be such that relief is negligible in influencing the movement of air masses, and the air masses governing rainfall and cloudiness are the same.

Much caution should be used when the method is to be applied to mountainous and coastal areas, where differences in exposure and altitude could be important or where rainfall is variable because of convective conditions.

Thus, for locations where a nearby weather station exists within the same homogeneous region, the best estimate for R_s is

$$R_s = (R_s/R_a)_{nt} R_a, \tag{5.43}$$

where R_a is the daily extraterrestrial radiation (MJ m^{-2} day^{-1}) for the same location and $(R_s/R_a)_{nt}$ is the ratio of shortwave to extraterrestrial radiation ( ) obtained from observations at the nearest (subscript nt) weather station.

This estimation method is recommended for monthly and daily calculations. However, when using it for daily estimates, a more careful analysis of weather data in the importing and exporting meteorological stations has to be performed to verify if both stations are in the same homogeneous climatic region. That analysis should include the comparison of daily data from both stations, particularly concerning $T_{\max}$, $T_{\min}$, $(T_{\max} - T_{\min})$, $RH_{\max}$,

RH_{min}, and $(RH_{max} - RH_{min})$. In fact, similar cloudiness (and similar n/N ratios) are related to similarities in temperature and humidity trends.

Use of a Relationship with the Temperature Difference ($T_{max} - T_{min}$). It is known that the difference between maximum and minimum air temperature is related to cloudiness conditions. Clear-sky conditions correspond to higher T_{max} because the atmosphere is transparent to the incoming radiation, and to low T_{min} because long-wave radiation is returned only in a small fraction. On the contrary, overcast conditions correspond to a lower T_{max} because a larger part of the radiation is absorbed and reflected upward by the clouds, whereas T_{min} is higher because, reflected downward, long-wave radiation is increased. Therefore, the difference $(T_{max} - T_{min})$ can be used as an indicator of the fraction of extraterrestrial radiation that becomes available at an evaporative surface. This principle has been utilized by Hargreaves and Samani [82] to develop estimates of ET_0 using only air-temperature data. Then [83],

$$R_s = K_r (T_{max} - T_{min})^{0.5} R_a, \tag{5.44}$$

where R_s is the daily shortwave incoming solar radiation (MJ m^{-2} day^{-1}); R_a is the daily extraterrestrial radiation (MJ m^{-2} day^{-1}), T_{max} and T_{min} are the maximum and the minimum air temperatures (°C), and K_r is an adjustment coefficient (°C$^{-1/2}$).

The adjustment coefficient K_r is empirical and takes different values for interior or coastal regions:

- For interior locations, where land mass dominates (i.e., where air masses are not strongly influenced by a large water body), $K_r = 0.16$.
- For coastal locations, situated on or adjacent to the coast of a large land mass that is greater than about 20 km wide in the inland direction (i.e., where air masses are influenced by a nearby water body but depend on the wind direction), $K_r = 0.19$.

This temperature difference method is recommended for locations where it is not appropriate to import R_s/R_a information from a nearby station [Eq. (5.43)], because no homogeneous climate conditions occur or because data are lacking in the region.

Caution is recommended when daily computations are required. Daily estimates should be used when summed or averaged to a several-day period (7 or 10 days, as in irrigation scheduling computations).

Case When Wind-Speed Data Are Missing

Two approaches can be utilized to overcome the problem of missing windspeed data.

Use of Wind-Speed Data from a Nearby Weather Station. Importing wind-speed data from a nearby station, as for R_s discussed earlier, relies on the fact that airflow above a homogeneous region may have relatively large space variations during the course of a day but small variation over longer periods or even over the day itself. This occurs when air masses are of the same origin or when the same fronts govern airflows in the region where both importing and exporting stations are located. Thus, when a location exists within a given region where wind-speed data are available (U_2), it may be possible to utilize the same U_2 at a nearby station where wind-speed data are missing.

The decision to import U_2 data from another station has to be taken after checking the regional climate and relief. In areas with relief, the nearest station may not necessarily be the best one for providing data but rather more distant a station with similar elevation and exposure to the dominant winds may be more appropriate. This may vary from one season to another.

Use of wind-speed data from a nearby weather station is acceptable for daily estimates if climate homogeneity is checked and if daily estimates are to be summed or averaged to several-day periods (7 or 10 days, or time intervals between irrigations) in irrigation scheduling computations. To check validity of imported wind-speed data, trends of other meteorological variables have to be checked. Strong winds often are associated with a decrease in air moisture, and low winds are common with high RH. Thus, trends in variation of daily RH_{max} and RH_{min} (or T_{dry}, T_{wet}, and $T_{dry} - T_{wet}$) should be similar in both locations.

Use of an Empirical Monthly Estimate of Wind Speed. When it is not possible to find a weather station within the climatic region from which to import wind-speed data, then the user may adopt monthly estimates of wind-speed data. Errors in the resulting monthly ET_0 estimates may not be very substantial, particularly when ET is energy driven.

Average wind estimates should be selected from the information available on the climate of the region [80] and may change with the seasonal changes in climate. Indicative values (m s^{-1}) are

Light wind	$U_2 \leq 1.0$
Low to moderate wind	$U_2 \sim 2.0$
Moderate to strong wind	$U_2 \sim 3.0$
Strong wind	$U_2 \geq 4.0$

$U_2 = 2$ m s^{-1} is an average value for much of the globe. Lower values are frequent in humid tropical regions. Caution in selecting an empirical wind-speed estimate is recommended.

Case When Humidity Data Are Missing

Humidity data are required to compute the actual vapor pressure e_a. The daily minimum temperature is suggested as an approximation to the dewpoint temperature:

$$e_a = e^0(T_{min}). \tag{5.45}$$

This approximation provides better estimates for low values of T_{min} and when the air is moist. It often can be expected that T_{min} should be decreased by 2°C to overcome problems resulting from aridity of the weather station [78], namely, outside rainy periods. T_{min} should be decreased by 2°C in semiarid climates and by 3°C in arid ones.

Case When Only T_{mean} Is Available

Because of nonlinearity of the e^0 (T) curve, the daily saturation vapor pressure e_s should be computed from T_{max} and T_{min} [Eq. (5.19)]. However, for some data sets, only the mean daily temperature T_{mean} is available. Computing e_s from T_{mean} would provide underestimation of e_s and of VPD, and thus of ET_0. It is then advisible to estimate T_{max} and T_{min} from T_{mean}.

Adopting the general relationship given by the Eq. (5.44), it is possible to compute

$$T_{max} = T_{mean} + \frac{1}{2}(5.9R_s/R_a)^2 \quad (5.46)$$

$$T_{min} = T_{mean} - \frac{1}{2}(5.9R_s/R_a)^2 \quad (5.47)$$

The empirical coefficient 5.9 corresponds in Eq. (5.44) to $K_r = 0.17$. This coefficient may be modified for specific regional climatic conditions.

Comments on Using ET_0 PM with Estimates of Missing Weather Data

Approaches suggested earlier intend to perform ET_0 calculations with only one standard equation, the FAO-PM, that is, Eq. (5.14), thus avoiding the problems of using alternative ET equations whose behavior could differ from that of the FAO-PM. Deviations from the approximations proposed to the full FAO-PM are expected to be in the range of those resulting from the use of an alternative ET equation or even less.

Use of the FAO-PM equation with only maximum and minimum temperature data, that is, estimating R_s from the nearest station [Eq. (5.43)], $T_d = T_{min}$ and $U_2 = 2$ m s^{-1}, provided an average standard error of estimate (SEE) equal to 0.60 mm day^{-1} for the monthly estimates of ET_0 relative to the FAO-PM equation with full data sets when applied to 918 locations in 20 countries, using the CLIMWAT database. Application of the Hargreaves-Samani equation (5.50) to the same data set resulted in a higher SEE = 0.85 mm day^{-1}. The FAO-PM equation with R_s estimated from $(T_{max} - T_{min})$, $T_d = T_{min}$, and $U_2 = 2$ m s^{-1} was compared with daily lysimeter data from Davis, California, resulting in SEE = 1.01 mm day^{-1}, which was only 6% higher than the SEE = 0.95 mm day^{-1} obtained from the Hargreaves-Samani equation (5.50). If imported wind could be used, a lower SEE could be expected. This indicates that a proper application of methodologies described in the preceding sections also can be used for estimating daily ET_0, mainly when these values are averaged over several-day periods, as in irrigation scheduling computations.

Approximations proposed earlier can be validated at the regional level (see "Calibration," below). Sensitivity analysis should be performed to check causes (and limits) for the method utilized to import the missing data.

Estimation of ET_0 Using Other Equations

Calibration

As presented above, the FAO-PM equation has been selected together with the updated definition of reference ET. All calculation procedures have been standardized according to the weather data observations and the timescale of computations to improve the accuracy of estimation. Therefore, it is highly recommended that the FAO-PM method be applied when full data are available and when appropriate estimation procedures are applied to missing data. Therefore, if a different method is to be used, it should first be validated and/or calibrated using the FAO-PM.

The validation or calibration of the method should be on a regional basis because different trends and ranges in the variation of meteorological variables induce different behavior of the ET equations. These are never linear and may respond with more sensitivity to one variable than to another. Thus, regional validation or calibration should be performed in large regions, where those trends and ranges in the variation of weather variables are similar. However, there is less need for climatic homogeneity as is required when variables have to be imported from one weather station to another to complete data sets.

For regional calibration of the selected equation, simple calibration factors can be computed on a monthly basis through linear regression as

$$b = \frac{ET_0}{ET_{\text{eqn}}} \quad \text{or} \quad b = \frac{ET_0 - a}{ET_{\text{eqn}}}, \tag{5.48}$$

where ET_0 is the reference ET as defined by the FAO-PM equation (5.14) and ET_{eqn} is ET predicted by the equation being calibrated. In general, it is appropriate to perform a regression through the origin. Regression coefficients b and a are calibration factors. ET_0 then is calculated with the selected equation as

$$ET_0 = b\,ET_{\text{eqn}} \quad \text{or} \quad ET_0 = a + b\,ET_{\text{eqn}} \tag{5.49}$$

The ET equation is considered calibrated when the SEE $\leq 20\%$ of the average ET_0 estimate.

For these regional validation or calibration studies, it is recommended that one use weather data observed above a green vegetation cover. Because the sensitivity of ET_0 estimates to changes in T and e_a are generally nonlinear, the ratios of sensitivities of two ET_0 equations will rarely equal the ratios of ET_0 estimates by the two equations. Therefore, use of nonreference weather data will introduce a bias in linear calibration coefficients.

When the vegetation becomes dry during any period, it is advisable to use only weather data from months during which precipitation or irrigation is sufficient (greater than ET_0) to ensure that the stations were surrounded by green, evaporating vegetation and that weather measurements were synchronized with the reference ET energy exchange.

ET_0 Equations

Methods for computing ET_0 from weather data of Doorenbos and Pruitl [55] may be utilized when validated or calibrated against the FAO-PM. However, they lose importance.

- The FAO-PM relies on a conceptual approach—the original Penman method—which has been modified into a new and more solid approximation of the evaporation from vegetated surfaces, the PM approach. The use of the original Penman requires a locally calibrated wind function [14], which would require appropriate data sets, available only in very few locations. Therefore, the Penman approaches definitely should be replaced by the PM (FAO-PM) method.
- The FAO-radiation (FAO-Rad) equation corresponds to an approximation of the radiation term $[\Delta(R_n - G)/(\Delta + \gamma^*)]$ in the FAO-PM equation, but uses an adjustment

factor c for daytime wind U_2 and mean RH. The FAO-Rad requires observations of temperature and sunshine, and the use of regional estimates for RH and U_2. The FAO-PM can be utilized when T_{min} replaces T_{dew} and when regional U_2 values are imported to the location where they are missing. Therefore, the use of FAO-Rad is not considered to be an alternative to FAO-PM when U_2 and RH data are missing.

- The FAO Blaney-Criddle (FAO-BC) method estimates ET_0 from temperature data only and requires correction for RH, wind speed, and relative sunshine using regional values of these variables. However, the FAO-BC has been shown to overestimate ET_0 for a large number of locations, particularly for the nonarid ones. Thus, as indicated earlier (in "Comments on Using ET_0 PM with Estimates of Missing Data"), when only temperature data are available, there is no advantage in using an alternative to the FAO-PM equation with estimates of missing weather data.
- The Hargreaves method [82, 84] not only has shown good results as a temperature method when applied in a variety of locations but, despite being empirical, it has the appropriate conceptual foundation to be utilized and represents a great improvement in relation to the FAO-BC method. It can be expressed as

$$ET_{0\,\mathrm{Harg}} = C_0(T_{max} - T_{min})^{0.5}(T_{mean} + 17.8)R_a, \tag{5.50}$$

where $ET_{0\,\mathrm{Harg}}$ is the estimate of grass reference ET (mm day^{-1}), T_{max} and T_{min} are the maximum and minimum daily air temperatures (°C), T_{mean} is the mean daily air temperature (°C), R_a is the extraterrestrial radiation (MJ m^{-2} day^{-1}), C_0 is the conversion coefficient $= 0.000939$ ($= 0.0023$ when R_a is in mm day^{-1}) and 17.8 is the factor for conversion of °C to °F.

The temperature difference ($T_{max} - T_{min}$) allows for an approximation to the amount of radiation available at the surface as analyzed for Eq. (5.44). The difference ($T_{max} - T_{min}$) is also an indicator for VPD, which is normally higher for clear sky (high $T_{max} - T_{min}$) and lower for overcast conditions (low $T_{max} - T_{min}$).

However, because the equation can underestimate ET for locations where high wind speed is associated with high VPD, and because it may overestimate if humidity is high, validation or calibration of the equation is recommended for such regions [see Eqs. (5.48) and (5.49)].

The Hargreaves equation can be computed on a monthly or a daily basis. However, the best use of daily estimates is made when summed or averaged for a week or 10-day period, as is usual for irrigation scheduling computations.

Pan-Evaporation Method

Evaporation pans provide a measurement of the integrated effect of radiation, wind, temperature, and humidity on evaporation from a specific open water surface. In a similar fashion, the plant responds to the same climatic variables, but several major factors may produce significant differences in loss of water. Reflection of solar radiation from a water surface is only 5%–8%, but from most vegetative surfaces is 20%–25%. Storage of heat within the pan can be appreciable and may cause high evaporation during the night; most crops transpire only during the daytime. Also, the difference in water losses from pans and from crops can be affected by differences in turbulence, temperature, and humidity of the air immediately above the surfaces. Heat transfer through the sides of the pan can occur, which may be severe for sunken pans. Also the color of the pan and the use of screens will affect water losses. The siting of the pan and the pan environment influence the measured results, especially when the pan is placed in fallow rather than cropped fields.

Notwithstanding these deficiencies, with proper siting the use of pans to predict CWR for periods of 10 days or longer is still warranted. The use of the U.S. Class A pan and the Colorado sunken is presented by Doorenbos and Pruitt [55]. To relate pan evaporation (E_{pan}) to reference crop ET (ET_0), empirically derived coefficients (K_p) are used to account for climate and pan environment. ET_0 can be obtained from

$$ET_0 = K_p E_{pan}, \tag{5.51}$$

where E_{pan} is the pan evaporation (mm day^{-1}) and represents the mean daily value for the period considered, and K_p is the pan coefficient ( ).

Values for K_p are given by Doorenbos and Pruitt [55]. Allen and Pruitt [58] developed polynomial equations to replace tables for the determination of the pan coefficients K_p according to the type of evaporation pan and upwind fetch. In general, local calibration of K_p coefficients can be performed as indicated for Eqs. (5.48) and (5.49) making $K_p = b$ in the linear regression through the origin. Polynominals for K_p can be obtained through stepwise multiregression analysis between the ratios ET_0/E_{pan} and the variables U_2 and RH_{mean} observed for the same period.

5.1.5 Crop Evapotranspiration

Introduction to Crop Evapotranspiration and Crop Coefficients

Crop ET (ET_c) is calculated by multiplying the reference crop ET (ET_0) by a dimensionless crop coefficient K_c:

$$ET_c = K_c ET_0. \tag{5.52}$$

The reference crop corresponds to a living, agricultural crop (clipped grass) and it incorporates the majority of the effects of variable weather into the ET_0 estimate. Therefore, because ET_0 represents an index of climatic demand on evaporation, the K_c varies predominately with the specific crop characteristics and only somewhat with climate. This enables the transfer of standard values for K_c between locations and between climates.

The crop coefficient K_c is the ratio of the crop ET_c to the reference ET_0, and it represents an integration of the effects of three primary characteristics that distinguish the crop from the reference ET_0. These characteristics are crop height (affecting roughness and aerodynamic resistance); crop–soil surface resistance (affected by leaf area, the fraction of ground covered by vegetation, leaf age and condition, the degree of stomatal control, and soil surface wetness); and albedo (reflectance) of the crop–soil surface (affected by the fraction of ground covered by vegetation and by the soil surface wetness).

The crop height h influences the aerodynamic resistance r_a [Eqs. (5.4) and (5.5)] and the turbulent transfer of vapor from the crop into the atmosphere. The combined crop–soil surface resistance influences the bulk surface resistance r_s [Eq. (5.6)], which represents the resistance to vapor flow from within plant leaves and from beneath the soil surface. The albedo of the crop–soil surface influences the net radiation of the surface R_n, which is the primary source of the energy exchange for the evaporation process.

Two K_c approaches are considered. The first uses a time-averaged K_c to estimate ET_c, which includes time-averaged (multiday) effects of evaporation from the soil surface. The second approach uses the basal crop coefficient, and a separate, daily calculation is made to estimate evaporation from the soil surface.

The time-averaged K_c is used for planning studies and irrigation system design where averaged effects of soil wetting are acceptable and relevant. For typical irrigation management, the time-averaged K_c is valid and currently is utilized in many irrigation-scheduling simulation models. The basal K_c approach, which requires more numerical calculations, is best for irrigation scheduling, soil water-balance computations, and for research studies where effects of day-to-day variation in soil surface wetness and the resulting impacts on daily ET_c, the soil moisture profile, and deep percolation fluxes are important.

The crop coefficient curve (Fig. 5.3) represents the changes in K_c over the length of the growing season. The shape of the curve represents the changes in the vegetation and ground cover during plant development and maturation that affect the ratio of ET_c to ET_0.

As shown in Fig. 5.3, shortly after planting of annuals or shortly after the initiation of new leaves for perennials, the value for K_c is small, often less than 0.4. The K_c begins to increase from the initial K_c value, $K_{c\,\mathrm{ini}}$, at the beginning of rapid plant development and reaches a maximum value, $K_{c\,\mathrm{mid}}$, at the time of maximum or near-maximum plant development. The period during which $K_c = K_{c\,\mathrm{mid}}$ is referred to as the midseason period. During the late-season period, as leaves begin to age and senescence due to natural or cultural practices occurs, the K_c begins to decrease until it reaches a lower value at the end of the growing period equal to $K_{c\,\mathrm{end}}$.

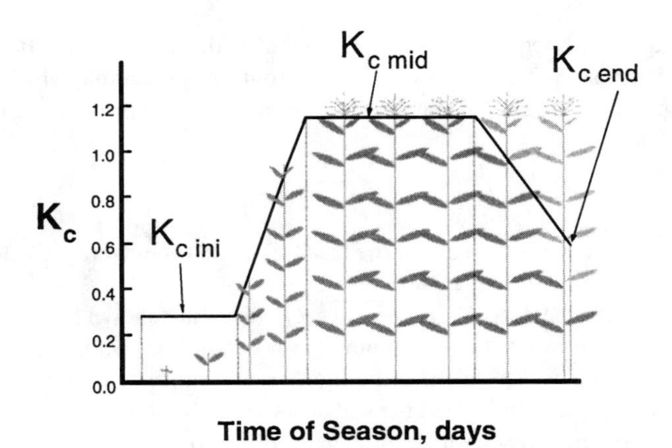

Figure 5.3. Generalized crop coefficient curve during a growing season.

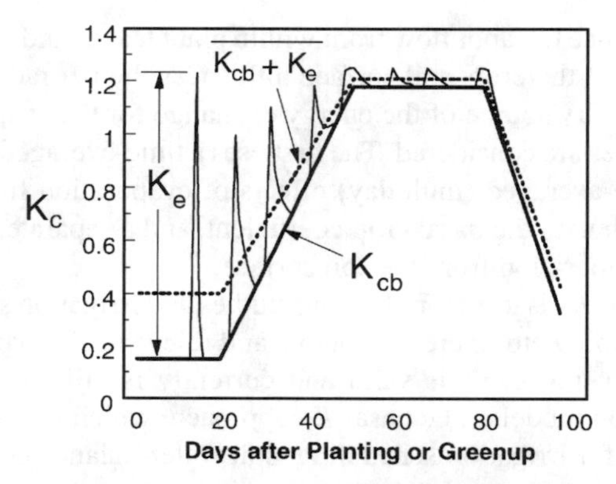

Figure 5.4. Crop coefficient definitions showing the basal K_{cb}, soil evaporation K_e, and time-averaged $\overline{K_{cb} + K_e}$ values.

The form for the equation used in the basal K_c approach is

$$K_c = K_s K_{cb} + K_e, \tag{5.53}$$

where K_s is the stress reduction coefficient (0 to 1), K_{cb} is the basal crop coefficient (0 to ~1.4), and K_e is the soil water evaporation coefficient (0 to ~1.4). K_{cb} is defined as the ratio of ET_c to ET_0 when the soil surface layer is dry, but where the average soil water content of the root zone is adequate to sustain full plant transpiration. As shown in Fig. 5.4, the K_{cb} represents the baseline potential K_c in the absence of the additional effects of soil wetting by irrigation or precipitation. K_s reduces the value of K_{cb} when the average soil water content of the root zone is not adequate to sustain full plant transpiration. K_e describes the evaporation component from wet soil in addition to the ET represented in K_{cb}.

Because Eq. (5.53) requires the calculation of a daily soil water balance for the surface soil layer, a simplification is required for routine application, which may involve long time steps or wetting effects averaged over many fields or over many years. The time-averaged K_c is

$$K_c = \overline{K_{cb} + K_e}, \tag{5.54}$$

where $\overline{K_{cb} + K_e}$ represents the sum of the basal K_{cb} and time-averaged effects of evaporation from the surface soil layer, K_e.

Typical shapes for the K_{cb}, K_e, and $\overline{K_{cb} + K_e}$ curves are shown in Fig. 5.4. The K_{cb} curve in the figure represents the minimum K_c for conditions of adequate soil moisture and dry soil surface. The K_e spikes in the figure represent increased evaporation when precipitation or irrigation has wetted the soil surface and has temporarily increased total ET_c. The spikes generally reach a maximum value of 1.0 to 1.2, depending on the climate, the depth of soil wetting, and the portion of soil surface wetted. When summed, the values for K_{cb} and for K_e represent the total crop coefficient, K_c, shown as the dashed line in

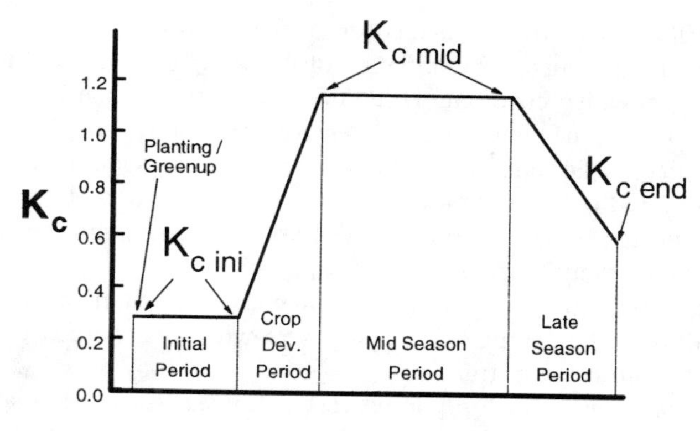

Figure 5.5. Crop coefficient curve and stage definitions.

Fig. 5.4. The K_c or $\overline{K_{cb} + K_e}$ curve lies above the K_{cb} curve, with potentially large differences during the initial and development stages, depending on the frequency of soil wetting. During the midseason period, when the crop is likely to be near full cover, the effects of soil wetness are less pronounced.

Average Crop Coefficients

The K_c Curve

A simple procedure is used to construct the K_c curve for the complete growing period. This procedure was first presented by Doorenbos and Pruitt [55]. The K_c curve, such as that shown in Fig. 5.5, is constructed using the following three steps:

1. Divide the growing period into four general growth stages that describe crop phenology or development, and determine the lengths of these stages, in days. The four stages are initial period, crop development period, midseason period, late-season period.
2. Identify the three K_c values that correspond to $K_{c\,\text{ini}}$, $K_{c\,\text{mid}}$ and $K_{c\,\text{end}}$.
3. Connect straight-line segments through each of the four growth-stage periods. Horizontal lines are drawn through $K_{c\,\text{ini}}$ in the initial period and through $K_{c\,\text{mid}}$ in the midseason period. Diagonal lines are drawn from $K_{c\,\text{ini}}$ to $K_{c\,\text{mid}}$ within the course of the development period and from $K_{c\,\text{mid}}$ to $K_{c\,\text{end}}$ within the course of the late-season period.

Lengths of Crop Growth Stages

The four crop growth stages are characterized in terms of benchmark crop growth stages or cultivation practices:

1. *Initial period*—planting date to approximately 10% ground cover. The length of the initial period is highly dependent on the crop, the crop variety, the planting date, and the climate. For perennial crops, the planting date is replaced by the green-up date, that is, the time when initiation of new leaves occurs.

2. *Crop development*—10% ground cover to effective full cover. Effective full cover for many crops occurs at the initiation of flowering or when LAI reaches 3. For row crops, effective cover can be defined as the time when leaves of plants in adjacent rows begin to intermingle or when plants reach nearly full size if no intermingling occurs. For tall crops, it occurs when the average fraction of the ground surface covered by vegetation is about 0.7 to 0.8. For densely sown vegetation, such as winter and spring cereals and grasses, effective full cover corresponds approximately to the stage of heading.
3. *Midseason*—effective cover to start of maturity. The start of maturity is often indicated by the beginning of the aging, yellowing, or senescence of leaves, leaf drop, or the browning of fruit.
4. *Late season*—start of maturity to harvest or full senescence. For some perennial vegetation in frost-free climates, crops may grow year round so that the date of termination may be taken as the same as the date of "planting."

The rates at which vegetation cover develops and the time at which it attains effective full cover are affected by weather conditions, and especially by mean daily air temperature [85]. Therefore, the length in time between planting and effective full cover will vary with climate, latitude, elevation, planting date, and cultivar (crop variety). Generally, once the effective full cover for a plant canopy has been reached, the rate of further phenological development (flowering, seed development, ripening, and senescence) is more dependent on plant genotype and is less dependent on weather [71]. In some situations, the time of emergence of vegetation and the time of effective full cover can be predicted using cumulative degree-based regression equations or by more sophisticated plant growth models [85–89].

The lengths of the initial and development periods may be relatively short for *deciduous trees and shrubs* that can develop new leaves in the spring at relatively fast rates. The $K_{c\,\mathrm{ini}}$ selected for trees and shrubs should reflect the ground condition prior to leaf emergence or initiation because $K_{c\,\mathrm{ini}}$ is affected by the amount of grass or weed cover, soil wetness, tree density, and mulch density.

The length of the late-season period may be relatively short (<10 days) for vegetation killed by frost or for agricultural crops that are harvested fresh. The value for $K_{c\,\mathrm{end}}$ after full senescence or following harvest should reflect the condition of the soil surface (average water content and any mulch cover) and the condition of the vegetation following plant death or harvest.

Indicative lengths of growth stages are given in FAO Papers [55, 90, 91]. Local observations of the specific stage of plant development should be used, whenever possible, to incorporate effects of plant variety, climate, and cultural practices. Local information can be obtained by interviewing farmers, ranchers, agricultural extension agents, and local researchers, by conducting local surveys, or by remote sensing. When determining stage dates from local observations, the preceding guidelines and visual descriptions may be helpful.

Crop Coefficients

Values for $K_{c\,\mathrm{ini}}$, $K_{c\,\mathrm{mid}}$, and $K_{c\,\mathrm{end}}$ are listed in Table 5.1 for various agricultural crops. Most of the K_c values in Table 5.1 were taken from Doorenbos and Pruitt [55] and Doorenbos and Kassam [90] after some modification. Additional information was obtained from Wright [71], Pruitt [92], Snyder *et al.* [93, 94], Jensen *et al.* [2], and other sources.

The K_c values in Table 5.1 are organized by crop group type. Usually there is close similarity in K_c within the same crop group because the plant height, leaf area, ground coverage, and water management are usually similar. For several group types, only one value for $K_{c\,\text{ini}}$ is listed and is considered to be representative of the whole group. The $K_{c\,\text{ini}}$ value in Table 5.1 is only approximate because $K_{c\,\text{ini}}$ can vary widely with soil wetting conditions, irrigation, and rainfall.

The K_c values in Table 5.1 represent potential water use by healthy, disease-free, and densely planted stands of vegetation and with adequate levels of soil water. When stand density, height, or leaf area are less than that attained under perfect or normal (pristine) conditions, the value for K_c should be reduced. The reduction in $K_{c\,\text{mid}}$ for poor stands can be as much as 0.3–0.5 and should be made according to the amount of effective (green) leaf area relative to that for healthy vegetation having normal planting densities.

K_c for Initial Stage for Annual Crops

ET during the initial stage for annual crops is predominately in the form of evaporation. Therefore, accurate estimates for $K_{c\,\text{ini}}$ must consider the frequency that the soil surface is wetted during the initial period. Numerical procedures to compute $K_{c\,\text{ini}}$ are given below. Graphical procedures are presented in Doorenbos and Pruitt [55] and in Allen *et al.* [91].

Evaporation from bare soil, E_s, can be characterized as occurring in two distinct stages [96, 97]. Stage 1 is termed the "energy-limited" stage. During this stage, moisture is transported to the soil surface at a rate sufficient to supply the potential rate of evaporation (E_{so}), which is governed by energy availability at the soil surface. E_{so} (mm day^{-1}) can be estimated from

$$E_{so} = 1.15\,ET_0, \tag{5.55}$$

where ET_0 is the the mean ET_0 during the initial period (mm day^{-1}).

Stage 2 is termed the "soil-limited" stage, where hydraulic transport of subsurface water to the soil surface is unable to supply water at the potential evaporation rate. During stage 2, the soil surface appears partially dry and a portion of the evaporation occurs from below the soil surface. The energy required for subsurface evaporation is supplied by transport of heat from the soil surface into the soil profile. The evaporation rate during stage-2 drying decreases as soil moisture decreases (Fig. 5.6). It therefore can be expressed as being proportional to the water remaining in the evaporation layer relative to the maximum depth of water that can be evaporated from the same soil layer during stage-2 drying. The maximum total depth of water that can be evaporated from the surface soil layer is termed W_x. If the evaporation rate during stage-2 drying is assumed to be linearly proportional to the equivalent depth of water remaining in the evaporation layer, as shown in Fig. 5.6, then the average soil water evaporation rate can be estimated as proposed by Allen *et al.* [76].

Table 5.1. Time-averaged crop coefficients K_c, maximum plant heights h, range of maximum root depths Z_r, and depletion fraction for no stress F_{ns}, well-managed crops in subhumid climates for use with the FAO-PM ET_0

Crop[a]	$K_{c\,ini}$[b]	$K_{c\,mid}$[c]	$K_{c\,end}$[c]	Maximum Plant Height h (m)	Maximum Root Depth Z_r (m)	F_{ns}[d]
a. Small Vegetables	**0.7**	**1.05**	**0.95**			
Carrots		1.05	0.95	0.3	0.5–1.0	0.35
Celery		1.05	1.00	0.6	0.3–0.5	0.20
Crucifers: cabbage, cauliflower, broccoli, Brussels sprouts		1.05	0.95	0.4	0.4–0.6	0.40
Garlic		1.00	0.70	0.3	0.3–0.5	0.30
Lettuce		1.00	0.95	0.3	0.3–0.5	0.30
Onions, dry		1.05	0.75	0.4	0.3–0.6	0.30
Onions, green		1.00	1.00	0.3	0.3–0.6	0.30
Spinach		1.00	0.95	0.3	0.3–0.5	0.20
Radishes		0.90	0.85	0.3	0.3–0.5	0.30
b. Roots and Tubers	**0.5**	**1.10**	**0.95**			
Beets, table		1.05	0.95	0.4	0.6–1.0	0.50
Cassava, year 1	0.3	0.80	0.30	1.0	0.5–0.8	0.60
Cassava, year 2	0.3	1.10	0.50	1.5	0.7–1.0	0.60
Parsnips	0.5	1.05	0.95	0.4	0.5–1.0	0.40
Sugar beets	0.35	1.20	0.70	0.5	0.7–1.2	0.55
Sweet potatoes		1.15	0.65	0.5	1.0–1.5	0.65
Turnips (and rutabagas)		1.10	0.95	0.6	0.5–1.0	0.50
c. Legumes	**0.4**	**1.15**	**0.55**			
Beans, green	0.5	1.05	0.90	0.4	0.5–0.7	0.45
Beans, dry and pulses	0.4	1.15	0.30	0.4	0.6–0.9	0.45
Chick peas		1.00	0.35	0.4	0.6–1.0	0.50
Fava beans (broad beans), fresh	0.5	1.15	1.10	0.8	0.5–0.7	0.45
Fava beans, dry/seed	0.5	1.15	0.30	0.8	0.5–0.7	0.50
Garbanzos	0.4	1.15	0.30	0.8	0.6–1.0	0.50
Green gram and cowpeas		1.05	0.60–0.35	0.4	0.6–1.0	0.50
Groundnuts		1.15	0.60	0.4	0.5–1.0	0.50
Lentils		1.10	0.30	0.5	0.6–0.8	0.50
Peas, fresh	0.5	1.15	1.10	0.5	0.6–1.0	0.35
Peas, dry/seed		1.15	0.30	0.5	0.6-1.0	0.40
Soybeans		1.15	0.50	0.5–1.0	0.6–1.3	0.50
d. Solanum crops	**0.6**	**1.15**	**0.80**			
Bell peppers, fresh		1.05	0.90	0.7	0.5–1.0	0.30
Eggplant		1.05	0.90	0.8	0.7–1.2	0.45
Potatoes		1.15	0.75–0.40	0.6	0.4–0.6	0.35
Tomatoes		1.15	0.70–0.90	0.6	0.7–1.5	0.40
e. Cucumber crops	**0.50**	**1.00**	**0.80**			
Cantaloupe	0.50	0.85	0.60	0.3	0.9–1.5	0.45
Cucumber, fresh market	0.60	1.00	0.75	0.3	0.7–1.2	0.50
Cucumber, machine harvest	0.50	1.00	0.90	0.3	0.7–1.2	0.50
Melons	0.50	1.05	0.75	0.4	0.8–1.5	0.40
Pumpkin, winter squash	0.50	1.00	0.80	0.4	1.0–1.5	0.35
Squash (zuchini and crookneck)	0.50	0.95	0.75	0.3	0.6–1.0	0.50
Watermelon	0.40	1.00	0.75	0.4	0.8–1.5	0.40

(Cont.)

Table 5.1. (Continued)

Crop[a]	$K_{c\,ini}$[b]	$K_{c\,mid}$[c]	$K_{c\,end}$[c]	Maximum Plant Height h (m)	Maximum Root Depth Z_r (m)	F_{ns}[d]
f. Tropical fruits						
Bananas, year 1	0.50	1.10	1.00	3.0	0.5–0.9	0.35
Bananas, year 2	1.00	1.20	1.10	4.0	0.5–0.9	0.35
Cacao	1.00	1.05	1.05	3.0	0.7–1.0	0.30
Coffee, bare soil	0.90	0.95	0.95	2.0–3.0	0.9–1.5	0.40
Coffee, with weeds	1.05	1.10	1.10	2.0–3.0	0.9–1.5	0.40
Dates	0.90	0.95	0.95	8.0	1.5–2.5	0.50
Palm trees	0.95	1.00	1.00	8.0	0.7–1.0	0.65
Pineapple, bare soil	0.50	0.30	0.30	0.6–1.2	0.3–0.6	0.50
Pineapples, vegetated soil	0.50	0.50	0.50	0.6–1.2	0.3–0.6	0.50
Rubber trees	0.95	1.00	1.00	10	1.0–2.0	0.60
Tea, nonshaded	0.95	1.00	1.00	1.5	0.9–1.5	0.40
Tea, shaded	1.10	1.15	1.15	2.0	0.9–1.5	0.45
g. Multiannual vegetables (with initially bare soil)						
Artichokes	0.50	1.00	0.95	0.7	0.6–0.9	0.45
Asparagus	0.50	0.95	0.30	0.2–0.8	1.2–1.8	0.45
Hops	0.30	1.05	0.85	5.0	1.0–1.2	0.50
h. Fiber crops	**0.35**					
Cotton		1.15–1.20	0.70–0.50	1.2–1.5	1.0–1.7	0.60
Flax		1.10	0.25	1.2	1.0–1.5	0.50
Sisal		0.4–0.7	0.4–0.7	1.5	1.0–2.0	0.80
i. Oil crops	**0.35**	**1.15**	**0.35**			
Castor beans (ricinus)		1.15	0.55	0.3	1.0–2.0	0.50
Rapeseed, canola		1.0–1.15	0.35	0.6	1.0–1.5	0.60
Safflower		1.0–1.15	0.25	0.8	1.0–2.0	0.60
Sesame		1.10	0.25	1.0	1.0–1.5	0.60
Sunflower		1.0–1.15	0.35	2.0	0.8–1.5	0.45
j. Cereals	**0.30**	**1.15**	**0.40**			
Barley		1.15	0.25	1.0	1.0–1.5	0.55
Oats		1.15	0.25	1.0	1.0–1.5	0.55
Wheat		1.15	0.25–0.40	1.0	1.0–1.5	0.55
Winter wheat	0.4–0.7	1.15	0.25–0.40	1.0	1.0–1.8	0.55
Maize field (grain) (field corn)		1.20	0.60–0.35	2.2	1.0–1.7	0.55
Maize, sweet (sweet corn)		1.15	1.05	1.5	0.8–1.2	0.50
Millet		1.00	0.30	1.5	1.0–2.0	0.55
Sorghum, grain		1.00–1.10	0.55	1.0–2.0	1.0–2.0	0.55
Sorghum, sweet		1.20	1.05	2.0–4.0	1.0–2.0	0.50
Rice	1.05	1.20	0.90–0.60	1.0	0.5–1.0	0.20[e]
k. Forages						
Alfalfa hay[f]	0.40	1.20	1.15	0.7	1.0–2.0	0.55
Clover hay[f], berseem	0.40	1.15	1.10	0.6	0.6–0.9	0.50
Grazing pasture, rotated grazing	0.40	0.85–1.05	0.85	0.15	0.5–1.5	0.60
Grazing pasture, extensive grazing	0.30	0.75	0.75	0.1	0.5–1.5	0.65
Turf grass, cool season	0.90	0.95	0.95	0.1	0.5–1.0	0.40
Turf grass, warm season	0.80	0.85	0.85	0.1	0.5–1.0	0.50
Bermuda spring crop for seed	0.35	0.90	0.65	0.4	1.0–1.5	0.60

(Cont.)

Table 5.1. (Continued)

Crop[a]	$K_{c\,ini}$[b]	$K_{c\,mid}$[c]	$K_{c\,end}$[c]	Maximum Plant Height h (m)	Maximum Root Depth Z_r (m)	F_{ns}[d]
Bermuda, summer for hay[f]	0.55	1.00	0.85	0.35	1.0–1.5	0.55
Rye grass, for hay[f]	0.50	1.05	1.00	0.35	0.5–1.0	0.60
Sudan grass for hay[f]	0.50	1.15	1.10	0.8–1.2	1.0–1.5	0.60
l. Sugar Cane	0.40	1.25	0.75	3–4	1.2–2.0	0.65
m. Nonwoody perennials						
Mint	0.60	1.15	1.10	0.6	0.4–0.8	0.40
Strawberries	0.40	0.85	0.75	0.2	0.2–0.3	0.20
n. Grapes and berries						
Berries (bushes)	0.30	1.05	0.50	1.5	0.6–1.2	0.50
Grapes for table	0.30	0.85	0.45	2.0	1.0–2.0	0.35
Grapes for wine	0.30	0.70	0.45	1.5–2.0	1.0–2.0	0.60
o. Fruit trees						
Almonds, no ground cover[g]	0.40	0.90	0.65	4.0	1.0–2.0	0.60
Avocado, no ground cover	0.60	0.85	0.75	3.0	0.5–1.0	0.70
Citrus, no ground cover[h]						
70% canopy	0.70	0.65	0.70	4.0	1.2–1.5	0.50.
50% canopy	0.65	0.60	0.65	3.0	1.1–1.5	0.50
20% canopy	0.50	0.45	0.55	2.0	0.8–1.1	0.50
Citrus, with active ground cover or weeds[g]						
70% canopy	0.75	0.70	0.75	4.0	1.2–1.5	0.50
50% canopy	0.80	0.80	0.80	3.0	1.1–1.5	0.50
20% canopy	0.85	0.85	0.85	2.0	0.8–1.1	0.50
Conifer trees	1.00	1.00	1.00	10.0	1.0–1.5	0.70
Deciduous orchard, killing frost						
Apples, pears, cherries, no ground cover	0.45	0.95	0.70	4.0	1.0–2.0	0.50
Peaches, stone fruit, no ground cover	0.45	0.90	0.65	3.0	1.0–2.0	0.50
Apples, pears, cherries, active ground cover	0.50	1.20	0.95	4.0	1.0–2.0	0.50
Peaches, stone fruit, active ground cover	0.50	1.15	0.90	3.0	1.0–2.0	0.50
Deciduous orchard, no killing frost						
Apples, pears, cherries, no ground cover	0.60	0.95	0.75	4.0	1.0–2.0	0.50
Peaches, stone fruit, no ground cover	0.55	0.90	0.65	3.0	1.0–2.0	0.50
Apples, pears, cherries, active ground cover	0.80	1.20	0.85	4.0	1.0–2.0	0.50
Peaches, stone fruit, active ground cover	0.80	1.15	0.85	3.0	1.0–2.0	0.50
Kiwi	0.40	1.05	1.05	3.0	0.7–1.3	0.35
Olives (40%–60% ground coverage by canopy)	0.65	0.70	0.70	3.0–5.0	1.2–1.7	0.70
Pistachios, no ground cover	0.40	1.10	0.45	3.0–5.0	1.0–1.5	0.60
Walnut orchard	0.50	1.10	0.65	4.0–5.0	1.7–2.4	0.50
p. Wetlands, temperate climate						
Cattails, bulrushes, killing frost	0.30	1.20	0.30	2.0		
Cattails, bulrushes, no frost	0.60	1.20	0.60	2.0		
Short vegetation, no frost	1.05	1.10	1.10	0.3		
Reed swamp, standing water	1.00	1.20	1.00	1–3		
Reed swamp, moist soil	0.90	1.20	0.70	1–3		

[a] When stand density, height, or leaf area are less than that attained under pristine conditions, the value for $K_{c\,mid}$ and, for most crops, for $K_{c\,end}$ should be reduced from 0.1 up to 0.5 according to the amount of effective, active leaf area relative to that for healthy vegetation under potential, pristine growing conditions.

Maximum plant heights are indicative and should be adapted for local conditions. Smaller values for maximum root depths refer to irrigated crops and the large ones to rainfed conditions, when the soil is nonrestrictive to root development.

[b] These are general values under typical irrigation management and soil wetting. For frequent wettings, such as with high-frequency sprinkle irrigation or rainfall, these values may increase substantially. Appropriate computation procedures are given in Section "K_c for Initial Stage for Annual Crops".

[c] Values represent those for a semihumid climate ($RH_{\min} \sim 45\%$) with moderate wind speed (averaging $2\ \mathrm{m\,s^{-1}}$). For more humid or arid conditions or for more or less windy conditions, values should be modified as described in Section "K_c for Midseason and End of Season".

[d] Fraction of available water that can be depleted before physiological, soil, and climatic influences reduce plant transpiration because of moisture stress. F_{ns} is similar to the term "management-allowed depletion" (MAD). However, the values for MAD are influenced by management and economic factors in addition to physical factors. Generally, MAD $< F_{ns}$ when there is risk aversion or uncertainty, and MAD $> F_{ns}$ when plant moisture stress is an intentional part of soil water management. Values listed for F_{ns} are for ET_c in the range of 5 to 6 mm day^{-1} and for nonsaline conditions. For ET_c outside this range, FAO [3, 90] suggests increasing F_{ns} for $ET_c < 5$ mm day^{-1} and decreasing F_{ns} for $ET_c > 6$ mm day^{-1}. The modifications to F_{ns} can be characterized by the equation $F_{ns} = F_{ns\text{ tab}} + 0.04\ (5 - ET_c)$, where $F_{ns\text{ tab}}$ is the F_{ns} value listed in the table and ET_c is the estimated crop ET (mm day^{-1}), from Eqs. (5.52) and (5.53) (both calculated using $K_s = 1$) and Eqs. (5.54).

[e] F_{ns} for rice is 0.20 of saturation [90].

[f] The three coefficients for hay crops represent the periods immediately following cutting, during full cover, and immediately before cutting, respectively.

[g] $K_{c\text{ end}}$ values are K_c prior to leaf drop. After leaf drop $K_{c\text{ end}} = 0.20$ for bare or dead ground cover, and K_c end $= 0.50$ to 0.80 for actively growing ground cover.

[h] For humid and subhumid climates, where there is less stomatal control, the K_c should be increased by 0.1–0.2.

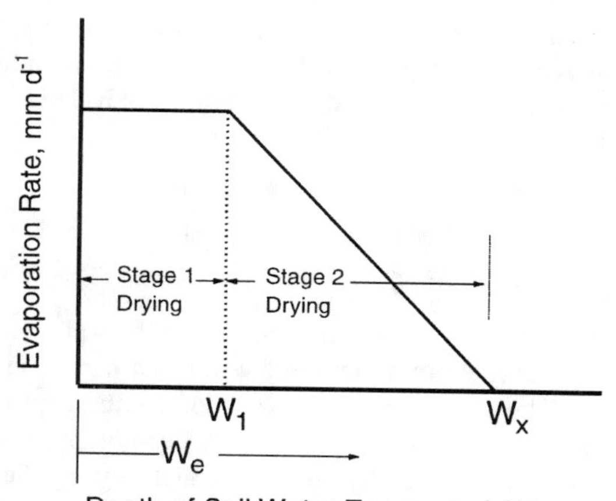

Figure 5.6. Two-stage model for soil evaporation.
***Source:* Adapted from [96].**

The basic equation for $K_{c\,\text{ini}}$ is

$$K_{c\,\text{ini}} = \frac{W_x - (W_x - W_1)\exp\left(\frac{-(t_w - t_1)E_{SO}\left(1 + \frac{W_1}{W_x - W_1}\right)}{W_x}\right)}{t_w ET_0} \tag{5.56}$$

for $t_w > t_1$, where W_1 is the cumulative depth (mm) of soil water evaporation when stage-1 drying is complete, W_x is the cumulative depth (mm) of soil water evaporation when soil evaporation effectively ceases (i.e., W_x = maximum potential soil water that can be evaporated from the soil surface layer between two wetting events), E_{so} is the potential evaporation rate during stage 1 (mm day^{-1}), t_w is the mean interval between wetting events (days), and t_1 is the time when stage-1 drying is completed ($t_1 = W_1/E_{\text{so}}$) (days). When $t_w < t_1$, that is, the entire process resides within stage 1, then

$$K_{c\,\text{ini}} = E_{SO}/ET_0. \tag{5.57}$$

Where furrow or trickle irrigation is practiced, and only a portion of the soil surface is wetted, the value calculated for $K_{c\,\text{ini}}$ in Eqs. (5.56) and (5.57) should be reduced in proportion to the average fraction of surface wetted, f_w (0.2–1.0). Indicative values for f_w are given in [91] and [98]: $f_w = 1.0$ for rain, sprinkling, and basin and border irrigation; $f_w = 0.4$ to 0.6 for furrows; $f_w = 0.3$ to 0.4 for irrigation with alternate furrows; $f_w = 0.2$ to 0.5 for trickle irrigation.

The value for the infiltration depth from irrigation (I_n) used in the following procedure also should be adjusted accordingly:

$$I_n = I_{ns}/f_w, \tag{5.58}$$

where I_n is the equivalent depth (mm) of infiltration from irrigation over the f_w fraction of the surface wetted, and I_{ns} is the depth (mm) of infiltration from irrigation, expressed as one-dimensional depth over the entire surface area.

W_1 (Fig. 5.6), is generally highest for medium-textured soils with high moisture retention and moderate hydraulic conductivity, and is lower for coarse soils having low moisture-retention capacity and for clays that have low hydraulic conductivity. Maximum values for W_1 (termed $W_{1\,\text{max}}$) are listed by Allen *et al.* [91]. Ritchie *et al.* [99] predicted potential values of W_1 (mm) according to soil texture:

$$\begin{aligned} W_{1\,\text{max}} &= 20 - 0.15\,(Sa) \quad \text{for } Sa > 80\%, \\ W_{1\,\text{max}} &= 11 - 0.06\,(Cl) \quad \text{for } Cl > 50\%, \\ W_{1\,\text{max}} &= 8 + 0.08\,(Cl) \quad \text{for } Sa < 80 \text{ and } Cl < 50\%, \end{aligned} \tag{5.59}$$

where *Sa* and *Cl* are the percentages of sand and clay in the soil.

The value for W_x is the maximum depth of water that can be evaporated from the soil following wetting. The value for W_x is governed by the depth of the soil profile contributing to soil water evaporation, by the soil water-holding properties within the evaporating layer, by the unsaturated hydraulic conductivity, by the presence of a hydraulically limiting layer beneath the evaporating layer, by the conduction of sensible heat into the soil to supply energy for subsurface evaporation, and by any root extraction of water from the soil layer. During winter and other cool season months, less radiation energy is available to penetrate the soil surface and to evaporate water from within a drying soil. An approximation for the maximum value of W_x for initial periods having $ET_0 \geq 5$ mm day^{-1} is

$$W_{x\,\text{max}} = z_e(\theta_{\text{UL}} - 0.5\theta_{\text{LL}}), \tag{5.60}$$

Table 5.2. Typical values for θ_{UL}, θ_{LL}, and θ_{AM} and sand, silt, and clay percentages for general soil classifications

	Mean Values of Soil Water Content[a] ($m^3\ m^{-3}$)			Soil Textures[b] (%)		
Soil Classification	θ_{UL}	θ_{LL}	θ_{AM}	Sand	Silt	Clay
Sand	0.12	0.04	0.08	92	4	4
Loamy sand	0.14	0.06	0.08	84	6	10
Sandy loam	0.23	0.10	0.13	65	25	10
Loam	0.26	0.12	0.15	40	40	20
Silt loam	0.30	0.15	0.15	20	65	15
Silt	0.32	0.15	0.17	7	88	5
Silty clay Loam	0.34	0.19	0.15	10	55	35
Silty clay	0.36	0.21	0.15	8	47	45
Clay	0.36	0.21	0.15	22	20	58

[a] Available moisture, $\theta_{AM} = \theta_{UL} - \theta_{LL}$.
[b] From the SCS soil texture triangle.
Source: [2].

and when $ET_0 < 5$ mm day^{-1} is

$$W_{x\ max} = z_e(\theta_{UL} - 0.5\theta_{LL})\sqrt{ET_0/5}, \tag{5.61}$$

where $W_{x\ max}$ is the maximum depth of water (mm) that can be evaporated from the soil when the upper 30 to 40 cm of the soil profile has been initially wetted completely, θ_{UL} is the soil water content at the drained upper limit of the soil (field capacity) ($m^3\ m^{-3}$), θ_{LL} is the soil water content at the lower limit of root extraction (wilting point) ($m^3\ m^{-3}$), and z_e is the mean total depth of soil that is subject to drying by evaporation [100–200 mm]. If unknown, a value of $z_e = 150$ mm is recommended. Typical values for θ_{UL} and θ_{LL} are given in Table 5.2. Soil water concepts are developed in Section 5.2.

The number of wetting (precipitation and irrigation) events N_w occurring during the initial period is determined by considering that two wetting events occurring on adjoining days should be counted as one event, and individual wetting events of less than 0.2 ET_0 can be ignored.

The mean time interval between wetting events, t_w (days), is determined by dividing the length of the initial growing stage t_{ini} (days) by the number of wetting events N_w, so that

$$t_w = t_{ini}/N_w. \tag{5.62}$$

P_{mean}, the average depth (mm) of water added to the evaporating layer at each wetting event, is determined by dividing the sum of the precipitation and irrigation infiltration ($\sum P_n + \sum I_n$) occurring during all wetting events by the number of events N_w thus,

$$P_{mean} = \left(\sum P_n + \sum I_n\right)\Big/ N_w. \tag{5.63}$$

However, each value of P_n and I_n must be limited to

$$P_n \leq W_{x\ \max} \text{ and } I_n \leq W_{x\ \max}.$$

The actual values for W_x and W_1 are calculated from $W_{x\ \max}$ and $W_{1\ \max}$ according to the average total water available during each drying cycle,

$$W_a = P_{\text{mean}} + M_{\text{ini}}/N_w, \tag{5.64}$$

where M_{ini} is the equivalent depth of water (mm) in the evaporation layer (of thickness z_e) at the time of planting. W_x then is calculated by comparing W_a and $W_{x\ \max}$:

$$W_x = \min(W_{x\ \max}, W_a) \tag{5.65}$$

and

$$W_1 = W_{1\ \max}\left[\min\left(\frac{W_a}{W_{x\ \max}}, 1\right)\right], \tag{5.66}$$

where min(·) is a function to select the minimum value of those in braces that are separated by the comma. Values W_1 and W_x then are used in Eq. (5.56).

K_c for Midseason and End of Season

Adjustment for Climate

The $K_{c\,\text{mid}}$ and $K_{c\,\text{end}}$ values in Table 5.1 represent $K_{cb}+K_e$ for irrigation management and precipitation frequencies typical of a subhumid climate where $RH_{\min} = 45\%$ and $U_2 = 2\ \text{m s}^{-1}$.

Under humid and calm conditions, the K_c values for full-cover agricultural crops generally do not exceed 1.0 by more than about 0.05. This is because full-cover agricultural crops and the reference crop of clipped grass both provide for nearly maximum absorption of shortwave radiation, which is the primary energy source for evaporation under humid and calm conditions. Generally, the albedos α are similar over a wide range of agricultural crops, including the reference. Because the VPD is small under humid conditions, differences in ET caused by differences in aerodynamic resistance r_a between the agricultural crop and the reference crop are also small, especially with low to moderate wind speed. In other words, under humid and calm wind conditions, the K_c values become less dependent on the differences between the aerodynamic components of ET_c and ET_0.

Under arid conditions, the effect of differences in r_a between the agricultural crop and the grass reference crop on ET_c become more pronounced because the VPD term may be relatively large. Hence, K_c will be larger under arid conditions when the agricultural crop has leaf area and roughness height that are greater than that of the grass reference. Because the $1/r_a$ term in the numerator of the PM equation (5.2) is multiplied by the VPD term $(e_s - e_a)$, ET from tall crops increases proportionately more relative to ET_0 than does ET from short crops when RH is low.

The values for $K_{c\,\text{mid}}$ and $K_{c\,\text{end}}$ in Table 5.1 represent conditions where $RH_{\min} \approx 45\%$ and $U_2 \approx 2\ \text{m s}^{-1}$. When climatic conditions are different from these values, the values for $K_{c\,\text{mid}}$ and $K_{c\,\text{end}}$ must be adjusted. Pereira *et al.* [14] presented a theoretical basis for quantifying the changes in K_c with changes in wind speed and humidity that result when the relatively smooth grass reference is used as the basis for K_c. Their "climatic coefficient" α_0, when simplified under conditions of relatively constant solar radiation and air temperature, can be reduced to an adjustment factor having the parameters U_2, $RH_{\min}$, and crop height h, thereby showing support for the climatic adjustment to $K_{c\,\text{mid}}$ and $K_{c\,\text{end}}$ provided by the following equation [76, 95]:

$$K_c = (K_c)_{\text{tab}} + [0.04(U_2 - 2) - 0.004(RH_{\min} - 45)](h/3)^{0.3}, \tag{5.67}$$

where $(K_c)_{\text{tab}}$ is the value for $K_{c\,\text{mid}}$ or $K_{c\,\text{end}}$ taken from Table 1, U_2 is the mean value for daily wind speed at 2-m height (m s^{-1}), $RH_{\min}$ is the mean value for daily minimum RH (%), h is the mean plant height (m), all mean values referring to the midseason or the late-season period. The values for U_2 and $RH_{\min}$ need only be approximate. Example values for maximum h are listed in Table 5.1 for various crops. However, h will vary with crop variety and with cultural practices. Therefore, when possible, h should be obtained from general field observations. However, an approximate value is generally adequate.

When $RH_{\min}$ are not available, they can be estimated from

$$RH_{\min} = 100e^0(T_{\min})/e^0(T_{\max}), \tag{5.68}$$

where $T_{\min}$ and $T_{\max}$ also refer to the midseason or the end-season periods [see Eq. (5.8)].

For frequent irrigation of crops (more frequently than every 3 days) and where $(K_{c\,\text{mid}})_{\text{tab}} < 1.1$, replace $(K_{c\,\text{mid}})_{\text{tab}}$ with approximately 1.1 to 1.3 to account for the combined effects of continuously wet soil.

Equation (5.67) is only applied for $K_{c\,\text{end}}$ when $(K_{c\,\text{end}})_{\text{tab}}$ exceeds 0.45. When crops are allowed to reach senescence and dry in the field ($K_{c\,\text{end}} < 0.45$), U_2 and $RH_{\min}$ have less effect on $K_{c\,\text{end}}$ and no adjustment is necessary.

Adjustments for Management Practices

The values given for $K_{c\,\text{end}}$ for vegetable and cereal crops reflect crop and water management practices particular to those crops. Vegetable crops often are irrigated frequently until they are harvested fresh, before the plants have reached senescence and died. Therefore, $K_{c\,\text{end}}$ values for these crops are relatively high (e.g., 0.90 for crucifers, green beans, and peppers). Cereal crops generally are allowed to reach senescence and to dry out in the field before harvest. Therefore, both the soil surface and the vegetation are drier with cereals and, consequently, the value for $K_{c\,\text{end}}$ is small (e.g., 0.35 for maize and grain if dried in the field, but about 0.60 if harvested at high grain moisture).

When the local water management and harvest timing practices are known to deviate from those typical of the values presented in Table 5.1, the user should make adjustments to the values for $K_{c\,\text{mid}}$ and $K_{c\,\text{end}}$, and to the length of the late-season period.

When stand density, height, or leaf area of the crop are less than that attained under perfect or normal crop and irrigation management conditions, the values for K_c are reduced. The reduction in $K_{c\,\text{mid}}$ for poor stands can vary from 0.1 up to 0.5. Adjustment of K_c is made according to the amount of effective (green) leaf area relative to that for healthy vegetation having normal plant density (pristine conditions). Then,

$$(K_c)_{\text{cor}} = K_c - A_{cm}, \tag{5.69}$$

where $(K_c)_{\text{cor}}$ is the crop coefficient to be used for poor stands, K_c is the $K_{c\,\text{mid}}$ for pristine conditions (from Table 5.1), and A_{cm} is the adjustment factor relative to the actual crop and irrigation management conditions (0–0.5). A_{cm} can be approximated through a green cover ratio of type:

$$A_{cm} = 1 - (\mathrm{LAI}_{\mathrm{actual}}/\mathrm{LAI}_{\mathrm{normal}}) \tag{5.70}$$

or similar ratio, referring to the midseason period.

A more complete approach to estimate K_c for nonpristine and unknown conditions is presented below in "K_c for Nonpristine and Unknown Conditions".

The Basal Crop Coefficient Approach

The basal crop coefficient, K_{cb} [see Eq. (5.53)] is defined as the ratio of ET_c/ET_0 when the soil surface is dry but transpiration is occurring at a potential rate (e.g., water is not limiting transpiration) [2, 71]. Therefore, K_{cb} represents primarily the transpiration component of ET. Its use provides for separate and special adjustment for wet soil evaporation immediately following specific rain or irrigation events. This results in more accurate estimates of ET_c when it must be computed on a daily basis.

Recommended values for K_{cb} are listed by Allen *et al.* [91, 95] but they can be obtained from K_c in Table 5.1 using the procedures indicated below.

K_{cb} for the Initial Period

K_{cb} during the initial period, denoted as $K_{cb\,\mathrm{ini}}$, is estimated for annual crops by assuming that a constant value can be used to represent evaporation from a mostly dry and bare surface soil layer that may occur during the initial period (see [2, 71, 92, 100]).

For *annual crops* having a nearly bare soil surface at the time of planting, it is recommended that

$$K_{cb\,\mathrm{ini}} = 0.15. \tag{5.71}$$

For *perennial crops having a nearly bare soil surface* at the beginning of the growing period before the initiation of new leaves, $K_{cb\,\mathrm{ini}}$ can be estimated as 0.15 to 0.20.

The $K_{cb\,\mathrm{ini}}$ for *perennial crops with ground cover* can be estimated by subtracting 0.10–0.20 from the $K_{c\,\mathrm{ini}}$ values in Table 5.1:

$$K_{cb\,\mathrm{ini}} = K_{c\,\mathrm{ini}} - 0.1 \quad \text{or} \quad K_{cb\,\mathrm{ini}} = K_{c\,\mathrm{ini}} - 0.2, \tag{5.72}$$

the second being recommended when crops are routinely irrigated frequently.

K_{cb} for Midseason and End Season

Most agricultural crop canopies nearly completely cover the soil surface during the midseason. As a result, soil wetness beneath the canopy has less effect on total ET_c. Therefore, $K_{cb\,\mathrm{mid}}$ is often nearly the same value as $K_{c\,\mathrm{mid}}$ listed in Table 5.1 and can be estimated by

$$K_{cb\,\mathrm{mid}} = K_{c\,\mathrm{mid}} - 0.05 \tag{5.73a}$$

for $C \geq \sim 0.8$, that is, for crops that have nearly complete ground cover, such as cereals; and

$$K_{cb\,\mathrm{mid}} = K_{c\,\mathrm{mid}} - 0.10 \tag{5.73b}$$

for $C < \sim 0.8$, where C is the average fraction of soil surface covered (or shaded) by vegetation (0–0.99).

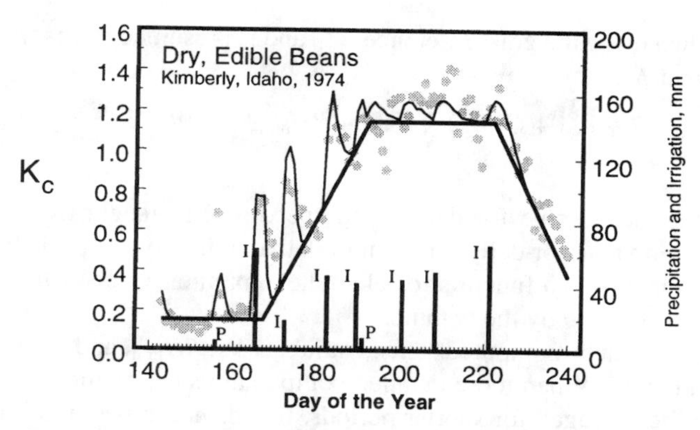

Figure 5.7. Application of the basal crop coefficient approach to edible beans in Kimberly, Idaho, 1974. ***Source:*** **J. L. Wright, personal communication.**

The basal K_c at the end of the growing season, $K_{cb\,\mathrm{end}}$, has a lower value than the $K_{c\,\mathrm{end}}$ values recommended in Table 5.1. In general, $K_{cb\,\mathrm{end}}$ values can be estimated from $K_{c\,\mathrm{end}}$ in this table:

$$K_{cb\,\mathrm{end}} = K_{c\,\mathrm{end}} - 0.10. \tag{5.74}$$

The $K_{cb\,\mathrm{mid}}$ and $K_{cb\,\mathrm{end}}$ need to be adjusted for climate using Eq. (5.67). The K_{cb} for the crop development period and for the late-season period are obtained by linear interpolation as indicated for the K_c curve (see "The K_c Curve," above). This is illustrated in Fig. 5.7.

Soil Evaporation Coefficient K_e

Evaporation from soil beneath a canopy, or from soil that is exposed to sunlight between plants, is governed by the amount of energy that is available at the soil surface to evaporate water. This energy is governed by the portion of total energy that has been consumed by plant transpiration [96, 98, 101, 102].

The maximum value for K_e corresponds to a period immediately following rain or irrigation. The K_e decays, depending on the cumulative amount of water evaporated from the surface soil layer. Thus, K_e can be calculated from

$$K_e = K_d(K_{c\,\max} - K_{cb}), \tag{5.75}$$

where K_d is the coefficient for soil evaporation decay (0–1), K_{cb} is the basal crop coefficient (see "K_{cb} for the Initial Period" and "K_{cb} for Midseason and End Season," above), and $K_{c\,\max}$ is the maximum value for K_c following rain or irrigation. However, K_e is limited by the fraction of wetted soil exposed to sunlight, f_s (0.01–1); then,

$$K_e \leq f_s K_{c\,\max}. \tag{5.76}$$

$K_{c\,\max}$ represents an upper limit on the evaporation and transpiration from any cropped surface and is imposed to reflect the natural constraints placed on available energy represented by the difference $R_n - G - H$ [see Eq. (5.1)]. $K_{c\,\max}$ ranges from about

1.05 to 1.35 when using the grass reference ET_0 and is presumed to change with climate similar to K_c and K_{cb}; thus

$$K_{c\,\max} = \max[\{1.2 + [0.04(U_2 - 2) - 0.004(RH_{\min} - 45)](h/3)^{0.3}\}, \quad \{K_{cb} + 0.05\}], \tag{5.77}$$

where U_2 is the mean value for daily wind speed at 2-m height over grass (m s^{-1}), $RH_{\min}$ is the mean value for daily minimum relative humidity (%), h is the mean plant height (m), and max[·] is a function to select the maximum value of the parameters in braces · that are separated by the comma.

A single $K_{c\,\max}$ can be computed. Then, variables U_2, $RH_{\min}$, and crop height refer to the midseason period. When more detailed computations are applied, the variables U_2 and $RH_{\min}$ can be averaged for shorter periods, five days for example. The mean crop height h also can be estimated for the same time intervals but, for the initial period, h can be artificially considered the same as for the grass reference crop ($h = 0.12$ m).

The simple model used to estimate evaporation from soil is similar to the one used to compute $K_{c\,\text{ini}}$ and is illustrated in Fig. 5.6, where the evaporation rate is at the maximum rate until the depth of water evaporated, W_e, equals W_1, the maximum cumulative depth of evaporation during stage 1. After W_e exceeds W_1, the evaporation process is in stage 2, and the evaporation rate begins to decrease in proportion to $(W_e - W_1)/(W_x - W_1)$. Therefore, the coefficient for soil evaporation decay, K_d, is calculated as

$$K_d = 1 \qquad (\text{for } W_e \leq W_1), \tag{5.78a}$$

$$K_d = \frac{W_x - W_e}{W_x - W_1} \qquad (\text{for } W_e > W_1). \tag{5.78b}$$

W_1 and W_x are estimated as described under "K_e for Initial Stage for Annual Crops," above. W_e is the current depth of water depleted from the f_s fraction of the soil (mm). W_e is computed from the water balance of the upper 100 to 150 mm of the soil profile. The daily soil water-balance equation recommended for the f_s fraction of wetted soil exposed to sunlight is

$$(W_e)_i = (W_e)_{i-1} - [P_i - (Q_r)_i] - \frac{(I_n)_i}{f_w} + \left(\frac{K_e ET_0}{f_s}\right)_i + (T_s)_i, \tag{5.79}$$

where $(W_e)_i$ and $(W_e)_{i-1}$ are the depths of water (mm) depleted from the evaporating soil upper layer at the end of days i and $i-1$ $[0 \leq (W_e)_i \leq W_x]$, P_i is the precipitation on day i (mm), $(Q_r)_i$ is the runoff from the soil surface on day i (mm) $[0 \leq (Q_r)_i \leq P_i]$, $(I_n)_i$ is the net irrigation depth on day i (mm) that infiltrates the soil [Eq. (5.58)], $(K_e ET_0/f_s)_i$ is the evaporation from the f_s fraction of the exposed soil surface on day i (mm), and $(T_s)_i$ is the transpiration from the f_s fraction of the evaporating soil layer on day i (mm).

To initiate the water balance [Eq. (5.77)], the user can assume that $(W_e)_i = 0$ immediately following a heavy rain or irrigation or, if a long period of time has occurred since the last wetting, the user can assume that $(W_e)_i = W_x$. An example of application of the basal crop coefficient approach is presented in Fig. 5.7.

When $P_i < 0.2\ ET_0$, P_i usually can be ignored. Estimation of $(Q_r)_i$ can be performed using the curve-number method as described by Martin and Gilley [74]. In most applications, $(Q_r)_i = 0$. For the majority of crops, except for very shallow rooted crops (maximum rooting depth <0.5 m), the amount of transpiration from the evaporating soil layer is small and can be neglected [i.e., $(T_s)_i = 0$]. This applies, for row crops, where little or none of $(T_s)_i$ depletes the f_s portion of the surface soil layer [$(T_s)_i = 0$]. When the complete soil surface is wetted as by precipitation or sprinkler, then f_s is essentially defined as $1 - C$, where C is the average fraction of ground covered by vegetation.

However, for irrigation systems where only a fraction of the soil surface is wetted, f_s is calculated as

$$f_s = \min(1 - C, f_w), \tag{5.80}$$

where C is the average fraction of ground covered (or shaded) by vegetation (0–0.99), and f_w is the average fraction of soil surface wetted by irrigation or precipitation (0.01–1) [see Eq. (5.58)]. In days when precipitation occurs, $f_w = 1$. The min(·) function selects the lowest value of the $1 - C$ and f_w.

Where it is not observed, the parameter C (0–0.99) can be estimated from

$$C = \left[\frac{K_{cb} - K_{c\,\min}}{K_{c\,\max} - K_{c\,\min}} \right]^{1+0.5h}, \tag{5.81}$$

where K_{cb} is the value for the basal crop coefficient for the particular day or period; $K_{c\,\min}$ is the minimum K_c for dry, bare soil with no ground cover (≈0.15–0.20), commonly the same as $K_{cb\,\text{ini}}$; $K_{c\,\max}$ is the maximum K_c immediately following wetting [Eq. (5.77)], and h is the mean plant height (m). The $1 + 0.5h$ exponent in Eq. (5.81) represents the effect of plant height on shading of the soil surface and on increasing the value for K_{cb}, given a specific value for C. The user should limit the difference $K_{cb} - K_{c\,\min}$ to ≥0.01 for numerical stability. The value for C will change daily as K_{cb} changes. Therefore, Eq. (5.81) is applied daily.

Reduction Coeficient for Limited Soil Water K_s

The K_{cb} in Eq. (5.53) and K_c in Eq. (5.54) are reduced when soil water content of the plant root zone is too low to sustain transpiration at potential levels. The reduction is made by multiplying K_{cb} or K_c by the water stress coefficient K_s. K_s can be predicted from [90]

$$K_s = \frac{\theta - \theta_{\text{LL}}}{\theta_t - \theta_{\text{LL}}} \tag{5.82}$$

when $\theta < \theta_t$, where K_s is the dimensionless ET reduction factor (0–1), θ is the mean volumetric soil water content in the root zone ($m^3\,m^{-3}$), θ_t is the threshold θ for the root zone, below which transpiration is decreased because of water stress ($m^3\,m^{-3}$), and θ_{LL} is the soil water content at the lower limit of water extraction by plant roots ($m^3\,m^{-3}$). By definition, $K_s = 1.0$ for $\theta > \theta_t$. Parameter θ_{LL} is synonymous with the term wilting point.

The threshold soil water content θ_t is predicted as

$$\theta_t = (1 - F_{ns})(\theta_{\text{UL}} - \theta_{\text{LL}}), \tag{5.83}$$

where F_{ns} is the average fraction of available soil water that can be depleted before water stress occurs (0–1), and θ_{UL} is the soil water content at the drained upper limit of the soil profile (field capacity) ($m^3 m^{-3}$). F_{ns} represents the depletion fraction for no stress, and is equivalent to the p term of Doorenbos and Pruitt [55] and Doorenbos and Kassam [90] and is similar to the term "management-allowed depletion" (MAD), introduced by Merriam [103]. However, values for MAD may be influenced by management and economic factors in addition to physical factors influencing F_{ns}. Generally, MAD $< F_{ns}$ when there is risk aversion or uncertainty, and MAD $> F_{ns}$ when plant water stress is an intentional part of soil water management.

Values for the θ parameters in Eqs. (5.82) and (5.83) represent averages for the effective root zone. F_{ns} normally varies from 0.30 for shallow-rooted plants at high rates of ET_0 (>8 mm day^{-1}) to 0.80 for deep-rooted plants at low rates of ET_0 (<3 mm day^{-1}). Values for F_{ns} are given in Table 5.1. Values listed need to be adjusted for climate as indicated in footnoted of the same table.

The determination of K_s requires a daily balance of soil water content.

K_c for Nonpristine and Unknown Conditions

For crops or vegetation where the K_c is not known, but where estimates of the fraction of ground surface covered by vegetation can be made, $K_{cb\,\mathrm{mid}}$ can be approximated as [76]

$$K_{cb\,\mathrm{min}} = K_{c\,\mathrm{min}} + [(K_{cb})_h - K_{c\,\mathrm{min}}]C_{\mathrm{eff}}^{\frac{1}{1+h}}, \tag{5.84}$$

where C_{eff} is the effective fraction of ground covered by vegetaton (0.01–1), $K_{c\,\mathrm{min}}$ is the minimum K_c for dry, bare soil (0.15–0.20), and $(K_{cb})_h$ is the maximum value for K_{cb} for vegetation having complete ground cover. $(K_{cb})_h$ is estimated for areas of vegetation >500 m^2 as

$$(K_{cb})_h = \min[(1.0+0.1h), 1.2] + 0.04(U_2-2) - 0.004(RH_{\mathrm{min}} - 45)(h/3)^{0.3}. \tag{5.85}$$

For small, isolated stand sizes, $K_{cb\,h}$ may need to be increased beyond the value given by the equation above.

The value for $K_{cb\,h}$ may be reduced for vegetation that has a high degree of stomatal control, such as some types of brush and trees. The exponent in Eq. (5.84) represents the effects of microscale advection (transfer) of sensible heat from dry soil surfaces between plants toward plant leaves, thereby increasing ET per unit leaf area and the effects of increased roughness as the value for C_{eff} decreases.

5.1.6 Soil Water Balance and Irrigation Water Requirements

Soil Water Balance

The soil water balance is made for the complete effective rooting depth, including the evaporation layer:

$$\theta_i = \theta_{i-1} + \frac{[P_i - (Q_r)_i] + (I_n)_i - (ET_c)_i - DP_i + GW_i}{1000(Z_r)_i}, \tag{5.86}$$

where θ_i is the mean volumetric soil water content in the root zone ($m^3 m^{-3}$) on day i, θ_{i-1} is soil water content on the previous day, P_i is depth of precipitation on day i (mm), $(Q_r)_i$ is the runoff from the soil surface on day i (mm), $(I_n)_i$ is net irrigation depth on day i (mm), (irrigation water infiltrating the soil), $(ET_c)_i$ is the crop ET from Eq. (5.52) on day i (mm), DP_i is any deep percolation on day i (mm), GW_i is any upward contribution of water from a shallow water table on day i (mm), and $(Z_r)_i$ is the rooting depth on day i (m). Estimation of GW_i is well described by Martin and Gilley [74]. $(Q_r)_i$ can be predicted using the SCS curve-number method [104].

Deep percolation is estimated as $DP_i = 0$ when $\theta_i \leq \theta_{UL}$, and $DP_i = 1{,}000(\theta_i - \theta_{UL})(Z_r)_i$ otherwise. In some applications, θ_i may be allowed to exceed θ_{UL} for one day before $DP_i \geq 0$ to account for some ET from excess soil water before it drains from the root zone.

The depth of the effective root zone for any day i can be predicted as

$$(Z_r)_i = (Z_r)_{\min} + [(Z_r)_{\max} - (Z_r)_{\min}]\frac{J_i - J_{\text{ini}}}{L_{rd}} \tag{5.87}$$

for $J_i \leq J_{\text{ini}} + L_{rd}$, where $(Z_r)_i$ is the effective depth of the root zone on day i (m), $(Z_r)_{\min}$ is the initial effective depth of the root zone (generally at $J = J_{\text{ini}}$), and $(Z_r)_{\max}$ is the maximum effective depth of the root zone (m). J_i is the day of the year [Eq. (5.30)] corresponding to day i and J_{ini} is the day of the year corresponding to the date of planting or initiation of growth (or January 1 if a perennial is growing through all months of the year). L_{rd} is the length of the root development period (days). When $J_i > J_{\text{ini}} + \text{length}$, $(Z_r)_i = (Z_r)_{\max}$. Indicative values for $(Z_r)_{\max}$ are given in Table 5.1.

The latest date for scheduling irrigation to avoid water stress is when θ_i equals θ_t [Eqs. (5.83) and (5.86)]. However, irrigations often are scheduled when the MAD fraction of water is depleted, where MAD may be higher or lower than F_{ns} [Eq. (5.83)]. In this case, irrigation is scheduled when

$$\theta_i = \theta_{\text{MAD}} = (1 - \text{MAD})(\theta_{\text{UL}} - \theta_{\text{LL}}) + \theta_{\text{LL}}. \tag{5.88}$$

The net irrigation depth to be applied then would be

$$(I_n)_i = 1{,}000(Z_r)_i(\theta_{\text{UL}} - \theta_i). \tag{5.89}$$

The soil water balance currently is computed through crop-water simulation models that allow the selection of best irrigation scheduling alternatives [49–51]. Irrigation scheduling principles and applications are described in Section 5.3.

Irrigation Water Requirement

The irrigation water requirement for a complete growing season is computed as

$$\text{IWR} = \frac{ET_c - P_e - GW - \Delta S}{1 - LR}, \tag{5.90}$$

where P_e is the effective precipitation, defined as gross precipitation less all runoff and deep percolation losses; GW is upward flow of groundwater into the root zone during the growing season; ΔS is the change in soil water storage in the root zone during the nongrowing season, the difference between θ at planting and at harvesting; and LR is the leaching requirement. Operational estimates of IWR are better done through crop-water simulation models.

The leaching requirement LR is estimated for noncracking soils as

$$LR = \frac{\mathrm{EC}_{iw}}{5\,\mathrm{EC}_e - \mathrm{EC}_{iw}}, \tag{5.91}$$

where EC_{iw} is the electrical conductivity of the irrigation water and EC_e is the desired average root zone salinity expressed as the electrical conductivity of the saturated soil extract. Values for EC_e are available in Ayers and Westcot [105]. Problems relative to salinity management are discussed in Sections 5.5–5.8.

The gross irrigation water requirement is computed as

$$\mathrm{GIWR} = \frac{\mathrm{IWR}}{Eff} \tag{5.92}$$

where *Eff* is the efficiency of the irrigation system. The value for Eff depends on the spatial scale for GIWR. For a single field, the Eff is the application efficiency of the field irrigation system. If GIWR is for a farm, Eff includes the product of application efficiencies and conveyence efficiencies within the farm. If GIWR is for an irrigation project or scheme, then Eff includes all system losses both on and off the farm. Note that a substantial portion of these losses eventually will return to the water resources system for reuse by other users. Values for Eff for application systems are described in Section 5.4.

References

1. Perrier, A. 1985. Updated evapotranspiration and crop water requirement definitions. *Crop Water Requirements*, eds. Perrier, A., and C. Riou, pp. 885–887. Paris: INRA.
2. Jensen, M. E., R. D. Burman, and R. G. Allen (eds.). 1990. *Evapotranspiration and Irrigation Water Requirements*. ASCE Manuals and Reports on Engineering Practices No. 70, New York.
3. Doorenbos, J., and W. O. Pruitt. 1975. *Guidelines for Predicting Crop Water Requirements*, FAO Irrigation and Drainage Paper 24, Rome: FAO.
4. Monteith, J. L. 1965. Evaporation and environment. *19th Symposia of the Society for Experimental Biology*, Vol. 19, pp. 205–234. Cambridge, UK: Cambridge University Press.
5. Monteith, J. L. 1985. Evaporation from land surfaces: Progress in analysis and prediction since 1948. *Advances in Evapotranspiration*, pp. 4–12. St. Joseph, MI: American Society of Agricultural Engineers.
6. Rijtema, P. E. 1965. Analysis of Actual Evapotranspiration. Agricultural Research Rept. No. 69, Wageningen: Centre for Agricultural Publications and Documents.
7. Alves, I., A. Perrier, and L. S. Pereira. 1996. The Penman-Monteith equation: How good is the "big leaf" approach? *Evapotranspiration and Irrigation Scheduling*, eds. Camp, C. R., E. J. Sadler, R. E. Yoder, pp. 599–605. St. Joseph, MI: American Society of Agricultural Engineers.
8. Matias P. G. M. 1992. SWATCHP, a model for a continuous simulation of hydrologic processes in a system vegetation–soil–aquifer–river. Ph.D. Dissertation, Technical University of Lisbon (in Portuguese).

9. Monteith, J. L., and M. H. Unsworth. 1990. *Principles of Environmental Physics*, 2nd ed., London: Edward Arnold.
10. Allen, R. G., W. O. Pruitt, J. A. Businger, L. J. Fritschen, M. E. Jensen, and F. H. Quinn. 1996. Evaporation and transpiration. *ASCE Handbook of Hydrology*, eds. Wootton, *et al.*, pp. 125–252. New York: American Society of Civil Engineers.
11. Perrier, A. 1982. Land surface processes: Vegetation. *Land Surface Processes in Atmospheric General Circulation Models*, ed. Eagleson, P. S., pp. 395–448. Cambridge, MA: Cambridge University Press.
12. Perrier, A., and A. Tuzet. 1991. Land surface processes: Description, theoretical approaches, and physical laws underlying their measurements. *Land Surface Evaporation: Measurement and Parameterization*, eds. Schmugge, T. J., and J.-C. Andre, pp. 145–155. Berlin: Springer-Verlag.
13. Shaw, R. H., and A. R. Pereira. 1982. Aerodynamic roughness of a plant canopy: A numerical experiment. *Agric. Meteorol.* 26:51–65.
14. Pereira, L. S., A. Perrier, R. G. Allen, and I. Alves. 1996. Evapotranspiration: Review of concepts and future trends. *Evapotranspiration and Irrigation Scheduling*, eds. Camp, C. R., E. J. Sadler, and R. E. Yoder, pp. 109–115. St. Joseph, MI: American Society of Agricultural Engineers.
15. Monin, A. S., and A. M. Obukhov. 1954. The basic laws of turbulent mixing in the surface layer of the atmosphere. *Akad. Nauk. SSSR Trud. Geofiz. Inst.* 24(151): 163–187.
16. Pruitt, W. O., D. L. Morgan, and F. J. Lourence. 1973. Momentum and mass transfers in the surface boundary layer. *Q. J. R. Meteorol. Soc.* 99:370–386.
17. Thom, A. S., and H. R. Oliver. 1977. On Penman's equation for estimating regional evaporation. *Q. J. R. Meteorol. Soc.* 103:345–357.
18. Businger, J. A. 1988. A note on the Businger-Dyer profiles. *Boundary-Layer Meteorol.* 42:145–151.
19. Monteith, J. L. 1973. *Principles of Environmental Physics*. London: Edward Arnold.
20. Brutsaert, W. H. 1982. *Evaporation into the Atmosphere*. The Netherlands: Deidel Dordrecht.
21. Allen, R. G., M. E. Jensen, J. L. Wright, and R. D. Burman. 1989. Operational estimates of reference evapotranspiration. *Agron. J.* 81:650–662.
22. Campbell, G. S. 1977. *An Introduction to Environmental BioPhysics*. New York: Springer-Verlag.
23. Garratt, J. R., and B. B. Hicks. 1973. Momentum, heat and water vapour transfer to and from natural and artificial surfaces. *Q. J. R. Meteorol. Soc.* 99: 680–687.
24. Verma, S. B. 1989. Aerodynamic resistances to transfers of heat, mass and momentum. *Estimation of Areal Evapotranspiration*, eds. Black, T. A., D. L. Spittlehouse, M. D. Novak, and D. T. Price, pp. 13–20. International Association for Hydrological Sciences (IAHS), Publ. No. 177.
25. Kustas, W. P., B. J. Choudhury, K. E. Kunkel, and L.W. Gay. 1989. Estimate of aerodynamic roughness parameters over an incomplete canopy cover of cotton. *Agric. For. Meteorol.* 46:91–105.

26. Kustas, W. P. 1990. Estimates of evapotranspiration with a one- and two-dimensional model of heat transfer over partial canopy cover. *J. Appl. Meteorol.* 29:704–715.
27. Szeicz, G., and I. F. Long. 1969. Surface resistance of crop canopies. *Water Resour. Res.* 5:622–633.
28. Ben-Mehrez, M., O. Taconet, D. Vidal-Madjar, and C. Valencogne. 1992. Estimation of stomatal resistance and canopy evaporation during the HAPEX-MOBILHY experiment. *Agric. For. Meteorol.* 58:285–313.
29. Garratt, J. R. 1992. *The Atmospheric Boundery Layer*. Cambridge. UK: Cambridge University Press.
30. Jordan, W. R., and J. T. Ritchie. 1971. Influence of soil water stress on evaporation, root absortion and internal water status of cotton. *Plant Physiol.* 48: 783–788.
31. Grant, D. R. 1975. Comparison of evaporation from barley with Penman estimates. *Agric. Meteorol.* 15:49–60.
32. Shuttleworth, W. J., and J. S. Wallace. 1985. Evaporation from sparse crops—an energy combination theory. *Q. J. R. Meteorol. Soc.* 111:839–853.
33. Wallace, J. S., J. M. Roberts, and M. V. K. Sivakuma. 1990. The estimation of transpiration from sparse dryland millet using stomatal condutance and vegetation area indices. *Agric. For. Meteorol.* 51:35–49.
34. Jarvis, P. G. 1976. The interpretation of the variations in leaf water potential and stomatal conductance found in canopies in the field. *Phil. Trans. R. Soc. London B* 273:593–610.
35. Stewart, J. B. 1988. Modelling surface conductance of pine forest. *Agric. For. Meteorol.* 43:19–35.
36. Kim, J., and S. B. Verma. 1991. Modeling canopy stomatal conductance in a temperate grassland ecosystem. *Agric. For. Meteorol.* 55:149–166.
37. Stewart, J. B., and S. B. Verma. 1992. Comparison of surface fluxes and conductances at two contrasting sites within the FIFE area. *J. Geophys. Res.* 97(D17): 18623–18628.
38. Itier, B. 1996. Measurement and estimation of evapotranspiration. *Sustainability of Irrigated Agriculture*, eds. Pereira, L. S., R. A. Feddes, J. R. Gilley, and B. Lesaffre, pp. 171–191. Dordrecht, The Netherlands: Kluwer.
39. Alves, I. L. 1995. Modelling crop evapotranspiration. Canopy and aerodynamic resistances. Ph.D. Dissertation, ISA, Technical University of Lisbon (in Portuguese).
40. McNaughton, K. G., and P. G. Jarvis. 1984. Using the Penman-Monteith equation predictively. *Agric. Water Manage.* 8:263–278.
41. Sharma, M. L. 1985. Estimating evapotranspiration. *Advances in Irrigation*, ed. Hillel, D., New York: Academic Press.
42. Hatfield, J. L., and M. Fuchs. 1990. Evapotranspiration models. *Management of Farm Irrigation Systems*, eds. Hoffman, G. J., T. A. Howell, and K. H. Solomon, pp. 33–59, St. Joseph, MI: American Society of Agricultural Engineers.
43. Burman, R., and L. O. Pochop. 1994. *Evaporation, Evapotranspiration and Climatic Data*. Amsterdam: Elsevier Science.

44. Gosse, G., A. Perrier, and B. Itier. 1977. Etude de l'évapotranspiration réelle d'une culture de blé dans le bassin Parisien. *Ann. Agron.* 28(5):521–541.
45. Perrier, A., N. Katerji, G. Gosse, and B. Itier. 1980. Etude "in situ" de l'évapotranspiration réelle d'une culture de blé. *Agric. Meterol.* 21: 295–311.
46. Katerji, N., and A. Perrier. 1983. Modélization de l'évapotranspiration réelle ETR d'une parcelle de luzerne: Rôle d'un coefficient cultural. *Agronomie* 3(6): 513–521.
47. Gash, J. H. C., W. J. Shuttleworth, C. R. Lloyd, J. C. André, J. P. Goutorbe, and J. Gelpe. 1989. Micrometeorological measurements in Les Landes forest during HAPEX-MOBILHY. *Agric. For. Meteorol.* 46:131–147.
48. Bastiaanssen, W. G. M. 1995. Regionalization of surface flux densities and moisture indicators in composite terrain. Ph.D. Thesis, Wageningen: Wageningen Agricultural University.
49. Pereira, L. S., A. Perrier, M. Ait Kadi, and P. Kabat (eds.). 1992. Crop water models. *ICID Bull.* 41(2):1–200.
50. Pereira, L. S., B. J. van den Broek, P. Kabat, and R. G. Allen (eds.). 1995. *Crop-Water Simulation Models in Practice*. Wageningen: Wageningen Pers.
51. Ragab, R., D. E. El-Quosy, B. J. van den Broek, L. S. Pereira (eds.). 1996. *Crop-Water-Environment Models*. Cairo: Egypt National Committee for International Commission on Irrigation and Drainage (ICID).
52. Shuttleworth, W. J. 1993. Evaporation. *Handbook of Hydrology*, ed. Maidment, D. R., pp. 4.1–4.53, New York: McGraw Hill.
53. Wright, J. L., and M. E. Jensen. 1972. Peak water requirements of crops in southern Idaho. *J. Irrig. Drain. Div. ASCE* 96(IR1):193–201.
54. Jensen, M. E. (ed.). 1974. *Consumptive Use of Water and Irrigation Water Requirements*. Rept. of the Technical Committee on Irrigation Water Requirements, Irrigation and Drainage Div. New York: American Society of Civil Engineers.
55. Doorenbos, J., and W. O. Pruitt. 1977. *Guidelines for Predicting Crop Water requirements*, Irrigation and Drainage Papers 24, 2nd ed. Rome: FAO.
56. Pruitt, W. O., and B. D. Swann. 1986. Evapotranspiration studies in N.S.W.: Daily vs. hourly meteorological data. *Irrigation '86*. Toowoomba, Queensland, Australia: Darling Downs Institute of Advanced Education.
57. Allen, R. G., W. O. Pruitt, and M. E. Jensen. 1991. Environmental requirements for lysimeters. *Lysimeters for Evapotranspiration and Environmental Measurements*, eds. Allen, R. G., T. A. Howell, W. O. Pruitt, I. A. Walter, and M. E. Jensen, pp. 170–181. New York: American Society of Civil Engineers.
58. Allen, R. G., and W. O. Pruitt. 1991. FAO-24 reference evapotranspiration factors. *J. Irrig. Drain. Eng. ASCE* 117(5):758–773.
59. Allen, R. G., M. Smith, A. Perrier, and L. S. Pereira. 1994. An update for the definition of reference evapotranspiration. *ICID Bull.* 43(2):1–34.
60. Allen, R. G., M. Smith, L. S. Pereira, and A. Perrier. 1994. An update for the calculation of reference evapotranspiration. *ICID Bull.* 43(2):35–92.
61. Perrier, A., P. Archer, and B. de Pablos. 1974. Etude de l'évapotranspiration réelle et maximele de diverses cultures: Dispositif et mesure. *Ann. Agron.* 25(3):229–243.

62. Pruitt, W. O., and F. J. Lourence. 1985. Experiences in lysimetry for ET and surface drag measurements. *Advances in Evapotranspiration*, pp. 51–69. St. Joseph, MI: American Society of Agricultural Engineers.
63. Meyer, W. S., and L. Mateos. 1990. Effects of soil type on soybean crop water use in weighing lysimeters. II. Effect of lysimeter canopy height discontinuity on evaporation. *Irrig. Sci.* 11:233–237.
64. Howell, T. A., A. D. Schneider, and M. E. Jensen. 1991. History of lysimeter design and use for evapotranspiration measurements. *Lysimeters for Evapotranspiration and Environmental Measurements*, eds. Allen, R. G., T. A. Howell, W. O. Pruitt, I. A. Walter, and M. E. Jensen. pp. 1–9, New York: American Society of Civil Engineers.
65. Grebet, P., and R. H. Cuenca. 1991. History of lysimeter design and effects of environmental disturbances. *Lysimeters for Evapotranspiration and Environmental Measurements*, eds. Allen, R. G., T. A. Howell, W. O. Pruitt, I. A. Walter, and M. E. Jensen. pp. 10–18. New York: American Society of Civil Engineers.
66. Pruitt, W. O. 1991. Development of crop coefficients using lysimeters. *Lysimeters for Evapotranspiration and Environmental Measurements*, eds. Allen, R. G., T. A. Howell, W. O. Pruitt, I. A. Walter, and M. E. Jensen. pp. 182–190. New York: American Society of Civil Engineers.
67. Blad, B. L., and N. J. Rosenberg. 1974. Lysimetric calibration of the Bowen-ratio energy balance method for evapotranspiration estimation in the Central Great Plains. *J. Appl. Meteorol.* 13(2):227–236.
68. Itier, B., and A. Perrier. 1976. Presentation d'une étude analytique de l'advection: II. Application à la mesure et à l'estimation de l'évapotranspiration. *Ann. Agron.* 27(4):417–433.
69. Perrier, A., B. Itier, J. M. Bertolini, and N. Katerji. 1976. A new device for continuous recording of the energy balance of natural surfaces. *Agric. Meteorol.* 16(1): 71–85.
70. Carrijo, O. A., and R. H. Cuenca. 1992. Precision of evapotranspiration estimates using neutron probe. *J. Irrig. Drain. Eng. ASCE* 118(6):943–953.
71. Wright, J. L. 1982. New evapotranspiration crop coefficients. *J. Irrig. Drain. Div. ASCE*, 108:57–74.
72. George, W., W. O. Pruitt, and A. Dong. 1985. Evapotranspiration modeling. California Irrigation Management Information System, Final Report, eds. Snyder, R., D. W. Henderson, W. O. Pruitt, and A. Dong. Land, Air and Water Resources Pap. 10013-A, pp. III-36 to III-59. Davis: University of Calfornia.
73. Feddes, R. A. 1987. Crop factors in relation to Makkink reference crop evapotranspiration. *Evaporation and Weather*, ed., Hooghart, J. C., pp. 33–45. The Hague: The Netherlands Organization for Applied Scientific Research.
74. Martin, D. L., and J. R. Gilley. 1993. Irrigation Water Requirements. SCS National Engineering Handbook, Chap. 2. Washington, DC: Soil Conservation Service.
75. Smith, M., R. G. Allen, J. L. Monteith, A. Perrier, L. Pereira, and A. Segeren. 1992. Report of the Expert Consultation on Procedures for Revision of FAO Guidelines for Prediction of Crop Water Requirements. Rome: FAO.

76. Allen, R. G., M. Smith, W. O. Pruitt, and L. S. Pereira. 1996. Modifications to the FAO crop coefficient approach. *Evapotranspiration and Irrigation Scheduling*, eds. Camp, C. R., E. J. Sadler, R. E. Yoder, pp. 124–132. St. Joseph, MI: American Society of Agricultural Engineers.
77. Smith, M., R. G. Allen, and L. Pereira. 1996. Revised FAO methodology for crop water requirements. *Evapotranspiration and Irrigation Scheduling*, eds. Camp, C. R., E. J. Sadler, R. E. Yoder, pp. 116–123. St. Joseph, MI: American Society of Agricultural Engineers.
78. Allen R. G., 1996. Assessing integrity of weather data for use in reference evapotranspiration estimation. *J. Irrig. Drain. Eng. ASCE* 122(2):97–106.
79. Allen, R. G., C. E. Brockway, and J. L. Wright. 1983. Weather station siting and consumptive use estimates. *J. Water Resour. Plan. Manage. Div. ASCE* 109(2): 134–146.
80. Smith, M. 1993. CLIMWAT for CROPWAT: A climatic database for irrigation planning and management. FAO Irrigation and Drainage Paper 49, Rome: FAO.
81. Allen, R. G. 1995. Evaluation of Procedures for Estimating Mean Monthly Solar Radiation from Air Temperature. Report prepared for FAO, Water Resources Development and Management Service, Rome: FAO.
82. Hargreaves, G. L., and Z. A. Samani. 1982. Estimating potential evapotranspiration. *J. Irrig. Drain. Eng. ASCE* 108(3):225–230.
83. Allen, R. G. 1987. Self calibrating method for estimating solar radiation from air temperature. *J. Hydrol. Eng. ASCE* 2(2):56–67.
84. Hargreaves, G. L., G. H. Hargreaves, and J. P. Riley. 1985. Agricultural benefits for Senegal River Basin. *J. Irrig. Drain. Eng. ASCE* 111:113–124.
85. Ritchie, J. T., and D. S. NeSmith. 1991. Temperature and crop development. *Modeling Plant and Soil Systems*, eds. Hanks, R. J., and J. T. Ritchie, pp. 5–29. Agronomy Series No. 31, Madison, WI: American Society of Agronomists.
86. Sinclair, T. R. 1984. Leaf area development in field-grown soybeans. Agron. J. 76: 141–146.
87. Flesch, T. K., and R. F. Dale. 1987. A leaf area index model for corn with moisture stress redutions. *Agron. J.* 19:1008–1014.
88. Ritchie, J. T. 1991. Wheat phasic development. *Modeling Plant and Soil Systems*, eds. Hanks, R. J., and J. T. Ritchie, pp. 31–54, Agronomy Series No. 31, Madison, WI: American Society of Agronomists.
89. Howell, T. A., J. L. Steiner, A. D. Schneider, and S. R. Evett. 1995. Evapotranspiration of irrigated winter wheat-southern high plains. *Trans. ASAE* 38(3):745–759.
90. Doorenbos, J., and A. H. Kassam. 1979: Yield response to water. FAO Irrigation and Drainage Paper 33, Rome: FAO.
91. Allen, R. G., L. S. Pereira, D. Raes, and M. Smith. 1999. Crop evapotranspiration. FAO Irrigation and Drainage Paper, Rome: FAO. (in print).
92. Pruitt, W. O. 1986. Traditional methods "Evapotranspiration research priorities for the next decade." ASAE Paper No. 86-2629.
93. Snyder, R. L., B. J. Lanini, D. A. Shaw, and W. O. Pruitt. 1987. Using reference evapotranspiration (ET_0) and crop coefficients to estimate crop evapotranspiration

(ET_c) for agronomic crops, grasses, and vegetable crops. Leaflet 21427, Cooperative Extension, Berkeley: University of California.
94. Snyder, R. L., B. J. Lanini, D. A. Shaw, and W. O. Pruitt. 1989. Using reference evapotranspiration (ET_0) and crop coefficients to estimate crop evapotranspiration (ET_c) for trees and vines. Leaflet 21428, Cooperative Extension. Berkeley: University of California.
95. Allen, R. G., M. Smith, L. S. Pereira, and W. O. Pruitt, 1996. Proposed revision to the FAO procedure for estimating crop water requirement. *Proceedings of 2nd International Symposium on Irrigation of Horticultural Crops*, ed. Chartzoulakis, K. S. No. 449, Vol. I:17–33. International Society for Horticultural Sciences (ISHS), AcTa Horticultural.
96. Ritchie, J. T. 1974. Evaluating irrigation needs for southeastern U.S.A. *Proceedings of the Irrigation and Drainage Special Conference, ASCE*, pp. 262–273. New York: ASCE.
97. Ritchie, J. T., and B. S. Johnson. 1990. Soil and plant factors affecting evaporation. *Irrigation of Agricultural Crops*, eds. Stewart, B. A., and D. R., Nielsen, Agronomy Series 30. pp. 363–390. Madison: WI American Society of Agronomists.
98. Martin, D. L., E. C. Stegman, and E. Fereres. 1990. Irrigation scheduling principles. *Management of Farm Irrigation Systems*. eds. Hoffman, G. J., T. A. Howell, and K. H. Solomon, pp. 155–203. St. Joseph, MI: American Society of Agricultural Engineers.
99. Ritchie, J. T., D. C. Godwin, and U. Singh. 1989. Soil and weather inputs for the IBSNAT crop models. *Decision Support System for Agrotechnology Transfer*, Part I, pp. 31–45. Honolulu: Dept. Agronomy and Soil Science, University of Hawaii.
100. Hanks, R. J. 1985. Crop coefficients for transpiration. *Advances in Evapotranspiration*, pp. 431–438. St. Joseph, MI: American Society of Agricultural Engineers.
101. Saxton, K. E., H. P. Johnson, and R. H. Shaw. 1974. Modeling evapotranspiration and soil moisture. *Trans. ASAE* 17(4):673–677.
102. Tanner, C. B., and W. A. Jury. 1976. Estimating evaporation and transpiration from a crop during incomplete cover. *Agron. J.* 68:239–242.
103. Merriam, J. L. 1966. A management control concept for determining the economical depth and frequency of irrigation. *Trans. ASAE* 9:492–498.
104. Soil Conservation Service. 1982. *National Engineering Handbook*. Washington DC: U.S. Government Printing Office.
105. Ayers, R. S., and D. W. Westcot. 1985. *Water quality for agriculture*. Rev. 1. FAO Irrigation and Drainage Paper 29. Rome: FAO.

5.2 Water Retention and Movement in Soil

N. Romano

5.2.1 Basic Concepts

Soil is a porous system made up of solid, liquid, and gaseous phases. The liquid phase (*soil solution*) consists of soil water, which usually contains a variety of dissolved minerals and organic substances. Water in soil may be encountered in three different states: as a liquid, a solid (ice), or a gas (water vapor).

The definitions and discussion in the following sections refer to a macroscopic description of an idealized continuous medium that replaces the actual complex geometry of a pore system. The various state variables (e.g., pressure potential and water content) and soil properties (e.g., bulk density and hydraulic conductivity) are considered to be continuous functions of position and time. They are viewed as macroscopic quantities obtained by volume averages over an appropriate averaging volume referred to as the *representative elementary volume* (REV) whose characteristic length should be much greater than that of a typical pore diameter but considerably smaller than a characteristic length of the porous system under study [1]. Within this effective continuum, the solid matrix is usually considered as rigid, the liquid phase is Newtonian and homogeneous, air is interconnected at the atmospheric pressure, and the analysis of flow regime is conducted by evaluating the flux density as volume of water discharged per unit time and per unit entire cross-sectional area of soil. Each point of the domain considered is the center of an REV. The macroscopic continuum approach represents a fertile tool for the development of theories applicable to the problem of water movement through porous media.

5.2.2 Soil Water Content

Soil water content generally is defined as the ratio of the mass of soil water to the mass of dried soil, or as the volume of water per unit volume of soil. In both cases, accuracy in calculating the water content depends on a clear and rigorous definition of the *dry soil* condition.

Because the interest of practical applications relies largely upon the determination of the magnitude of relative time changes in water content in a certain point, the condition of *dry soil* refers by tradition to a standard condition obtained by evaporating the water from a soil sample placed in an oven at 100–110°C until variations in sample weight are no longer noticed. The choice of these temperatures is somewhat arbitrary and does not result from scientific outcomes. Rather, within the above range of temperatures the evaporation of *free water* from the sample is guaranteed and the standard condition can be attained easily using commercial ovens.

It is useful to define the water content of soil on a volumetric basis θ (m^3/m^3) as the dimensionless ratio

$$\theta = V_w / V_t \tag{5.93}$$

of water volume V_w (m^3) to total soil volume V_t (m^3). Especially when subjecting a soil sample to chemical analyses, the soil water content is expressed usually on a mass basis as

$$w = M_w / M_s, \tag{5.94}$$

where M_w is the mass of water (kg) and M_s is the mass of dry soil particles (kg).

If $\rho_b = M_s / V_t$ denotes the oven-dry bulk density (kg/m^3) and $\rho_w = M_w / V_w$ is the density of liquid water (kg m^{-3}), the volumetric soil water content θ and the gravimetric soil water content w have the following relationship:

$$\theta = w(\rho_b/\rho_w). \tag{5.95}$$

The maximum water content in the soil is denoted as the water content at saturation θ_s. In some cases the amount of water in the soil is thus computed as a percentage of saturated water content s (%) and is expressed in terms of degree of saturation with respect to the water, $s = \theta/\theta_s \times 100$. However, evaluating the volumetric water content at saturation is highly uncertain, if not impossible, in swelling shrinking soils or as phenomena of consolidation of the soil matrix occur. In such cases, a useful expression of soil water content is the *moisture ratio* ϑ, defined as

$$\vartheta = V_w/V_s, \tag{5.96}$$

where V_s (m^3) is the volume of solids.

Measurement of Water Content of Soil

The measurement of water content in the soil is of great importance in many investigations and applications pertaining to agriculture, hydrology, meteorology, hydraulic engineering, and soil mechanics. In the fields of agronomy and forestry, the amount of water contained in the soil affects plant growth and diffusion of nutrients toward the plant roots, as well as acting on soil aeration and gaseous exchanges, with direct consequences for root respiration. Also, continuous monitoring of soil water content can support the setting up of optimal strategies for the use of irrigation water. In hydrology, moisture condition in the uppermost soil horizon plays an important role in determining the amount of incident water—either rainfall or irrigation water — that becomes runoff. Evapotranspiration processes, transport of solute and pollutants, and numerous hydraulic (e.g., retention, conductivity) or mechanical (e.g., consistency, plasticity, strength) soil properties depend on soil water content.

Several methods have been proposed to determine water content in soil, especially under field conditions. Soil water content can be measured by direct or indirect methods [2, 3].

Direct Methods

Direct methods involve removing a soil sample and evaluating the amount of water that it contains. Their use necessarily entails the destruction of the sample and hence the inability to repeat the measurement in the same location.

The most widely used direct method is the thermogravimetric method, often considered as a reference procedure because it is straightforward, accurate, and inexpensive in terms of equipment. This method consists of collecting a disturbed or undisturbed soil sample (usually of about 100–200 g taken with an auger or sampling tube) from the appropriate soil depth, weighing it, and sealing it carefully to prevent water evaporation or the gaining of moisture before it is analyzed. Then, the soil sample is placed in an oven and dried at 105–110°C. The residence time in the oven should be such that a condition of stable weight is attained, and it depends not only on the type of soil and size of the sample but also on the efficiency and load of the oven. Usual values of the residence time in the oven are about 12 h if a forced-draft oven is used, or 24 h in a convection oven.

At completion of the drying phase, the sample is removed from the oven, cooled in a desiccator with active desiccant, and weighed again. The gravimetric soil water content is calculated as follows:

$$w = [(w_w + t_a) - (w_d + t_a)]/[(w_d + t_a) - t_a], \tag{5.97}$$

where w_w and w_d represent the mass of wet and dry soil (kg), respectively, and t_a is the tare (kg).

The major source of error using the thermogravimetric method together with (5.95) is related to sampling technique. The fact that the soil cores may contain stones, roots, and voids, as well as certain unavoidable disturbances during sampling, may affect the precision in determining the value of the volumetric water content in soil.

Indirect Methods

Basically, indirect methods consist of measuring some soil physical or physicochemical properties that are highly dependent on water content in the soil. In general, they do not involve destructive procedures and use equipment that also can be placed permanently in the soil, or remote sensors located on airborne platforms and satellites. Thus, indirect methods are well suited for carrying out measurements on a repetitive basis and in some cases also enable data to be recorded automatically, but require the knowledge of accurate calibration curves.

The main indirect methods are gamma attenuation, neutron thermalization, electrical resistance, time-domain reflectometry (TDR). Other indirect methods are low-resolution nuclear magnetic resonance imaging and remote-sensing techniques.

A typical nondestructive laboratory method for monitoring water contents in a soil column is based on the attenuation and backscattering of a collimated beam of gamma rays emitted by a radioactive source, such as cesium-137. In case of shrinking/swelling porous materials, a dual-energy gamma-ray attenuation system (usually employing cesium-137 and americium-241 as the radioactive sources) can be used to measure simultaneously bulk density and water content in a soil sample.

Instead, the neutron method often is used for field investigations and enables soil water contents to be determined by the thermalization process of high-energy neutrons colliding with atomic nuclei in the soil, primarily hydrogen atoms [2]. Because hydrogen is the major variable affecting energy losses of fast neutrons, the count rate of thermalized (slow) neutron pulses can be related to soil water content. Actually, the calibration curve linearly relates the volumetric soil water content θ to the relative pulse count rate N/N_R; that is,

$$\theta = n_1(N/N_R) + n_2, \tag{5.98}$$

where N is the measured count rate of thermalized neutrons, N_R is the count rate under a "reference" condition, and n_1 and n_2 are parameters. Some manufacturers suggest that the reference count rate N_R be obtained in the same protective shield supplied for the probe transportation, but this value can be highly affected by humidity and temperature of the surrounding environment and by the relatively small size of the shield. More effectively, the value of the reference count rate should be taken in a water-filled tank (e.g., a cylinder of 0.6 m in height and 0.5 m in diameter) on a daily basis during the investigation. Parameter n_1 chiefly depends on the presence of substances that play a basic role in the thermalization process, such as boron, cadmium, iron, and molybdenum, whereas the value of parameter n_2 is strongly affected by soil bulk density and is nearly zero for very low values of bulk density.

Employing a factory-supplied calibration curve can be inadequate in most situations. It thus is recommended that the calibration curve be obtained experimentally in the field by relating the measured count ratio N/N_R for a soil location to simultaneous measurements of soil water content with the thermogravimetric method. Oven-dry bulk densities are also to be measured. One drawback of the neutron method is the low spatial resolution under certain conditions associated with the thermalization process. Close to saturation, the measuring volume is approximately a sphere 0.15 m in diameter, but under dry condition the diameter of this sphere is about 0.50–0.70 m. Therefore, larger uncertainties are to be expected when the soil profile consists of several alternating layers of highly contrasting soil texture, as well as when measurements are performed close to the soil surface. Moreover, because of the influence of the size of the sphere, the neutron method is not very useful for distances between the measuring depths less than 0.10 m.

In the past decade, indirect estimation of water content by measuring the propagation velocity of an electromagnetic wave is becoming increasingly popular. One method that exploits this principle and that can be employed in laboratory and field experiments, is TDR, which actually determines the apparent dielectric permittivity of soil by monitoring the travel time for an electromagnetic signal (TDR pulse) to propagate along a suitable probe inserted in the soil at the selected measuring depth. Dielectric properties of a substance in the presence of an electromagnetic field depend on the polarization of its molecules and are described by the apparent relative dielectric permittivity ε, which is a dimensionless variable always greater than unity and conveniently defined by a complex relation as the sum of a real part, ε', and an imaginary part ε'' of ε. The real part of the dielectric permittivity mainly accounts for the energy stored in the system due to the alignment of dipoles with the electric field, whereas the imaginary part accounts for energy dissipation effects [4]. In a heterogeneous complex system, such as soil, essentially made of variable proportions of solid particles, air, water, and mineral organic liquids, it is extremely difficult to interpret dielectric behavior, especially at low frequencies of the imposed alternated electrical field. However, within the frequency range from about 50 MHz to 2 GHz, the apparent relative dielectric permittivity of soil, $\varepsilon'_{\text{soil}}$, is affected chiefly by the apparent relative dielectric permittivity of water ($\varepsilon'_{\text{water}} \cong 80$ at 20°C) because it is much larger than that of air ($\varepsilon'_{\text{air}} = 1$) and of the solid phase ($\varepsilon'_{\text{solid}} \cong 3-7$). Within the above range, it is therefore possible to relate uniquely the measurements of soil relative permittivity $\varepsilon'_{\text{soil}}$ to volumetric water content by means of a calibration curve. Moreover, the employed measurement frequency makes the soil relative permittivity rather invariant with respect to the frequency and hence usually it also is referred to as the dielectric constant of soil.

By examining a wide range of mineral soils, Topp and his colleagues [5] determined the empirical relationship

$$\theta = -a_1 + a_2\varepsilon'_m - a_3\left(\varepsilon'_m\right)^2 + a_4\left(\varepsilon'_m\right)^3 \tag{5.99}$$

between the volumetric soil water content θ and the TDR-measured dielectric permittivity of the porous medium ε'_m. The regression coefficients a_i are, respectively, $a_1 = 5.3 \times 10^{-2}$, $a_2 = 2.92 \times 10^{-2}$, $a_3 = 5.5 \times 10^{-4}$, and $a_4 = 4.3 \times 10^{-6}$.

Even though the calibration equation (5.99) does not describe accurately the actual relation $\theta(\varepsilon'_m)$ when ε'_m tends toward 1 or to the value of the dielectric permittivity of free water; however, it is simple and allows good soil water content measurements within the range of $0.05 < \theta < 0.6$, chiefly if only relative changes of θ are required. For non-clayey mineral soils with low-organic-matter content, absolute errors in determining water content by Eq. (5.99) can be even less than $\pm 0.015 \text{m}^3/\text{m}^3$, whereas an average absolute error of about $\pm 0.035 \text{m}^3/\text{m}^3$ was reported for organic soils.

When absolute values of θ and a greater level of accuracy are required, a site-specific calibration of the TDR-measured dielectric permittivity ε'_m to soil water content θ should be evaluated. In this case, and especially if measurements are to be carried out close to the soil surface, a zone where soil temperature fluctuations can be relatively high during the span of the experiment, one also should take into account the dependence of temperature T(°C) upon $\varepsilon'_{\text{water}}$:

$$\varepsilon'_{\text{water}} = b_1 - b_2 T + b_3 T^2 - b_4 T^3, \tag{5.100}$$

where values of the constants b_i of this polynomial are, respectively, $b_1 = 87.74$, $b_2 = 0.4001$, $b_3 = 9.398 \times 10^{-4}$, and $b_4 = 1.410 \times 10^{-6}$. However, relation (5.100) strictly holds for free water only and can be considered as acceptable for sandy soils, but it cannot be used for clayey and even for loamy soils.

Finally, note that this device does not lead to point measurements, but rather it averages the water content over an averaging volume that mainly depends on the length and shape of the TDR probe employed.

5.2.3 Soil Water Potential

Water present in an unsaturated porous medium such as soil is subject to a variety of forces acting in different directions. The terrestrial gravitational field and the overburden loads due to the weight of soil layers overlying a nonrigid porous system tend to move the soil water in the vertical direction. The attractive forces occurring between the polar water molecules and the surface of the solid matrix and those coming into play at the separation interface between the liquid and gaseous phases can act in various directions. Moreover, ions in the soil solution give rise to attractive forces that oppose the movement of water in the soil.

Because of difficulties in describing such a complex system of forces and because of the low-velocity flow field within the pores, so that the kinetic energy can be neglected, flow processes in soil are referred instead to the potential energy of a unit quantity of water resulting from the force field. Thus, flow is driven by differences in potential energy, and soil water moves from regions of higher to regions of lower potential. In particular, soil water is at equilibrium condition if potential energy is constant throughout the system.

Because only differences in potential energy between two different locations have a physical sense, it is not necessary to evaluate soil water potentials through an absolute scale of energy, but rather they are referred to a standard reference state. This standard reference state usually is considered as the energy of the unit quantity of pure water (no solutes), free (contained in a hypothetical reservoir and subject to the force of gravity only), at atmospheric pressure, at the same temperature of water in the soil (or at a different, specified temperature), and at a fixed reference elevation. The concept of soil water potential is of fundamental importance for studies of transport processes in soil and provides a unified way of evaluating the energy state of water within the soil–plant–atmosphere system. To consider the different field forces acting upon soil water separately, the potential is used, defined thermodynamically as the difference in free energy between soil water and water at the reference condition.

A committee of the International Soil Science Society [6] defined the total soil water potential as "the amount of work that must be done per unit quantity of pure water in order to transport reversibly and isothermally an infinitesimal quantity of water from a pool of pure water at a specified elevation at atmospheric pressure to the soil water (at the point under consideration)." This definition, though a really formal one and not useful for effectively making measurements [7, 8], allows us to consider the total potential as the sum of separate components, each of which refers to an isothermic and reversible transformation that partly changes water state from the reference condition to a final condition in the soil.

Following the Committee's proposal, the total potential of soil water, ψ_t, can be broken down as follows:

$$\psi_t = \psi_g + \psi_p + \psi_o, \tag{5.101}$$

where the subscripts g, p, and o refer to the gravitational potential, the pressure potential, and the osmotic potential, respectively. Different units can be employed for the soil water potentials and they are reported under "Units of Potential," below.

The potentials ψ_g and ψ_o account for the effects of elevation differences and dissolved solutes on the energy state of water. The pressure potential ψ_p comprises all the remaining forces acting upon soil water and accounts for the effects of binding to the solid matrix, the curvature of air–water menisci, the weight of overlying materials, the gas-phase pressure, and the hydrostatic pressure potential if the soil is saturated. Thus, strictly speaking, the gravitational and pressure potentials refer to the soil solution, whereas the osmotic potential refers to the water component only.

However, the above definition of pressure potential ψ_p generally is not used because, in the realm of soil physics, the energy changes associated with the soil water transport from the standard reference state to a certain state in the soil at a fixed location are traditionally split up into other components of potential that separately account for the effects of pressure in the gaseous phase, overburdens, hydrostatic pressure, and links between water and the solid matrix.

The component of pressure potential that accounts for adsorption and capillary forces arising from the affinity of water to the soil matrix is termed matric potential ψ_m. Under fully saturated conditions, $\psi_m = 0$. In nonswelling soils (for which the solid matrix is rigid) bearing the weight of overlying porous materials and in the presence of an interconnected gaseous phase at atmospheric pressure, the matric potential ψ_m coincides with pressure potential ψ_p. Under relatively wet conditions, in which capillary forces predominate, the matric component of the pressure potential can be expressed by the capillary equation written in terms of the radius of a cylindrical capillary tube, R_{eff} (effective radius of the meniscus, in meters):

$$P_c = (2\sigma \cos \alpha_c)/R_{\text{eff}}, \tag{5.102}$$

where $P_c = P_{nw} - P_w$ is the capillary pressure (N/m^2), defined as the pressure difference between the nonwetting and the wetting fluid phases, σ is the interfacial tension between wetting and nonwetting fluid phases (N/m), and α_c is the contact angle. If the nonwetting phase is air at atmospheric pressure, the capillary pressure is equal to $-\psi_m$.

Moreover, the sum of the matric and osmotic potentials often has been termed water potential ψ_w, and it provides a measure of the hydration state of plants, as well as affecting the magnitude of water uptake by plant roots.

Units of Potential

The total soil water potential and its components are defined as energy per unit quantity of pure water. Therefore, their relevant units vary if reference is made to a unit mass, unit volume, or unit weight of water.

When referring to unit mass, the dimensions are L^2/T^2, and in the SI system the potential units are J/kg. Although it is better to use the unit mass of water because it does not change with temperature and pressure, this definition of the potential energy is widely used only in thermodynamics. If one considers that, in the most practical applications, water can be supposed incompressible and its density is independent of potential, energy can be referred to the unit volume instead of unit mass. Hence, the dimensions are those of a pressure $M{\cdot}L^{-1}T^{-2}$ and in the SI system the units of potential are J/m^3 = N/m^2 (pascal, Pa). By expressing the potential as energy per unit weight of water, the units are J/N = m, and the relevant dimensions are those of a length L. This latter way to evaluate the water energy potential is more useful and effective.

When effects of the presence of solutes in the soil solution can be neglected, as applies to most cases, instead of using the symbol ψ, it is customary in analogy to hydraulics to define the total soil water potential per unit weight, H (m), in terms of head units as

$$H = H_g + H_p = z + h, \tag{5.103}$$

where z (m) is the elevation of the point under consideration, or gravitational head, and h (m) is the soil–water pressure head.

Measurement of Soil Water Potential

Knowing soil water potential values in soil is of primary importance in studies of transport processes within porous media as well as in evaluating the energy status of water in the crops.

The most widely known direct method for measuring the pressure potential in soil is that associated with use of a tensiometer [2]. The matric potential then can be calculated from measurements of the gas-phase pressure, if different from the atmospheric pressure, and of the overburden potentials in the case of swelling soils. Schematically, a tensiometer consists of a porous cup (or disk), mostly made of a ceramic material, connected to a pressure sensor (e.g., mercury manometer, vacuum gage, pressure transducer) by means of a water-filled tube. The porous cup in inserted in the soil at the selected measurement depth. In spite of specific limitations, chiefly related to a good contact between the porous cup and the surrounding soil and to the measurement range because of vaporization of liquid water, and related failure of liquid continuity, when pressure potential reaches about −85 kPa, this device is widely used in both laboratory and field experiments.

To monitor matric potential at numerous locations within a field, as necessary for automatic irrigation scheduling or environmental monitoring studies, various methods have been proposed employing sensors that measure a certain variable (e.g., electric resistance, heat dissipation) strongly affected by soil water content. An empirical calibration curve then is required for evaluating matric potential in soil. However, uncertainties when using this methodology can be relatively large and chiefly associated with the hysteretic behavior of the porous sensor and with the validity of the selected empirical calibration curve for all or most of the sensors installed in the area of interest. All of these methods are indirect methods for matric potential measurements.

Measurements of water potential (sum of matric potential and osmotic potential) are particularly useful for evaluating crop water availability. These measurements usually are carried out employing thermocouple psychrometers [9], which actually measure the relative humidity of the vapor phase in equilibrium with the liquid phase of the soil.

5.2.4 Soil Water Retention Characteristics

The relationship between volumetric soil water content θ and pressure potential h (expressed here as an equivalent height of water) is called the water retention function. The relationship $\theta(h)$ is strongly and chiefly affected by soil texture and structure. With reference to drying conditions, Fig. 5.8 reports typical shapes of the water retention curve for a sandy and a clayey soil. Starting from an equilibrium condition at saturation ($h = 0$), a slight reduction in h may not cause reductions in θ until pressure potential in the soil reaches a critical value h_E (air-entry potential head), which depends mainly on the pore-size distribution, especially that of larger soil pores. Thus, for $h_E \leq h \leq 0$, the soil should not necessarily be under unsaturated conditions. However, the presence of an air-entry potential is particularly evident only for coarse-textured soils. Beyond this critical value, decreases in h will result in a more or less rapid decrease in θ. When water content is reduced so that soil conditions are very dry (residual water-content conditions), a slight reduction in soil water content may cause the pressure potential to decrease even by orders of magnitude.

The hysteretic nature of soil water retention characteristic $\theta(h)$ under nonmonotonic flow conditions has been demonstrated both theoretically and experimentally. In practice, the θ values are related to the h values in different ways, depending on the drying or wetting scenarios to which the porous medium is subjected and specifically even on the ($\theta - h$) value when the time derivative $\partial\theta/\partial t$ changes its sign. One could say that soil has "memory" of the drying and wetting histories that precede the setting up of a new equilibrium condition and this phenomenon of hysteresis is more evident when approaching fully saturated conditions. Figure 5.9 shows the excursion of water content and pressure potential (main wetting, main drying, and scanning soil water

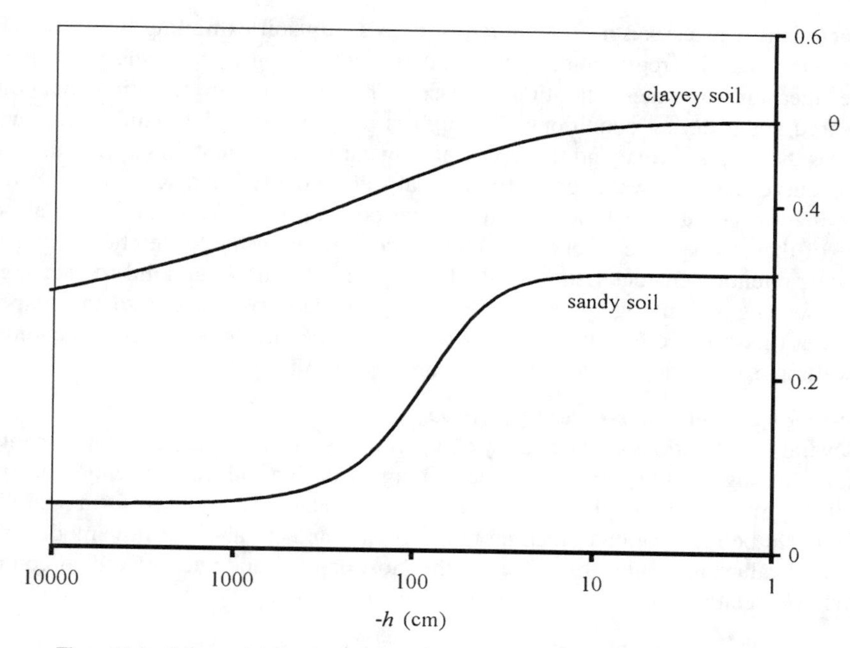

Figure 5.8. Schematic water retention characteristics for a sandy and a clayey soil during drainage.

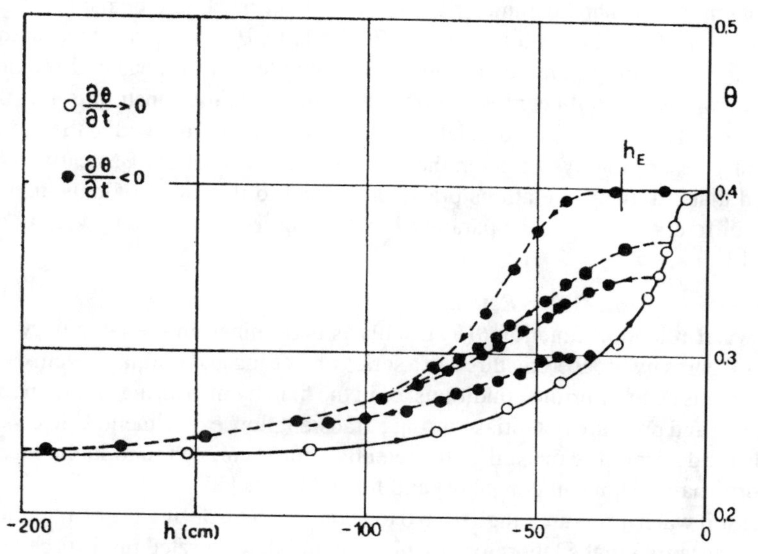

Figure 5.9. Measured soil water retention characteristics for a sandy soil exhibiting hysteresis. *Source:* From [10], with permission.

characteristics) measured in the laboratory on a sandy soil exhibiting hysteresis. The main wetting curve is represented by the solid line, with the open dots being experimental values measured as water saturation increases. The close dots are retention data points measured under drying conditions. The highest dashed line is the main drying curve, whereas the others correspond to a few scanning curves. Note that, in general, the main drying curve starts at a value equal to the total soil porosity, whereas the main wetting curve may reach, at $h = 0$, an effective water content that is less than the total pore space of the soil because of entrapped air. Major causes of the hysteretic behavior of the water retention characteristics are the following: different water-solid contact angles during wetting and drying cycles, as well as high variability in both size and shape of soil pores (ink-bottle effects); the amount of air entrapped in the pore space; phenomena of swelling or shrinking of the individual particles [2, 3].

Description of Soil Water-Retention Curve

Several investigations carried out by comparing a great deal of experimental retention data sets highlighted the possibility of describing the drying soil water-retention function reasonably well by employing empirical analytical relations. A closed-form analytical relation can be incorporated much more easily into numerical water flow models than measured values in tabular form. One of the most popular and widely verified nonhysteretic $\theta(h)$ relations has been proposed by van Genuchten [11]:

$$S_e = [1 + (\alpha|h|)^n]^{-m} \tag{5.104}$$

where $S_e = (\theta - \theta_s)/(\theta_s - \theta_r)$ is effective saturation; θ_s and θ_r are saturated and residual water content, respectively; h (m) is the soil water pressure head; and α (1/m), n, and m represent empirical shape parameters. However, attention should be paid to the concept and definition of residual saturation [8, 12]. Usually, θ_s and θ_r are measured values, whereas the remaining parameter values are computed from measured retention data points by employing nonlinear regression techniques, with the constraints $\alpha > 0, n > 1$, and $0 < m < 1$. A few empirical relations also have been introduced in the literature for analytically describing hysteresis in the water retention function. Basically, it has been proposed that parametric relations practically equal to that of van Genuchten be used, but with different values for the parameters when describing the main wetting or drying curves [13].

Determination of Soil Water-Retention Curve

The water retention function $\theta(h)$ usually is determined in the laboratory on undisturbed soil cores by proceeding through a series of wetting and drainage events and taking measurements at equilibrium conditions, or in the field by measuring simultaneously water contents and pressure potentials during a transient flow experiment. Reviews of direct methods for determining the soil water-retention curve can be found in the literature for laboratory analyses on soil cores [14] and for field soils [15].

In the soil water-pressure range from 0 to about −2.5 m, drying water-retention values often are measured in the laboratory by placing initially saturated undisturbed soil cores on a porous material (e.g., sand-kaolin bed, mixture of glass and diatomaceous powders), which then is subjected to varying soil water-pressure heads. The selected porous bed is held in a container, which usually is made of ceramics or Perspex and is provided with a cover to prevent evaporative losses. After reaching conditions of water equilibrium at a fixed pressure head, water content in each core is measured gravimetrically. Before removing a soil core from the porous bed for weighing, it is thus important to know whether equilibrium has been reached. This condition can be monitored conveniently by placing on the upper surface of the soil core a tensiometer, consisting, for example, of a sintered glass slab connected to a pressure transducer.

In-situ $\theta(h)$ data points can be obtained readily at different soil depths from simultaneous measurements of θ and h using field tensiometers for pressure potentials and a neutron probe, or TDR probes, for water content.

Indirect methods to determine the soil water-retention curve also have been proposed and are discussed in Section 5.2.8.

5.2.5 Flow Within the Soil

Generally, the hydrodynamic description of a fluid-flow problem requires knowledge of the momentum equation, the law expressing the conservation of mass (continuity equation), and a state relationship among density, stress, and temperature. Mathematically the flow problem therefore is defined by a more or less complicated system of partial differential equations whose solution requires specification of boundary conditions and, if the flow is unsteady, initial conditions describing the specific flow situation.

From a merely conceptual point of view, the flow of water within the soil should be analyzed on the microscopic scale by viewing the soil as a disperse system and applying the Navier-Stokes equations. Such a detailed description of flow pattern at every point in the domain is practically impossible because actual flow velocities vary greatly in both magnitude and direction due to the complexity of the paths followed by individual fluid particles when they move through the interconnected pores. On the other hand, in many applications, greater interest is attached to the knowledge of flux density. Therefore, flow and transport processes in soils typically are described on a macroscopic scale by defining a REV and an averaged set of quantities and balance equations.

Basic Flow Equations

The above conceptualization of a porous medium allows description of water movement through soil, either saturated or unsaturated, by the experimentally derived Darcy's law,

$$\boldsymbol{q} = -K\nabla H, \tag{5.105}$$

written in vectorial form for a homogeneous isotropic medium under isothermal conditions. This equation relates macroscopically the volumetric flux density (or Darcy velocity) $\boldsymbol{q}$ to the negative vector gradient of the total soil water-potential head H by means of the parameter K, called the soil hydraulic conductivity. For an anisotropic porous medium, this parameter becomes a tensor. The hydraulic conductivity is assumed to be independent of the total potential gradient but may depend on other variables. Because the term ∇H is dimensionless, both $\boldsymbol{q}$ and K have dimensions of L/T and generally units of meters per second or centimeters per hour.

Note that Darcy's law can be derived from the Navier-Stokes equations for viscous-flow problems because it practically describes water flow in porous media when inertial forces can be neglected with respect to the viscous forces. Therefore, the range of validity of Darcy's law depends on the occurrence of the above condition; readers wishing further details are directed to the literature [3].

The soil water-flow theory based on the Darcy flux law provides only a first approximation to the understanding and description of water-flow processes in porous media. Apart from the already-cited nonlinear proportionality between the Darcy velocity and the hydraulic potential gradient at high flow velocity due to the increasing weight of the inertial forces with respect to the viscous forces in determining the magnitude of the stresses acting on soil water and the presence of turbulence, allowances are made for possible deviations from Darcy's law even at low flow velocities [16]. Other causes of deviations from the Darcy-based flow theory can be attributed chiefly to the occurrence of macropores (such as earthworm holes, cracks, and fissures), nonisothermal conditions, nonnegligible effects of air pressure differences, and solute–water interactions. However, these causes may become more important when modeling transport processes under field-scale conditions with respect to laboratory-scale situations.

The description of mass conservation is still made using the concept of REV and usually with the assumption of a rigid system. The principle of mass conservation requires that the change with time of mass stored in an elemental soil volume must equal the difference between the inflow- and the outflow-mass rates. Therefore, the basic mass balance for water phase can be written as

$$\frac{\partial(\rho_w \theta)}{\partial t} = -\nabla \cdot (\rho_w \boldsymbol{q}), \tag{5.106}$$

where ρ_w is water density (kg/m^3), θ is volumetric water content, and t is time (s). If one should take the presence of source or sink terms into account (e.g., recharging well, water uptake by plant roots), the equation of continuity is

$$\frac{\partial(\rho_w \theta)}{\partial t} = -\nabla \cdot (\rho_w \boldsymbol{q}) + \rho_w S, \tag{5.107}$$

where S is a function representing sources (positive) and sinks (negative) of water in the porous system and has dimensions of $1/T$ and units of $1/\text{s}$.

5.2.6 Water Flow in Saturated Soil

When the pore system is completely filled with water, and hence pressure potential h is positive throughout the system, the coefficient of proportionality in Darcy's law (5.105) is called the saturated hydraulic conductivity K_s. The value of K_s is practically a constant, chiefly because the soil pores are always filled with water, and it depends not only on soil physical properties (e.g., bulk density, soil texture), but also on fluid properties (e.g., viscosity).

When water is incompressible and the solid matrix is rigid (or, of course, when the flow is steady), the flux equation (5.105) and the continuity equation (5.106) reduce to

Laplace's equation for H:

$$\nabla^2 H = 0. \tag{5.108}$$

Values of saturated hydraulic conductivity K_s are obtained in the laboratory using a constant-head permeameter (basically, a facility reproducing the original experiment carried out by Darcy to demonstrate the validity of his flux law) or a falling-head permeameter [2]. Field measurements of hydraulic conductivity of a saturated soil are commonly made by the augerhole method [17].

One alternative to direct measurements is to use theoretical equations that relate the saturated hydraulic conductivity to other soil properties. By assuming an equivalent uniform medium made up of spherical particles and employing the Hagen-Poiseuille equation for liquid flow in a capillary tube, the following Kozeny-Carman relation holds between saturated hydraulic conductivity K_s and soil porosity p:

$$K_s = cp^3/A^2, \tag{5.109}$$

where p is defined as the dimensionless ratio of the pore volume to the total soil volume, A is the specific surface area of the porous medium per unit volume of solid (m^2/m^3), and c is a constant (m^3 s^{-1}) [18]. Mishra and Parker [19] used van Genuchten's water retention curve [VG retention curve, Eq. (5.104)] to derive the following expression:

$$K_s = c'(\theta_s - \theta_r)^{2.5}\alpha^2, \tag{5.110}$$

where θ_s, θ_r, and α are parameters as defined by Eq. (5.104), and c' is equal to 108 cm^3 s^{-1} if K_s is expressed in cm s^{-1} and α in 1/cm.

In layered soils, it is relatively simple to determine the equivalent saturated hydraulic conductivity of the whole porous system by analogy with the evaluation of the equivalent resistance of electrical circuits arranged in series or parallel. For soil layers arranged in series to the flow direction (the more common case), the flow rate is the same in all layers, and the total potential gradient equals the sum of the potential gradient in each layer. Conversely, in the parallel-flow case, the potential gradient is the same in each layer, and the total flow is the sum of the individual flow rates.

5.2.7 Water Flow in Unsaturated Soil

Water movement in a porous material whose interconnected pores are filled only partially with water is defined as unsaturated water flow. Important phenomena occurring in the hydrological cycle, such as infiltration, drainage, redistribution of soil water, water uptake by plant roots, and evaporation, all involve flow of water in unsaturated soil.

Historically, the development of the physical theory of water flow in unsaturated porous media was promoted by Richards [20], who considered the original Darcy law for saturated flow, and therefore its underlying physical meaning, still valid under unsaturated conditions. In the unsaturated zones, the gaseous phase (generally, air and water in the vapor phase) is assumed to be continuous and interconnected at a constant pressure value, usually at atmospheric pressure. Moreover, the flow of the interconnected air or gas is neglected because it is a nearly frictionless flow. The presence of the gaseous phase reduces the hydraulic conductivity of the system in different ways from point to point of the flow domain, depending on the local values of water content. Therefore, the proportionality factor K becomes a function of volumetric water content θ and is called the unsaturated hydraulic conductivity function.

According to the Richards approximation and neglecting sinks, sources, and phase changes, equating Darcy's law (5.105) and the equation of continuity (5.106) yields the following partial differential equation governing unsaturated water flow:

$$\frac{\partial(\rho_w \theta)}{\partial t} = \nabla \cdot (\rho_w K \nabla H). \tag{5.111}$$

An alternative formulation for the flow equation can be obtained by introducing the soil water diffusivity $D = K\mathrm{d}h/\mathrm{d}\theta$ (dimension of L^2/T and units of m^2/s):

$$\frac{\partial(\rho_w \theta)}{\partial t} = \nabla \cdot [\rho_w (K \nabla z + D \nabla \theta)] \tag{5.112}$$

where z (m) is the gravitational component of the total soil water-potential head.

The use of θ as dependent variable seems more effective for solving flow problems through porous media with low water content. However, when the degree of saturation is high and close to unity, employing Eq. (5.112) proves difficult because of the strong dependency of D upon θ. In particular, in the saturated zone or in the capillary fringe region of a rigid porous medium, the term $\mathrm{d}h/\mathrm{d}\theta$ is zero, D goes to infinity, and Eq. (5.112) no longer holds. Also, an unsaturated-flow equation employing θ as dependent variable hardly helps to model flow processes into spatially nonuniform porous media, in which water content may vary abruptly within the flow domain, thereby resulting in a nonzero gradient $\nabla\theta$ at the separation interface between different materials. The selection of h as dependent variable may overcome such difficulties as soil water potential is a continuous function of space coordinates, as well as it yields Eq. (5.111) that is valid under both saturated and unsaturated conditions.

Water transport processes in the unsaturated zone of soil are generally a result of precipitation or irrigation events which are distributed on large surface areas relative to the extent of the soil profile. The dynamics of such processes is driven essentially by gravity and by predominant vertical gradients in flow controlled quantities. These features thus allow us the opportunity to mathematically formulate most practical problems involving flow processes in unsaturated soils as one- dimensional in the vertical direction. The equation governing the vertical, isothermal unsaturated-soil water flow is written traditionally as

$$C\frac{\partial h}{\partial t} = \frac{\partial}{\partial z}\left[K\left(\frac{\partial h}{\partial z} - 1\right)\right] \tag{5.113}$$

known as the Richards equation. This equation uses soil water-pressure head h as the dependent variable and usually is referred to as the pressure-based form of the governing unsaturated water-flow equation. In Eq. (5.113), z denotes the vertical space coordinate (m), conveniently taken to be positive downward; t is time (s); K is the unsaturated hydraulic conductivity function ($\mathrm{m\,s^{-1}}$); and $C = \mathrm{d}\theta/\mathrm{d}h$ is the capillary hydraulic storage function (1/m), also termed specific soil water capacity, which can be computed readily by deriving the soil water-retention function $\theta(h)$.

From the numerical modeling viewpoint, Celia and coworkers [21] have conducted several studies to show that the so-called mixed form, which originates by applying the temporal derivative to the water content and the spatial derivative to the pressure potential, hence avoiding expansion of the time derivative term, should be preferred to both the θ-based and h-based formulations of the Richards equation.

Unsaturated-Soil Hydraulic Conductivity Curve

The functions $\theta(h)$ and $K(\theta)$, the unsaturated-soil hydraulic properties, are highly nonlinear functions of the relevant independent variables and they characterize a soil from the hydraulic point of view. The unsaturated hydraulic conductivity also can be viewed as a function of pressure head h, because water content and pressure potential are directly related through the water retention characteristic. If one uses this relationship, strictly speaking, unsaturated hydraulic conductivity should be considered as a function of matric potential only but, according to the Richards approximation, the terms matric potential or pressure potential can be used without distinction. The presence of a hysteretic water retention function will cause the $K(h)$ function to be hysteretic as well. However, some experimental results have shown that hysteresis in $K(\theta)$ is relatively small and negligible in practice.

Typical relationships between unsaturated hydraulic conductivity K and pressure head h under drying conditions are illustrated in Fig. 5.10 for a sandy and a clayey soil. The water retention functions $\theta(h)$ for these two types of soil are depicted in Fig. 5.8. As evident from Fig. 5.10, size distribution and continuity of pores have a strong influence on the hydraulic conductivity behavior of soil. Coarser porous materials, such as sandy soils, have high hydraulic conductivity at saturation K_s and relatively sharp drops with decreasing pressure potentials. This behavior can be explained readily if one considers that coarse soils are made up primarily of large pores that easily transmit large volumes

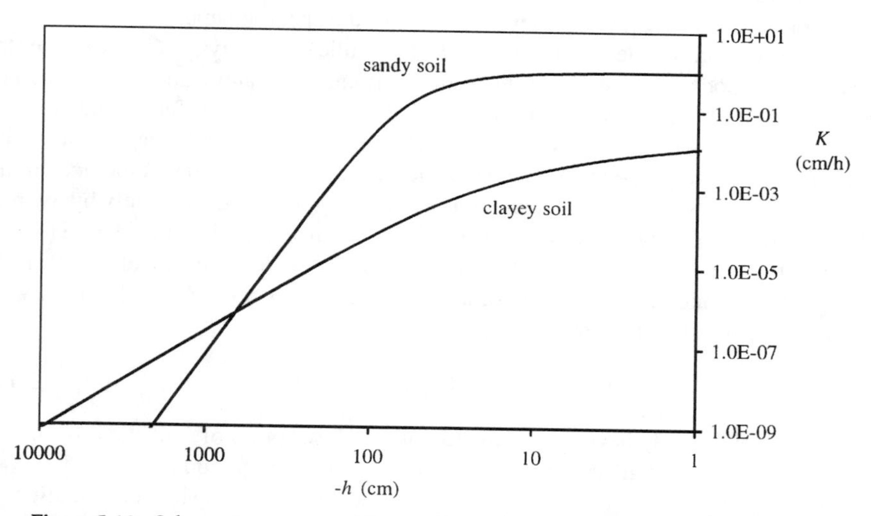

Figure 5.10. Schematic unsaturated hydraulic conductivity functions for a sandy and a clayey soil during drainage.

of water when filled, but are emptied even by small reductions in pressure potentials from saturation conditions. On the other hand, fine porous materials such as clay soils have lower K_s values and then unsaturated hydraulic conductivity decreases slowly as pressure potentials decrease from $h = 0$. In fact, the higher proportion of fine pores that characterizes clay soils makes the water transport capacity of this type of porous materials still relatively high even under dry conditions. Intermediate situations can occur for porous materials having soil textures between these two extreme cases. Even if coarse soils are more permeable than fine soils at saturation or close to this condition, the reverse can be true when the porous materials are under unsaturated conditions. Therefore, knowledge of texture, structure, and position of the different layers in a soil profile is of primary importance to accurately assess the evolution of water movement in soil.

Determination of Unsaturated-Soil Hydraulic Properties

The unsaturated hydraulic conductivity function $K(\theta)$, or $K(h)$, is somewhat difficult to determine accurately, insofar as it cannot be measured directly and, in any case, it varies over many orders of magnitude not only among different soils, but also for the same soil as water content ranges from saturation to very dry conditions. Even though, at present, no proven specific measuring devices are commonly available to determine the hydraulic conductivity function, the numerous proposed methods usually involve measurements of water content and pressure potential for which widespread and well-known commercial devices do exist.

One of the more common and better-known methods for determining $K(\theta)$, or $K(h)$, is the instantaneous profile method [22]. Although the related procedure is quite tedious, one of the main advantages of this method is that it can be applied with minor changes under both laboratory and *in situ* conditions. The crust method often is used as a field method [23, 24]. For information on operational aspects, applicability, and limitations of these or other methods, the reader is referred to the specific papers cited.

However, evaluating the dependence of the hydraulic conductivity K on water content θ by direct methods is time-consuming, requires trained operators, and is therefore very costly. Several attempts have been made to derive models for the function $K(\theta)$ from knowledge of the soil water-retention characteristic $\theta(h)$, which is easy to determine and reflects well the pore-size geometry, which in turn strongly affects the unsaturated hydraulic properties. A hydraulic conductivity model that is used frequently by hydrologists and soil scientists was developed by van Genuchten, combining the relation (5.104), subject to the constraint $m = 1 - 1/n$, with Mualem's statistical model [25, 26]. According to this model, unsaturated hydraulic conductivity is related to volumetric water content by the following expression:

$$K(\theta) = K_0 S_e^{\lambda}\left[1 - \left(1 - S_e^{1/m}\right)^m\right]^2, \tag{5.114}$$

where S_e and m already have been defined under "Description of Soil Water-Retention Curve," λ is a dimensionless empirical parameter on average equal to 0.5, and K_0 is the hydraulic conductivity when $\theta = \theta_0$. The advantages of using this relation are that it is a closed-form equation, has a relatively simple mathematical form, and depends mainly on the parameters describing the water retention function. To obtain the hydraulic conductivity curve for a certain soil under study, at least one value of K should be measured at a fixed value of θ. It is customary that the prediction be matched to saturation conditions, such that $\theta_0 = \theta_s$ and $K_0 = K_s$. However, this criterion is not very effective, mainly because hydraulic conductivity at saturation may be ill-defined since it is strongly affected by macroporosity, especially in the case of structured soils. It thus has been suggested that the $K(\theta)$ curve be matched at a point K_0 measured under unsaturated conditions ($\theta_0 < \theta_s$) [26].

5.2.8 Indirect Estimation of Water-Retention and Hydraulic-Conductivity Functions

Much work is being directed toward the evaluation of the soil water-retention and hydraulic-conductivity functions from related soil physical properties. The increasingly complex computer models employed in environmental studies require a large amount of input data, especially those characterizing the soil from the hydraulic viewpoint, which in turn are notoriously difficult to determine. Therefore, when simulating hydrological processes in large areas, the possibility of deriving hydraulic parameters from soil data (such as bulk density, organic-matter content, and percentage of sand, silt, and clay), which are relatively simple to obtain or already available, is highly attractive. This task is carried out by using the *pedotransfer functions* (PTFs), which transfer basic soil physical properties and characteristics into fixed points of the water-retention function or into values of the parameters describing an analytical $\theta(h)$ relationship [27]. Unsaturated hydraulic conductivity characteristics then usually are evaluated by PTF predictions of hydraulic conductivity at saturation [28]. PTFs appear to provide a promising technique to predict soil hydraulic properties, and they are highly effective for deriving soil water-retention characteristics; however, there is still some debate in the literature on the accuracy and reliability of unsaturated hydraulic conductivity evaluated by this predictive method [29]. To date, most of the research relating to PTFs usually have been directed toward comparisons between measured and estimated hydraulic properties for different types of soils [30], but a few studies have investigated the effects of PTF predictions on some practical applications [31].

More recently, many authors have come to be interested in the feasibility of simultaneously estimating the water-retention and hydraulic-conductivity functions from transient flow experiments by employing the inverse-problem methodology in the form of the parameter optimization technique. By using this approach, only a few selected variables need to be measured during a relatively simple transient flow event obtained for prescribed but arbitrary initial and boundary conditions. Data processing assumes that the soil hydraulic properties $\theta(h)$ and $K(\theta)$ are described by analytical relationships with a small number of unknown parameters, which are estimated by an optimization method minimizing deviations between the real system response measured during the experiment and the numerical solution of the governing flow equation for a given parameter vector. Assuming homoskedasticity and lack of correlation among measurement errors, the optimization problem reduces to a problem of nonlinear ordinary least squares. For soil hydrology applications, however, the observations usually consist of quantities (e.g., water contents, pressure potentials, water fluxes) that show differences in measurement units and accuracy. Thus, a less restrictive reasonable hypothesis accounting for this situation is that one assumes uncorrelated errors but unequal error variances among the different variables. The method becomes a weighted least-squares minimization problem. Another advantage of the parameter estimation methods is that they also can provide information on parameter uncertainty.

Such a methodology is suitable for characterizing hydraulically a soil either in the laboratory (employing, for example, multistep outflow experiments [32] or evaporation experiments [33]) or in the field (generally by inversion of data from transient drainage experiments [34, 35]). In many cases, the inverse-problem methodology based on the parameter estimation approach allows experiments to be improved in a way that makes test procedures easier and faster. The variables to be measured, the locations of the sensors, the times at which to take measurements, as well as the number of observations used as input data for the inverse problem, may exert a remarkable influence on reliability of parameter estimates. Therefore, the experiment should be designed to ensure that the relevant inverse problem allows solution without significantly compromising accuracy in parameter estimates. In fact, parameter estimation techniques are inherently ill-posed problems. Ill-posedness of the inverse problem is associated mainly with the existence of a solution that can be unstable or nonunique, as well as to the fact that model parameters can be unidentifiable. Even if problems relating to ill- posedness of the inverse solution can arise, parameter optimization techniques are undoubtedly highly attractive for determining flow and transport parameters of soil and have proved to be effective methods, especially when a large amount of data needs to be analyzed.

Nomenclature

Basic dimensions are M = mass, L = length, T = time, and K = temperature. F (force) = $M{\cdot}L{\cdot}T^{-2}$ is a derived dimension.

Roman Letters

a_i = coefficients in Eq. (5.99)
A = specific surface area of the medium, L^{-1}
b_i = coefficients in Eq. (5.100)
c = constant in Eq. (5.109), L^3/T
c' = constant in Eq. (5.110), L^3/T
C = capillary hydraulic storage function, L^{-1}
D = soil water diffusivity, L^2/T
h = soil-water pressure head, L
h_E = air-entry potential head, L
H = total soil-water potential head, L
K = soil hydraulic conductivity, L/T
K_s = saturated soil hydraulic conductivity, L/T
m = parameter in the VG retention curve, dimensionless
M_s = mass of solids, M
M_w = mass of water, M

n = parameter in the VG retention curve, dimensionless
n_i = parameters in Eq. (5.98), dimensionless
N = thermalized neutron count rate, dimensionless
N_R = "reference" count rate, dimensionless
p = soil porosity, dimensionless
P_c = capillary pressure, F/L^2
P_{nm} = pressure of nonwetting fluid phase, F/L^2
P_w = pressure of wetting fluid phase, F/L^2
$\boldsymbol{q}$ = volumetric water flux density, L/T
R_{eff} = effective radius of the meniscus in a capillary tube, L
s = degree of saturation, dimensionless
S = sources or sinks of water in the system, T^{-1}
S_e = effective saturation, dimensionless
t = time, T
t_a = tare, M
T = temperature, K
V_s = volume of solids, L^3
V_t = total volume of the soil body, L^3
V_w = volume of water, L^3
w = gravimetric soil water content, dimensionless
w_d = mass of the dry soil sample, M
w_w = mass of the wet soil sample, M
z = elevation, gravitational head, L
∇ = gradient operator, L^{-1}
∇^2 = Laplace operator, L^{-2}

Greek Letters

α = parameter in the VG retention curve, L^{-1}
α_c = contact angle, dimensionless
$\varepsilon'_{\text{air}}$ = apparent relative dielectric permittivity of air, dimensionless
ε'_m = TDR-measured apparent relative dielectric permittivity of the medium, dimensionless
$\varepsilon'_{\text{soil}}$ = apparent relative dielectric permittivity of the soil body, dimensionless
$\varepsilon'_{\text{solid}}$ = apparent relative dielectric permittivity of solids, dimensionless
$\varepsilon'_{\text{water}}$ = apparent relative dielectric permittivity of water, dimensionless
λ = parameter in Eq. (5.114)
θ = volumetric soil water content, dimensionless
θ_r = residual soil water content, dimensionless
θ_s = saturated soil water content, dimensionless
ρ_b = oven-dry bulk density, M/L^3
ρ_w = density of water, M/L^3
σ = surface tension between wetting and nonwetting fluid phases, F/L
ψ_g = gravitational potential, L^2/T^2, or F/L^2 or L
ψ_m = matric potential, L^2/T^2, or F/L^2 or L
ψ_o = osmotic potential, L^2/T^2, or F/L^2 or L

ψ_p = pressure potential, L^2/T^2, or F/L^2 or L
ψ_t = total soil water potential, L^2/T^2, or F/L^2 or L
ψ_w = water potential L^2/T^2, or F/L^2 or L

References

1. Bear, J., and Y. Bachmat. 1990. *Introduction to modeling of transport phenomena in porous media*. Norwell, MA: Kluwer.
2. Hillel, D. 1980. *Fundamentals of Soil Physics*. San Diego, CA: Academic Press.
3. Kutílek, M., and D. R. Nielsen. 1994. *Soil Hydrology*. Cremlingen-Destedt, Germany: Catena-Verlag.
4. Heimovaara, T. J. 1994. Frequency domain analysis of time domain reflectometry waveforms: 1. Measurement of the complex dielectric permittivity of soils. *Water Resour. Res.* 30:189–199.
5. Topp, G. C., J. L. Davis, and A. P. Annan. 1980. Electromagnetic determination of soil water content: Measurement in co-axial transmission lines. *Water Resour. Res.* 16:574–582.
6. Bolt, G. H. 1976. Soil physics terminology. *Bull. Int. Soc. Soil Sci.* 49:26–36.
7. Corey, A. T., and A. Klute. 1985. Application of the potential concept to soil water equilibrium and transport. *Soil Sci. Soc. Am. J.* 49:3–11.
8. Nitao, J. J., and J. Bear. 1996. Potentials and their role in transport in porous media. *Water Resour. Res.* 32:225–250.
9. Rawlins, S. F., and G. S. Campbell. 1986. *Water Potential: Thermocouple Psychrometry. Methods of Soil Analysis*. Part I (2nd ed.) ed. Klute, A., Agronomy Monograph 9, pp. 597–618. Madison, WI: American Society of Agronomy Inc.
10. Santini, A. 1981. Natural replenishment of aquifers. CNR Paper No. 72, pp. 53–89, Milan: National Research Council of Italy (in Italian).
11. van Genuchten, M. T. 1980. A closed form equation for predicting the hydraulic conductivity of unsaturated soils. *Soil Sci. Soc. Am. J.* 44:892–898.
12. Nimmo, J. R. 1991. Comment on the treatment of residual water content in "A consistent set of parametric models for the two-phase flow of immiscible fluids in the subsurface", by L. Luckner, *et al. Water Resour. Res.* 27:661–662.
13. Kool, J. B., and J. C. Parker. 1987. Development and evaluation of closed-form expressions for hysteretic soil hydraulic properties. *Water Resour. Res.* 23: 105–114.
14. Klute, A. 1986. Water retention: Laboratory methods. *Methods of Soil Analysis*. Part I (2nd ed.) ed. Klute, A., Agronomy Monograph 9, pp. 635–662. Madison, WI: American Society of Agronomy Inc.
15. Bruce, R. R., and R. J. Luxmoore. 1986. Water retention: Field methods. Methods of Soil Analysis, Part I (2nd ed.) ed. Klute, A., Agronomy Monograph 9, pp. 663–686. Madison, WI: American Society of Agronomy Inc.
16. Swartzendruber, D. 1966. Soil-water behavior as described by transport coefficients and functions. *Adv. Agron.* 18:327–370.

17. Amoozegar, A., and A. W. Warrick. Hydraulic conductivity of saturated soils: Field methods. *Methods of Soil Analysis*. Part I (2nd ed.) ed. Klute, A. Agronomy Monograph 9, pp. 735–798, Madison, WI: American Society of Agronomy Inc.
18. Carman, P. C. 1956. Flow of gases through porous media. New York: Academic Press.
19. Mishra, S., and J. C. Parker. 1990. On the relation between saturated hydraulic conductivity and capillary retention characteristics. *Ground Water* 28:775–777.
20. Richards, L. A. 1931. Capillary conduction of liquids through porous mediums. *Physics* 1:318–333.
21. Celia, M. A., E. T. Bouloutas, and R. L. Zarba. 1990. A general mass-conservative numerical solution for the unsaturated flow equation. *Water Resour. Res.* 26:1483–1496.
22. Dirksen, C. 1991. Unsaturated hydraulic conductivity. *Soil Analysis—Physical Methods*, eds. Smith, K. A., and C. E. Mullins, pp. 209–269. New York: Marcel Dekker.
23. Bouma, J., D. Hillel, F. D. Hole, and C. R. Amerman, 1971. Field measurement of hydraulic conductivity by infiltration through artificial crusts. *Soil Sci. Soc. Am. Proc.* 35:362–364.
24. Hillel, D., and W. R. Gardner. 1969. Steady infiltration into crust-topped profiles. *Soil Sci.* 107:137–142.
25. Mualem, Y. 1976. A new model for predicting the hydraulic conductivity of unsaturated porous media. *Water Resour. Res.* 12:513–522.
26. van Genuchten, M. T., and D. R. Nielsen. 1985. On describing and predicting the hydraulic properties of unsaturated soils. *Ann. Geophys.* 3:615–628.
27. Bouma, J. 1989. Using soil survey data for quantitative land evaluation. *Adv. Soil Sci.* 9:177–213.
28. Vereeken, H., J. Maes, P. Darius, and J. Feyen. 1989. Estimating the soil moisture retention characteristics from texture, bulk density and carbon content. *Soil Sci.* 148:389–403.
29. Tietje, O., and V. Hennings. 1996. Accuracy of the saturated hydraulic conductivity prediction by pedo-transfer functions compared to the variability within FAO textural classes. *Geoderma* 69:71–84.
30. Tietje, O., and M. Tapkenhinrichs. 1993. Evaluation of pedo-transfer functions. *Soil Sci. Soc. Am. J.* 57:1088–1095.
31. Romano, N., and A. Santini. 1997. Effectiveness of using pedo-transfer functions to quantify the spatial variability of soil water retention characteristics. *J. Hydrol.* 202:137–157.
32. van Dam, J. C., J. N. M. Stricker, and P. Droogers. 1994. Inverse method to determine soil hydraulic functions from multistep outflow experiments. *Soil Sci. Soc. Am. J.* 58:647–652.
33. Santini, A., N. Romano, G. Ciollaro, and V. Comegna. 1995. Evaluation of a laboratory inverse method for determining unsaturated hydraulic properties of a soil under different tillage practices. *Soil Sci.* 160:340– 351.

34. Kool, J. B., and J. C. Parker. 1988. Analysis of the inverse problem for transient unsaturated flow. *Water Resour. Res.* 24:817–830.
35. Romano, N. 1993. Use of an inverse method and geostatistics to estimate soil hydraulic conductivity for spatial variability analysis. *Geoderma* 60:169–186.

5.3 Irrigation Scheduling Techniques

F. Martín de Santa Ollala, C. Fabeiro Cortés, A. Brasa Ramos, and A. Legorburo Serra

5.3.1 Introduction

Irrigation scheduling is the process of defining the most desirable irrigation depths and frequencies. Scheduling provides for the optimal profit on yield of a crop, taking into consideration crop, farming, water, and environmental restrictions [1].

The use of irrigation scheduling is becoming more and more necessary because of the continuous increase in water demand, both in agriculture and in other sectors (industry, recreation, and urban use), when water resources are becoming scarcer all over the planet [2]. These techniques are useful not only to prevent water waste, but also to avoid negative effects of overirrigation on crops, both on yield and on the environment.

Although it may seem simple, irrigation scheduling is a complex problem indeed, because the satisfaction of crop water requirements must consider all the restrictions imposed on farm management [3]. Among the particular determining factors of each farm are water availability, manpower and energy availability, characteristics of the existing irrigation system and equipment, legal factors affecting the farmland, and user training [3]. Other determinants to be considered when designing irrigation scheduling are soil factors (texture, water-retention capacity, and depth), climatic factors (temperature, solar radiation, humidity, wind speed), factors related to the crop (crop type and variety, characteristics of root system, susceptibility to water stress and salinity), and cropping factors (sowing time, length of the growth cycle, critical growth stages, tillage, fertilizing, control of pests, diseases, and weeds). Many of these factors are interdependent and may vary both in space and time, thus confirming the complexity in accomplishing good irrigation scheduling. In practice and to be operative, the technique can be simplified with the aid of field sensors, computers, and automation [3].

The objectives may be of a different nature, ranging from technical (including yield) to economic and environmental aims, although they usually are combined. The selection of a particular objective for irrigation scheduling depends on specific needs in each situation, but four main strategies can be noted [4].

The first strategy consists of maximizing yields per unit of irrigated surface. To obtain this yield optimization, the user must fully satisfy crop evapotranspiration demand. This approach is becoming increasingly difficult to justify economically because of increasing water scarcity, high energy costs, and changes in agricultural policies. Its achievement could be justified in small farms where land is the limiting factor.

The second objective is to maximize yield per unit of water applied. This requires adoption of strategic irrigations, the water being applied during the critical periods of the crop cycle, when the effects of water stress could strongly affect yields. This optimization strategy is justified when water is the limiting resource and when its cost is high.

Another objective is the economic optimization by maximizing benefits in the farm production. This approach takes into account each of the farm's limitations and may be achieved when there are no marginal benefits, that is, when the cost of the last unit of water applied is equal to the benefit produced.

Other objectives relate to the environment. One strategy would be to minimize the use of energy. Its achievement is related to the application of penalties during the periods of peak energy demand. This requires not only the scheduling of irrigation but also the selection of crops, that have lower water requirements. Other strategies consist of reducing the water return flows, thus ensuring a better use of the resources available, avoiding groundwater pollution, and controlling soil salinity. This requires avoiding overirrigation and integrating irrigation scheduling with other cultivation techniques, such as soil tillage and fertirrigation (fertilizer application through irrigation system).

Adopting these strategies implies in most cases an accurate knowledge of the functions that relate yield to the volumes of water applied. These production functions usually adopt different forms and may be related to the growth and development cycles of the crop [5].

By not applying water during a given period of crop development, there may be no negative effects on final yield, particularly when the water shortage affects only nonproductive plant organs. This is the case with cereals when complete grain development has occurred. These aspects are dealt with in Section 5.5.

Production functions represent a key element for successful irrigation scheduling [4]. At present, few production functions are well defined in relation to the amount of water applied . Therefore, they must be used very carefully (see Section 5.5).

As more and better production functions become available, irrigation scheduling will be come easier to use in response to economic criteria, as well as to environmental and technical ones, including saving water during some specific periods of plant development.

5.3.2 Methods

Irrigation strategies for different crops and varying soil and climatic conditions can be determined using longterm data representing average weather conditions or short-term predictions based on real-time information. Irrigation scheduling methods are set out below, focusing on their applicability and limitations. Aspects peculiar to the different irrigation methods are dealt with in Section 5.4. However, the challenge to agricultural research is to develop successful methods that are simple to implement and easy to understand from the farmer's and the project management's standpoints [6]. Most of the practiced applications can be found in the literature [1, 7]. Common approaches generally include measurement of soil water, plant-stress indicators, and soil water balance.

Methods based on soil water measurements and on plant-stress indicators present some difficulties, particularly for farmers, related to improper handling, malfunctioning, or miscalibration of instrumentation, or lack of understanding of equipment functioning principles. Methods that follow the soil water balance also are not readily adoptable by the farmers because of difficulties in obtaining quality weather data, appropriate information about soils and crops, or about updating schedules.

Trained staff can apply these procedures on a real-time basis to produce irrigation scheduling calendars based on fixed intervals for long-term usage and adoption of very simple rules [8]. In more complicated situations, these can be based on sophisticated models. Irrigation scheduling models currently are used worldwide since microcomputer capability has improved and the required input variables are better defined [7, 9, 10]. Water-balance models consider that irrigation should start when a threshold value of water content is reached. Some models neglect water fluxes that are more difficult to determine, such as the capillarity rise. The effects of weather conditions on plant growth, combined with models predicting seasonal yields as influenced by soil water content, usually are simulated using mechanistic models [10, 11]. A review of crop–water–yield models for water management is given in Section 5.5.

Irrigation models can be used for large areas that include a variety of crops, providing farmers with the daily information needed to make timely decisions. Although these models seem to be difficult to implement in small-scale farming or in developing countries, irrigation scenarios originating from them could be easier to use for definition of irrigation strategies, choosing crops, planning crop according to farm constraints, and allocation of irrigation water supplies. However, the best use of these models is real-time irrigation scheduling [12].

Measuring or Estimating Soil Water

Observation of the soil water content and the soil water potential (definitions and measurement techniques are described in Section 5.2) can be used to schedule irrigations.

The main difficulties consist in the spatial variability of the soil water properties and in appropriately exploring the soil water-measurement cadences by the irrigating farmer. Spatial variability requires careful selection of measurement locations along the cropped field. Difficulties in using these techniques by the farmers can be overcome by specialized support services.

Soil water measurements currently are used for irrigation scheduling by establishing the soil water balance of the cropped field when measuring the soil water by use of a neutron probe [13, 14]. Time-domain reflectometry also is becoming popular, given improvements in measurement techniques [15]. A threshold value for the soil water content is selected and observations are made periodically that are used to forecast the day when the threshold will be attained. The soil water-depletion rate utilized for this forecast is established from two or three observations spaced several days apart. A simple graphical procedure often is used to forecast the irrigation date. The application depth is calculated from the difference between the soil water at field capacity and at the threshold [1].

Tensiometers are also very popular to schedule irrigations. Tensiometers measure the soil water potential, thus providing only information on the irrigation date. Also, threshold soil water potential is defined with this objective, and successive tensiometer readings allow forecasting of when the threshold will be attained. The application depths are either previously fixed or are computed from the soil water-retention curve (see Section 5.2). Applications are discussed elsewhere [1].

The soil water conditions and the evapotranspiration can be estimated using remote-sensing techniques [16–20]. These techniques overcome problems of spatial variability, especially when used with a geographic information system. The use of remote-sensing techniques requires an irrigation scheduling support service to provide information and advice to farmers. Irrigation scheduling using soil water measurements can be performed at the farm scale on large commercial farms [21], but other solutions are required for small farms.

Plant-Stress Indicators

It is possible to get messages from the plant itself or from the plant canopy as a whole indicating that it is time to irrigate. But water stress is not the only stress affecting the plant; diseases, parasite attacks, thermal and light deficits, or even water excess can produce the same type of effect on the crop. Observation of apparent symptoms of water stress cannot be considered a practical guideline for irrigation scheduling because generally it is too late when symptoms appear (see Section 5.5).

Measurements collected on individual plants will require a accurate sampling. Micromorphometric methods allow detection of changes in the dimension of vegetative organs and give interesting information, mainly on trees. Respective sensor devices are not difficult to place on branches or to connected to a logging system. The main problem encountered is that the same response is obtained with excess or lack of water [22]. Methods based on leaf water-potential measurement are difficult, expensive, and cannot be automated. Scholander's chamber and the thermocouple psychrometer are used [23]. The main interest is in linking values of predawn leaf water potential to evapotranspiration [24]. When threshold values are selected, successive measurements allow forecasting of the irrigation date.

Sap-flow measurements are based on the heat capacity of water. Two techniques are available: the sap-flux density technique, which is limited by the need to determine the cross-sectional area of the water-conducting tissue; or the mass flux technique, which is restricted to estimations of small-tree transpiration [4]. However, much progress is to be expected.

Surface-temperature measurements [25] performed by means of infrared hand-held thermometers have proved feasible and are used in practice. Some indicators, such as the Stress Degree Day (SDD) and the Crop Water Stress Index (CWSI), can be applied but they are crop specific. Efforts to simplify field applications have been carried out [16–20] but they can be used only if weather conditions are somewhat stable because indicators are based on the differences in temperature between the crop and the air. As a result, indicators also are related to specific environmental conditions and require local calibration.

5.3.3 Water Balance

The water balance is dealt with in Section 5.1.6, where information on computing the soil water-threshold values also is provided.

The water balance for irrigation scheduling requires information on soil hydraulic properties (soil water content (θ) at field capacity and at wilting point for the different soil layers, or the water retention and hydraulic conductivity curves in case of mechanistic models); weather data on rainfall and reference evapotranspiration (see Section 5.1.4);

and crop characteristics relative to crop growth stages, root depths, plant heights, and crop coefficients (see Section 5.1.5). In contrast, some models use other evapotranspiration models, requiring specific inputs.

The water balance can be performed using only weather data to compute the soil water depletion. The majority of models use this approach. Models can be applied in real time to plan the seasonal irrigation calendar, or to study alternative strategies for irrigation scheduling [7, 9, 10]. When models are applied in real time, they often use information on actual soil water observations to adjust the forecast of the irrigation date and depth.

The threshold for water stress can be defined by a crop-specific factor of soil water depletion, namely the p factor [26, 27]. This approach is described in Section 5.1.6, where this factor is related to the management-allowed deficit (MAD) [11]. This approach is utilized in more common models [7, 10, 28]. Mechanistic models usually incorporate a plant-growth submodel. This allows an appropriate response of the model to water deficits, not requiring the definition of a soil water depletion [10, 29].

The yield impacts of irrigation scheduling strategies are considered through a simple yield-evapotranspiration function [26] in less-sophisticated models. On the contrary, yield is an output of most mechanistic models, namely the CERES-type models [10]. Some models also incorporate any function of crop responses to salinity, resulting from parameters and information provided in the literature [30, 31].

5.3.4 Irrigation Programming Under Adverse Conditions

Irrigation Management Under Water-Shortage Conditions

The lack of water has a negative effect on the whole set of vital plant processes such as photosynthesis, respiration, absorption of nutrients, and assimilate translocation. It also affects growth, reproduction, and the development of the seed. From the agronomical point of view, the main interest is focused on the effect of water deficit on yield. The effect of water stress on plant development and productivity has been the object of several studies [32–35] and is dealt with in Section 5.5.

The most visible effects of water deficit are decreased size of the plant and of its foliar surface, as well as decreased yield. Other agronomic consequences of water deficit are modifications in the chemical composition of agricultural products (such as sugar, oil, and protein) and in other quality features of yield (such as diameter and color of fruits). Negative impacts brought about by stress essentially will depend on the stage of the cycle during which it takes place, on the intensity of the deficit involved, and on its duration.

To illustrate some aspects of water stress on yield and yield components, an experiment with soybeans is referenced [36]. The treatments were established by fixing six different levels of ET_c restitution through irrigation: 20%, 40%, 60%, 80%, 100%, and 120%. Deficits or excess water applications were kept constant throughout the crop cycle. The seasonal depths of irrigation water applied were, respectively, 84, 171, 253, 324, 435, and 504 mm. In addition, all plots received 87 mm of effective rainfall.

The yield obtained was shown to have been clearly affected by the water treatment received, with values ranging from 1135 to 4660 kg/ha of grain. Among the components of the yield, the one that seemed to be the most sensitive to the different depths applied was the number of grains per plant. The number of plants per unit surface and the number of pods per plant were affected, but the only significant difference was between those treatments receiving less water and the remaining treatments. The quality of the grain was assessed through oil and protein content. The oil content grew with the seasonal quantity of water (from 24.4% to 27%), whereas the protein content decreased (45.5% to 35.1%). The treatments with the greatest water supply, which kept the largest active leaf area over the longest time, achieved the largest quantity of oil. However, the yield of both oil and protein tended to increase with the depth of water applied.

In areas with a shortage of water, a water-saving irrigation strategy thus increasingly is being implemented, is known as controlled-deficit irrigation (CDI). This is based on reducing water applications during those phenological periods in which controlled water deficit does not significantly affect the production and quality of the crop involved, while satisfying crop requirements during the remainder of the crop cycle. CDI may produce economic returns higher than those obtained when irrigating for maximum production [37].

Research carried out in several regions in the world [38–42] into the response of CDI in fruit trees, such as peach, pear, almond, and citrus, shows that this technique can lead to up to 50% water savings with a small decrease in yield, while keeping, or even improving, the quality of the production. The main problem is to determine the phenological periods when the impact of water deficit does not significantly affect the production and/or quality of the crop. Existing literature on the subject does not provide homogeneous information; hence, appropriate data validation needs to be performed on local conditions before planning CDI.

Planning for deficit irrigation entails a greater challenge than for full irrigation, because it requires the knowledge of the appropriate MAD and how this deficit will affect the yield. Deficit irrigation may call for changes in cropping practices, such as moderation of plant population density, decreased application of fertilizers and agrochemicals, flexible sowing dates, and the selection of shorter-cycle varieties.

The relationship between water deficit and yield has to be well known when planning deficit irrigation. To determine when irrigation is to take place (and the amount of water to be applied), suitable water-stress indicators should be used. These indicators may refer to the depletion of soil water, soil water potential, plant water potential, or canopy temperature. The last indicator seems to be the most suitable, but the most widely used, for practical reasons, concerns soil water content and soil water potential. However, the spatial variability of the soil and of the spatial irrigation give rise to uneven soil water storage, which causes problems in analyzing indicator information when these originate from point measurements. Indicators resulting from areal observation or measurement, such as the CWSI [24], are the most appropriate.

There are different ways to carry out deficit-irrigation management. The irrigator can reduce the irrigation depths, refilling only a part of the soil reserve capacity in the root zone. Or the irrigator may reduce the frequency of irrigations, but pay attention to the timing and depth of a limited number of irrigation sessions. In surface irrigation, deficit irrigation can be carried out by moistening furrows alternately or digging them farther apart [43].

Irrigation Management Under Variable Rainfall Conditions

In regions where rainfall may be significant, the effective rainfall must be considered to reduce irrigation requirements and minimize the negative effects of overirrigation. These impacts include deep percolation, transport of nutrients and solutes into groundwater, surface runoff, and soil erosion. The scheduling of irrigation therefore requires data on potential future rainfall. This information can be in the form of a short-term weather forecast, or using information relative to previous rainfall records covering a significant number of years or using random generated rainfall events [1].

Short-term forecasts have been used successfully in wet regions, where there is a high rainfall probability, its quantity being the only unknown factor. In these cases, the net irrigation quantity is reduced to allow for rainwater storage in the soil without generating percolation. This type of strategy is highly adaptable to crops with a deep-root system, on soils with moderate to high water-retention capacity, and for systems applying light and frequent irrigations. In semiarid regions, the rainfall during irrigation periods is usually from local storms, and both their frequency and quantity are highly variable. In such cases, irrigators must rely on actual rainfall data and delay irrigation dates by a number of days as a function of the quantity fallen and the actual evapotranspiration rate.

There are several irrigation scheduling models that program irrigation by using short-term rainfall forecasts [10]. The EPICPHASE model [29] is an example of a mechanistic crop growth model, capable of simulating different irrigation strategies using the weather forecast. The application of models having such capabilities may help to achieve water savings, as well as reduce the risk of percolation and the resulting leaching of nutrients, while not affecting yields.

Irrigation Planning Under Saline Conditions

The problem of salinity is important in arid and semiarid areas because the rain contributes little to the lixiviation of salts, in addition to which the quality of the irrigation water available is usually poor.

To avoid reductions in yield when the concentration of salts exceeds crop tolerance, the excess salts must be lixiviated below the root zone. Therefore, an additional fraction of water must be added to the net irrigation requirement. This quantity, called a leaching requirement, is calculated as a function of the soil and water salinity levels, to allow salt leaching through drainage (see Sections 4.1, 5.7, and 5.8).

When leaching requirements are applied, it must be remembered that excessive leaching can lead to the lixiviation of nutrients. When irrigation is managed under saline conditions, the concentration of salts in the irrigation water, crop tolerance to salts, dependable rainfall, depth of the groundwater, and ease of drainage all must be taken into consideration.

The leaching frequency has been studied by several researchers. The results indicate that the most tolerant crops allow for a delayed leaching and for a relatively high concentration of salts in the root zone, with a minimum effect on yield as long as a low salt content is kept in the zone where most of the root water uptake occurs. It also has been found that plants are capable of making up for low water absorption in a more saline zone by increasing absorption in a less saline zone within their reach without decrease in yield. How much salt can be stored in the root zone before leaching is required and how often leaching requirements are to be applied are questions still unanswered [44], despite progress of research.

Crop tolerance to salts grows throughout the season. If the salinity level is sufficiently low during the seedling stage and the proper quantities of water with low salt content are applied, leaching may not be necessary and the salt level may be allowed to continue to increase until the end of the cycle. Winter rainfall or preplanting irrigation may saturate the soil and leach the salts accumulated so that there may be no need for leaching in the following irrigation.

On the contrary, when the water is saline, winter rainfall is scarce, and preplanting irrigation is light, leaching throughout the season will be necessary to prevent yield from being affected. Remember, however, that leaching is only required when the concentration of salts exceeds a threshold value [44].

For a long time, it has been assumed that very frequent irrigation would reduce the impact of salinity. A high water potential may partly reduce osmotic tension, but no increase in yield has been observed that would corroborate this assumption [45]. If the surface of the soil is moistened frequently, evapotranspiration will be high most of the time and salts will be concentrated in shallow upper layers of the soil. In addition, water absorption by roots will tend to take place in shallow layers when they are moistened frequently, whereas if the surface of the soil is allowed to dry with less frequent irrigation, absorption also will take place in deeper layers. Both water absorption and evapotranspiration processes tend to concentrate salts near the surface under frequent irrigation conditions. Trickle irrigation systems are an exception because localized water displaces salts beyond the limits of the wet bulb. In this case, lixiviation prevails over evapotranspiration and water absorption, but a leaching fraction also has to be considered. The only acceptable measure to control salinity seems to be a controlled increase in the quantity of water applied, associated with drainage of the leachates.

When water resources are limited and nonsaline water costs become prohibitive, irrigation with saline water can be carried out on crops with moderate to high tolerance to salts, particularly during the latest growing stages. A mixture of saline and nonsaline water can be used for irrigation purposes, but this is a questionable practice. It may be more advisable to use nonsaline water during the most sensitive phases of the crop cycle and saline or mixed water during the rest of the time.

The use of saline water for irrigation generally calls for the selection of crops tolerant to salts, the development of appropriate irrigation scheduling, and the maintenance of the relevant soil physical properties to ensure better hydraulic conductivity.

Under saline conditions, irrigation scheduling requires better accuracy for the estimation of the components in the soil water balance, to better estimate the leaching requirements. With regard to irrigation management, it is advisable to consider how uniformly the irrigation water will be distributed so as to decide which part of the field is receiving the required leaching fraction.

The indiscriminate use of saline water for irrigation should be avoided because it leads to soil salinization. With regard to the various irrigation systems, trickle irrigation is the one that offers the best advantages under saline conditions [46]. Sprinkler irrigation affects the leaves of sensitive crops but it is appropriate for leaching with water of good quality. Basin irrigation has the advantage, when compared to other surface irrigation systems, of ensuring evenly distributed leaching as long as the basins are sized and leveled correctly. In furrow irrigation, salts tend to accumulate near the seed, because leaching takes place down from the furrow. To control salinity in these irrigation systems, special attention has to be paid to the depth applied and to the uniformity of the application. Subsurface irrigation produces a continuous upward flux from the water table, which leads to the accumulation of salts on the surface and is not appropriate to control salts. Salts cannot be lixiviated with these systems and regular leaching will therefore be needed in the form of rainfall or surface irrigation [46].

5.3.5 Irrigation Advisory Systems

The practical application of irrigation scheduling techniques requires appropriate technology transfer and support to farmers. Irrigation extension or consulting services may provide this support and help in the transfer of technologies from research to practice. Such services should not only provide information but also stress how this information is being used and what the impacts are for improving irrigation systems and management.

Farmer support must be adjusted to the technology level of the user. De Jager and Kennedy [21] define three levels:

1. *Top technology*. This situation means that there is a weather station in the farmland with a support system for decision making that allows individualized complete scheduling.
2. *Intermediate technology*. The station is strategically located for a group of farms that receive common information. This general information may be complemented individually with specific data regarding each field, such as soil water properties, crop phenology, and irrigation systems. This information allows for the individual scheduling of irrigation. There is also a collective technical support system available.
3. *Low technology*. In this situation, fixed irrigation periods and volumes are scheduled for the whole campaign, based on large series of climatic parameters and crop average water requirements. Hill and Allen [8] present an interesting scheduling system for this situation.

Irrigation-scheduling advising systems commonly use an irrigation scheduling system based on the soil-crop water balance. When other methods are used, such as an evaluation of soil or plant water status, they are usually a complement of the water balance and often are used to address the results obtained.

An irrigation-scheduling advising system generally is composed of the following elements:

- A data collection system, including at least an automated weather station and a monitoring system relative to crop phenological evolution. Ideally, it also should include monitoring and evaluation of the irrigation systems used by the farmers.
- A system for information processing able to provide actual reference evapotranspiration data, the ET_0 for the climatic conditions prevailing in the area, crop coefficient data relative to crops in the area, as well as information on soils and soil water. With

the help of models, indicative information on irrigation depths and frequencies also may be provided.

- An information transmission system. Traditional methods for information spreading were represented by local press, radio, and telephone . At present, using personal computers, one has access to information systems and can access individual information in real time [7, 9, 10].
- A system for information validation and comparison of the results obtained. This is a fundamental piece of the system and may be supported by monitoring selected fields in the study area. Results can be evaluated and compared with those from other fields that have not adopted the systems advice or that are following different irrigation schedules.

The evaluation of irrigation systems should be considered as a main monitoring task of irrigation-scheduling advising services. Knowing the crop water requirements is useless unless the way that water is applied to each field is known. Often, irrigation systems installed a few years ago apply water differently from the original design. This is because supply conditions, such as discharge and pressure, may have changed over time, or the hydraulic structures are not functioning as designed, or the systems have deteriorated. It is necessary not only to evaluate the off-farm systems, but to evaluate the performances of on-farm systems, for uniformity and efficiency, and to provide for improving on-farm irrigation [7, 28] (see Section 5.4).

Consulting services may be provided to individual farmers or to users' associations. Farmers' participatory activities should be encouraged, including the funding.

The transfer of irrigation technologies is a main task of services providing irrigation support to farmers [47]. Burt [48] points out that effective use of irrigation scheduling models requires a process of preparation and training of the users. The lack of such support may be one cause for low adoption of irrigation scheduling advice and for the limited use of irrigation scheduling techniques and models [7].

Several examples of users' participatory activities indicating positive future trends can be provided. An interesting example is presented by Blackmore [49] in the Murray-Darling hydrographic basin in Australia. Tollefson [6] analyzes the role being developed in Canada by the Prairie Farm Rehabilitation Administration in the development of users' participation in water management in collaboration with universities and different institutions. In Castilla-La Mancha (Spain) [23], plans for the exploitation of mined aquifers have been initiated with the scientific and technical support of the university in cooperation with the regional government and with an intense participation of users' organizations.

In developing countries, interesting experiences are also taking place, such as those in West Java and Bihar, India [50], with the Damodar Valley Corporation in India [33], in the Jingtaichuan district in China [51], and in Bangladesh [52].

Other aspects that may be of interest for achieving successful use of irrigation scheduling by farmers are the allocation of water to the farmers in volumetric terms. At the beginning of the campaign, the farmer knows the total volume available and the adequate rate of use; water prices then are established such that the farmer is induced to make the best use of allocated water [6].

References

1. Martin, D. L., E. C. Stegman, and E. Fereres. 1990. Irrigation scheduling principles. *Management of Farm Irrigation Systems*, eds. Hoffman, G. J., *et al.*, pp. 155–203. St. Joseph, MI: American Society of Agricultural Engineers.
2. Evans, T. E. 1996. The effects of changes in the world hydrological cycle on availability of water resources. *Global Climate Change and Agriculture Productions*. Coord. Bazzaz, F., and W. Sombroeck, pp. 15–48, Chichester: Wiley-Rome: FAO.
3. De Juan Valero, J. A., F. J. Martín de Santa Olalla Mañas, and C. Fabeiro Cortés. 1992. La Programación de Riegos (I): Los objetivos y los métodos. *Riegos y Drenajes XXI* 66:19–27.
4. Martín de Santa Olalla, F., and J. A. de Juan. 1993. *Agronomía del Riego*. Madrid: Mundi-Prensa-UCLM.
5. Hunt, R. 1982. Plant growth curves. *The Functional Approach to Plant Growth Analysis*. London: Edward Arnold.
6. Tollefson, L. 1996. Requirements for improved interactive communications between researchers, managers, extensionists and farmers. *Irrigation Scheduling: From Theory to Practice*, eds. Smith, M., *et al.*, FAO Water Reports, Vol. 8, pp. 217–226. Rome: International Commission on Irrigation and Drainage and FAO.
7. Smith, M., L. S. Pereira, J. Berengena, B. Itier, J. Goussard, R. Ragab, L. Tollefson, and P. van Hoffwegen (eds.). 1996. *Irrigation Scheduling: From Theory to Practice*. FAO Water Reports, Vol. 8. Rome: International Commission on Irrigation and Drainage and FAO.
8. Hill, R. W., and R. G. Allen. 1996. Simple irrigation calendars: a foundation for water management. *Irrigation Scheduling: From Theory to Practice*, eds. Smith, M., *et al.*, FAO Water Reports, Vol. 8, pp. 69–74. Rome: International Commission on Irrigation and Drainage and FAO.
9. Camp, C. R., E. J. Sadler, and R. E. Yoder (eds.). 1996. *Evapotranspiration and Irrigation Scheduling*. St. Joseph, MI: American Society of Agricultural Engineers.
10. Pereira, L. S., B. J. van der Broek, P. Kabat, and R. G. Allen (eds.). 1995. *Crop-Water Simulation Models in Practice*. Wageningen: Wageningen Pers.
11. Hoffman, G. J., T. A., Howell, and K. H. Solomon (eds.). 1990. *Management of Farm Irrigation Systems*. St. Joseph, MI: American Society of Agricultural Engineers.
12. Itier, B., F. Maraux, P. Ruelle, and J. M. Deumier. 1996. Applicability and limitations of irrigation scheduling methods and techniques. *Irrigation Scheduling: From Theory to Practice*. FAO Water Reports, Vol. 8, pp. 19–32. Rome: International Commission on Irrigation and Drainage and FAO.
13. Burman, R. D., R. H. Cuenca, and A. Weiss. 1983. Techniques for estimating irrigation water requirements. *Advances in Irrigation*, ed. Hillel, D., pp. 335–394. Orlando: Academic Press.
14. Cuenca, R. H. 1989. Hydrologic balance model using neutron probe data. *J. Irrig. Drain. Eng.* 144:645–663.
15. Hook, W. R., N. J. Livingstone, Z. J. Sun, and P. B. Hook. 1992. Remote diode shorting improves measurement of soil water by time domain reflectometry. *Soil Sci. Soc. Am. J.* 56:1384–1391.

16. Brasa, A., F. Martín de Santa Olalla, and V. Caselles. 1996. Maximum and actual evapotranspiration for barley (*Hordeum vulgare L.*) through NOAA satellite images in Castilla-La Mancha, Spain. *J. Agric. Eng. Res.* 64:283–293.
17. Caselles, V., J. Delegido, J. A. Sobrino, and E. Hurtado. 1992. Evaluation of the maximum evapotranspiration over the La Mancha region, Spain, using NOAA AVHRR data. *Int. J. Remote Sensing* 13:939–946.
18. Lagouarde, J. P. 1991. Use of NOAA AVHRR data combined with an agrometeorological model for evaporation mapping. *Int. J. Remote Sensing* 12:1853–1864.
19. Sandholt, I., H. S. Andersen. 1993. Derivation of actual evapotranspiration in the Senegalese Sahel using NOAA-AVHRR data during the 1987 growing season. *Remote Sensing Environ.* 46:164–172.
20. Vidal, A., and A. Perrier. 1989. Analysis of a simplified relation for estimating daily evapotranspiration from satellite thermal IR data. *Int. J. Remote Sensing* 10:1327–1337.
21. De Jager J. M., and J. A. Kennedy. 1996. Weather-based irrigation scheduling for various farms (commercial and small-scale). *Irrigation Scheduling: From Theory to Practice*, eds. Smith, M., *et al.*, FAO Water Reports, Vol. 8, pp. 33–38. Rome: International Commission on Irrigation and Drainage and FAO.
22. Huguet, J. G. 1985. Appréciation de l'état hydrique d'une plante á partir des variations microétriques de la dimension des fruits on des tigues en cours de journée. *Agronomie* 5:733–741.
23. Martín de Santa Olalla, F., Brasa Ramos, A., Fabeiro Cortés, C., Fernández González, D., López Córcoles, H. (1999). Improvement of irrigation management towards the sustainable use of groundwater in Castilla-La Mancha, Spain. *Agri. Water Manage.* Special Issue.
24. Jackson, R. D., S. B. Idso, R. J. Reginato, P. J. Printer. Jr. 1981. Canopy temperature as a crop water indicator. *Water Resour. Res.* 17:1133–1138.
25. Jackson, R. D. 1983. Canopy temperature and crop water stress. *Advances in Irrigation*, Vol., pp. 43–85, New York: Academic Press.
26. Doorenbos, J., and A. H. Kassam. 1979. Yield Response to Water. Irrigation and Drainage Paper 33. Rome: FAO.
27. Doorenbos, J., and W. O. Pruitt. 1977. Guidelines for Predicting Crop Water Requirements. Irrigation and Drainage Paper 24. Rome: FAO.
28. Tarjuelo, J. M. 1995. *El Riego por Aspersión y su tecnología*. Madrid: Mundi-Prensa.
29. Cabalguenne, M., J. Puech, P. Debaeke, N. Bosc, and A. Hilaire. 1996. Tactical irrigation management using real time EPIC-phase model and weather forecast. *Irrigation Scheduling: From Theory to Practice*, eds. Smith, M., *et al.*, FAO Water Reports, Vol. 8, pp. 185–194. Rome: International Commission on Irrigation and Drainage and FAO.
30. Food and Agriculture Organization of the United Nations. 1992. Wastewater treatment and use in agriculture. Irrigation and Drainage Paper 47. Rome: FAO.
31. Food and Agriculture Organization of the United Nations. 1992. The use of saline waters for crop production. Irrigation and Drainage Paper 48. Rome: FAO.
32. Hsiao, T. C. 1973. Plant responses to water stress. *Ann. Rev. Plant Physiol.* 24:519–570.

33. Paleg, L. G., and D. Aspinall. 1981. *The Physiology and Biochemistry of Drought Resistance in Plants*. New York: Academic Press.
34. Bradford, K. J., and T. C. Hsiao. 1982. *Encyclopedia of Plant Physiology*, Vol. 12B; pp. 263–324. New York: Lange.
35. Kramer, P. J. 1983. *Water Relations of Plants*. New York: Academic Press.
36. Martín de Santa Olalla, F., J. A. de Juan, and C. Fabeiro. 1993. Growth and yield analysis of soybean under different irrigation schedules in Castilla-La Mancha, Spain. *Eur. J. Agron.* 33:187–196.
37. Hargreaves, G. H., and Z. A. Samani. 1984. Economic considerations of deficit irrigation. *J. Irrig. and Drain. Eng.* 110:343–358.
38. Mitchell, P. D., and D. J. Chalmers. 1982. The effect of reduced water suppy on peach tree growth and yields. *J. Am. Soc. Hort. Sci.* 107:853–856.
39. Mitchell, P. D., P. H. Jerie, and D. J. Chalmers. 1984. Effects of regulated water deficits on pear tree growth, flowering, fruit growth and yield. *J. Am. Soc. Hort. Sci.* 109:604–606.
40. Goldhamer, D. A., and K. Shackel. 1989. Irrigation cutoff and drought irrigation strategy effects on almond. *17th Annual Almond Research Conference*, pp. 35–37. Modesto: Almond Res. Conf.
41. Li, S. H., J. G. Huguet, P. G. Schoch, and P. Orlando. 1989. Response of peach tree growth and cropping to soil water deficits at various phenological stages of fruit development. *J. Hort. Sci.* 64:541–552.
42. Domingo, R. 1995. Respuesta de los cítricos al riego deficitario. Limonero. *Riego Deficitario Controlado*. Cuaderno Value 1, pp. 119–171. Madrid: Mundi-Prensa.
43. Ragab, R. 1996. Constraints and applicability of irrigation scheduling under limited water resources, variable rainfall and saline conditions. *Irrigation Scheduling: From Theory to Practice*, eds. Smith, M., *et al.*, FAO Water Reports, Vol. 8, pp. 149–166. Rome: International Commission on Irrigation and Drainage and FAO.
44. Hoffman, G. J., J. D. Rhoades, J. Letey, and F. Sheng. 1990. Salinity management. *Management of Farm Irrigation Systems*. eds. Hoffman, G. J., T. A. Howell, and K. H. Solomon, pp. 667–715. St. Joseph, MI: American Society of Agricultural Engineers.
45. Shainberg, I., and J. Shalhevet. 1984. *Soil Salinity Under Irrigation. Processes and Management*. Berlin: Springer-Verlag.
46. Shalhevet, J. 1994. Using water of marginal quality for crop production: Major issues. *Agric. Water Manage.* 25:233–269.
47. World Bank. 1985. Agriculture Research and Extension: An Evaluation of the World Bank's Experience, pp. 3–100. Washington, DC: World Bank.
48. Burt, C. M. 1996. Essential water delivery policies for modern on-farm irrigation management. *Irrigation Scheduling: From Theory to Practice*, eds. Smith, M., *et al.*, FAO Water Reports, Vol. 8, pp. 273–278. Rome: International Commission on Irrigation and Drainage and FAO.
49. Blackmore, D. J. 1994. Integrated catchement management. The Murray-Darling Basin experience. *Water Down Under Conf.* Adelaide, Vol I, 1–7.
50. Vermillion, D. L., and J. D. Brewer. 1996. Participatory action research to improve irrigation operations: Examples from Indonesia and India. *Irrigation Scheduling:*

From Theory to Practice, eds. Smith, M., *et al.*, FAO Water Reports, Vol. 8, pp. 241–250. Rome: International Commission on Irrigation and Drainage and FAO.
51. Xianjun, C. 1996. Introduction of water-saving irrigation scheduling through improved water delivery. *Irrigation Scheduling: From Theory to Practice*, eds. Smith, M., *et al.*, FAO Water Reports, Vol. 8, pp. 257–260. Rome: International Commission on Irrigation and Drainage and FAO.
52. Harun-ur Rashid, M. 1987. A research-development process for improving on-farm irrigation water management programmes in Bangladesh. *ADAB News* 14(4):18–21.

5.4 Irrigation Methods

L. S. Pereira and T. J. Trout

5.4.1 Irrigation Systems and Irrigation Performance

General Aspects

Irrigation has been practiced for millennia. Until the 20th century, all irrigation depended on gravity to deliver water to the fields and to spread water across the surface of the land. Development of efficient engines, pumps, and impact sprinklers in the first half of this century allowed farmers to mimic rainfall with sprinkler irrigation. In the past 20 years, microirrigation has become more common where water is scarce, crop values are high, and the technology is practical. Surface irrigation remains the dominant system in use worldwide, although sprinkler irrigation has become widely used in some areas.

Total irrigated areas in the World are given in Table 5.3. An estimate of the percentages of the three primary methods of irrigation in four regions is presented in Table 5.4. Data in the table cannot be extrapolated to world figures because the data are not available for most regions, particularly Asia, which represents 69% of the world's irrigated area and where surface irrigation is predominant.

The many types of irrigation systems usually fall into one of three categories. Surface irrigation systems are those that depend on gravity to spread the water across the surface of the land. These systems also are referred to as gravity or flood irrigation systems. The shape of the soil surface and how the water is directed across the surface determine the types of surface systems (i.e., furrow, border, or basin). Sprinkler systems attempt to mimic rainfall by spraying the water evenly across the soil surface. The water is

Table 5.3. Estimated irrigated areas in the world

Region	Irrigated Area (ha)
Asia	174,300,000
Europe	25,150,000
Africa	11,480,000
South/Central America	17,650,000
United States and Canada	22,100,000
Australia	2,300,000
World	252,990,000

Source: Food and Agriculture Organization AQUASTAT survey.

Table 5.4. Estimated use of the three primary irrigation methods

Region	Use (%)		
	Surface	Sprinkle	Micro
Africa[a]	85.0	12.5	2.5
Near East[b]	87.6	11.0	1.4
United States[c]	51.6	44.1	4.3
Former Soviet Union[d]	58	42	0.1

[a] For 20 countries representing 44.7% of the total irrigated area in the region [1].
[b] For 18 countries representing 45.8% of the total irrigated area in the region [2].
[c] [3].
[d] [4].

pressurized with a pump, distributed to areas of the fields through pipes or hoses, and sprayed across the soil surface with rotating nozzles or sprayers. Types of sprinkler systems depend on the layout of the distribution pipelines and the way they are moved (i.e., solid set, hand move, center pivot, or rain gun). Microirrigation systems, also called drip or trickle systems, use small tubing to deliver water to individual plants or groups of plants. These systems use regularly spaced emitters on or in the tubing to drip or spray water onto or into the soil. Microirrigation systems are categorized by the type of emitters (i.e., drip or microspray).

Some systems do not clearly fit into these three categories. For example, subirrigation uses gravity to distribute water below the soil surface; it is uncommon and is not discussed here. Low-energy precision application (LEPA) systems use center-pivot machines to spray or dribble water onto small areas. They are discussed with sprinkler systems. Overhead or undertree impact sprinklers on fruit trees are considered sprinkler systems, whereas undertree sprays are considered microsprays.

Quantifying Performance

Irrigation performance often is described in terms of the water application efficiency and water distribution uniformity. Because there are many aspects to irrigation and several irrigation methods, a wide range of performance parameters have been proposed but there is no consensus for standardization. Next some basic performance parameters that have been widely used are discussed.

Efficiency

The definition of water application efficiency is not well established despite the fact it is used worldwide. Reviews of various efficiency terms are provided elsewhere [5–7].

The classical definition of irrigation efficiency introduced by Israelsen in 1932 is the ratio between the irrigation water consumed by the crops of an irrigated farm or project during crop growth and the water diverted from a river, groundwater, or other source into farms or project canals [8]. However, this definition is the cause of much misinterpretation [8]. It is necessary to make a clear distinction between water consumption—which includes evaporation, transpiration, and water embodiment in a product—and water use, which also includes nonconsumptive components that may be of practical necessity (such as leaching salts) and may be available for reuse. Irrigation performance evaluation depends on the point of view.

The term "efficiency" is restricted to output/input ratios of the same nature, such as the ratio of delivered/diverted water volumes or infiltrated/applied water depths. In both cases it is possible to identify the nonconsumed fraction that can be recovered or reused for agriculture or other purposes. Water-use efficiency (WUE) should be used to represent plant or crop output per unit water use (i.e., the photosynthetic WUE, the biomass WUE, or the yield WUE, as proposed by Steduto [9]). This concept is not used as a measure of irrigation performance.

The efficiency terms used in on-farm irrigation to measure the performance of water application are the following:

Application Efficiency e_a. Measured as percentage (%), this is defined as

$$e_a = 100(Z_r/D), \tag{5.115}$$

where Z_r is the average depth of water (mm) added to the root zone storage, and D is the average depth of water (mm) applied to the field. The condition

$$Z_r \leq SWD \tag{5.116}$$

must be met everywhere on the field; SWD is the soil water deficit (mm), at time of irrigation.

Application Efficiency of Low Quarter (e_{lq}). Measured as a percentage (%), e_{lq} is defined by Merriam and Keller [10] as

$$e_{lq} = 100(Z_{r,lq}/D), \tag{5.117}$$

where $Z_{r,lq}$ is the average depth of water (mm) added to root-zone storage in the quarter of the field receiving the least water. This indicator differs from e_a (Eq. 5.115) by the fact that it allows consideration of the nonuniformity of water application when underirrigation is practiced. As for e_a, the numerator cannot exceed the SWD. The e_{lq} is used for surface, sprinkler, and microirrigation.

Potential Efficiency of Low Quarter (PELQ). Measured as a percentage (%), is used for design and corresponds to the system performance under good management when the correct depth and timing are being used. $PELQ$ is given by [10]

$$PELQ = 100(Z_{lq,\mathrm{MAD}}/D_{\mathrm{MAD}}), \tag{5.118}$$

where $Z_{lq,\mathrm{MAD}}$ is the average low-quarter depth infiltrated (mm) when equal to the management-allowed deficit (MAD) and D_{MAD} is the average depth of water applied (mm) when $SWD = \mathrm{MAD}$ (about MAD, see Section 5.1).

Uniformity.

Several parameters are used as indicators of the uniformity of water application to a field. The most commonly used [5] are described below.

Distribution Uniformity (DU). Calculated as a percentage (%), DU is defined by the ratio

$$DU = 100(Z_{lq}/Z_{av}), \tag{5.119}$$

where Z_{lq} is the average infiltrated depth (mm) in the quarter of the field receiving the least water and Z_{av} is the average infiltrated depth (mm) in the entire field.

The infiltrated depths may be replaced by the application depths in sprinkler irrigation (assumes no surface runoff or redistribution), and by the emitter discharges in microirrigation. Some authors prefer to replace the low-quarter averages in the numerator with the minimum observed values. This indicator then becomes the absolute distribution uniformity (DU_{abs}).

Coefficient of Uniformity (CU). Calculated as a percentage (%) and also known as the Christianson uniformity coefficient, CU is defined as

$$CU = 100\left(1 - \frac{\sum X}{n\bar{X}}\right), \tag{5.120}$$

where X are the absolute deviations of application (or infiltrated) depths from the mean (mm), $\bar{X}$ is the mean of observed depths (mm), and n is the number of observations. This indicator often is used in sprinkler irrigation.

Both DU and CU are related. The approximate relationship between them is [11]

$$CU = 100 - 0.63(100 - DU) \tag{5.121}$$

or

$$DU = 100 - 1.59(100 - CU). \tag{5.122}$$

The parameter CU relates to the standard deviation (sd) and the mean ($\bar{X}$) of the individual observations by

$$CU = 100\left(1.0 - \frac{sd}{\bar{X}}\left(\frac{2}{\pi}\right)^{0.5}\right), \tag{5.123}$$

which can be rearranged to give [12]

$$sd = \frac{\bar{X}}{(2/\pi)^{0.5}}\left(1.0 - \frac{CU}{100}\right). \tag{5.124}$$

The relation between the percentage of the surface area receiving the target water depth and the depth of water applied at different CU (or DU) values can be calculated if the data distribution is known. For sprinkler irrigation, a normal distribution is assumed by Keller and Bliesner [11] and a uniform distribution is adopted by Mantovani *et al.* [13]. Several distribution functions are analyzed by Seginer [14]. The use of these relationships for irrigation design and management is discussed later under "Set Sprinkler Systems."

Statistical Uniformity Coefficient U_s. Calculated as a percentage (%), this indicator is defined as

$$U_s = 100(1 - V_q) = 100\left(1 - \frac{S_q}{q_a}\right), \qquad (5.125)$$

where V_q is the coefficient of variation of emitter flow, S_q is the standard deviation of emitter flow ($L\,h^{-1}$), and q_a is the average emitter flow rate ($L\,h^{-1}$).

This indicator often is used in microirrigation. Bralts *et al.* [15] adopted U_s for design and defined V_q as the geometric mean of the coefficients of variation relative to emitter clogging, manufacturing variation, and pressure variation.

5.4.2 Surface Irrigation

Description of Systems

The practice of surface irrigation is ancient and is used on more than 90% of the world's irrigated area. The sustainability of irrigated agriculture depends on improvements and innovations in surface irrigation methods, their appropriateness for the different systems, and their adoption in field practice.

Surface irrigation methods include several processes of water application to irrigated fields [16]: furrow, basin, border, contour ditches (wild flooding), and water spreading. The two last processes have several variants. They basically consist of directing water diverted from ditches or watercourses onto sloping fields. They are primarily used to irrigate pasturelands and are generally very inefficient. The main surface methods are basin, furrow, and border irrigation.

Basin Irrigation

Basin irrigation is the most commonly used system worldwide. It consists of applying water to leveled fields bounded by dikes. Two different types are considered: one for paddy rice irrigation, where ponded water is maintained during the crop season; and the other for other field crops, where ponding time is short—until the applied volume infiltrates.

In traditional rice irrigation, small basins are flooded before planting and are drained only before harvesting. The depth of water in the basins can be very large in case of floating rice but usually it should be kept between 5 and 10 cm [17]. Very frequent water applications are used. In sloping lands, the basin dikes usually are built on the contour and fields often irrigate and drain from field to field in cascade. In flat areas, the basins are commonly rectangular and often have independent supply and drainage. When water-saving irrigation is practiced in tropical areas where water ponding is not required for temperature regulation, water is applied only to keep the soil near saturation [18]. In modern rice basins, laser leveling is used, basin size often exceeds 1 ha, and each basin has independent supply and drainage facilities. Rice transplanting may be replaced by mechanical seeding, sometimes in dry soil.

For nonrice crops, basin irrigation can be divided into two categories: traditional basins, of small size and with traditional leveling; and modern precision-leveled basins, which are laser leveled and large, and have regular shapes (Fig. 5.11). Especially with

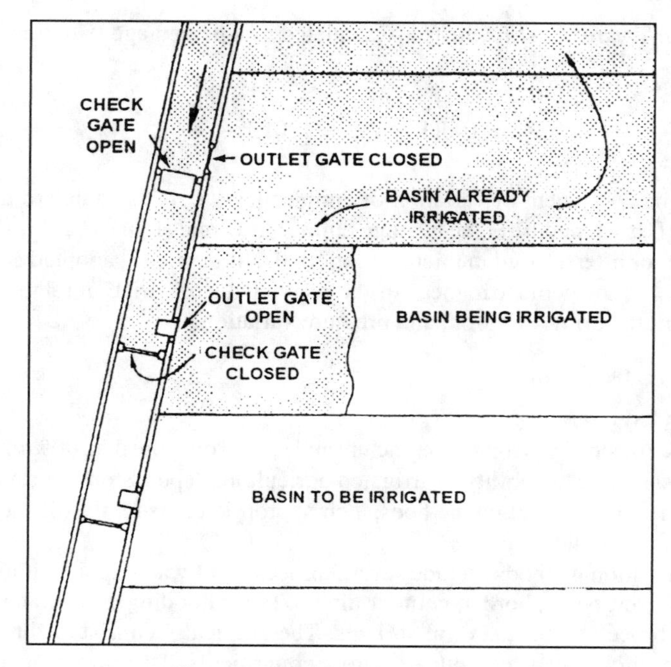

Figure 5.11. Basin irrigation system.

traditional basins, shape depends on the land slope and may be rectangular in flat areas and follow natural land contours in steep areas. For row crops, and especially horticultural crops, the basins often are furrowed, with the crops being planted on raised beds or ridges. For cereals and pastures, the land is commonly flat inside the basin. Tree crops sometimes have raised beds around the tree trunks for disease control.

Basin irrigation is most practical where soil infiltration rates are moderate to low so that water spreads quickly across the basin and water-holding capacity of the soil is high so that large irrigations can be given. Basin irrigation depths usually exceed 50 mm. Inflow rates for basin irrigation have to be relatively high (>2 L s^{-1} per meter width) to achieve quick flooding of the basin and therefore provide for uniform time of opportunity for infiltration along the basin length. Basins must be leveled precisely for uniform water distribution because basin topography determines the recession of the ponded water. Figure 5.12 shows the irrigation phases for basins and furrows. In basins, because of large inflow rates, the advance time, as well as the time of cutoff, is short. The depletion phase is large, to allow for the infiltration of the ponded water.

Surface drainage often is not provided with basin irrigation. This simplifies the layout of the fields and the water delivery channels, but can result in waterlogging and soil aeration problems if soil infiltration is low and rainfall is high or irrigations are large. Where rainfall may be high during cropping, a network of surface drainage channels should be provided. Because there is normally no runoff from basins, quantifying irrigation

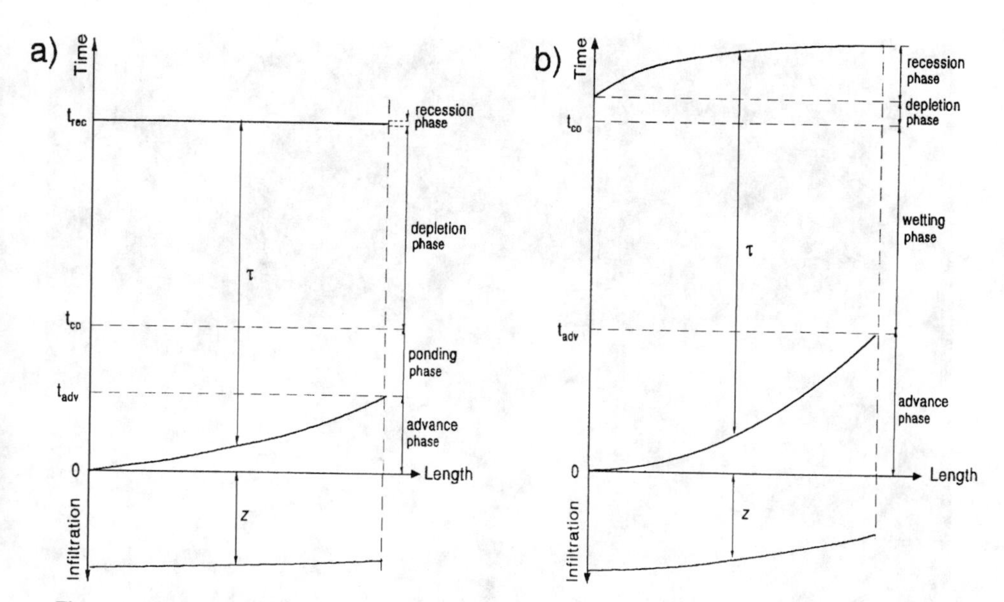

Figure 5.12. Irrigation phases, times of advance t_{adv}, cutoff t_{co}, and recession t_{rec}, infiltration opportunity time τ, and infiltrated depths Z for (a) level basin irrigation and (b) sloping furrow irrigation.

applications requires only measurement of inflows and irrigation time. A desired net application can be preset if flow rate is known.

Water usually is delivered to basins through open channels. In traditional systems, water is diverted from the earthen channels into the fields through cuts in the banks. Improved systems with concrete or steel outlet and check gates greatly improve water control and reduce the labor requirement. Concrete lining of field channels also improves water control and reduces seepage losses and maintenance. The labor requirements of basin irrigation can be low if the basins are large and precisely leveled and water control is good.

Furrow Irrigation

In furrow irrigation, small regular channels direct the water across the field (Fig. 5.13). These channels, called furrows or corrugations, serve both to convey water across the field and as the surface through which infiltration occurs. Because conveyance and infiltration are two opposing purposes, designing and operating furrow systems always requires balancing the trade-off between quickly conveying the water across the field and maintaining the flow long enough to infiltrate adequate water. Efficient furrow irrigation nearly always requires irrigation times longer than advance times, and thus drainage of tailwater runoff at the downstream end. Runoff typically varies from 10% to 40% of the applied water and should be collected, stored, and reused as indicated in ASAE EP408.2 [19]. Without tailwater reuse, furrow irrigation efficiency is unlikely.

Irrigation furrows usually are directed along the predominant slope of the field. Cross-slope or contour furrows reduce the furrow slope but, when the cross slope exceeds the

Figure 5.13. Furrow irrigation with siphon tubes from a concrete-lined ditch.

furrow slope, precise leveling and well-constructed furrows are necessary to prevent water crossover and the resulting uncontrolled channeling and gullying across the field. Furrows are used on slopes varying from 0.001 to 0.05 $\mathrm{m\,m^{-1}}$. Low slopes require soils with low infiltration rates. Slopes greater than 0.01 usually result in soil erosion.

Because furrow slope and soil infiltration rate are usually beyond the control of the designer, furrow design primarily involves determining appropriate furrow length and flow rate. Lengths are commonly between 200 and 400 m. Short lengths are required when infiltration rates are high, and may be required where slopes are steep and soils are erodible. Long lengths allow larger, more efficient, fields. Subdividing long fields with gated pipes or temporary ditches allows efficient irrigation on large fields.

Furrow irrigation is used most commonly for row crops planted on beds or ridges. Furrows may be formed between each plant row or bed, or between alternate rows. Alternate furrows may be used for any given irrigation. This increases the effective MAD because of the wider spacing and thus the greater area irrigated from each furrow. Furrow spacing is limited by the amount of lateral movement of infiltrated water. Fine-textured soils and deep root zones allow widely spaced furrows. Furrow spacing in broadcast and row crops varies from 0.6 to 1.6 m. Furrows in orchards may be up to 3 m apart. Small furrows or corrugations also are used in close-growing crops such as small grains, pastures, and forage. These small furrows often do not completely contain the flows but help to direct them more quickly and uniformly across the field.

Furrow inflow rates are selected to balance the runoff loss that results from high inflow rate and rapid advance against the nonuniform distribution that results from low inflow rate and slow advance. The soil infiltration rate is the most important factor in the relationship between inflow rate and irrigation performance. Because infiltration varies seasonally and is difficult to predict, furrow irrigation systems must be designed to operate over a fairly wide range of inflow rates, and efficient furrow irrigation requires monitoring of the irrigation process and adjusting the inflow rates as required. Consequently, furrow irrigation is relatively labor intensive. Inflow cutback can improve performance greatly by allowing both high initial flows for rapid advance and low final flows to minimize runoff loss. Cutback irrigation seldom is practiced because most irrigation water is supplied at a constant rate and managing the remaining water after the flow cutback is difficult. Cutback also increases labor. Tailwater reuse reduces the need for cutback.

Water is supplied to furrows from ditches or gated pipes along the upper end of the field. In some traditional systems, water is supplied to furrows from earthen ditches through periodic cutouts through the bank that supplies a small ditch that delivers water to 5 to 10 furrows. This feed-ditch method is labor intensive, difficult to control, and results in nonuniform water application. The preferred and more common way to deliver water to furrows from ditches is with siphon tubes (Fig. 5.13). Siphon tubes are typically about 1.7 m long and made from rigid aluminum or polyethylene. Siphon tube use requires that the water level in the supply ditch be at least 10 cm above the field elevation. Water in the ditch is dammed or "checked up" to the required height with flexible (canvas or reinforced polyethylene) dams across the earthen ditches or rigid dams or weirs in concrete-lined ditches. Siphon-tube flow rate is adjusted by changing the elevation of the downstream end of the tube, by switching to a larger or a smaller tube, or by using two or more tubes together. The advantage of siphon tubes over spiles through the bank is that they do not require plugs or gates and do not leak.

Gated pipes (Fig. 5.14) are laid on the surface at the head end of the furrows and have adjustable outlets for each furrow. The most common type of outlet gate is a small rectangular slide that covers a rectangular slot in the pipe. Round outlets that adjust by rotating the circular cover and spigot outlets are also available. Rigid gated pipe usually is made from either 4-mm-thick aluminum or about 8-mm-thick PVC. PVC pipe must have ultraviolet inhibitors to prevent deterioration in the sun. Both have low-pressure, gasketed ends that are easy to connect and disconnect. Common pipe diameters are 15, 20, 25, and 30 cm and lengths up to 9 m are easy to maneuver. Advantages of gated pipe over open ditches are that they requires little or no field area and they are portable, and thus can be moved for tillage and harvesting operations.

Thin-walled (lay-flat) tubing made from PVC sheeting is used like rigid gated pipe. Both fixed and adjustable outlet gates for lay-flat tubing are available. Lay-flat tubing can hold only about 5 kPa of water pressure and generally lasts only one year. It is made in a range of diameters and wall thicknesses. It is a low-initial-cost alternative to rigid gated pipe but often has a higher annual cost because of its short useful life.

Figure 5.12b shows the irrigation phases for furrows. In basins, because of large inflow rates, the advance and ponding phases are relatively short and the depletion phase

Figure 5.14. Gated pipes in polyvinyl chloride (PVC).

is large to allow for the infiltration of the ponded water. In furrow irrigation, inflows are often selected (0.2 to 1.2 $L\,s^{-1}$) so that the advance phase is 20% to 50% of the irrigation time. Desirable advance-to-cutoff time ratios depend on the soil infiltration characteristics and whether runoff is reused. Depletion time is often short and sometimes may be ignored. Recession time is usually much shorter than advance time, and so, wetting time must be sufficient to adequately irrigate the tail end (low quarter) of the furrow.

Border Irrigation

In border irrigation, the field is divided into sloping strips of land separated by parallel border dikes or ridges. Water is applied at the upstream end and moves as a sheet down the border (Fig. 5.15). Border irrigation is used primarily for close-growing crops, such as small grains, pastures, and fodder, and for orchards and vineyards. The method is best adapted to areas with low slopes, moderate soil infiltration rates, and large water supply rates. These conditions allow large borders that are practical to farm.

Borders are most common and practical on slopes less than 0.005 $m\,m^{-1}$. They can be used on steeper slopes if infiltration is moderately high and crops are close growing. Irrigation to establish new crops on steep borders is difficult because water flows quickly, is difficult to spread evenly, and may cause erosion. Design and management of very flat borders approximates conditions for level basins.

Border width is determined by cross slope and available flow rates. The elevation difference across a border should be less than 30% of the flow depth to ensure adequate

Figure 5.15. Border irrigation.

water coverage. Thus, border width is limited by field cross slope or by the amount of land movement required to eliminate cross slope. Deep flow levels, resulting from low slopes, high flow rates, or high crop roughness, allow more cross slope. Land leveling is critical to efficient border operation. Within these limitations, border widths should be multiples of the width of the machinery that will be used, to allow efficient machinery operation. In orchard and vineyards, width is determined by row spacing. Borders typically vary from 5 to 60 m wide.

Border length affects advance time and thus irrigation cutoff time. Longer borders require longer irrigation times and result in greater irrigation depths. Borders up to 400 m long are used where infiltration rates are moderately low and MAD is high.

In the ideal border irrigation, the recession curve parallels the advance curve, giving equal infiltration opportunity time all along the border. Thus, efficient border irrigation, like basin irrigation, requires large flow rates per unit width to advance the water quickly down the field. On sloping borders, this requires that inflow cutoff, and often even recession at the top end of the field, occur before advance completion at the tail end. If soil infiltration or crop roughness changes, inflow rates must change to maintain this balance. The downstream end of borders may be closed or open. Closed borders require accurate flow cutoff times to prevent excessive ponding at the tail, and may require drainage of excessive rainfall. Open borders require drainage systems and, preferably, tailwater collection and reuse. Water application to borders can be from cutouts or side gates from ditches or from large siphon tubes.

Border irrigation is gradually declining in use. Laser leveling allows some border irrigated fields to be converted to easier-to-manage basins. Some fields also have been converted to sprinkler irrigation, and orchards and vineyards may be converted to microirrigation.

Analyses in the following sections are oriented to these main surface irrigation methods, focusing on progress for modernizing surface irrigation systems with improved performance and labor and energy savings. This can contribute to the competitiveness of surface irrigation systems compared to pressurized systems.

Governing Equations and Modeling

Flow Equations

The process of surface irrigation combines the hydraulics of surface flow in the furrows or over the irrigated land with the infiltration of water into the soil profile. The flow is unsteady and varies spatially. The flow at a given section in the irrigated field changes over time and depends upon the soil infiltration behavior. Performance necessarily depends on the combination of surface flow and soil infiltration characteristics.

The equations describing the hydraulics of surface irrigation are the continuity and momentum equations [20]. In general, the continuity equation, expressing the conservation of mass, can be written as

$$\frac{\partial A}{\partial t} + \frac{\partial Q}{\partial x} + I = 0, \tag{5.126}$$

where t is time (s), Q is the discharge ($m^3\,s^{-1}$), x is the distance (m) along the flow direction, A is the flow cross-sectional area (m^2), and I is the infiltration rate per unit length ($m^3\,s^{-1}\,m^{-1}$).

The momentum equation, expressing the dynamic equilibrium of the flow process, is

$$\frac{v}{g}\frac{\partial v}{\partial x} + \frac{v}{gA}\frac{\partial Q}{\partial x} + \frac{v}{gA}\frac{\partial A}{\partial t} + \frac{1}{g}\frac{\partial v}{\partial t} = S_o - S_f - \frac{\partial y}{\partial x}, \tag{5.127}$$

where g is the gravitational acceleration ($m\,s^{-2}$), S_o is the land (or furrow) slope ($m\,m^{-1}$), S_f is the friction loss per unit length or friction slope ($m\,m^{-1}$), v is the flow velocity ($m\,s^{-1}$), and y is the flow depth (m).

These equations are first-order nonlinear partial differential equations without a known closed-form solution. Appropriate conversion or approximations of these equations are required. Several mathematical simulation models have been developed.

Infiltration Equations

Several infiltration equations are used in surface irrigation studies. Most common are the empirical Kostiakov equations,

$$I = ak\tau^{a-1} \tag{5.128}$$

and

$$I = ak\tau^{a-1} + f_0, \tag{5.129}$$

where I is the infiltration rate per unit area (mm h^{-1}), a and k are empirical parameters, f_0 is the empirical final or steady-state infiltration rate (mm h^{-1}), and τ is the time of opportunity for infiltration (h). The latter is commonly used in furrow infiltration where cutoff times are long and infiltration tends to approach a steady rate. When initial preferential flow occurs, as is the case with swelling or cracking soils, an initial "instantaneous" infiltration amount must be added to the cumulative infiltration.

Other infiltration equations used are the empirical Horton equation,

$$I = f_0 + (I_i - f_0)e^{\beta\tau}, \tag{5.130}$$

where β is an empirical parameter, I_i is the initial infiltration rate (mm h^{-1}), and f_0 is the final infiltration rate (mm h^{-1}); the semiempirical equation of Philips,

$$I = 0.5S\tau - 0.5 + A_s, \tag{5.131}$$

where S is soil sorptivity (mm h^{-2}) and A_s is soil transmissibility (mm h^{-1}); and the Green-Ampt equation,

$$I = K\left[1 + \frac{(\theta_s - \theta_i)h'}{Z}\right], \tag{5.132}$$

where K is the saturated hydraulic conductivity (mm h^{-1}), θ_s is the saturated-soil water content (m^3 m^{-3}), θ_i is the soil initial water content (m^3 m^{-3}), h' is the matric potential at the wetting front (mm), and Z is the cumulative depth of infiltration (mm).

However, the equation that more precisely describes the flow in porous media is the Richards equation. For border or basin irrigation, the one-dimensional form is appropriate, whereas, for furrow irrigation, the two-dimensional form would be required:

$$C(h)\frac{\partial h}{\partial t} = \frac{\partial}{\partial x}\left[K(h)\frac{\partial h}{\partial x}\right] + \frac{\partial}{\partial z}\left[K(h)\left(\frac{\partial h}{\partial z} - 1\right)\right], \tag{5.133}$$

where $C(h)$ is $\partial\theta/\partial h$, θ is soil water content, $h(\theta)$ is the pressure head, $K(h)$ is the hydraulic conductivity, t is time, x is the horizontal distance, and z is the vertical distance from soil surface (positive downward).

The use of Eq. (5.133) in the continuity equation (5.126) not only increases the complexity of the solution of the flow equations but also requires much more detailed and accurate information on the hydraulic soil properties. However, the information provided by the corresponding model would be more detailed and, hopefully, better represent the dynamics of the irrigation process [21–23].

Computer Models

A great deal of effort has been put forth to develop numerical solutions for both the continuity and momentum equations (5.126) and (5.127) and the Richards and Green-Ampt infiltration equations (5.132) and (5.133). The current approaches to solutions of Eqs. (5.126) and (5.127) are the method of characteristics, converting these equations into ordinary differential ones; the Eulerian integration, based on the concept of a deforming control volume made of individual deforming cells; (c) the zero-inertia approach, assuming that the inertial and acceleration terms in the momentum equation (5.127) are negligible in most cases of surface irrigation; and the kinematic-wave approach, which assumes that a unique relation exists to describe the $Q = f(y)$ relationship. A consolidated review and description of those solutions, using the Kostiakov equations (5.128) or (5.129) for the infiltration process, are given by Walker and Skogerboe [20].

These solutions are incorporated in the computer programs SIRMOD [24] and SRFR [25, 26]. SRFR solves the nonlinear algebraic equations adopting time-space cells with variable time and space steps and also includes a full hydrodynamic model adopting the Kostiakov infiltration equation (5.129). The model adapts particularly well to describe level furrows and basins as well as the impacts of geometry of furrows on irrigation performance. With the same origin [27], a menu-driven program, BASIN, for design of level basins has been developed [28]. These user-friendly programs correspond to the present trends in software development, which make complete design tools available to users.

These computer models can be used for most cases in irrigation practice, for both design and evaluation. Nevertheless, there are many other developments in modeling recently reported in literature, mainly relative to improvements in zero-inertia and kinematic wave models (comments in [29]).

For many problems in the irrigation practice, the simple volume balance equation [20] can be appropriate for sloping furrows and borders:

$$Q_0 t = V_y(t) + V_z(t), \tag{5.134}$$

where Q_0 is the inflow rate at the upstream end ($m^3\,h^{-1}$), t is the time since irrigation started (h), V_y is the volume of water on the soil surface (m^3), and V_z is the volume of water infiltrated (m^3). This describes the mass conservation. Related models are particularly useful to study the advance phase and, consequently, to derive the infiltration characteristics from observations of advance. In general, empirical time-based infiltration equations (5.128) and (5.129) are used. Results from volume balance models are sensitive to the shape coefficients describing the surface water stream and the infiltration pattern.

The volume balance model is appropriate for real time (automatic) control because, under these circumstances, the advance phase is the most critical [30]. Simplified models proposed by Mailhol [31] and Eisenhauer *et al.* [32] also use the volume balance approach to optimize irrigation parameters—flow rate and time for cutoff—when a target irrigation depth is known. These models, in combination with irrigation scheduling programs, could be used for real-time furrow irrigation management.

Field Evaluation

Field evaluation of farm irrigation systems, described in detail by Merriam and Keller [10], plays a fundamental role in improving surface irrigation. Evaluations provide information used to advise irrigators on how to improve their system design and/or operation, as well as information on improving design, model validation and updating, optimization programming, and developing real-time irrigation management decisions. Basic field evaluation includes observation of

- inflow and outflow rates and volumes (volume balance);
- timing of the irrigation phases, particularly advance and recession;
- soil water requirements and storage;

- slope, topography, and geometry of the field; and
- management procedures used by the irrigator.

More thorough evaluations require independent measurement of the infiltration using a furrow or basin infiltrometer, as well as estimates of water stored on the soil surface and/or surface roughness. When soils are erodible, erosion observations also should be performed [33].

The most difficult to measure and important parameter affecting surface irrigation design and performance is infiltration. Field data from evaluations can be used to estimate the infiltration parameters for the Kostiakov equation. The application of the volume balance method to the advance phase of sloping furrows and borders led to the development of a well-proved methodology for estimating infiltration parameters—the two-point method [34]. Several authors have reported on successful use of this method, and usefulness to design and evaluation is well established [20]. Smerdon *et al.* [35] provide an interesting evaluation of methodologies, and Blair and Smerdon [36] analyze several forms of advance and infiltration equations. A standard engineering practice for furrow evaluation (ASAE EP419.1) has been developed [19].

Estimation of the infiltration parameters and the roughness coefficient for surface flow also can be done through the inverse surface irrigation problem by numerical simulation models [37, 38]. Techniques to optimize the infiltration and roughness parameters by using the simulation models interactively are available [39, 40]. Of particular interest are the methodologies aimed at real-time control of irrigation. The examples offered by Mailhol [31] and Eisenhauer *et al.* [32] describe simplified, easy-to-implement approaches.

The two main performance parameters—the distribution uniformity DU [Eq. (5.119)] and the application efficiency e_a [Eq. (5.118) or (5.117)] can be computed from field data. Distribution uniformity primarily depends on the parameters characterizing the irrigation event; e_a also is influenced by the irrigation scheduling decision, i.e., the irrigation timing (SWD) and depth. DU can be functionally described by

$$DU = f_1(q_{\text{in}}, L, n, S_0, I_c, F_a, t_{\text{co}}), \tag{5.135}$$

where q_{in} is the unit inflow rate (per furrow or per unit width of the border or basin); L is the length of the furrow, border, or basin; n is the roughness coefficient; S_0 is the longitudinal slope of the field; I_c represents the intake characteristics of the soil; F_a represents of the cross-sectional characteristics of the furrow, border, or basin; and t_{co} is the time of cutoff.

The water application efficiency can be described by the same factors above together with the SWD when irrigation starts:

$$e_a = f_2(q_{\text{in}}, L, n, S_0, I_c, F_a, t_{\text{co}}, SWD). \tag{5.136}$$

Design

Traditionally, design of surface irrigation systems has been based on past experiences in an area. This method met the practical needs of the farmers as long as irrigation water was plentiful and conditions were similar. As the need for improved design and performance grew, engineers began combining past experiences with simple hydraulic relationships to develop semiempirical design procedures in the forms of guidelines, equations, and nomographs [16].

Recent advances in simulation modeling, computing, and user-friendly software for personal computers provide good tools for design of surface irrigation systems. However, the use of these tools lags behind the potential. Among the main reasons are the lack of required input data, the need for using information from field evaluations as design criteria or as input parameters, influences of land leveling and soil management on actual field performance, dependence of performance parameters on many farming factors including the irrigator decisions, and temporal and spatial variability of soil characteristics.

The derivation of infiltration from field evaluation data, as mentioned above, can play an essential role for design, either using simulation models or adopting optimization techniques. Simulation models are used for design in an iterative process of searching the best geometric and inflow parameters that provide optimum values for the performance parameters.

The input parameters include the infiltration characteristics; the length and the cross-sectional characteristics of the furrows, borders, or basins; the desired irrigation depth; and the roughness coefficient n. The slope, S_0, can be fixed or adjusted together with the other output parameters: the inflow rate q_{in} and the time duration of the irrigation t_{co}. Input parameters can vary from one simulation to another until the best solution is obtained. The capabilities of simulation provide for alternative best solutions concerning different processes of water application and different irrigation management conditions.

Land Leveling Requirements

The functional relationships for DU and e_a [Eqs. (5.135) and (5.136)] indicate that, for each set of field characteristics, there exists an optimal value for the slope S_0. However, a given average slope may correspond to a precisely leveled field or one with uneven microtopography. Consequences of irrigating an uneven basin are shown in Fig. 5.16. Differences in advance time are small because of the high inflow rate, but differences in recession time are large, giving widely varying infiltration opportunity time. This results in uneven infiltrated depths, greatly affecting irrigation performance, particularly for basins and furrows.

Recent land leveling advances have been made in both computational procedures [41–43] and the use of laser control of land grading equipment.

The leveling precision can be estimated from the standard deviation of field elevation differences to the target elevation:

$$S_d = \left[\sum_{i=1}^{n} (h_i - h_{ti})^2 / (n-1) \right]^{1/2}, \qquad (5.137)$$

where S_d is the standard deviation of field elevation (m), h_i is the field elevation at point i (m), and h_{ti} is the target elevation at the same point (m). When precision laser leveling is used, it is possible to achieve $S_d < 0.012$ m whereas conventional equipment does not provide better than $S_d = 0.025$ m.

Poor land leveling particularly affects the distribution uniformity in furrowed level basins supplied through an earthen ditch at the upstream end. Differences in furrow

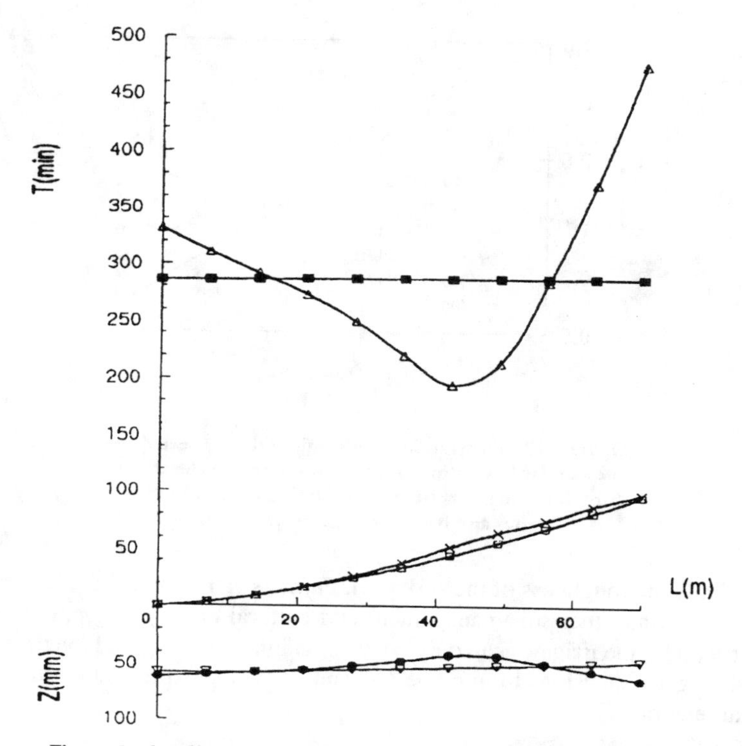

Figure 5.16. Effects of microtopography on irrigation characteristics: advance (□–□) and recession times (■–■) and infiltrated depths (∇–∇) for a precise level basin compared with advance (x–x) and recession times (Δ–Δ) and infiltration depths (●–●) for an uneven field.

entrance elevations cause differences in inflow rates and volumes entering each furrow. Even though the water may be redistributed after the advance is completed through a ditch at the downstream end, DU is highly influenced by S_d. Combined effects of individual inflow rates (q_{in}) and basin lengths (L) are shown in Fig. 5.17, relating the S_d of furrow entrance elevations with the ratio between the actual DU and the maximum value for DU expected when land leveling would be optimal; DU_{max}.

Maintenance of precise laser leveling requires tillage equipment that conserves the landform after land grading and the adoption of appropriate tools for furrow opening. An economic analysis of land leveling impacts given by Sousa *et al.* [45] shows how poor land grading leads to lower yields and higher maintenance costs.

Management

Control of Inflow Rates and Time of Cutoff

For a given field, two of the parameters in the performance functional relationships (5.135) and (5.136) remain constant—the length L and the slope S_0. Several others may change from season to season and from irrigation to irrigation as a consequence of irrigation and farming practices—soil infiltration characteristics I_c, the shape of the

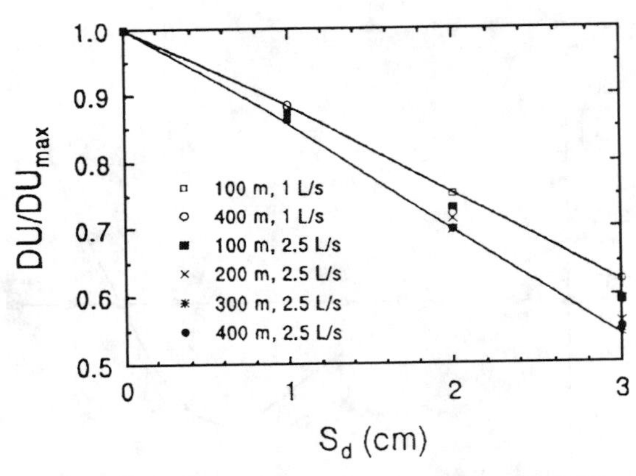

Figure 5.17. Distribution uniformity DU as affected by the standard deviation S_d of furrow entrance elevations for a furrowed level basin, considering different inflow rates and basin lengths. *Source:* [44].

furrow F_a, and the roughness of the surface n. Farmers can change these to some extent through tillage and other soil management and cultural or cropping practices. At the time of irrigation, the irrigator can only control the inflow rate q_{in} and the time of cutoff t_{co}. Thus, irrigator decisions to improve DU and e_a can be expressed by the simplified functional relations

$$DU = f(q_{\text{in}}, t_{\text{co}}) \tag{5.138}$$

and

$$e_a = f(q_{\text{in}}, t_{\text{co}}, SWD). \tag{5.139}$$

Appropriate control of inflow rates and duration of irrigation is particularly relevant where overirrigation has to be avoided. This is the case when deep percolation of excess water contributes to water table rise, salinization, leaching of fertilizers and other agrochemicals, and excessive drainage volumes. Such control is also essential when irrigating saline soils or using saline water. Therefore, the improved control of q_{in} and t_{co} becomes not only a matter of irrigation performance but environmental protection.

When the irrigator selects the best time for irrigating, and thus the best SWD, he may achieve the best performance when appropriate inflow rate and time of cutoff are selected and applied. Selection of the appropriate rates and times requires either simulation of the event, with appropriate input parameters, or experience. Applying the desired rates and times requires an ability to measure and control the flows in rate and time. Lack of water control at the farm and field level often constrains an irrigator's ability to do the job well.

Good farm water control requires good water conveyance systems, such as buried pipelines, and often on-farm water storage so that irrigation rates and timings are not completely determined by water supply rates and timings. Good field control requires the ability to adjust and control field inflow rates and to provide uniform discharges along a furrow-irrigated field. Good field application systems include the use of concrete channels with siphon tubes for furrows or side gates for borders and basins; the use of surface gated pipes or lay-flat, flexible tubing equipped with appropriate regulation valves or orifices for furrows; and buried pipelines with valved risers for basins and borders. This equipment, although more expensive than traditional earthen ditches, is easy to operate, less labor intensive, and provides more precise control. Further advances in control of q_{in} and t_{co} are obtained through automation.

Automation

A primary disadvantage of surface irrigation is the labor requirement. Engineers have worked for many years to develop automated, or at least mechanized, surface irrigation systems. Humpherys [46] describes many of the systems that have been developed and tested in the past 50 years.

Two recently developed mechanization systems for surface irrigation that can reduce labor and improve water control have seen modest adoption: cablegation (Fig. 5.18) and surge flow. Cablegation [47] automatically applies a gradually decreasing inflow rate to furrows from gated pipes. Considering that infiltration rates also decrease with time, this can be a great advantage in decreasing both deep percolation and tailwater runoff losses.

When properly designed, installed, and managed, cablegation systems can provide good irrigation performance. An example with blocked furrows is presented in Fig. 5.19. Field evaluation during irrigation provided information to reduce the inflow rate and increase the time duration of the subsequent irrigation. This resulted in increasing DU and e_a.

Experiences with furrow cablegation systems have shown that careful placement of the gated pipe on a uniform slope at the upstream end of the field is very important to

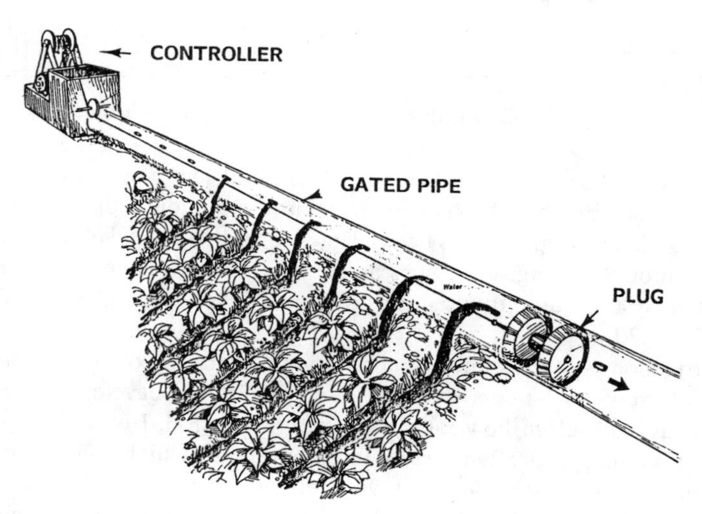

Figure 5.18. Cablegation system for furrow irrigation.

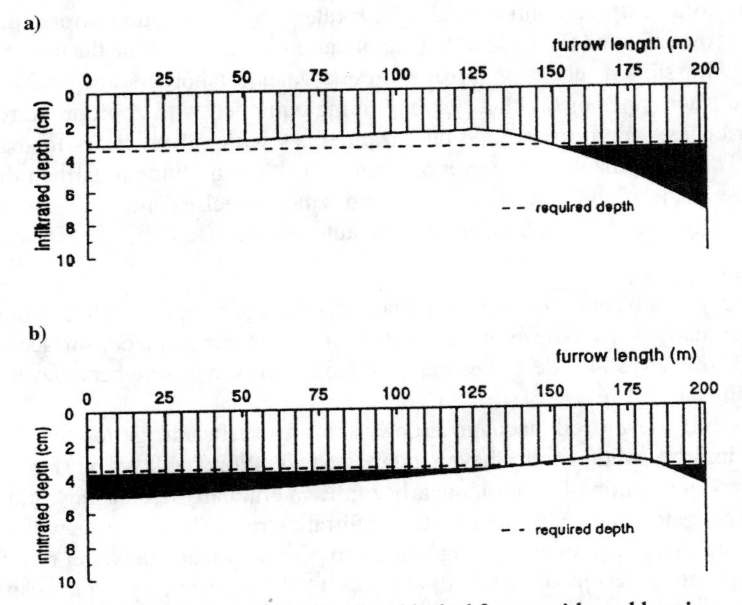

Figure 5.19. Infiltration profiles in a blocked furrow with a cablegation system: (a) irrigation on 08/21/92, with DU = 73.7% and e_a = 73.4%; (b) irrigation on 09/15/92 after decreasing the inflow rate and increasing the cutoff time, with DU = 80% and e_a = 80%. *Source:* [40].

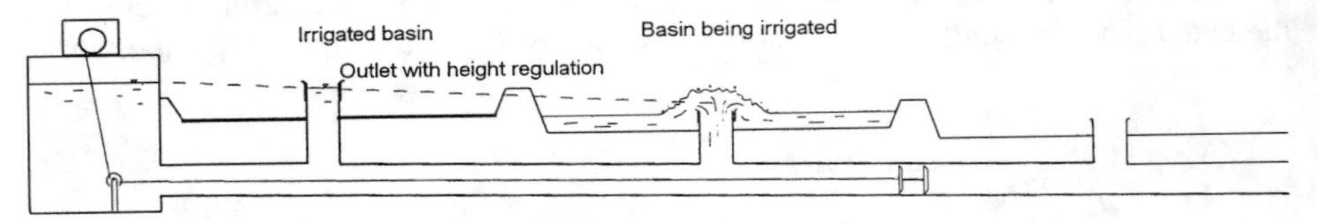

Figure 5.20. Cablegation for automated level-basin irrigation.

obtain uniform applications along the pipe. Because of low-pressure flow in the pipe, the outlet flows are greatly influenced by the relative elevation of the outlets.

The cablegation concept also has been adapted to borders and level basins [49]. Water is delivered through underground pipes and applied through large open-topped risers. A cablegation system currently used for level-basin irrigation at Coruche, Portugal, is shown in Fig. 5.20.

Surge irrigation is the intermittent application (alternating on-and-off flow cycles) of water to furrows or borders [50]. Surge originally was developed as a means to provide a quasi-cutback in flow rates to reduce runoff [51]. However, engineers discovered that surging also often results in a reduction in high infiltration rates and faster stream advance when compared with continuous water application. The mechanisms by which intermittent application can reduce furrow infiltration rates are relatively well known [52, 53], but the amount of reduction varies with the soil

Figure 5.21. Programmable surge valve.

conditions and cannot be accurately predicted. Commercial programmable surge valves are available to automatically cycle the water between two irrigation sets (Fig. 5.21). Thus, surge irrigation can decrease labor (irrigate two sets at one setting), improve distribution uniformity (faster advance), and reduce runoff (quasi-cutback flows) [50, 54–56].

For both cablegation and surge irrigation, appropriate combinations of inflow rate and cutoff time are required for good performance. Thus, at least some field observations are required. Specific evaluation techniques are available for both cablegation [57], and surge irrigation [58]. Surge irrigation performance also can be simulated with several surface hydraulics models, if the surge effect on infiltration can be predicted. Results of surge field evaluations often report improved advance time and infiltration rates [55] and decreased deep percolation and tailwater runoff [59, 60], at least for the first irrigation. Surge flow also may apply to graded borders.

Appropriate filters should be used when automated irrigation is used in areas where water carries trash and algae, which could cause plugging of outlets and malfunctioning of the system [61].

Several automation devices have been developed for level-basin irrigation, mainly for timer-activated automatic field outlet gates [62–64].

Very little surface irrigation mechanization has been adopted, for two primary reasons: First, because infiltration varies throughout the season, and surface irrigation performance depends very much on the infiltration rate, it is difficult to accurately predict the required inflow rates and irrigation times. Thus, good performance requires either feedback control from field sensors or human monitoring and adjustment. Feedback control is expensive and complex. Human monitoring reduces the labor-saving benefits of mechanization. Second, few private companies have been willing to invest in developing, testing, and marketing surface irrigation mechanization devices. Surface irrigation systems generally do not require sufficient system-specific hardware to merit large private-sector investment.

Soil Management

Because surface irrigation performance is very dependent on infiltration [65], soil management plays a crucial role in efficient irrigation. Management practices can be used to increase infiltration of soils of low infiltrability (deep tillage, cultivation, controlled traffic, residue incorporation, residue management, reduced tillage, gypsum amendment, polyacrylamide amendment); reduce infiltration of soils with excessive infiltration (compaction, surge irrigation); improve soil water storage to allow larger, more uniform irrigations (organic-matter incorporation); improve soil aeration (drainage, deep tillage); control soil crusting and sealing (residue management, gypsum amendment); and reduce soil erosion (irrigation management, residue management, reduced tillage, polyacrylamide amendment) [66].

A common problem in furrow irrigation is high infiltration rates after primary tillage. Several studies have shown that furrow compaction of newly formed furrows can reduce infiltration by up to 50% [66–68]. Furrow performance also may be affected by furrow shape, uniformity, and roughness. Therefore, the selection of furrow opener systems should consider the capabilities for furrow forming, smoothing, and firming. Laser control of the depth of the furrowing tool has been tested as a means to improve furrow uniformity [69].

Conventional tillage practices can destroy precision leveling. Alternative tillage systems, including reduced tillage and no-till can extend the life of a level field surface. However, modifying soil tillage practices impacts infiltration. Sealing may have negative or positive impacts according to soil type [53, 70]. Reduced till may have contradictory effects. Both a slight decrease and a large increase of advance time are reported when rotary till or minimum till is compared with moldboard plow. Contradictory results also have been obtained when application efficiencies were compared. Because of the variation in soil types, it is sometimes difficult to predict the effect of changes in tillage and other soil management practices.

Fertigation

Fertigation (and chemigation) currently is applied with microirrigation and sprinkler irrigation, particularly with automated systems [72] but it is rare with surface irrigation. However, fertigation may be used for both furrow and basin irrigation when good irrigation uniformity and performance can be achieved. When chemicals are applied in surface irrigation water, backflow prevention devices must be used at any groundwater or public water supply, and any tailwater should be captured and reused, preferably on the same field.

Recent studies show the technical feasibility of fertigation with surface irrigation. An experimental study [73] to analyze the variation in concentration of urea fertilizer in an irrigation supply channel show that injection must be done about 7 m upstream of the first field outlet when flow velocities are low (near 0.3 m s^{-1}). Best mixing occurs when the injection is made in the stream centerline. Results also indicate that, for current irrigation practices, the fertilizer concentrations are uniform along the supply conduit. These conclusions also may be valid for gated pipes or lay-flat tubing.

Santos *et al.* [74] report that distribution uniformity of irrigation water is closely followed by the distribution uniformity of nitrogen fertilizer in level-basin-irrigated maize in a silty loam soil. The application of fertilizers with each irrigation improves fertilizer use by the crop and decreases the amount of nitrate in the soil after harvesting [75]. Although more studies are required, there is evidence that surface irrigation fertilization can reduce the environmental impacts of nitrates.

Surface Irrigation Scheduling

The purpose of irrigation scheduling (see Section 5.3) is to apply the appropriate irrigation depth when the *SWD* is less than or equal to the MAD. Scheduling of irrigations is essential to attain high application efficiencies [Eq. (5.139)], to avoid deep percolation and runoff losses of water and chemicals, and to prevent stress of the crop. Scheduling is of particular importance for management of saline soils or where low-quality water is used, so that the appropriate leaching fraction is given; for conditions of water scarcity, where wasted water is costly; and where drainage is poor and excess applications cause a high water table and poor soil aeration.

Scheduling surface irrigation is difficult for three reasons: (1) It is often difficult to quantify the amount of water applied, (2) it is difficult to quantify the uniformity of the application, and (3) water deliveries from irrigation water suppliers often are not under the farmer's control. Quantifying water application requires flow measurement. If water runs off the field, both inflows and outflows must be measured. Flow measurement of canal water supplies is often inadequate or completely lacking. Runoff is seldom measured. Without flow measurement to quantify applications, schedulers usually assume that the root zone is completely refilled with each irrigation. This may be a poor, and costly, assumption. Flow measurement is critical to good irrigation scheduling and irrigation performance. Methods to measure irrigation water are widely described [76, 77].

Because distribution uniformity in surface irrigation is dependent on complex interactions of parameters, and on inherently variable infiltration characteristics, water application uniformity is difficult to predict. Knowledge of *DU* is necessary to adjust total applications to the effective (usually low quarter) application amounts used for scheduling. Knowledge of the spatial application variability is important to select locations for soil water measurements. Because uniformity and variability are difficult to predict or measure, precise scheduling of surface irrigation is not possible. Overirrigation, although costly in terms of water wastage, effectively reduces water storage variability and facilitates scheduling.

Most surface-irrigated areas are supplied from collective irrigation canal systems. Farm irrigation scheduling depends upon the delivery schedule, for example, rate, duration, and frequency, which are dictated by the system operational policies. Flow rate and duration may impose constraints on the volume of application, and supply frequency may constrain, or even determine, the irrigation timing. In general, surface irrigation delivery systems are rigid and the time interval between successive deliveries is too long. Irrigators tend to compensate for this by applying all the water they are entitled to use. For deep-rooted crops and soils with high water-holding capacity, this strategy can be appropriate; however, for shallow-rooted crops and/or coarse-textured soils, percolation losses can be substantial and crops may be stressed by alternating drought and waterlogging. Under rigid delivery schedules, it is difficult both to modernize the irrigation methods and to implement irrigation scheduling programs. Several papers in Smith *et al.* [78] discuss technical, social, and organizational problems of rigid water-supply schedules and possible solutions.

In general, it is difficult to apply small irrigation depths by surface irrigation, especially early in the season when the infiltration rate is high and roots are shallow. In furrow irrigation, the time of cutoff usually should be longer than the advance to provide for sufficient infiltration opportunity time. This implies that surface irrigation applications are usually large, which makes this method more suitable for situations where *SWD* and MAD are relatively large. Under these circumstances, scheduling can be simplified when adopting simple irrigation calendars [79]. This is also advantageous for collective system management and for introducing improved tools for controlling the flow rates at the farm level. An example of development of irrigation scheduling calendars for basin irrigation that takes into consideration the system constraints is presented by Liu *et al.* [80].

When the user is in control of the delivery timings, or when an arranged delivery schedule is used, and when irrigation applications can be quantified adequately, it is appropriate to use irrigation scheduling simulation models to determine the soil water status using soil, crop, and meteorological data. Numerous examples of such models are available in the literature (e.g. [81–83]). The use of scheduling models in real time requires periodic field validation of soil water status.

Irrigation also can be scheduled through regular monitoring of the soil-water or plant-water status. Soil samples analyzed with the feel method are appropriate for any situation. The use of soil water-monitoring devices is generally economical only for large fields and crops which can justify the costs of purchase and operation. The use of these systems requires the selection of an irrigation threshold corresponding to MAD [see Section 5.1, Eq. (5.88)]. Observations or estimations of the *SWD* have to be performed regularly to track changes in the *SWD* and allow the prediction of the irrigation date when *SWD* = MAD. Furrow irrigation poses a special problem for soil water monitoring in that water distribution varies with distance from each furrow, and so, representative sampling locations and depths must be determined and assumptions must be made about the effective MAD.

A further degree of sophistication can be achieved by combining soil water-balance simulation with soil water measurements. This can be done in two ways: (1) soil water information is used to validate the model predictions, and (2) continuous soil water and meteorological monitoring are coupled in an irrigation simulation model. An example of the latter approach is presented by Malano *et al.* [40].

5.4.3 Sprinkler Irrigation

Introduction

Sprinkler irrigation in agriculture began with the development of impact sprinklers and lightweight steel pipe with quick couplers. In the 1950s, improved sprinklers, aluminium pipe, and more efficient pumping plants reduced the cost and labor requirements and increased the usefulness of sprinkler irrigation. In the 1960s, the development of moving systems, namely the center pivot, provided for moderate-cost, mechanical, high-frequency irrigation. Additional sprinkler innovations are continually being introduced that reduce labor, increase the efficiency of sprinkling, and adapt the method to a wider range of soils, topographies, and crops. Shortages of labor and water are resulting in more widespread use of mechanization and automation, including self- propelled sprinkler systems, automatic valves, and computer monitoring and control.

There are many types of sprinkler systems, but all have the following basic components:

- The *pump* draws water from the source, such as a reservoir, borehole, canal, or stream, and delivers it to the irrigation system at the required pressure. It is driven by an internal combustion engine or electric motor. If the water supply is pressurized, the pump may not be needed.
- The *mainline* is a pipe that delivers water from the pump to the laterals. In some cases the mainline is placed below ground and is permanent. In others, portable mainline laid on the surface can be moved from field to field. Buried mainlines usually are made of coated steel, asbestos-cement, or plastic. Portable pipes usually are made of lightweight aluminium alloy, galvanized steel or plastic. In large fields the mainline supplies one or more submains that deliver the water to the laterals.
- The *lateral* pipeline delivers water from the mainline to the sprinklers. It can be portable or permanent and may be made of materials similar to those of the mainline, but is usually smaller. In continuous-move systems, the lateral moves while irrigating.
- *Sprinklers* spray the water across the soil surface with the objective of uniform coverage.

Sprinklers irrigation systems can be divided broadly into set and continuous-move systems. In set systems, the sprinklers remain at a fixed position while irrigating; in continuous-move systems, the sprinklers operate while the lateral is moving in either a circular or a straight path. Set systems include solid set or permanent systems as well as periodic-move systems, which are moved between irrigations, such as hand-move and wheel-line laterals and hose-fed sprinklers. The principal continuous-move systems are center-pivot and linear-move laterals, and traveling raingun sprinklers.

Adaptability

Sprinklers are available in a wide range of characteristics and capacities and are suitable for most crops and adaptable to most irrigable soils. Set systems can apply water at any selected rate down to 3 mm h^{-1}. This extends the use of sprinkling to fine-textured soils with low infiltration rates. High-application-rate systems, such as the center-pivot and traveling rainguns, are not applicable to low-infiltration-rate conditions. Care is required to select the proper sprinklers for the existing conditions.

Periodic-move systems can be used where the crop-soil-climate conditions do not require irrigations more than every 5 to 7 days and the crop is not too tall or delicate. Where shallow-rooted crops are grown on soils with low water-holding capacities, light, frequent irrigations are required and fixed (solid set) or continuously moving systems are more suitable. Fixed systems also can be designed and operated for frost and freeze protection, blossom delay, and crop cooling.

Sprinklers can be adapted to most climatic conditions, but high wind conditions decrease distribution uniformity and increase evaporation losses, especially when combined with high temperatures and low air humidities. Although sprinkling is adaptable to most topographic conditions, large elevation differences result in nonuniform application unless pressure regulation devices are used. Other aspects relative to adaptability and suitability of sprinkler irrigation and its advantages and limitations are given in the literature [11, 84].

Advantages

Sprinkler irrigation has the following advantages compared to surface irrigation:

- Properly designed and operated sprinkler irrigation systems can give high seasonal irrigation efficiencies and save water.
- Sprinkler irrigation performance is not dependent on soil infiltration (as long as application rate does not exceed infiltration rate), and thus is dependable and predictable.
- Soils with variable textures and profiles can be efficiently irrigated.
- Land leveling is not required; shallow soils that cannot be graded for surface irrigation without detrimental results can be irrigated.
- Steep and rolling topography can be irrigated without producing runoff or erosion.
- Light, frequent irrigations, such as for germination of a crop, can be given.
- Sprinkler systems can effectively use small, continuous streams of water, such as from springs and small-tube or dug wells.
- Mechanized sprinkler systems require very little labor and are relatively simple to manage.
- Fixed sprinkler systems require very little field labor during the irrigation season and may be fully automated.
- Periodic-move sprinkler systems require only unskilled labor; irrigation management decisions are made by the manager.
- Fixed sprinkler systems can be used to control weather extremes by increasing air humidity, cooling the crop, and reducing freeze damage.
- Sprinklers can be managed to supplement rainfall.
- Sprinklers can leach salts from saline soils more effectively than surface or microirrigation methods.
- Cultural practices such as conservation tillage and residue management can be used easily under sprinkler irrigation.

Limitations

Sprinkler irrigation has the following limitations:

- Initial costs are higher than for surface irrigation systems unless extensive land grading costs are required.
- Energy costs for pressurizing water is a significant expense, depending on the pressure requirements of sprinklers used and power costs.
- When water is not continuously available at a sufficient, constant rate, the use of a storage reservoir is required.

- Soil infiltration rate of less than 3–5 mm h^{-1} will constrain system selection and operating procedures and may result in runoff; center pivots require initial infiltration rates above 20 mm h^{-1}.
- Windy and dry conditions cause water loss by evaporation and wind drift.
- Irregular field shapes are more expensive and less convenient, especially for mechanized sprinkler systems.
- Certain waters are corrosive to the metal pipes used in mainline and laterals.
- Water containing trash or sand must be cleaned to avoid clogging and nozzle wear.
- Sprinkler irrigation water containing salts may cause problems because salts drying on the leaves affect some crops. High concentrations of bicarbonates in irrigation water may affect the quality of fruits. Sodium or chloride concentration in the irrigation water exceeding 70 or 105 parts per million (ppm), respectively, may injure some fruit crops.
- The high humidity and wet foliage created by sprinkling is conducive to some fungal and mold diseases.

Sprinklers

Types of Heads and Nozzles

Sprinkler heads are the most important component in the system because their performance determines the effectiveness and efficiency of the whole system. A sprinkler operates by forcing water under pressure through a small hole or nozzle and into the air. Nozzle size and water pressure determine the flow rate. Most sprinklers are designed to give a circular wetting pattern. The distance from the sprinkler to the outer edge of that circle is called the throw or wetted radius. Nozzle design, size, and pressure determine the pattern wetted diameter. Sprinklers may have special features that allow them to irrigate only a part of the circle.

In *rotary sprinkler heads*, the water jet commonly discharges at an angle above horizontal between 22° and 28°. The jet breaks up into small drops as it travels through the air, and falls to the ground like natural rainfall. Good water distribution is dependent on maintaining water pressure (jet velocity) within the range that produces the proper droplet sizes. The sprinkler rotates in a horizontal plane to produce a circular wetting pattern.

Sprinklers may have two nozzles discharging in opposite directions. The larger provides for larger throw and creates the sprinkler rotation. The small one provides for wetting the inner circle.

Impact sprinkler rotation is caused by the water jet impinging on a spring-loaded swing arm (Fig. 5.22). The water jet impulse forces the spring arm sideways. The spring returns the arm, which impacts the body of the sprinkler, rotating it a few degrees. Then the cycle repeats. The rotational speed (1 to 3 rotations per minute) is controlled by the swing-arm weight and spring tension. It is important that the sprinkler rotates correctly so that no area is left underirrigated.

In *gear-driven rotary sprinklers* the pressurized water entering the sprinkler rotates a small water turbine which, through reducing gears, provides for slow, continuous sprinkler rotation. Gear-drive mechanisms require clean water to prevent clogging and wear.

Figure 5.22. Impact sprinkler.

Spray heads discharge a jet of water vertically from a nozzle onto an impingement plate that redirects it into a circular pattern. This plate may be smooth or serrated and have a flat, convex, or concave surface, depending on the desired pattern shape (throw and droplet sizes). Sprayers generally operate at low pressure and have smaller pattern diameters than impact sprinklers. They are commonly used on continuous-move lateral systems. Recent adaptations of spray heads use rotating or wobbling plates with curved grooves that turn the plate by jet reaction. These variations are designed to increase throw at low pressures and improve water distribution patterns.

Characteristics

A sprinkler is characterized by

- operating pressure P (kPa) required to provide good water distribution,
- discharge or flow rate q_s ($m^3\ h^{-1}$), and
- effective diameter of the wetted circle D_w (m).

The same sprinkler head may be used for different flow rates and diameters by changing the operating pressure and/or the nozzle diameter d_n (mm). Sprinkler charts given by the manufacturer should provide the information on the best combination $P–q_s–D_w$ for each d_n (Table 5.5). These characteristics are interrelated by

$$q_s = K_d P^{0.5} \tag{5.140}$$

Table 5.5. Discharge q_s ($m^3 h^{-1}$) and wetted diameter D_w (m) for typical rotary sprinklers with trajectory angles between 22° and 28° and standard nozzles without vanes

Sprinkler Pressure (kPa)	Nozzle Diameter d_n (mm) 2.4		3.2		4.0		4.8	
	q_s	D_w	q_s	D_w	q_s	D_w	q_s	D_w
140	0.26	19.2	—	—	—	—	—	—
170	0.29	19.5	0.51	23.2	0.80	25.0	—	—
205	0.32	19.8	0.56	23.5	0.88	25.9	1.20	27.8
240	0.34	20.1	0.61	23.8	0.94	26.5	1.36	28.7
275	0.37	20.4	0.65	24.1	1.01	26.8	1.45	29.3
310	0.39	20.7	0.69	24.4	1.07	27.1	1.54	29.9
345	0.41	21.0	0.73	24.7	1.13	27.4	1.63	30.5
380	0.43	21.3	0.77	25.0	1.19	27.7	1.71	30.8
415	0.45	21.6	0.80	25.3	1.24	28.0	1.78	31.1
450	—	—	0.83	25.6	1.29	28.3	1.86	31.4
485	—	—	—	—	1.34	28.6	1.93	31.7

Source: Adapted from [11].

and, more empirically,

$$D_w = K_r q_s^{0.5} \tag{5.141a}$$

or

$$D_w = K_r P^{0.25}. \tag{5.141b}$$

Parameters K_d and K_r primarily depend on the nozzle diameter but they also vary with nozzle design and shape. The exponent in Eq. (5.141b) may be slightly different from 0.25. The value 0.272 is given by Rochester and Hackwell [85] for small rainguns. The wetted diameter can be increased by increasing the water pressure and the nozzle angle (within limits) or using nozzles with vanes [11].

Application Rate

The rate at which sprinklers apply water when operating is called the application rate i_a ($mm\,h^{-1}$). The application rate depends on the sprinkler discharge q_s ($m^3 h^{-1}$) and, for set systems, the spacing between the sprinklers, which determines the irrigated area a_w (m^2):

$$i_a = 1000(q_s/a_w). \tag{5.142}$$

Increasing the nozzle size or the pressure, or spacing the sprinklers closer together will increase the application rate. Manufacturers supply the information required to compute i_a for their sprinklers. The application rate should always be less than the rate at which the soil can infiltrate water to avoid surface water ponding, redistribution, and runoff. Redistribution lowers application uniformity. Runoff not only wastes water but also may cause soil erosion. Tillage management practices including surface residue management and reservoir tillage are used to increase infiltration and/or surface storage

Table 5.6. Suggested maximum continuous sprinkler application rates ($mm\,h^{-1}$) for average soil, slope, and tilth

	Slope (%)			
Soil Texture and Profile	0–5%	5–8%	8–12%	12–16%
Deep coarse sandy soils	50	38	25	13
Coarse sandy soils over more compact soils	38	25	19	10
Deep light sandy loams	25	20	15	10
Light sandy loams over more compact soils	19	13	10	8
Deep silt loams	13	10	8	5
Silt loams over more compact soils	8	6	4	2.5
Heavy textured clays or clay loams	4	2.5	2	1.5

Source: Adapted from [11].

and reduce runoff [86]. Table 5.6 provides suggested maximum continuous application rates according to soil infiltration conditions and slope.

Drop Sizes

A sprinkler normally produces a wide range of drop sizes from 0.5 mm up to 4.0 mm in diameter. The small drops usually fall close to the sprinkler whereas the large ones travel much farther. Information about drop-size distributions along the wetted radius as influenced by pressure is given by Kincaid *et al.* [87]. The range of drop sizes can be controlled by the size and shape of the nozzle and its operating pressure. At low pressures, drops tend to be large. At high pressures they are much smaller and misting may occur. Noncircular nozzle shapes have been developed to produce smaller drops at low pressures. A detailed analysis is provided by Li *et al.* [88, 89].

Large drops have high kinetic energy and can damage delicate crops. They also break down the surface structure of some soils, resulting in reduced infiltration rate and crusting. Sprinklers producing large drops should not be used on soils that tend to crust.

Water Distribution

Sprinklers generally cannot produce an even water distribution over the whole of the wetted radius. Often the application is highest close to the sprinkler and decreases toward the edge, resulting in a radial pattern of distribution shaped like a triangle. To make the distribution more uniform over the field, several sprinklers must operate close enough together that their distribution patterns overlap. The sprinkler pattern determines the desired spacing between sprinklers. Uniformity usually is improved by putting sprinklers close together, but this increases water application rates and cost of the system.

Fixed set sprinklers usually are placed in a square or rectangular grid, although triangular grids improve pattern overlap and distribution uniformity (Fig. 5.23). In continuous move systems, only spacing along the lateral affect distribution (assuming the movement is adequately continuous). Continuous move systems usually produce better uniformity than set systems.

Sprinkler Classification

Sprinklers can be classified according to several factors including

- type—rotary or spray;

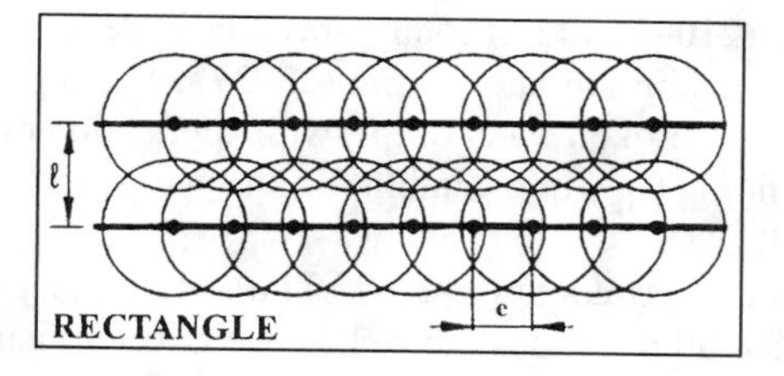

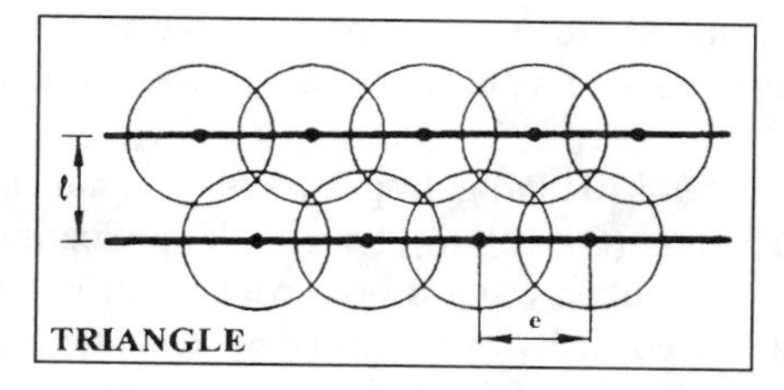

Figure 5.23. Rectangular and triangular sprinkler grids, with spacing *e* along the lateral and *l* between laterals.

- area shape wetted—full or part of the circle;
- throw or wetted diameter—small (<5 m) to large (>50 m);
- operating pressure—low (<150 kPa) to high (>350 kPa);
- discharge—small (<0.5 $m^3 h^{-1}$) to large (>50 $m^3 h^{-1}$);
- application rate—low (<5 mm h^{-1}) to high (>15 mm h^{-1});
- number of nozzles—one or two;
- angle of the jet with the horizontal—very low (<10°) for undertree irrigation, low (18°–21°) to reduce wind drift, or normal (25°–28°); and
- drop-size distribution.

In general, these classifications should be combined and considered under the perspective of adaptability of the sprinkler to the field conditions. According to Keller and Bliesner [11],

- Low-pressure (35–140 kPa) impact or spray sprinklers have a small wetted diameter (6–15 m), produce large water drops with a fair water distribution and relatively high application rate (>10 mm h^{-1}). They are mostly suitable for small areas or continuous-move systems.
- Rotary sprinklers with low to moderate pressure (105–210 kPa) produce a medium wetted diameter (18–24 m), water drops are fairly well broken up, and application rates can be selected over a wide range (>3 mm h^{-1}). Water distribution is good when the pressure is near 200 kPa. They are suitable for most crops, including vegetables and undertree irrigation and are also suitable for continuous-move systems.
- Low- to medium-pressure (70–245 kPa) spinners or sprayers for undertree orchard irrigation produce wetted circles with moderate diameters (12–27 m), moderate-size drops, and fairly good water distribution. A large range of application rates is obtainable (>5 mm h^{-1}). Ideal for orchards in windy areas.

- Medium pressure (210–410 kPa) rotary sprinklers are available with one or two nozzles, irrigate a medium to large circle (23–37 m), and produce excellent water distribution with well-broken water drops, with application rates also in a very wide range (>2.5 mm h^{-1}). They are suitable for all type of soils, including those with low intake, and all crops.
- High-pressure rotary sprinklers (340–690 kPa), either single or dual nozzle, wet large diameters (34–90 m), drops are well broken, water distribution is good when wind speed does not exceed 6 km h^{-1}, but application rates are relatively high (>10 mm h^{-1}). They are suitable for field crops, soils with nonlimiting infiltration rate and regions without excessive wind. They can be used as center-pivot end guns and as traveller guns.
- Very high pressure (550–830 kPa) gun sprinklers, generally single nozzled, irrigate circles of large diameters (60–120 m), have high application rates (>15 mm h^{-1}) and produce very well broken water drops. Water distribution is good under calm conditions but is distorted easily by wind. They are suitable for field crops in soils with good infiltration characteristics and are mostly used as traveling rainguns.

Water Distribution Profiles and Recommended Spacing

In choosing a sprinkler, the aim is to find the combination of sprinkler spacing, operating pressure, and nozzle size that provides the desired application rate with the best distribution uniformity. The uniformity obtainable with a set sprinkler system depends largely on the water distribution pattern and spacing of the sprinklers. The uniformity is strongly affected by wind and operating pressure.

Small droplets from sprinklers are blown easily by wind, distorting wetting patterns and reducing irrigation uniformity. Losses due to wind effects are analyzed by Yazar [90]. The distortion of the precipitation distribution patterns caused by the wind is analyzed by Han *et al.* [91]. Although 15 km h^{-1} is only a gentle breeze, it seriously disrupts the operation of a sprinkler system [92]. Sprinklers need to operate close together under windy conditions to distribute water evenly. In prevailing wind conditions, the orientation of the laterals should be at right angles to the wind direction and the sprinkler spacing along the lateral should be reduced.

A sprinkler performs best within a pressure range that normally is specified by the manufacturer. If the pressure is too low, the water jet does not break up adequately, and most of the water falls in large drops near the outer diameter of the pattern. If the pressure is too high, the jet breaks up too much, causing misting, and most of the water falls close to the sprinkler. Both of these patterns have a reduced throw.

Manufacturers of sprinklers specify a wetted diameter for all nozzle-size and operating-pressure combinations for each type of sprinkler. These diameters, together with the water distribution profile are used when making sprinkler spacing recommendations.

Indicative spacing recommendations based on the wetted diameter D_w are given in Table 5.7 for the most common water distribution profiles. Triangular and elliptic profiles are characteristic of sprinklers operating at the recommended pressure. A donut profile generally is produced with sprinklers operating at pressures lower than those recommended and by sprinklers with straightening vanes just upstream from the nozzle [11].

Table 5.7. Water distribution profiles and suggested spacings (% D_w) for fixed set sprinklers

	Sprinkler Spacing Grid (% D_w)		
Profiles	Square	Triangular Equilateral	Rectangular
Triangular	55	66	40 × 60
Elliptic	60	66	40 × 60 or 65
Donut	40 80 (fair)	80	40 × 80

Source: Adapted from [11].

The wetted diameters listed in manufacturers' brochures usually are based on tests under essentially no wind conditions. Under field conditions with up to 5 km h^{-1} wind, such diameters should be shortened by 10% from the listed figure to obtain the effective diameter. A reduction of 1.5% for each 1 km h^{-1} over 5 km h^{-1} is proposed for the usual range of wind conditions under which sprinklers are operated [11].

Detailed information on water distribution profiles produced by several agricultural sprinklers with different nozzles and operating at various pressures is provided by Tarjuelo *et al.* [93]. Information also includes expected uniformity (CU) at various spacings and as influenced by wind.

Set Sprinkler Systems

Set systems, using many small rotary sprinklers operating together, are the most commonly used sprinkler system. The sprinklers operate at medium to high pressures. Application rates vary from 3 to 35 mm h^{-1}. Single laterals can irrigate an area 9 to 24 m wide and up to 400 m long at one setting. Set systems are described in detail by several authors [11, 94–99].

Fixed systems

When sufficient laterals and sprinklers are provided to cover the whole irrigated area so that no equipment needs to be moved, the system is called a *solid-set* system. For annual crops, the portable pipes and sprinklers are laid out after planting and remain in the field throughout the irrigation season. The equipment is removed from the field before harvesting. In perennial crops such as orchards, laterals and sprinklers often are left in place from season to season. The system then is called *permanent*. Permanent systems often are buried below ground but they also may be laid out on the posts over the top of the crop in case of overtree irrigation for frost protection and chemigation.

Because of the large flow requirements, most fixed systems have only part of the system irrigating at one time. Flow is diverted from one part of the system to another by hydrants or valves that may be automated. However, for special conditions, such as crop cooling or frost protection, it is essential to have sufficient capacity to operate the whole of the system at the same time.

Fixed systems are expensive initially because of the amount of pipes, sprinklers and fittings, and valves required, but labor costs are low. These systems are particularly suited to automation and are useful in areas where labor is a limiting factor.

Semipermanent Systems

Sprinkler systems have been developed with the advantages of both portable and fixed equipment to combine both low capital costs and low labor requirements. These often are referred to as semipermanent systems and the most commonly used are the pipe-grid and hose-pull systems. These systems are designed to reduce the number or size of laterals.

Pipe-Grid Systems

These are similar in many aspects to fixed systems. Small-diameter laterals (about 25 mm) are used to keep system costs low. Laterals are laid out over the whole field and they remain in place throughout the irrigation season. In general, two sprinklers are connected to each lateral, one near the end, the other near the middle. When the irrigation depth has been applied, each sprinkler is disconnected and moved along the lateral to the next position. This procedure is repeated until the whole field has been irrigated. A typical system would involve at least two sprinkler moves on every lateral each day.

Hose-Pull Systems

Originally developed for orchard undertree irrigation, these systems now are being used for some row crops (Fig. 5.24). The mainline and laterals usually are permanently installed, either on or below the ground surface, but also can be portable. Small-diameter plastic hoses supply water from the lateral to one or two sprinklers. The hose length is normally restricted to about 50 m because of friction losses. Initially, the sprinkler is placed in the farthest position and remains there until the irrigation depth is applied. Then it is pulled along to the next position and so on until irrigation is complete.

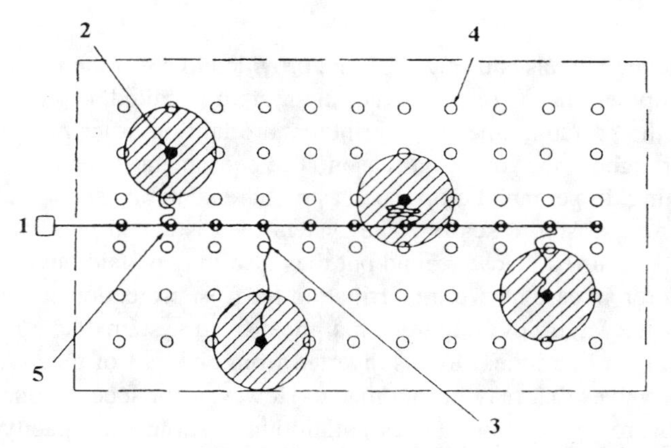

Figure 5.24. Hose-pull system: 1) pump, 2) sprinkler, 3) flexible hose, 4) sprinkler positions, 5) lateral.

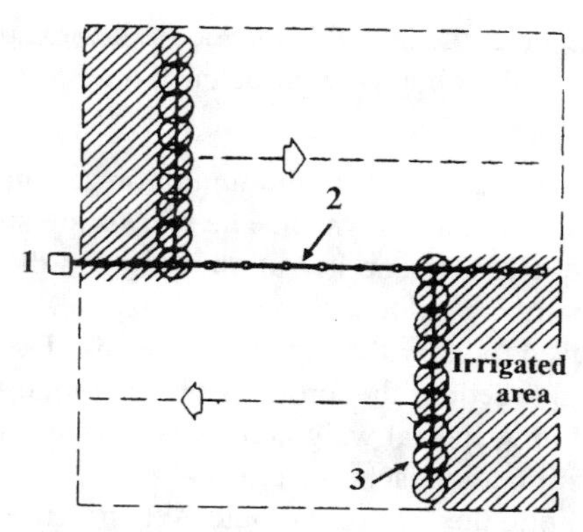

Figure 5.25. Hand-moved system: 1) pump, 2) mainline, 3) lateral.

The use of hoses reduces the number of laterals that are needed, and so, the system costs are less than for permanent systems. Problems can arise with the plastic hoses. The system is reliable for orchards but operators may have difficulties in managing the system in field crops.

Portable Systems

Hand-Moved Systems. These are designed to be moved by hand. The lateral is usually in aluminum or plastic pipe between 50 mm and 100 mm in diameter and 9 to 12 m long, so that it can be moved easily by one person. The laterals remain in position until irrigation is complete. The pump then is stopped and the lateral disconnected from the mainline, drained, dismantled, moved by hand to the next point on the mainline, and reassembled. Usually, the lateral is moved between one and four times each day. It gradually is moved around the field until the whole field is irrigated (Fig. 5.25).

Systems may have two or more laterals to irrigate large areas. They are connected to the mainline using valve couplers. This allows irrigation to continue while one of the laterals is being moved. In some cases, when the sprinklers are used to germinate new plantings, to leach salts, or to supplement rainfall, the whole system including pump and mainline is moved from field to field.

Hand-moved sprinklers are used to irrigate a wide range of field and orchard crops. Their capital cost is low and they are simple to use. However, they require a large labor force, often working in wet, muddy, and uncomfortable conditions.

Towed Systems. To alleviate labor requirements, laterals (aluminium or plastic) can be mounted on wheels or skids and towed across the submain to their new settings. Towed systems are used for large fields.

Sprinkler-Hop Systems. Sprinklers are placed only at every second or third position along the laterals. When the irrigation depth has been applied, the sprinklers are disconnected and moved or "hopped" along the lateral to the next position. This is done without stopping the flow in the lateral because each sprinkler connection is fitted with a special valve that automatically stops the flow when the sprinkler is removed. After the two or three hops are complete, the lateral is moved to the next position. Normally only one lateral move or one sprinkler hop is required each day.

Side-Roll Systems

Side-roll or wheel-line systems use an aluminium or galvanized steel lateral as the axle of a large (1.5 to 2.0 m diameter) wheel. The wheels are spaced 9–12 m apart and allow the lateral to be rolled from one irrigation setting to the next. A small internal combustion engine normally is used to roll the whole lateral. The pipes must be strong, and rigid couplings are used to carry the high torque loads. The engine often is located in the middle of the lateral to reduce the torque. The small rotary sprinklers, spaced 9 to 12 m apart, are mounted on a special weighted swivel assembly to make sure they are always in an upright position after each move.

The mainline is laid along the side of the field. When irrigating, the lateral remains in one place until the water has been applied. The pump then is stopped and the lateral uncoupled from the mainline and drained (to reduce the weight) and rolled to the next position using the engine. A flexible hose connection to the hydrant allows the lateral to be moved over two or three sets with supply from the same hydrant.

This system is best suited to large flat rectangular areas growing low field crops. In heavy soils, the wheels may become bogged down in the mud.

Mobile Raingun Systems

Mobile raingun systems (also called *traveling sprinklers* or *travelers*) use a large rotary sprinkler operating at high pressure. The term raingun is used because of the large size of sprinkler used and its ability to throw large quantities of water over wide areas. They have become popular because of their relatively low capital cost and low labor requirements. They are well adapted to supplemental irrigation. Because of the high pressure requirements, they have high energy costs.

Rainguns normally operate at high pressure from 400 to 800 kPa with discharges ranging from 30 to 200 $m^3\,h^{-1}$. They can irrigate areas up to 100 m wide and 400 m long (4 ha) at one setting. Application rates vary from 7.5 to 25 $mm\,h^{-1}$. Information below should be complemented with other background literature [11, 94–99].

Hose-Reel Systems

The hose-reel machine has a raingun mounted on a sledge or wheeled carriage. Water is supplied through a semirigid hose that is flexible enough to be wound onto a large reel. The 200- to 400-m-long hose is used to pull the raingun toward the hose reel.

In a typical layout for a hose-reel system (Fig. 5.26), the mainline is across the center of the field. The hose-reel is placed close to the mainline at the start of the first run and connected to the water supply. The raingun is slowly pulled out across the field by a tractor and the hose is allowed to uncoil from the reel. The pump is started and the valve coupler is opened slowly to start the irrigation. The raingun then is pulled back slowly across the field by winding the hose onto the hose reel. Power to drive the hose reel can be provided by a water motor or, more often, by an internal combustion engine. At the end of a run, the hose reel automatically stops winding and shuts down the water supply.

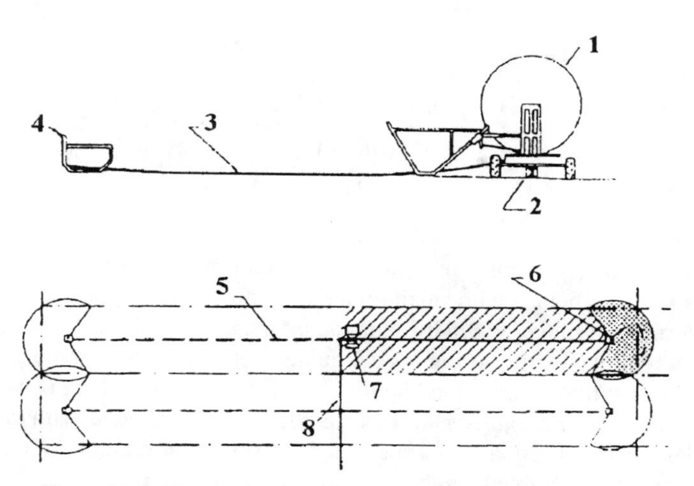

Figure 5.26. Hose-reel system: 1) hose reel, 2) valve coupler on the mainline, 3) semirigid hose, 4) raingun mounted on a sledge carriage, 5) towpath of the raingun, 6) initial position of the raingun to irrigate the dashed area, 7) field location of the hose reel, 8) mainline.

When the hose reel is located in the center of the field, it is rotated 180° and the raingun is pulled out in the opposite direction to start the next irrigation run. When irrigation is completed in this position, the hose reel and the raingun are towed by tractor to the next outlet along the mainline.

Hose-Pull or Cable-Drawn Systems

The hose-pull machine has a raingun mounted on a wheeled carriage. Water is supplied through a flexible hose up to 200 m long and 50–100 mm in diameter, which is pulled along behind the machine. The mainline is laid across the center of the field. A strip up to 400 m long can be irrigated at one setting of a 200 m long flexible hose.

The raingun carriage is positioned at the start of its first run at a distance equal to $\frac{1}{3}D_w$ from the field edge. The flexible hose is laid along the travel lane and connected to the raingun and the valve coupler on the mainline.

A steel guide cable on the sprinkler carriage is pulled out to the other end of the field and firmly anchored. The valve coupler is opened slowly to start the irrigation. The raingun carriage is moved either by a "water motor" powered from the water supply using a piston or turbine drive, or, more often, an internal combustion engine.

At the end of a run the carriage stops automatically and shuts down the main water supply to the raingun. Labor is required only to reposition the hose, cable, and machine to start the next run.

The pressure at the raingun determines the application rate. The forward speed of the machine controls the depth of water applied. Typical machine speeds vary from 10 to 50 m h^{-1}. The faster the machine travels, the smaller the depth of water applied.

The required machine speed, V_{tg} (m h^{-1}) can be calculated from

$$V_{tg} = 1000\frac{q_s}{DW}, \qquad (5.143)$$

where q_s is the sprinkler flow rate (m^3 h^{-1}), D is the desired irrigation depth (mm), and W is the width of the irrigated strip (mm).

The duration of operation for each set, t_i (h) is

$$t_i = L_f / V_{tg}, \tag{5.144}$$

where L_f is the length of the irrigated field (m).

Side-Move Systems

These are traveling systems that combine hose-reel machines with moving laterals. In place of a raingun, booms with small rotary sprinklers or sprayers extend out to each side of the portable carriage. When irrigating, the carriage is positioned at one end of the field and slowly pulled across using a steel guide cable and winch in the same way as the hose-pull raingun. A strip of land up to 70 m wide and 400 m long can be irrigated at one setting (2.8 ha). For light systems, a rigid hose can be used as with the hose-reel raingun. These systems require less operating pressure than raingun systems, but the instantaneous application rate is higher.

Rainguns

There are two types of these large rotary sprinklers: *swing-arm rainguns*, which are large-impact sprinklers and *water-turbine rainguns*, which are gear-driven sprinklers. Rainguns have sector stops to adjust for the desired circular arc to be irrigated.

Rainguns are fitted with either taper or ring nozzles. Taper nozzles normally produce a good water jet that is less affected by wind, and they have a slightly greater throw than ring nozzles. Ring nozzles, however, provide better stream breakup at low operating pressures. They are less expensive and provide greater flexibility in size selection. Typical nozzle diameters vary from 15 to 50 mm. Typical discharges and wetted diameters corresponding to common nozzle sizes are in Table 5.8. The trajectory angle for rainguns varies between 15° and 28°. Generally, the higher the angle, the larger the throw for a given operating pressure. Low angles (<20°) are preferred under windy conditions (>15 km h^{-1}). Because rainguns operate at high pressures, it is important that the jet of water leave the nozzle relatively undisturbed. Turbulence reduces the throw of

Table 5.8. Typical discharges q_s (m^3 h^{-1}) and wetted diameters D_w (m) for raingun sprinklers with 24° trajectory angles and tapered nozzles

Sprinkler Pressure (kPa)	Nozzle Diameter (mm) 20.3		25.4		30.5		35.6		40.6	
	q_s	D_w	q_s	D_w	q_s	D_w	q_s	D_w	q_s	D_w
415	32	87	51	99	75	111	—	—	—	—
480	35	91	55	104	81	116	109	133	—	—
550	37	94	59	108	86	120	117	139	153	146
620	40	97	63	111	92	125	123	143	162	151
690	42	100	66	114	96	128	130	146	171	155
760	44	104	69	117	101	131	137	149	179	158
825	46	107	73	120	105	134	143	152	187	163

Source: Adapted from [11].

the sprinkler. Modern rainguns have vanes, which "straighten" the flow and suppress turbulence.

Rainguns irrigate only part of a circle behind the machine. This ensures that the machine always moves on a dry towpath. The application depth profile is not uniform and varies across the strip irrigated by a traveling sprinkler as influenced by the effect of changing the wetted sector angle. The most uniform profile is with a sector angle $\omega = 240°$. The most commonly used is $\omega = 270°$, which is still fairly uniform. As ω is increased further, the uniformity of the profile decreases.

Towpath Spacing and Application Rate

The application uniformity of raingun sprinklers is affected by wind velocity and direction, jet trajectory, nozzle type, wetted sector angle, sprinkler profile characteristics, and overlap. Variations in operating pressure also affect uniformity. Under calm wind conditions (0–3.5 km h^{-1}), a towpath spacing of 80% to 90% of the wetted diameter produces good uniformity. Towpath spacing should be reduced about 5% for each 2 km h^{-1} of wind-speed increase, resulting in a towpath spacing of 55% when wind speed is expected to average 16 km h^{-1}.

For a part-circle gun sprinkler spaced to give sufficient overlap between towpaths, the application rate i_a (mm h^{-1}) is approximately

$$i_a = \frac{1}{1000} \frac{q_s}{\pi (0.9 D_w/2)^2} \frac{360}{\omega}, \qquad (5.145)$$

where q_s is the sprinkler discharge ($m^3\ h^{-1}$), D_w is the wetted diameter (m), and ω is the wetted sector angle (degrees).

Constant travel speed is required for uniform water distribution over the irrigated area. Traveler speed should vary no more than 10%.

Moving Lateral Systems

These systems have laterals that move continuously while applying water. There are three main types of systems: center pivot, lateral move, and side move (for complementary information see [11, 94–96, 99]).

Center-Pivot Systems

These systems consist of a single galvanized steel lateral that rotates in a circle about a fixed pivot point in the center of the field (Fig. 5.27). Lateral pipe diameters range from 100 to 250 mm. The lateral is supported using cables or trusses as much as 3 m above the ground on A-shaped steel frames mounted on wheels (Fig. 5.28). The frames are spaced approximately 30 m apart. Laterals vary in length from 100 to 800 m. A common lateral length is 400 m, which irrigates up to 50 ha.

Water is supplied to the center pivot by a buried mainline or directly from a well located near the pivot point. Water flows through a swivel joint to the rotating lateral and sprinklers. When irrigating, the lateral rotates continuously about the pivot, wetting a circular area. One revolution can take from 20 to 100 h depending on the lateral length and the amount of water to be applied. The slower that the lateral rotates, the more water that is applied. Typical applied depths vary from 5 to 30 mm. A center-pivot lateral is therefore a system that can effectively apply light, frequent irrigations.

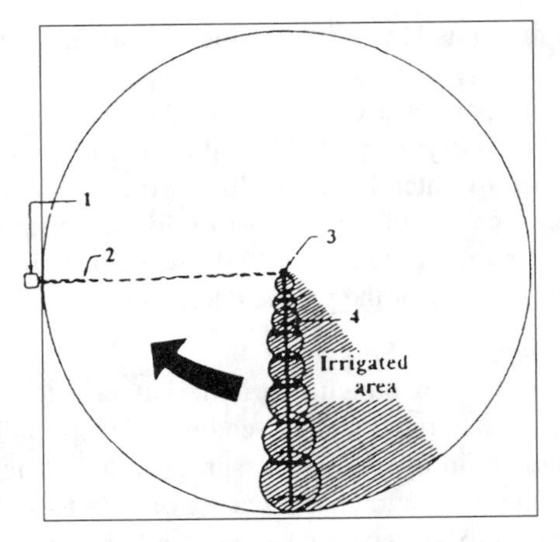

Figure 5.27. Center-pivot system: 1) pump, 2) mainline, 3) pivot center, 4) lateral.

Figure 5.28. Center-pivot lateral equipped with spray heads on top.

Because the lateral moves in a circle, uniform watering is achieved by linearly increasing the application rate toward the outer end of the lateral. This is done by varying either the nozzle size or the spacing of sprinklers. The first method uses equally spaced sprinklers with small nozzles close to the pivot and larger ones toward the outer end. The second method uses the same size of sprinkler but the sprinklers are placed closer together toward the outer end. This method can simplify maintenance because all of the sprinklers require the same spare parts.

A large gun sprinkler can be used at the outer end to extend the effective length of the lateral. End guns require a pressure of 350 to 700 kPa, which often requires a booster pump mounted near the end gun. When low-pressure spray or impact sprinklers are used, the booster pump is essential.

Both impact and spray sprinklers are used on center pivots. Impact sprinklers give a longer throw and thus fewer are required, and the instantaneous application rate is lower. Sprayers require 25% to 50% less pressure and thus require less energy. Often, spray nozzles are suspended from long tubes to spray water close to the canopy and out of the wind (Fig. 5.29).

The main objection to spray nozzles is that the wetted diameter and thus the instantaneous application width are much smaller than with rotary sprinklers. This results in very high application rates at the outer end (up to 100 mm h^{-1}), which often exceeds the soil infiltration rate. To prevent surface-water runoff, special cultivation practices can

Figure 5.29. Single-row drop sprayers on a center-pivot lateral showing a small wetted width.

Table 5.9. Range of normal operating pressures and associated wetted diameter D_w for sprinkler types and spacing configurations most commonly used on center-pivot laterals

Sprinkler Type and Spacing Configuration	Pressure Range (kPa)	D_w Range (m)
Low-pressure spray		
Single-row drop	70–200	3–9
Single-row top	70–200	6–14
On short booms	70–140	12–18
On long booms	100–170	20–26
Low-pressure impact		
Variable spacing	140–240	18–23
Semiuniform spacing	200–275	21–24
Medium-pressure impact		
Variable spacing	275–350	27–34
Semiuniform spacing	275–380	30–37
High-pressure impact		
Uniform spacing	380–450	40–50

Source: Adapted from [11].

be used. Conservation practices such as ridge tillage on row crops that preserves plant residue on the surface improves infiltration and slows runoff. However, this often cannot adequately prevent runoff and erosion in sloping fields. Reservoir tillage creates small basins that store water on the surface until it can infiltrate [86]. To decrease application rates, the sprayers can be mounted on booms extending out from the lateral. Summary information on pressure ranges and pattern widths for different sprinkling configurations is given in Table 5.9. A procedure to select the spray or sprinkler device that has a wetted diameter capable of satisfying the infiltration capacity of the soil and the surface storage is proposed by Allen [100].

Sprinkler pressure variations that occur as a lateral rotates on a sloping field cause discharge variations that are proportional to the square root of the operating pressure. Therefore, the water distribution uniformity from center pivots with low-pressure sprinklers operating on uneven topography may be poor unless the sprinklers are fitted with flexible-orifice flow-control nozzles or pressure regulators.

The required system supply rate Q ($m^3\,h^{-1}$) depends on the area irrigated A (ha) and the water requirements of the crop expressed as daily application depth D (mm day^{-1}):

$$Q = 0.42D \times A \tag{5.146}$$

with $A = \pi R_p^2/10{,}000$, where R_p is the center-pivot wetted radius (m). Center pivots often are designed with application rates less than peak crop water requirements. Thus, they operate continuously for much of the irrigation season.

Irrigation intervals and water applications per revolution depend upon the water-holding capacity of the soil, the rooting depth of the crop, and the infiltration rate of the soil. Light, frequent irrigations maintain more uniform soil water but result in higher evaporation losses and greater wear on the pivot drive mechanisms. The infiltration capacity of the soil often limits the allowed water application per pass. A review of infiltration under center-pivot irrigation is presented by von Bernuth and Gilley [101]. For soils with adequate infiltration and water storage capacity, 20 to 30 mm often is applied each rotation, resulting in typical irrigation intervals of about 4 days.

Each tower is driven by its own electrical motor (usually 0.5–1.0 kW). Hydraulic-powered systems seldom are used because speed control and maintaining tower alignment are difficult, and the system will only move when irrigating. The rotating speed of a center-pivot lateral is controlled by regulating the speed of the end tower V_p (m h^{-1}) given by

$$V_p = \frac{2\pi R_e}{t_r}, \tag{5.147}$$

where R_e is the length of the lateral (m) and t_r is the desired duration of one lateral revolution (h). The speed usually is set as a percentage on time because the constant-speed tower drive motors cycle on and off over short intervals to maintain the desired average speed. The on-time for each tower varies with the distance from the pivot. Typically the on/off cycle is 1 min. The lateral is kept in alignment between the end tower and the pivot point by special control devices activated by deflections created by misalignment. When a tower falls behind, the deflections activate the drive motor until the tower catches up.

One main advantage of this system is that it can be fully automated and controlled from a panel near the pivot or remotely from some office nearby. Time clocks are used to start and stop the machine and several safety devices are used for protection. For example, if the water pressure drops or one of the tower drives breaks down, the system will automatically stop irrigating and an alarm will alert the operator. Several center pivots covering large areas can be easily controlled and maintained by a few people, particularly when automated remote monitoring and control are used. This allows easy scheduling of irrigation, fertigation, and chemigation. The application of fertilizers (mainly nitrogen) and chemigation with center pivots is becoming popular.

Center pivots operate best on sandy soils that infiltrate water quickly and can support the heavy wheel loads from the towers. Traction problems may occur when irrigating heavy soils, especially when laterals are equipped with sprayers that give high application rates. Main problems in management relate to the light application depths, which allow neither for refilling the soil for large root depths of crops nor for leaching of salts. Low-infiltration-rate soils also limit use of center pivots, especially if equipped with high-application-rate spray heads. The use of center pivots in medium to heavy soils in arid or semiarid areas is marginal. They should not be used in saline arid conditions.

A primary limitation of center pivots is that they cannot completely irrigate a rectangular or square area. A pivot can irrigate only about 80% of a square field with no obstructions. The remaining corners must be irrigated by some other method, or left fallow. Pivot lateral extensions (corner systems) and end guns, designed to irrigate corners, reduce the unirrigated area but increase system cost and operational complexity.

Linear-Move Systems

Linear-move (or lateral-move) systems are similar in construction to the center pivot, with the lateral supported over the crop on towers. The primary difference is that the complete lateral continuously moves in a linear direction. The main advantage over center pivots is that they can completely irrigate a rectangular field. Also, water application rates are uniform along the lateral, resulting in simpler sprinkler design and lower peak application rates.

Water is supplied to the moving control tower by a flexible hose attached to a pressurized water supply or by pumping water from a small canal along the edge or the center of the field. The control tower is equipped with an engine, pump, and generator that pressurizes the water and supplies the lateral wheel motors with electrical power.

Guidance is provided by signals emitted by buried electrical wire, by an aboveground guide cable stretched along one of the field edges, or by a guide wheel that follows a small furrow. Antennas in the control tower sense the buried wire and transmit the signal to a guidance control box. Levers on the control tower are activated by the guidance cable or wheel and align the lateral to follow the line.

Linear-move laterals are equipped with sprayers or impact sprinklers but usually not end guns. Sprinklers are at uniform spacing and should have similar characteristics along the lateral. Application depth varies with the lateral speed. This system can be automated in the same way as a center pivot.

When the lateral reaches the far end of the field, it has to be moved back to the beginning. This means moving a heavy machine over recently irrigated land. On sandy soils this may not be a problem, but on fine-textured soils the towers may sink into the soil even when crawler tracks are used. It may be necessary to wait a few days to reposition the system. One operational sequence to avoid this problem is to divide the field into two parts. Irrigation starts on one edge and continues to the center of the field. The lateral then is moved dry to the other end where irrigation starts again toward the center of the field. Upon reaching the center, the lateral again is moved without irrigation to the edge to start the next irrigation from the initial position.

Linear moving systems, like center-pivot laterals, are mainly appropriate to apply light and frequent irrigations and thus have similar problems when irrigating heavy soils, saline soils, and in arid or semiarid climates. However, because of low peak application rates, they adapt to a wider range of soil conditions than pivots.

LEPA Systems

An adaptation of moving lateral systems that is becoming popular is LEPA [102]. The overhead sprinklers are replaced with bubblers on drops closely spaced [103]. Usually, a bubbler is positioned close to the ground (0.3–0.6 m) between alternate rows of the crop and the crop is precisely planted so that the rows follow the bubbler paths. The purpose of the system is to eliminate all wind drift losses and part of the surface evaporation loss. It essentially eliminates nonuniformity caused by sprinkler distribution patterns, but nonuniformity resulting from lateral start-stop can be substantial. LEPA systems can operate at very low pressures.

The main disadvantage of the system is that instantaneous water application rates are very high. Consequently, LEPA requires either very flat land with very high infiltration rates, or special reservoir tillage to create sufficient surface storage to store much of the applied water until it can infiltrate [86].

A variety of moving lateral sprinkler types and configurations have been developed that operate similarly to LEPA systems. These commonly use small spray heads on drops. Like the LEPA bubblers, the sprayers are positioned between alternate crop rows. The heads are higher off the ground than the bubblers, but still designed to be below the top of the crop. The purpose is to gain the advantages of LEPA systems with slightly lower instantaneous application rates. Reservoir tillage usually is required.

Issues on Irrigation Performance

The distribution uniformity DU [Eq. (5.119)] is determined by several design variables and can be expressed functionally by the following relation:

$$DU = f(P, \Delta P, S, d_n, \text{WDP}, \text{WS}), \tag{5.148}$$

where P is the pressure (kPa) available at the sprinkler, ΔP is the variation of the pressure (kPa) in the operating set or along the moving lateral, S represents the spacings (m) of the sprinklers along the lateral and between laterals or spacings between travelers, d_n is the nozzle diameter (mm), which influences the sprinkler discharge and the wetted diameter for a given P, WDP represents the water distribution pattern of the sprinkler, and WS is the average wind speed (m s^{-1}).

All of the above variables are set at the design stage. The designer, together with the farmer, first has to select the system according to the field, farming characteristics, and crops to be irrigated. Then the designer selects the sprinkler characteristics and spacings. At this stage, variables S, d_n, and WDP are set, and the average wind speed WS during operation has been forecasted. Then, the hydraulics calculations are performed to select pipe sizes, pump characteristics, and other system equipment. These computations produce values for the average pressure at the sprinklers and the respective variation along the laterals.

When sprinklers are selected, and thus the variables P, q_s, D_w, and d_n are defined, the main variables governing the distribution uniformity are the spacing S and the variation of pressure ΔP in the operating set. Excessive spacings often are observed as a cause of low system performance. Excessive spacing was observed in 65% of traveling guns and 70% of solid-set systems evaluated in France [104]. The same problem was observed in California, where other causes for low uniformity also include inappropriate nozzles and high pressure variation within the system [105].

It is commonly accepted that the variation in sprinkler discharges in an operating system should not exceed 10%; thus the variation in pressure must be no more than 20%. Lower ranges are desirable when high-value crops are irrigated. To achieve this, the hydraulics design has to be performed carefully.

The application efficiency e_a [Eqs. (5.115) and (5.117)] depends not only on the design but on management variables and it can be functionally described by

$$e_a = f(P, \Delta P, S, d_n, \text{WDP}, \text{WS}, I_c, i_a, t_i, SWD), \tag{5.149}$$

where, besides variables defined above, I_c are the intake characteristics of the soil (mm h^{-1}), i_a is the application rate of the sprinkler (mm h^{-1}), t_i is the duration of the irrigation event (h), and SWD is the soil water deficit before the irrigation event (mm).

Equation (5.149) shows that the application efficiency depends not only on design and distribution uniformity but also on the management of the system by the farmers, including irrigation scheduling.

Most of the parameters in Eq. (5.149) are controlled by the designer. The pressure head P also depends on the functioning of the pressurized supply system. Dubalen [104], referring to Midi-Pyrénnées, France, reports that only 52% of raingun systems and 58% of solid-set systems evaluated had the correct pressure at the sprinkler. This largely affects sprinkler discharges and makes it difficult to apply appropriate water depths. The duration of the application and the SWD are management variables controlled by the irrigator. However, when the farmer does not know the discharge or the application rate, he may not be able to correctly set the irrigation depth. The same study [104] shows that actual irrigation depths deviate by more than 20% from those claimed by the farmers in 46% of cases for raingun systems and 34% for solid-set systems.

Design Issues

High distribution uniformities and efficiencies can be achieved when hydraulic design produces the pressure required for the selected sprinkler discharge, with acceptable variation within the system, and when the layout design is based on optimal sprinkler spacings. The design of the system layout, including the selection of sprinklers and respective spacings is dealt with in several publications [11, 94, 96, 98, 99], as are the hydraulics of sprinkler systems [11, 96, 106–109]. Several engineering standards also support appropriate design (ASAE S261.7, S263.3, S376.1, S394, S395) [19].

Uniformity [DU or CU, Eq. (5.119) and (5.120)] resulting from adjacent sprinklers can be estimated by simulation of the overlap of four adjacent sprinklers in a rectangle, or three in the case of a triangle. The effect of wind can be considered by changing the original circular pattern into an ellipse with radii that vary with the wind speed [91].

The design also should aim at achieving high efficiencies. Considering that the application efficiency depends upon the uniformity, Keller and Bliesner [11] propose to compute the distribution efficiency DE_{p_a} (%) corresponding to a desired percentage p_a of the irrigated area receiving the target irrigation depth (or area adequately irrigated):

$$DE_{p_a} = 100 + (606 - 24.9p_a + 0.349(p_a)^2 - 0.00186(p_a)^3)\left(1 - \frac{CU}{100}\right), \quad (5.150)$$

where CU is the coefficient of uniformity [see Eqs. (5.120) and (5.123)].

The design application efficiency E_{p_a} (%) for any percentage p_a of the area adequately irrigated then is given by

$$E_{p_a} = DE_{p_a} R_e O_e, \quad (5.151)$$

where R_e is the effective fraction (0.1–1.0) of water applied, that is, after estimating the losses by evaporation and wind drift; and O_e is the effective fraction (0.9–1.0) of water discharged, that is, after estimating the losses from leakage.

R_e expresses how the application efficiency is affected by evaporation and wind drift as influenced by the size of the droplets. R_e can be computed from the foreseen values of reference evapotranspiration ET_0 (mm day^{-1}), wind speed WS (km h^{-1}), and the coarseness index CI:

$$R_e = 0.976 + 0.005\,ET_0 - 0.00017\,ET_0 + 0.0012\,\text{WS} - CI(0.00043\,ET_0 - 0.0018\,\text{WS} + 0.000016\,ET_0\,\text{WS}) \quad (5.152)$$

for CI such that, if $CI < 7$, let $CI = 7$; if $CI > 17$, let $CI = 17$.

CI is an empirical estimate for the size of droplets produced by the sprinkler as a function of the operating pressure P (kPa) and of the nozzle diameter d_n (mm):

$$CI = 0.032 P^{1/3}/d_n. \quad (5.153)$$

The gross application depth D then is calculated from the net irrigation depth I_n (see Section 5.1.6) and the efficiency E_{p_a}. When irrigating with saline water or in saline soils, a leaching fraction LF must be added (see Section 5.6):

$$D = \frac{I_n}{(E_{p_a}/100)(1 - LF)}, \quad (5.154)$$

where D is the gross application depth (mm), I_n is the net irrigation depth (mm), and E_{p_a} is the design application efficiency (%).

Field Evaluation

Field evaluation of sprinkler systems in operation can play a major role in improving irrigation performance. Field evaluations of set systems include the following observations [10, 11, 19, 110]:

- sprinkler spacings;
- pressure at the sprinkler nozzles at different locations;
- pressure variation ΔP along a lateral and within the operating set;
- sprinkler discharges and respective variations along a lateral and within the operating set;
- applied depths in a sample area on both sides of a lateral or between two laterals (using a grid of containers);
- duration of the irrigation;
- *SWD* at the time of irrigation;
- wind speed and direction;
- irrigator management practices, including MAD and the target application depth; and
- pressure and flow rate at the inlet of the system.

These observations provide for the computation of the actual distribution uniformity and coefficient of uniformity of the system, the actual and potential application efficiences, effective depth applied, percentage of area adequately irrigated, and system evaporation and wind drift losses. This information is used to provide recommendations to the farmer relative to

- improvements in the irrigation scheduling practices;
- changes needed in the system (spacings, sprinkler nozzles, number of laterals or sprinklers operating simultaneously, pressure regulation);

- improvements concerning the pumping system; and
- maintenance of the system and components.

Field evaluations for traveling guns [104] mainly focus on the traveler velocity, applied depths perpendicular to the traveler direction, and pressure available at the gun sprinkler and upstream. They also provide for computation of performance and management parameters and allow recommendations relative to irrigation scheduling and to improve the management and maintenance of the system.

In the case of center-pivot laterals [111], applied water depths are observed along a radius and, when possible, along a travel path. Observations of the pressures and discharge rates of the sprinklers along the lateral and of the end gun are also desirable. The velocity of the lateral and the wetted width also should be observed, as well as the ocurrence of runoff and erosion. Computed performance and management indicators generally do not include the efficiency because depths applied are very small. Recommendations should pay particular attention to the rotation velocity because velocity determines the applied depths. They also should include any need for changing sprinklers along the lateral, and the working conditions of the end gun.

Field evaluations are extremely important for helping farmers improve operation and management of sprinkler systems, to achieve higher performance and decrease water losses. When sprinkler systems are used to apply liquid fertilizers and agrochemicals, field evaluations also help farmers to improve fertigation and chemigation [112].

Sprinkler Irrigation Management, Fertigation, and Chemigation

As discussed above, the uniformity of sprinkler water application [Eq. (5.148)] essentially depends on the conditions fixed at the design phase and how close actual conditions are to those assumed at design. The only exception is wind speed WS, an intermittant problem that is avoided by not irrigating during windy periods. Good maintenance is needed to maintain potential uniformities. Pumps must maintain their intended discharge and pressure, pipes and joints must be maintained leak-free, and nozzle wear and sprinkler operation must be monitored and problems corrected. When conditions deviate from those at design (i.e., pumping-depth decline or sprinkler-nozzle changes), the performance of the whole system should be reanalyzed.

The application efficiency can be influenced greatly by management. When neglecting the variables relative to uniformity, Eq. (5.149) simplifies to

$$e_a = f(I_c, i_a, t_i, SWD). \tag{5.155}$$

The infiltration rate I_c may be influenced by soil management practices that avoid soil sealing and crusting or increase the intake rate of the soil. Among soil management practices are reduced tillage and direct seeding techniques [66]. High infiltration rates are particularly important in orchards and noncovering crops cultivated in soil where runoff and erosion can be critical, as is the case for tropical soils. The use of straw mulches and undertree vegetation are useful techniques.

The application rate i_a is fixed at the design phase but can be modified somewhat after field evaluations. However, in case of center-pivot and linear-move systems, where i_a is not constant but varies over time during the water application, the moving speed of the lateral can be adjusted to minimize the time during which i_a exceeds I_c [113]. Field observations are required to appropriately decide on such adjustments. In row crops, where i_a often exceeds I_c, furrow diking or reservoir tillage can be used to prevent runoff [86].

Variables t_i and SWD in Eq. (5.155) depend on the irrigation scheduling practices. The time duration of irrigation t_i is the variable controlling the irrigation depth. In set systems, the irrigation depth D (mm) is directly controlled by t_i (h) when the application rate i_a (mm h^{-1}) [Eq. (5.142)] is known ($t_i = D/i_a$).

For mobile rainguns and mobile lateral systems, the irrigation depth is controlled by the system velocity [Eqs. (5.143) and (5.147)], and thus indirectly by the time duration of water application.

Sprinkler irrigation scheduling methods depend on two main factors: the frequency of irrigations and the size and technological level of the farm. Large farms are more often in control of pressure and discharge, timing of applications, and duration of irrigations and have better conditions to invest in both control systems and irrigation-scheduling sensing tools [114]. Small farms are also in control of irrigation timing and duration when delivery is made on demand or when they manage their own water source. However, they may have less control over pressure and discharge at the farm hydrant when the delivery system is not fully responsive to the demand during peak periods. Under these circumstances, both pressure and discharge drop, inducing low performances and requiring an adjustment of the irrigation time [115].

System controls [116] include timers and/or volumetric control valves, which enable automatic regulation of the application duration in each unit of the system; solenoid valves, which allow the selection of the portion of the area to be irrigated next; and pressure and/or discharge regulation devices, which permit maintainance of uniform flow in the operating system. Where farm pumps are used, control systems also are applied to the pumping system, mainly to automatically control the "on" and "off" conditions and to protect the motor against short circuits and overcurrent.

Control systems may be connected with irrigation sensing tools to automatically start irrigation when sensed variables reach a preset threshold value. Information from field sensors may be stored and handled at a control center and irrigation systems controlled remotely from there. This technology mainly applies to farms having several center-pivot systems where remote control is applied to irrigation, as well as to reduce energy costs. On the contrary, irrigation control systems are uncommon in small and medium-size farms, where irrigation valves are operated manually.

Because portable systems are labor consuming, they should be used to apply large depths, often 50–90 mm. Fixed systems may be used for variable frequency of irrigations, for D varying from 10 to 90 mm or more. Depths applied by mobile raingun systems are limited by the speed range of the system, with D commonly ranging from 15 to 50 mm. Mobile laterals are appropriate for frequent, daily up to 4-day applications (7–25 mm), and seldom are used for infrequent applications.

For set systems and traveling guns, when the MAD is large, the methods for irrigation scheduling are the same as for surface irrigation: water balance simulation models, monitoring the soil water status, or a combination of both. In areas with a large number of small farmers, the use of simple irrigation calendars may be more appropriate. The successful use of a simple simulation model through a videotel system for advising farmers within a large irrigation project has been reported by Giannerini [117]. Other examples of sprinkler irrigation scheduling practices are given by Smith *et al.* [78].

When frequent irrigations are applied, scheduling strategies should be based on the replacement of the volume of water consumed during the preceding irrigation interval. For center pivots, special irrigation scheduling models have been developed with modules for water application, fertigation, chemigation, and energy management. An illustration is model SCHED [118]. The estimation of ET_c using the basal-crop-coefficient approach (see Section 5.1) is appropriate for performing the soil water balance for scheduling frequent irrigations.

Fertilizers and other chemicals can be applied with sprinkler systems and are commonly applied through center-pivot laterals. Systems must be equipped with a fertilizer and/or a chemigation tank and an injection pump. When nozzle diameters are very small (sprayers), a filter should be placed downstream of the injection point to prevent clogging by large fertilizer particles.

Fertigation and chemigation require high water-application uniformities [112]. Otherwise, the applied products are distributed unevenly in the field with negative impacts on crops where they are insufficient and, for the environment, where they are in excess. Details on application of fertilizers and agrochemicals with sprinkler irrigation are given in [72] and [119].

5.4.4 Microirrigation

General Aspects

Microirrigation, also called trickle or drip irrigation, applies water to individual plants or small groups of plants. Application rates are usually low to avoid water ponding and minimize the size of distribution tubing. The microirrigation systems in common use today can be classified in two general categories:

- *Drip irrigation*, by which water is applied slowly through small emitter openings from plastic tubing. Drip tubing and emitters may be laid on the soil surface, buried, or suspended from trellises.
- *Microspray irrigation*, also known as microsprinkling, by which water is sprayed over the soil surface. Microspray systems are used for widely spaced plants such as fruit trees.

A third type of localized irrigation, bubbler systems, uses small pipes and tubing to deliver a small stream of water to flood small basins adjacent to individual trees. Bubbler systems may be pressurized with flow emitters or may operate under gravity pressure without emitters. They are not common and will not be discussed here. Descriptions and design procedures for gravity bubbler systems are given by Rawlins [120] and Reynolds *et al.* [121].

Drip Irrigation

Drip irrigation systems are designed to slowly apply water to individual points. The spacing of the emitters, and thus the layout and cost of the system, depends on the crop spacing and rooting pattern and the soil characteristics. For closely spaced, water-sensitive crops with small root systems, the emitters may be as close as 20 cm apart along each crop row. For row crops with extensive root systems in fine-textured soil, the emitter spacing may be up to 1 m on alternate crop rows. Two emitters per vine is common in vineyards. For tree crops, four to eight emitters per tree may be sufficient. Thus, emitter and tubing requirements for drip systems vary from 2500 emitters and 3000 m of tubing per hectare for widely spaced trees to 20,000 emitters and 15,000 m of tubing per hectare for closely spaced vegetable crops.

Drip tubes are normally laid out in, or parallel to, crop rows. The tubing often is laid on the soil surface. In crops with trellising, such as vineyards, the tubing may be suspended from the trellising to keep it out of the way of tillage operations. In horticultural crops, thin-wall tubing (drip tape) may be placed a few centimeters below the soil surface and/or under plastic mulch to help hold it in place.

Drip tubing also can be placed up to 60 cm below the soil surface. Subsurface drip irrigation (SDI), when placed below tillage depths, allows the tubing to be left in place for several seasons. It also minimizes wetting of the soil surface and thus weed germination and surface evaporation. Subsurface drip usually requires specialized tillage operations and equipment, and also requires special equipment or management to prevent roots from growing into and plugging the emitters. It may require sprinkling to germinate new crops or to periodically leach salts. Specialized equipment with a hollow shank is used to inject drip tubing to the desired depth. A disadvantage of subsurface drip is that plugged emitters are not evident until the crop is damaged. Special care must be taken to prevent plugging.

Drip tubing is made from polyethylene. Tubing wall thickness varies from 0.1 mm to 1.3 mm. Thin-walled (0.1–0.4 mm) tubing, sometimes called drip tape, lays flat when not pressurized and usually is used for only one season. Thick-walled tubing can be used for several years and may be removed and replaced between crops. Drip-tubing diameters (outside) vary from 6 to 35 mm, but 16, 18, and 21 mm are the most common sizes. With moderate emitter discharge rates (4 L h^{-1} or 6 L h^{-1} m^{-1}), 18-mm thick-wall tubing or 16 mm thin-wall tubing can be used for run lengths up to 100 m without excessive pressure loss (<30 kPa). With thick-wall tubing, it is often less expensive to subdivide a long field with submains rather than use large-diameter tubing. Large-diameter thin-wall tubing recently has become available that allows long (400 m) run lengths.

Drip emitters can be either manufactured into the tubing or attached onto the line during field installation. In thin-wall tubing, the emitters are often an integral part of the tubing wall. In thick-wall tubing, and a few thin-wall products, in-line emitters are injection molded from rigid plastic and then integrally incorporated into the tubing during manufacture (Fig. 5.30). Drip tubing is available with in-line emitters spaced from 0.2 to 1.2 m apart. Drip tubing with porous walls is available but generally is not suited for

Figure 5.30. Drip tubing in-line emitter (upper) and on-line button pressure-compensating emitter (lower).

agriculture because of plugging and poor water distribution. Tubing with small holes punched or laser-drilled through the walls is not recommended for the same reasons. Tubing with in-line emitters can be installed quickly and easily.

On-line emitters are molded from rigid plastic. Their barbed inlets are inserted into small holes punched into the tubing in the field. Small-diameter tubing can be attached to on-line emitters to deliver the water away from the lateral tubing. An advantage of on-line emitters is that they can be spaced as desired, additional emitters can be added as

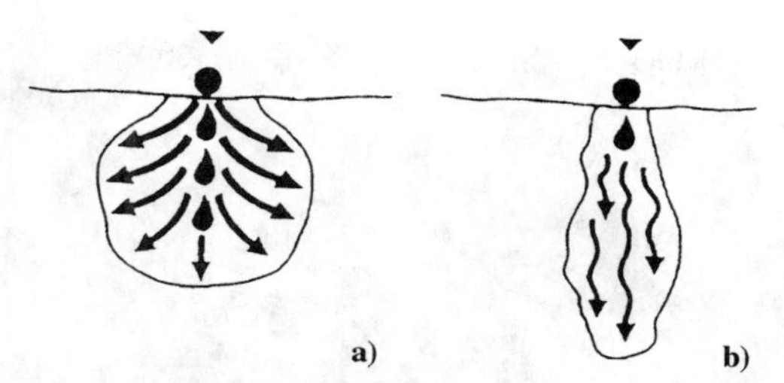

Figure 5.31. Wetted bulbs for (a) fine-textured soil and (b) coarse-textured soil.

perennial crops mature, and plugged emitters can be replaced. On-line emitters usually are not buried and generally are used only for perennial crops.

Drippers are point sources of water with small discharges, typically 1–8 L h^{-1} with 2 and 4 L h^{-1} being most common for thick-wall tubing and 1–2 L h^{-1} being common for closely spaced emitters in thin-wall tubing. The water enters the soil profile and percolates downward and outward. The result is a bulbous-shaped volume of moist soil. The size and shape of the "bulb" depend on the discharge rate of the emitter, the duration of application, and the soil type. Other conditions being constant, the bulb is more circular in fine-textured soils, in which the major water-moving force is capillary, whereas, in coarse-textured soils, the movement of water is caused largely by gravity and a narrower, deeper bulb results (Fig. 5.31). High discharge rate may increase the relative horizontal wetting. On the ground surface, wetted circles form, which may coalesce into continuous wetted strips. The soil surface between the tubing is kept dry.

Active plant roots are concentrated inside the wetted bulbs and therefore may be restricted in volume. Nevertheless, yields often (but not always) exceed those obtained by other irrigation methods. This is because, inside the bulb, light, frequent irrigations and fertilizer applications (fertigation) can maintain optimum growth conditions. Irrigation frequency varies from daily to every three or four days.

Salts in the soil move with the water toward the periphery of the wetted zone. Inside the wetted bulb, where the main root activity occurs, the salt concentration is generally low and not harmful to plants. However, lack of periodic leaching from irrigation or rainfall can result in harmful soil salt concentrations near the edges. These salt concentrations can be especially damaging during germination of new crops, or if rainfall moves the accumulated salts back into the active rooting area. Periodic large water applications are required to leach out salts. This must be done often with sprinklers.

Microspray Irrigation

Microspray emitters spray water over 2 to 6 m diameter circles or partial circles. Microspray was developed to wet a larger percentage of the rooting area of tree crops than was practical with drip irrigation. Microspray heads are small versions of low-pressure sprinkler heads. In the most common type, a small, vertical water jet from a small orifice nozzle impacts a deflector plate that diverts the jet into a horizontal pattern. The nozzle size determines the flow rate (at a given pressure) and the deflector-plate shape determines

Figure 5.32. Microspray irrigation showing a half-circle pattern sprayer and the microtubing connection.

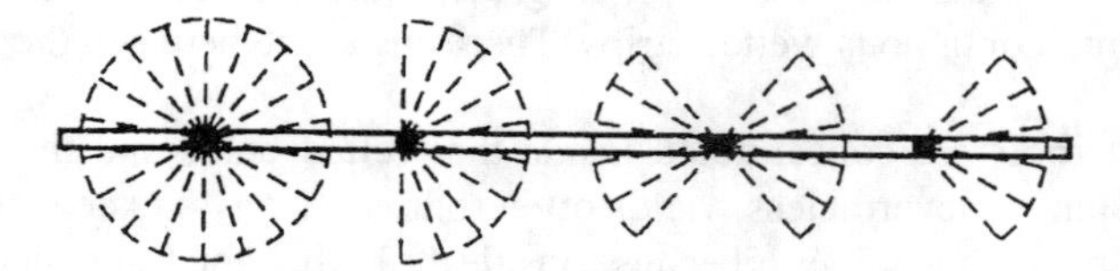

Figure 5.33. Typical microspray wetting patterns.

the spray pattern. A variety of flow rates and patterns are available. Microsprayers usually are inserted into the end of short pieces of 6 mm tubing and held upright on a stake (Fig. 5.32) or hung from suspended tubing. The small tubing is attached to the drip-tubing laterals with small plastic barbed connectors. In addition to sprayers, various types of spinner heads also are made, on which the jet impacts a small groove or channel that rotates, rapidly throwing the water in a circular pattern. Spinners usually produce a larger wetted pattern.

Microsprays are commonly used for widely spaced tree crops. They also may be used in greenhouses. In tree crops, one or two sprayers are used on each tree. The pattern usually is oriented so that the water does not spray on or at the base of the trunk, to reduce disease problems (Fig. 5.33). Flow rates from microsprayers are generally much higher than with drippers—typically 20 to 80 L h^{-1}. Although the spacing between sprayers may be 2 to 4 m, flow rate per unit length is usually higher than with drip systems, and so, larger tubing or shorter runs are required.

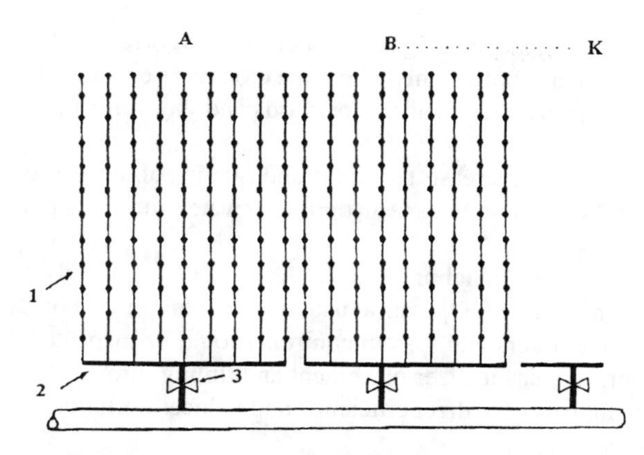

Figure 5.34. Layout of a microirrigation system with automatic valves to control irrigation for several sets: 1) lateral with emitters, 2) manifold, 3) automation valve.

Water Distribution, Filtration, and Control

Water is supplied to one end of the drip or microspray tubing through a manifold. These are usually buried PVC or polyethylene pipes, but in temporary systems, also may be flexible tubing laid on the surface. Flexible tubing or hose often is used to connect the surface drip tubing to buried manifold pipes. In large systems, main and submain pipeline networks deliver water to the manifolds. Manual or automatic valves are used to distribute water to the desired sets (Fig. 5.34). Pressure regulation may be required in the distribution system, as well as air release valves for SDI systems. The downstream ends of all drip or microspray lines must be accessible for flushing. Individual lines can be flushed manually by opening the downstream ends, or by pressure-activated flush valves. With SDI, sets of lines usually are connected to a flushing manifold so they can be flushed simultaneously.

Microirrigation systems require a pump to pressurize the water, one or more filters to clean the water, and valves to regulate pressure. Microirrigation water is applied under low pressure, usually in the range of 50 to 200 kPa. Thin-walled drip tubing usually allows no more than 80 kPa. For other systems, pressures are usually at least 100 kPa to reduce the effects of varying pressure on uniformity. Because emitter orifices or paths are small, filtration is nearly always required. Several types of filters are available for different types of contaminants. Fertilizer injection is recommended to gain the full benefit of precise management. Chemical injection to clean emitters is required with some water qualities and is recommended with SDI. Automation, through use of automatic valves and controllers, reduces labor requirements.

Adaptability and Suitability of Microirrigation

Based upon analyses provided by Pair *et al.* [96], Keller and Bliesner [11], and Papadopoulos [122], advantages and disadvantages of microirrigation systems include:

Advantages

Microirrigation has the following advantages:

- It has the potential to reduce irrigation water use and corresponding operating costs because water can be applied only where the crop roots develop. This is particularly true for widely spaced crops such as orchards and vineyards or for shallow-rooted crops.
- It has been shown to increase the yield and quality of some crops. This is most likely the result of maintaining near-optimum water and fertility conditions in the root zone.
- It can reduce the cost of labor because the systems need only to be maintained and managed, not tended. Operation usually is accomplished by automatic timing devices, but the emitters and system controls should be inspected frequently.
- A greater control over fertilizer placement and timing through fertigation with microirrigation improves fertilizer efficiency and reduces pollution hazards associated with fertilizers.
- It can reduce weed growth and the incidence of some diseases because foliage and much of the soil surface are not wetted. This reduces costs of labor and chemicals to control weeds and diseases and reduces related pollution hazards.
- It is less disruptive to field operations because the noncropped soil between crop rows is not wetted.
- Frequent irrigations maintain soil water content and keep the salts in the active root zone more dilute, making it possible to use more saline water than with other irrigation methods.
- Well-designed microirrigation systems can operate efficiently on almost any topography.
- Problem soils with low infiltration rates, low water-holding capacity, and variable textures and profiles can be irrigated efficiently.
- It usually requires lower operating pressure and thus less energy than sprinkler systems.

Disadvantages

Microirrigation has the following disadvantages:

- Equipment costs usually are higher than for surface irrigation systems and may be higher than for sprinkler systems.
- Equipment often is complex and requires frequent monitoring to ensure good performance.
- Energy costs to pressurize the system are higher than with surface irrigation.
- Because emitter outlets are very small, they can become clogged by particles of mineral or organic matter. Clogging reduces discharge rates and the water distribution uniformity; thus filtration is required in most cases. Iron oxide, calcium carbonate, algae, and microbial slimes may be problems requiring chemical treatment of the water to prevent clogging.
- Because microirrigation systems operate at low pressures, varying field topography can result in significant pressure variations and irrigation nonuniformity. Careful design and pressure regulation are required on undulating land.

- Some crops do not germinate well with drip irrigation and especially with subsurface drip. A second method of irrigation, usually portable sprinklers, may be required for germination.
- Salts may concentrate at the soil surface and between emitters and become a potential hazard. Localized salt accumulation can hinder crop germination. Light rains can leach accumulated salts downward into the root zone. Irrigation should continue on schedule unless adequate rain has fallen to ensure leaching of salts below the root zone.
- Salts also concentrate below the surface at the perimeter of the wetted bulbs. Too much drying of the soil between irrigations may allow the movement of water and salts back toward the inner bulb. To avoid this damage, irrigation must be frequent under saline conditions.
- If unexpected events (equipment failure, power failure, or water-supply interruption) interrupt irrigation, crop damage may occur quickly because roots use only a small volume of wetted soil. At least 33% of the total potential root zone should be wetted. Careful system maintenance and a secure water supply are a must.
- When a main supply line breaks or the filtration system malfunctions, contaminants may enter the system, resulting in emitter clogging. Secondary filters can be used to protect against these problems.
- Rodents and other small animals sometimes chew and damage polyethylene tubing. In some cases, animal damage disallows laying tubing or emitters on the soil surface. Burrowing rodents also may damage subsurface tubing. Rodent control may be necessary to reduce the problem.

Emitters

Knowledge about the emitter characteristics and functioning is important for design, operation, and maintenance of microirrigation systems. The information given below should be supplemented with other background literature [11, 98, 123, 124].

General

Emitters are designed to dissipate pressure and to discharge a small uniform flow of water at a constant rate. The ideal emitter

- has a large opening to minimize plugging,
- has low discharge rate to maximize lateral length,
- is relatively insensitive to pressure and temperature variations,
- is simple and relatively inexpensive to manufacture to high uniformity,
- is durable and long-lived (sunlight resistant; minimal wearing of orifices or moving parts), and
- is resistant to insect and rodent damage.

All of these criteria cannot be met with a single design. In fact, several criteria create opposing conditions (i.e., large opening and small discharge rate). Manufacturers have developed many emitter designs. Table 5.10 summarizes some of these basic designs and their advantages and disadvantages. A large majority of the emitters in use are labyrinth or pressure-compensating for drip and orifices for microspray.

Table 5.10. Typical emitter types and their characteristics

Path Type	Description	Merits[a]	Problems[b]	Typical Discharge Exponent x
Microtube	Long, small-diameter spaghetti tube; laminar flow.	a	2,5,6	0.7–0.8
Molded, long, smooth	Long, smooth, coiled or spiral passageway in a molded emitter body; laminar flow.		2,4	0.7
Vortex	Water enters tangentially into a chamber in which it spins and then exits through a hole on the opposite side.	a,b,c	3,4	0.4
Tortuous	Labyrinth or zigzag path; turbulent flow at some points in the passageways.	c,d,e		0.5–0.55
Porous pipe	Very small holes in the tubing itself sweat or emit water.		2,3,5,6	≥1.0
Pressure-compensating	Some type of flexible membrane, O-ring, or other design is used to reduce the path size at higher pressures; quality is highly variable among manufacturers.	Possibly b,c,d,e	Possibly 1,4,6,7	0–0.5
Multiple flexible orifice	Water passes through several orifices in flexible membranes; particles caught in one orifice will create backpressure, expanding the orifice and moving the particle through.	d,e	1, possibly 7	0.7
Orifice	A single, simple hole, typical of microsprayers.	a,b,c		0.5

[a] a, inexpensive; b, flow rate is insensitive to temperature changes; c, low manufacturing C_v (little variation between emitters); d, typically a large hole; e, less susceptible to plugging than other emitters with the same hole size.

[b] 1, expensive; 2, flow rate is sensitive to temperature changes; 3, typically a small hole; 4, relatively sensitive to plugging; 5, very sensitive to plugging; 6, large manufacturing C_v with some makes and models; 7, discharge characteristics of some makes and models may change after a few years.

Source: Adapted from [11].

Hydraulic Characteristics of Emitters

Hydraulically, most emitters can be classified as long-path, orifice, vortex, pressure-compensating, or porous-pipe emitters. The hydraulic characteristics of each emitter are related directly to the flow regime inside the emitter as characterized by the Reynolds number (Re). These flow regimes usually are characterized as laminar, $Re < 2{,}000$; unstable, $2{,}000 < Re \leq 4{,}000$; partially turbulent $4{,}000 < Re \leq 10{,}000$; and fully turbulent $10{,}000 > Re$.

The flow regime in an *orifice emitter* is fully turbulent. The flow rate is given by

$$q = 3.6AC_o\sqrt{2gH}, \tag{5.155}$$

where q is the emitter flow rate (L h^{-1}), A is the orifice area (mm^2), C_o is the orifice coefficient (usually about 0.6), H is the pressure head at the orifice (m), and g is the acceleration of gravity, 9.81 m s^{-2}. Because orifice flow is usually fully turbulent, small changes in fluid viscosity caused by fluid temperature changes usually do not affect emitter performance. *Short-path emitters* generally behave like orifice emitters. For twin-chamber tubing, with n_o external orifices for each orifice in the inner chamber, Eq. (5.155) also applies with appropriate modifications.

The flow in a *long-flow-path emitter* is through a small microtube. When the flow regime is turbulent, the emitter flow rate can be expressed as

$$q = 113.8A\sqrt{2gHd/fL}, \tag{5.156}$$

using the Darcy-Weisbach equation with q(L h^{-1}), d is inside diameter (mm), L is microtube length (m), and f is the friction factor (dimensionless). The cross-sectional shape of the conduit will affect the hydraulic characteristics. For laminar flow, the emitter discharge becomes proportional to H and the effects of fluid temperature changes on viscosity can cause significant flow variation.

One of the most popular types of emitters is the *tortuous path* or labyrinth emitter. It allows the maximum opening size for a given flow rate. For example, one manufacturer's tortuous-path drip emitter that discharges 2 L h^{-1} at 100 kPa has a minimum path dimension of 1.4 mm and a path length of 160 mm. An orifice to give the same flow would have an opening diameter of only 0.3 mm (Eq. 5.155) and be much more susceptible to plugging. Flow in tortuous-path emitters is usually turbulent and thus is insensitive to water temperature, and discharge is given by a equation similar in form to Eq. (5.156).

The *vortex emitter* (or *sprayer*) has a flow path containing a round cell that causes circular flow. The circular motion is achieved by having the water enter tangentially to the outer wall. This produces a fast rotational motion, creating a vortex at the center of the cell. Consequently, both the resistance to the flow and the head loss in the vortex emitter are greater than for a simple orifice having the same diameter. Flow rate is usually given by

$$q = 3.6AC_o\sqrt{2gH^{0.4}}. \tag{5.157}$$

Large openings that are less susceptible to clogging can be used. Variations in emitter operating pressures due to elevation differences and pipe friction cause smaller variations in the discharge from vortex emitters.

Pressure-compensating emitters attempt to overcome the hydraulic constraints imposed by orifice or long-flow-path emitters and to provide a constant emitter flow rate. Usually, an elastic material, which changes shape as a function of pressure, is used separately or in combination with orifices or small-diameter conduits. These emitters usually allow only small changes in emitter flow rate as pressure is changed within a given design range. Pressure-compensating emitters allow the use of smaller lateral pipe diameters, longer laterals, and/or fewer manifolds. Pressure-compensating emitters may be the only way to achieve uniform water application when slopes are steep or when the topography is hilly and uneven.

Automatic flushing emitters are less susceptible to clogging. In these drippers, the flow passages open more widely at low pressures than at the normal operating pressures. This results in high flow rates, which flush the system and wash away any deposits that may otherwise clog the emitters. *On/off flushing emitters* flush for only a few seconds each time the system is started and again when it is shut off. *Continuous flushing emitters* are constructed so that they can eject relatively large particles during operation. They do this by using relatively large-diameter flexible orifices in series to dissipate pressure. Particles larger than the diameter of the orifices are ejected by a local increase of pressure as the particles reach each flexible orifice.

Emitter flow equation

Emitters flow rates are described for design purposes by experimentally determining flow rate as a function of operating pressure (Fig. 5.35). This empirical *emitter-flow equation* is

$$q = K_e P^x, \tag{5.158}$$

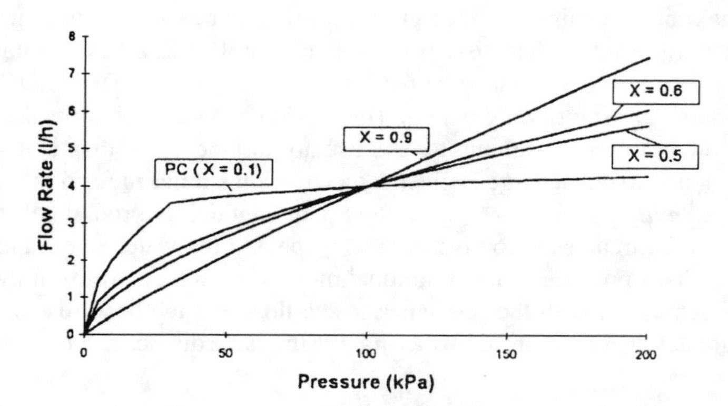

Figure 5.35. Discharge-pressure curves for typical types of emitters.

where q is the emitter flow rate (L h^{-1}), K_e is the proportionality factor that characterizes the emitter dimensions, P is the operating pressure (kPa), and x is the *emitter discharge exponent*, which characterizes the flow regime. The coefficients K_e and x are determined by plotting q versus P on a log-log plot. The slope of the straight line is x, and the intercept at $P = 1$ is K_e:

$$x = \frac{\log(q_1/q_2)}{\log(P_1/P_2)}. \tag{5.159}$$

Actual discharge/pressure relationships may differ significantly from those given by the manufacturers [125]. Reliable information on x and K_e is often available from laboratories where irrigation equipment is tested.

Low values of x (low sensitivity to pressure variations) allow the use of long laterals or small lateral diameters. In addition, the performance of drippers laid along steep slopes is improved. As x approaches zero (pressure-compensating drippers), the discharge varies little with variations in pressure. Good pressure-compensating emitters should have x values below 0.1 over the expected pressure range. For *laminar flow emitters*, x is close to 1; therefore, the variations in operating pressure should be held within about ±5% of the desired average. For *turbulent-flow emitters*, $x = 0.5$, and the pressure-head variation should be within about ±10% of the desired average.

Sensitivity to Temperature

Emitter flow will vary with temperature if the flow cross sections vary with the thermal expansion and contraction of the emitter material. With long-path, laminar-flow emitters, flow also varies with the viscosity of the water, which changes with temperature. Temperature effects can be important because the temperature of water flowing slowly through polyethylene laterals lying in the sun can increase substantially (>20°C) from the head to the tail end. Manufacturers should give information on the temperature effects on emitter flows. Laminar-flow emitters should not be used where temperatures vary through the system. Information on the temperature discharge ratio (TDR), relating the emitter discharge at high temperature to the standard emitter discharge at 20°C, is given by Keller and Bliesner [11].

Sensitivity to Clogging

Two critical parameters affecting emitter clogging susceptibility are the minimum flow-passage dimension and the velocity of the water through the passage. The relation between the passage cross section and the susceptibility to clogging is *very sensitive* (<0.7 mm), *sensitive* (0.7–1.5 mm), and *relatively insensitive* (>1.5 mm) for continuously flushing emitters. Keller and Bliesner [11] give information on minimum flow-passage dimension for main types of emitters. For microsprinklers in Florida citrus orchards, Boman [126] found that plugging decreased about 50% when orifice diameter was increased about 30%. Velocities of water through the emitter passage ranging from 4 to 6 m s^{-1} generally result in reduced clogging.

The manufacturer's recommendations for filtration also give an indication of the emitter's sensivity to clogging. The greater the sensivity, the finer the recommended filtration. The following classifications and filter size (μm) recommendations used in France [98] are

Extremely sensitive	filters < 80 μm
Very sensitive	filters ~ 80 μm
Sensitive	filters ~ 100 μm
Low sensitive	filters ~ 125 μm
Very low sensitivity	filters ~ 150 μm

Coefficient of Manufacturing Variation

The *coefficient of manufacturing variation for an emitter*, C_v, is used as a measure of the anticipated variations in discharge for new emitters. The value of C_v should be available from the manufacturer. It also might be available from independent testing laboratories and can be measured from the discharge data of a sample set of at least 50 emitters operated at a reference pressure:

$$C_v = \frac{\sqrt{\left(q_1^2 + q_2^2 \cdots + q_n^2 - nq_a^2\right)/(n-1)}}{q_a} \tag{5.160}$$

where $q_1, q_2, \ldots, q_n$ are individual emitter discharge rates (L h^{-1}), n is the number of emitters in the sample, and q_a is the average emitter discharge rate for the sample (L h^{-1}). Manufacturing variability can be classified in accordance with Table 5.11. Significant differences between C_v values given by the manufacturers and those obtained in independent tests often occur [125]. Many emitters are available with C_v in the range of 0.03 to 0.05.

System Layout and Components

Basic components of a microirrigation system are the pump, filtration equipment, controllers, main pressure regulators, control valves, water-measuring devices, and chemical injection equipment, which usually are centrally located at the *pump/filtration station*; the *delivery system*, including the main and submain pipelines that transfer water from

Table 5.11. Classification of emitter coefficient of manufacturing variation, C_v

Classification	C_v Range
Point source	
Excellent	<0.05
Average	0.05 to 0.07
Marginal	0.07 to 0.11
Poor	0.11 to 0.15
Unacceptable	>0.15
Line source	
Good	<0.10
Average	0.10 to 0.20
Marginal to unacceptable	>0.20

Source: ASAE EP405.1 [19].

the source to the manifolds, which also may have filters, pressure regulators, and control valves; the *manifolds*, which supply water to the laterals; and the *laterals*, which carry water to the emitters. Figure 5.34 shows the layout of a typical system.

Pump/Filtration Station

Elements that may be required at the central supply point include (see [11, 98, 123, 127–130]):

- a reservoir to store water if the water supply varies in rate or time from the system requirements;
- a pump to pressurize the water (unless the water supply is adequately pressurized);
- a backflow prevention valve to prevent the backward flow of water that may contain chemicals or fertilizers into the pump or main supply system or well;
- a chemical injection system for injecting fertilizer and chemicals into the microirrigation system;
- primary, and possibly secondary, filters to clean particulates from the water;
- control valves to manually or automatically regulate flows or pressures;
- an air-release/vacuum relief valve, located at a high point, used to release any air before it enters the delivery system and relieve vacuums following pump shutdown or valve closure; and
- a flow measurement device.

Filtration. Filters are critical to reduce clogging of drippers by solid particles suspended in the water. Prefiltration also may be required to prevent damage to the pump and valves. Burt and Styles [124] give details about filtration needs and designs.

Prefiltration usually is accomplished by settling in a reservoir and/or some type of prescreening. Water residence time in reservoirs should be 1–2 h to settle out all sand-sized particles and a portion of the silts. Reservoirs may introduce organic contaminants such as algae into the water supply. Prescreens include manual or automatically brushed trash racks and coarse screens, self-cleaning gravity overfall screens such as turbulent fountain screens [61], and manual and self-cleaning cylindrical screens for pump intakes.

Four main types of filters are used with microirrigation systems. The *vortex sand separator* (hydrocyclone) can be used to remove dispersed solid material, particularly sand. The hydrocyclone is a conical container, wider at the top. Water enters tangentially at the top and flows down at a high rotational velocity. This pushes the solid particles against the wall of the container from where they are carried downward toward the collecting chamber. The clean water at the bottom reverses direction and flows upward axially through the center of the container and out the discharge line. The accumulated sand is discharged periodically from the lower-end chamber. Sand separators should not be used as the final filter because it removes only a portion of the particulates.

Cylindrical *screen filters* use finely woven screen to prevent passage of particles. Screens work best for mineral particles (sands and silts) and for water without a large particulate load. The screens usually are cleaned by flushing water through the upstream chamber. However, the flow velocity across the screen surface is seldom sufficient to clean the screen. More thorough cleaning requires backflushing clean water through the screen, or by manually removing the screen cylinder and cleaning it. Screen filters are available with an automatic vacuuming mechanism that cleans the screen.

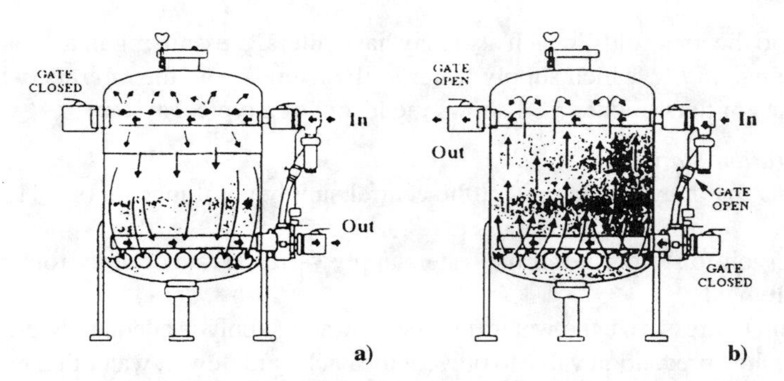

Figure 5.36. Media fillters: (a) Normal operation; (b) reverse flow for filter cleaning.

Disc filters comprise a series of grooved discs that are stacked together. When pressed together, the rings form a cylindrical filtering body. They work well for removing mineral and organic matter. They are cleaned by separating the disks and washing them off. Self-cleaning disc filters are available that automatically release the tension on the discs and backflush water through them.

Media filters use sand and gravel in tanks to filter out both mineral and organic particles (Fig. 5.36). The sand is sized to achieve the required filtration. Media filters are recommended when the water is relatively dirty. They are cleaned by diverting the inlet and backflushing clean water through the media. Effective cleaning requires sufficient backflush water velocity to lift the sand media without washing it away.

Filters are designed on the basis of flow rate and allowable pressure drop, the amount and type of suspended particles in the flow, and the size of particulates that are allowed to pass. Some self-cleaning screen and disc filters require as much as 220 kPa pressure to operate effectively, and may require downstream pressure-sustaining valves.

Emitter manufacturers should specify the filtration size required. Typically, the system must filter to at least one-fifth, and preferable to one-tenth, the smallest opening in the emitter. For example, water for a microspray with a 0.4-mm orifice would need to be filtered to at least 80 μm. Common filters for microirrigation filter to between 80 and 150 μm.

When water carries a large amount of suspended material, the filter should clean automatically. Automatic cleaning can occur in response to pressure loss across the filter and/or at preset time intervals. In media and disk filters, the flow through the filter must be reversed, and so, multiple filters must be used in parallel so that, as the filters are cleaned in turn, the others can operate normally to continue to discharge to the fields and to supply clean backflush water.

Where the water is relatively dirty, use of two filters in series is desirable. This prevents the microirrigation system from being ruined if one of the filters fails. A common combination is to have an automatic, self-cleaning media filter backed up by a manual screen filter. The backup filters also can be placed at the inlets to the manifolds in the field, to protect against particulates that might enter the pipelines during repairs or from scale or algae that might originate in the pipelines.

Filters do not, as a rule, completely solve the problem of clogging. Microirrigation laterals should be flushed periodically to remove accumulated fine particles. Calcium carbonate deposits in the system can be dissolved with a solution of nitric or hydrochloric acid (2 to 5 L acid per m^3 water). Deposits of organic material can be removed by flushing the system with water for about 15 min and then with compressed air under high pressure. When surface waters are used, microorganisms and small algae may pass through the filter and later accumulate and grow within the system. Treatment with chlorine (1 to 5 mg L^{-1}) continuously or periodically will prevent algae growth.

Some substances in the water, such as iron and manganese, pose special clogging problems [124]. An analysis of irrigation water for inorganic constituents is recommended before design of the filtration/chemigation system.

Chemical Injection. Fertilizer and other soluble chemicals such as acids and chlorine can be added to the pressurized water by several methods. The simplest method is to bypass a small portion of the flow around a fitting that creates a small head loss or pressure differential (<5 kPa), through a chemical storage tank. As the water flows into the tank, the solution is forced out the opposite end and into the system. A valve regulates the flow rate. The tank must be able to hold the maximum expected system pressure. Depending on the density and mixing of the solution, the chemical concentration of the tank outflow will decrease with time. This simple method is used only if the concentration of the added chemical can vary with time. A variation on this pressure-tank concept is to pressurize the tank with a compressed gas (usually N_2) to force the solution into the irrigation system.

Electrically driven injection pumps are an accurate method for adding chemicals at a constant rate. The flow rate of these diaphragm or piston pumps is regulated with a motor speed controller, or by mechanical adjustment of the piston displacement. The pump materials must be compatible with the chemicals used. Good-quality injection pumps are relatively expensive and require good maintenance, but they deliver an adjustable, constant flow and require no pressure loss in the water delivery system. Hydraulically driven injection pumps are also available. Some hydraulic pumps can automatically adjust flow rate proportional to the system flow rate to maintain constant chemical concentrations.

The third method to inject chemicals is through a venturi. All or a portion of the flow is diverted through a venturi. High-velocity flow through the venturi throat, such that the flow velocity head exceeds the water pressure, creates suction that draws solution from the chemical tank. Chemical injection venturis are available in a range of sizes and configurations. A large venturi in the delivery line creates a substantial (15%–30%) pressure loss. A valve in the delivery line to create the bypass flow through a small venturi is simple and inexpensive, but requires more than 70 kPa pressure drop in the water delivery system, and thus can be very expensive in terms of energy cost. A less-energy-consuming alternative is to use a small booster pump to bypass a small portion of the water through the small venturi.

Fertilizers and chemicals that will not cause filter damage should be injected upstream of the final filter in case particulates are introduced with the solution. Chemical application should begin after the lateral lines are fully pressurized and cease with sufficient irrigation time remaining to flush the entire system with clean irrigation water.

Control/Monitoring Devices. *Pressure regulators* are required if the inlet pressure to the system may vary beyond an acceptable range. Reasons for varying pressure include fluctuating water supply pressures, such as fluctuating well-water depths, and varying system flow rates. Constant system pressure will ensure constant irrigation application rates to all manifolds. Pressure regulation is critical if the pressure can exceed the burst pressure of system components. A pressure relief valve can protect against system overpressure.

If system hydraulics or topography result in nonuniform pressure at the various manifolds, regulators may be required at the manifold inlets. Pressure-compensating emitters may eliminate the need for system pressure regulation. Pressure regulators normally require 40 to 80 kPa pressure loss to operate properly and thus may increase the system pressure requirements [116].

Microirrigation allows precise water application. *Automatic control* of microirrigation provides precise water application without high labor. Automatic control of irrigation times can be accomplished in a number of ways. Commonly, a master-control unit is installed at the central pump/filtration station and is connected via buried wires to field stations located along the main pipeline. Each field station in turn controls a number of automatic valves (fitted with solenoids) at the inlets to submains or manifolds. The system automatically activates successive groups of manifolds to apply any desired amount of water and fertilizer. The control unit records any problems, such as pump failures, that may occur.

Flow measurement is important to ensure that the system is operating as designed. Decreased flow rate is often the first indication that emitters are plugged. Increased flow may indicate system failures such as lateral breaks. Flow totalizers document power outages. Flow should be measured at the pump/filtration station, and may be measured at submains. Because microirrigation water is clean, several types of mechanical and electronic devices can be used. Pressure measurement also is recommended to help verify proper system operation.

Distribution-Pipe Network

A manifold with its attached laterals is the basic system subunit and the last control point in the system. Manual valves, automatic valves, pressure regulators, filters, and flow measurement can be provided at manifold inlets. These controls also can be provided upstream at bifurcations between the main and submain units. Upstream from pressure control points, the allowed pressure variation for the subunit does not affect pipe-size selection. Therefore, the selection of pipe sizes for the main water supply lines should be based primarily on the economic trade-off between energy costs and costs for pipe and installation.

Normal operating pressures being low, the pipe pressure class rating will be determined more by structural strength than by system requirements. Air-vacuum release and pressure-relief valves should be considered and incorporated into the main system at the appropriate locations. Means to flush and drain the pipelines also should be provided.

Design Principles

Emitter Placement

The first decision in designing a microirrigation system is selecting the type and placement of emitters. This will depend on the crop (annual or perennial, spacing, rooting extent, sensitivity to water and nutrient stress, economics), the soil (permeability and water-holding capacity), and the cultural practices (tillage and harvesting operations, availability and cost of labor). For example, for annual row crops, either portable surface drip, subsurface drip, or disposable thin-wall tubing may be used, depending on the economics and practices. For perennial crops, the microirrigation system can be above- or belowground and permanent. For trellised perennial crops, such as vineyards, the drip laterals can be hung from the trellising.

The amount of the soil that must be wetted is the most important factor in choosing emitter spacing. In general, less wetting of the soil surface reduces system costs and evaporation losses. However, experience with microirrigation systems increasing shows that, with some crops, a substantial portion of the potential rooting extent should be wetted to reduce risk and achieve maximum yields. This has motivated the recent trend from drip to microspray irrigation of tree crops. A reasonable objective of design for widely spaced crops, such as vines, bushes, and trees, is to wet between one-third and two-thirds of the cropped area. For water-stress-sensitive row crops, at least two-thirds of the area should be wetted, whereas less than half may be adequate for drought-tolerant crops, especially if significant rainfall is expected. If the soil or water contains significant salts, salinity management may be a critical factor in emitter placement. Salts cannot be allowed to build up to damaging levels in the active root zone.

It is difficult to accurately predict the horizontal water movement from a point drip source. Keller and Bliesner [11] propose the following empirical equation to compute the wetted width or diameter from a point source emitter:

$$D_w = C_2(V_w)^{0.22}\left(\frac{K_s}{q}\right)^{-0.17}, \tag{5.161}$$

where D_w is the subsurface wetted diameter (m), V_w is the volume of water applied (L), K_s is the saturated hydraulic conductivity of the soil (m s^{-1}), q is the point-source emitter discharge (L h^{-1}), and C_2 is an empirical coefficient, 0.031, for these units.

Mechanistic models have been developed to simulate the soil water distribution around a point source. The two-dimensional and three-dimensional Richards equation often is utilized. However, these models require knowledge of the $K(h)$ and $h(\theta)$ curves (see Section 5.2), which are not available for most of applications. An analysis comparing several modeling approaches is presented by Angelakis *et al.* [132].

Simple field tests are the most reliable way to determine wetting diameter. They consist of selecting operating emitters at a few representative sites and then measuring the resulting wetting patterns.

The diameter of soil surface wetted by an emitter usually is about three-quarters as large as that measured at a depth of 30 cm. On sloping fields the wetted pattern may be

Table 5.12. Estimated wetted diameter D_w from a 4 L h^{-1} drip emitter by degree of soil stratification

Soil or Root Depth and Soil Texture[a]	Wetted Diameter (m)		
	Degree of Soil Stratification		
	Homogeneous	Stratified[b]	Layered[c]
0.75-m Depth			
Coarse	0.5	0.8	1.1
Medium	0.9	1.2	1.5
Fine	1.1	1.5	1.8
1.5-m Depth			
Coarse	0.8	1.4	1.8
Medium	1.2	2.1	2.7
Fine	1.5	2.0	2.4

[a] *Coarse* includes coarse to medium sands; *medium* includes loamy sands to loams; *fine* includes sandy clay to loam to clays (if clays are cracked, treat as coarse-to-medium soils).
[b] Soil of relatively uniform texture, but having some particle orientation or some compaction layering that gives higher horizontal than vertical permeability.
[c] Soil that changes in texture with depth and in particle orientation as well as being moderately compacted.
Source: Adopted from [11].

distorted in the downslope direction, but the actual wetted diameter will be similar to that on flat ground.

Table 5.12. gives estimates of the wetted diameters D_w of a standard 4 L h^{-1} emitter for different soil conditions and wetting depths. Values in Table 5.12 are based on daily or every-other-day irrigations that apply volumes of water sufficient to slightly exceed the crop's water-use rate. Almost all soils are either stratified or layered to some extent. However, assuming stratification or layering is risky. This must be determined by actual field checks.

Once the desired wetting pattern is established, the lateral spacing and placement and emitter spacing must be determined. In row crops, the lateral spacing decision generally will be between a few alternatives, for example, one or two laterals per bed or a lateral for each or alternate rows. In tree and vine crops, the decision usually will be between one or two lateral lines per crop row. Many growers have found that one drip lateral per vine row is adequate, but two drip laterals or one lateral with microsprays per tree row give better production. Emitter spacing then should provide the required wetting. The spacing of emitters (S_e) along the lateral should be about 80% of the wetted diameter given in Table 5.12 ($S_e = 0.8D_w$) to ensure full contact between adjacent wetting bulbs.

Lateral depth placement will depend on crop, soil, and cultural practices. Laterals often are suspended in vineyards and laid on the surface for trees. The disadvantage of on-ground or aboveground laterals is that they may be damaged by rodents or other animals, or by farming activities. The advantages are ease of placement, monitoring, and repair.

Permanent subsurface drip must placed below tillage depth. Deep placement may necessitate use of sprinklers to germinate the crop. In general, disposable thin-wall tubing is placed 40 to 100 mm below the surface, whereas permanent tubing is placed 200 to 500 mm deep.

Uniformity

Several parameters have been used to describe water distribution uniformity of microirrigation systems [see Eq. (5.119) to (5.125)]. Emission uniformity (EU, %), is used primarily to describe the predicted emitter flow variation along a lateral line:

$$\mathrm{EU} = 100[1.0 - 1.27(C_v/\sqrt{N})](q_n/q_a), \tag{5.162}$$

where C_v is the coefficient of manufacturing variation [Eq. (5.160)], N is the number of emitters per plant, q_n is the minimum emitter discharge rate (L h^{-1}) computed [Eq. (5.158)] from the minimum pressure along a lateral in the subunit or in a system, and q_a is the average or design discharge rate (L h^{-1}). The ratio q_n/q_a expresses the relationship between the minimum and average emission rates resulting from pressure variations within the subunit or system. The factor $(1 - 1.27C_v/\sqrt{N})$ adjusts for the additional nonuniformity caused by manufacturing variation between individual emitters. The EU values recommended in ASAE Standard EP405.1 [19] for different site conditions are presented in Table 5.13.

Some authors [15] use the statistical uniformity U_s [Eq. (5.125)], defining the coefficient of variation of emitter flow, V_q, as

$$V_q = \left[\sum_{i-1}^{n} \frac{(q_i - q_a)^2}{n-1}\right]^{1/2} \Bigg/ q_a, \tag{5.163}$$

where q_i are the observed or simulated emitter discharges and q_a is the average discharge. It is assumed that this coefficient of variation results from different causes: manufacturing variation, pressure variation, and emitter plugging or clogging. The various sources of nonuniformity can be combined such that the relative importance of each on the overall

Table 5.13. Recommended ranges of design emission uniformities EU [19]

Emitter Type	Spacing (m)	Topography	Slope (%)	EU Range (%)
Point source on perennial crops	>4	Uniform	<2	90 to 95
		Steep or undulating	>2	85 to 90
Point source on annual or semipermanent crops	<4	Uniform	<2	85 to 90
		Steep or undulating	>2	80 to 90
Line source on annual or perennial crops	All	Uniform	<2	80 to 90
		Steep or undulating	>2	70 to 85

Source: [19].

variability can be determined:

$$V_q = \frac{\left(V_p^2 + V_k^2 + x^2 V_h^2\right)^{1/2}}{N^{1/2}}, \tag{5.164}$$

where N is the number of emitters per plant, V_p is the coefficient of variation due to emitter plugging, V_k is the coefficient of variation of the emitter discharge coefficient K_e [Eq. (5.158)], x is the characteristic exponent [Eq. (5.158)], and V_h is the coefficient of variation of the hydraulic pressure.

The distribution uniformity DU [Eq. (5.119)], based on the low-quarter application relative to the average application; also can be used for microirrigation. It is practical when based on field measurements for complete systems and is useful to predict irrigation requirements based on meeting low-quarter needs.

Application Depths and System Capacity

As with other irrigation systems, the microirrigation system must be designed with sufficient capacity to meet the water requirements of the crops to be irrigated, plus allowances for leaching salts, nonuniform water application, and system downtime for repairs and cultural practices. Crop water requirements can be determined as indicated in Section 5.1. For design purposes, they should be based on peak monthly water-use values. The basal-crop-coefficient approach (Section 5.1.5) should be used because it appropriately takes into consideration the effects of wetting only a portion of the soil surface on the soil evaporation component. Once the daily net irrigation requirement I_n (mm day^{-1}), is determined for the local climate and expected crop mix, it must be adjusted for other requirements before calculating system capacity.

For moderate to low water salinity values, leaching usually can be applied early or late in the season when it will not impact system capacity requirements. For high-salinity waters, a continuous leaching fraction (LF) between 5% and 20% may be required.

To take into consideration the unavoidable percolation, a transmission coefficient T_r must be considered [11] for shallow, coarse-textured soils. The daily gross application depth D (mm day^{-1}) is then

$$D = \frac{I_n}{(\mathrm{EU}/100)(1 - \mathrm{LF})} T_r, \tag{5.165}$$

where I_n is the daily net irrigation requirement (mm day^{-1}), EU is the emission uniformity (%), LF is the leaching fraction (0.05–0.20), and T_r is the transmission coefficient (1.0–1.15).

Because microirrigation is based on frequent irrigations, and the entire root zone often is not wetted, a substantial allowance must be made for equipment failure or other potential downtime. A 25% downtime allowance DT is reasonable. With this information, the required system capacity Q (m^3 h^{-1}) can be calculated as

$$Q = 10 \frac{24}{1 - DT/100} DA, \tag{5.166}$$

where DT is the downtime allowance (%), D is the daily application depth (mm day^{-1}), and A is the irrigated area (ha).

Once the system capacity is set, irrigation times for the various crops are determined from their individual daily water requirements and the application rates of the emitters selected. Irrigation frequency for drip systems is often much higher than for other systems in order to maintain constant soil water contents. Frequency typically varies from daily to twice per week.

Hydraulic Design

The hydraulic design of microirrigation systems requires particular attention because the pressure at the emitters is low, often below 130 kPa and sometimes below 60 kPa. Relatively small friction head losses and variations in elevation, which would be acceptable in sprinkler irrigation, cause high relative variations in the operating pressure of the emitters.

The principles for hydraulic design are well presented in manuals [11, 124, 127] and are not presented here. In recent developments with microirrigation modeling [133–136], the hydraulic computations are combined with calculations of the expected uniformity indicators.

System pressure control, mainly pressure regulators, often must be used to maintain acceptable pressures. The type and position of control will influence the system layout and the length and diameters of submains and manifolds. For large systems, controls may be required at submain inlets. When the system is irrigating sloping land, pressure regulators may be required at manifold inlets. When laterals are along a slope or are very long, the pressure regulation may be required for each emitter (pressure-compensating emitters).

Microirrigation Performance

Similar to the approach used for the other methods, the performance indicators can be described by functional relationships with selected system variables. The distribution uniformity or emission uniformity in microirrigation [Eq. (5.119)] is determined by a combination of design parameters that can be functionally expressed as follows:

$$DU = f(P,\ \Delta P,\ x,\ E_c,\ C_v,\ FI), \tag{5.167}$$

where P is the pressure at emitters, ΔP is the variation in pressure along the unit or system, x is the characteristic exponent of the emitter, representing the sensitivity to variations in pressure, E_c represents the emitter characteristics related to variations in discharge, mainly representing the sensitivity to clogging and to temperature, C_v is the coefficient of manufacturing variation for the emitter, and FI represents the filtering capabilities of the system.

As for sprinkler systems, the irrigator has little control over the uniformity performance of the system except for the standard of system maintenance. Variables P and ΔP are set when the hydraulic design is made. Variables x, E_c, and C_v are set when emitters are selected, and FI is established when the filtration station is designed.

When the design is performed carefully, uniformities lower than foreseen should not be expected. However, lower DU may occur when emitters selected by the user do not have the quality expected or the system is not properly maintained. Filter maintenance, lateral flushing, and the periodic chemical treatment against clogging are required to maintain the design uniformity. Evaluations in California, reported by Pitts *et al.* [105] and Hanson *et al.* [137], confirm the importance of these issues and show that actual field DU values are usually between 65% and 80%.

As with surface and sprinkler irrigation systems, the application efficiency e_a also is determined by the combination of design and management variables, as indicated in the following functional relationship:

$$e_a = f(P,\ \Delta P,\ x,\ E_c,\ C_v,\ FI,\ K_s,\ SW,\ t_i,\ \Delta t_i), \tag{5.168}$$

where, besides variables defined for DU, K_s is the hydraulic conductivity of the soil, SW represents soil water conditions at the time of irrigation, t_i is the duration of the irrigation, and Δt_i is the time interval between irrigations. The irrigator's control basically is reduced to the timing of irrigation and the volumes and frequency of applications.

Variable K_s expresses a potential for deep percolation losses. For very permeable soils, emitter spacing must be small, irrigation must be very frequent, and applied volumes must be small to avoid percolation below the root zone for highly permeable soils. For soils with very low hydraulic conductivity, spacings can be larger, irrigation can be less frequent, and emitter discharge can be smaller. These conditions have to be set at design when selecting the emitters and lateral layout.

Soil water conditions at the time of irrigation (*SW*) play a major role in defining the irrigation timing and volume in humid climates, when rainfall meets part of the water requirements, and in irrigation of deep-rooted crops, mainly in orchards, where the soil water reserve can be large. Under arid conditions and most applications when fertigation with a nutrient solution is applied, light, frequent irrigations are used because the soil water reserve plays a minor role under these circumstances. *SW* also is considered as a variable related to soil cracking that occurs in many soils when the soil dries. Soil cracking favors percolation of water and solutes through preferential flow. Irrigation always must start when the soil is wet enough to avoid cracking. This condition also corresponds to the need to maintain soil water above the stress threshold.

Field Evaluation

The evaluation of operating systems can play a major role in improving their performance ([10, 11, 98] and ASAE EP458 [19]). Observations include

- discharges q_i of emitters along various laterals in the irrigation unit;
- pressure at several locations within the unit, and thus P and ΔP;
- the area wetted by an emitter, around a tree or along a crop row;
- the soil water conditions before and after irrigation;
- working conditions of filters and control devices; and
- the fertigation strategies being applied.

This information allows calculation of the distribution uniformity [Eq. (5.119)] determination of the fraction of area wetted, assessment of the quality of design, evaluation of the actual conditions of functioning system, strategies for irrigation scheduling, and the maintenance conditions of the system. The information can be used to advise the farmer about improvements needed in the system, irrigation and fertigation scheduling, and in the maintenance of filters, controllers, and fertigation devices.

Irrigation and Fertigation Management

Microirrigation systems should be used to achieve the highest returns and yields while optimizing the use of water and other production inputs. Microirrigation systems may use less water when not all of the area is irrigated and when system and application losses are minimized, but they should not be managed with the sole intent of saving water; instead, they should be managed to supply the amount of water required by the crop with high frequency.

When very frequent irrigations are applied, the essential information for irrigation scheduling is a forecast of the crop water use. An estimation of ET_c may be sufficient, either using meteorological information (see Section 5.1) or specific sensors. An example using automated pan evaporation is given by DeTar *et al.* [138]. For tree crops, the use of soil and/or plant sensors may be useful [114]. Minimizing percolation losses should be a main objective of scheduling [139].

Microirrigation systems allow precise control of flow rate, duration, and frequency of irrigations. However, in collective pressurized systems, unexpected pressure variations may occur in the supply system. These variations can be high during periods of peak demand. The application duration has to be increased when pressure drops. Flowmeters are critical to show that the desired amount of water has in fact been applied. These problems are reduced when pressure regulators or pressure-compensating emitters are used.

Fertigation is discussed by many authors [98, 122, 124, 128]. It is recommended that fertigation not be limited to the application of primary fertilizers (N, P, and K) but include other nutrients according to the soil conditions and plant requirements. Fertigation may cause contamination of soils and groundwater when excessive quantities of chemical ions are applied. Thus the composition of the nutrient solution has to be based on soil analysis and nutrient requirements. Concentration and composition of the nutrient solution should change during the season as the nutrient requirements change with the crop development.

5.4.5 System Selection

The first decision that an irrigation designer must make is the selection of the irrigation method. This choice depends on both physical and socioeconomic factors, including the cost, availability, and quality of the water supply; the soil type; the field topography and geometry; the crop type and value; the labor cost and availability, material costs, energy costs; and the practicability and availability of the various technologies [140]. Table 5.14 gives a brief summary of how these factors affect selection of the irrigation system. The listed characteristics are not strict guidelines. Economics and innovative design often can supersede physical limitations.

Table 5.14. Characteristics that favor each of the irrigation methods

Factor	Surface Irrigation	Sprinkler Irrigation	Microirrigation
Water cost	Low	Medium	High
Water availability	Periodic or irregular	Regular	Continuous
Water cleanliness	Any	Minimal trash or sand	Clean
Soil infiltration	Medium to low rate; uniform	Medium to high	Any
Soil water storage	High	Medium	Low
Surface topography	Uniform, low slope	Uniform to somewhat irregular	Irregular
Crop geometry	Any	Low-growing	Wide-spaced
Water stress sensitivity	Low	Moderate	High
Crop value	Low	Moderate	High
Cost of labor	Low	Varies with the system	High
Cost of energy	High	Low	Moderate
Capital availability	Low	Medium to high	High
Technology availability	Low	Medium to high	High

Table 5.15. Estimated seasonal water application efficiency by method of irrigation

Efficiency	Surface	Sprinkler	Micro
Potential (%)	60–80	75–90	90–95
Actual (%)	30–80	50–80	65–90

The water application efficiency of the various irrigation methods varies with conditions and system type, and it is difficult to estimate (Table 5.15). Surface irrigation is often relatively inefficient because of lack of water control and the dependence on inherently variable soils. The potential efficiency of microirrigation is very high but requires good system design, maintenance, and operation. Sprinkler irrigation also can be efficient, especially when used under low-wind conditions. Measurements in the United States show that these systems often do not reach their potentials [105].

To support system selection, Table 5.16, containing indicative values for initial investment costs, economic equipment life, and maintenance costs, is provided (complementary information is given by Keller [141]). The table also includes the expected range for the seasonal application efficiencies of the systems. This information is helpful for estimating the gross irrigation water requirements [Section 5.1, Eq. (5.92)]. However, Table 5.16 does not refer to traditional surface irrigation, which has very low investment in equipment and often produces uniformity and efficiency near the lower values in Table 5.15.

Table 5.16. Costs and efficiencies of different types of modern on-farm irrigation systems

Irrigation Method and Type	Equipment Initial Cost (U.S. dollars/ha)[a]	Economic Life (years)	Annual Maintenance (% of cost)	Application Efficiency (%)
Surface precision				
Basin (level)	370–1,085	10–15	10	70–90
Border	370–1,085	10–15	10	70–85
Furrow	150–750	10–15	3–5	65–85
Conveyance				
Lined	400–1,250	15	3	—
Piped	800–2,500	20	1	—
Automation	300	10	5	—
Sprinkle				
Lateral				
Hand-move	450–675	15	2	65–80
End-tow	600–950	10	3	65–75
Side-roll	800–1,100	15	2	65–80
Side-move	950–1,350	15	4	65–80
Hose-fed	450–675	5–20	3	60–80
Traveling gun	950–1,200	10	6	55–70
Center-pivot[b]				
Standard (400 m)	1,100	15	5	70–85
w/Corner	1,200	15	6	65–85
Long (500 m)	700	15	5	65–85
Linear Move[b]				
Ditch-feed	1,100–1,300	15	6	65–85
Pipe-feed	1,600–2,050	15	6	65–85
Solid-Set				
Portable	2,700–3,250	15	2	65–75
Permanent	2,300–3,500	20	1	65–75
Microirrigation				
Orchard				
Drip/spray	1,500–3,500	10–20	3	75–90
Bubbler	2,500–4,000	15	2	60–85
Row-crop				
Drip Tubing	2,000–5,000	10–20	3	65–90
Thin-wall tubing	1,650–3,000	1–20	20	60–80

[a] 1992 prices.
[b] Costs more than double when systems irrigate smaller areas (<30 ha).
Source: Adapted from [141].

References

1. Food and Agriculture Organization of the United Nations. 1995. *Irrigation in Africa in Figures*. FAO Water Reports 7. Rome: FAO.
2. Food and Agriculture Organization of the United Nations. 1997. *Irrigation in the Near East Region in Figures*. FAO Water Reports 9. Rome: FAO.
3. Irrigation Association. 1995. Farm and ranch survey verifies irrigation efficiency. *Irrig. Bus. Tech.* III(6):18–22.

4. Food and Agriculture Organization of the United Nations. 1998. *Irrigation in Countries of the Former Soviet Union in Figures*. FAO Water Report 15. Rome: FAO.
5. Heermann, D. F., W. W. Wallender, and G. M. Bos.1990. Irrigation efficiency and uniformity. *Management of Farm Irrigation Systems*, eds. Hoffman, G. J., T. A. Howell, and K. H. Solomon, pp. 125–149. St. Joseph, MI: American Society of Agricultural Engineers.
6. Wolters, W. 1992. *Influences on the Efficiency of Irrigation Water Use*. ILRI Publication. No. 51, Wageningen: ILRI.
7. Bos, M. G., D. H. Murray-Rust, D. J. Merrey, H. G. Johnson, and W. B. Sneller. 1994. Methodologies for assessing performance of irrigation and drainage management. *Irrig. Drain. Sys.*, 7(4):231–261.
8. Jensen, M. E. 1996. Irrigated agriculture at the crossroads. *Sustainability of Irrigated Agriculture*, eds. Pereira, L. S., R. A. Feddes, J. R. Gilley, and B. Lesaffre, pp. 19–33. Dordrecht: Kluwer.
9. Steduto, P. 1993. Water use efficiency. *Sustainability of Irrigated Agriculture*. eds. Pereira, L. S., R. A. Feddes, J. R. Gilley, and B. Lesaffre, pp. 193–209. Dordrecht: Kluwer.
10. Merriam, J. L., and J. Keller. 1978. *Farm Irrigation System Evaluation: A Guide for Management*. Logan: Department of Agricultural and Irrigation Engineering, Utah State University.
11. Keller, J., and R. D. Bliesner. 1990. *Sprinkle and Trickle Irrigation*. New York: Van Nostrand Reinhold.
12. Hart, W. E., W. N. Reynolds. 1965. Analytical design of sprinkler systems. *Trans. ASAE* 8(1):83–85, 89.
13. Mantovani, E. C., F. J. Villalobos, F. Orgaz, and E. Fereres. 1995. Modelling the effects of sprinkler irrigation uniformity on crop yield. *Agric. Water Manage.* 27:243–257.
14. Seginer, I. 1987. Spatial water distribution in sprinkler irrigation. *Advances in Irrigation*, Vol. 4, ed. Hillel, D., pp. 119–168. Orlando: Academic Press.
15. Bralts, V. F., D. M. Edwards, and I-P. Wu. 1987. Drip irrigation design and evaluation based on the statistical uniformity concept. *Advances in Irrigation*, Vol. 4, ed. Hillel, D., pp. 67–117. Orlando: Academic Press.
16. Hart, W. E., H. G. Collins, G. Woodward, and A. S. Humpherys. 1980. Design and operation of gravity or surface systems. *Design and Operation of Farm Irrigation Systems*, ed. Jensen, M. E., pp. 501–580. St. Joseph, MI: American Society of Agricultural Engineers.
17. Paulo, A. M., L. A. Pereira, J. L. Teixeira, and L. S. Pereira. 1995. Modelling paddy rice irrigation. *Crop-Water-Simulation Models in Practice*, eds. Pereira, L. S., B. van den Broek, P. Kabat, and R. G. Allen, pp. 287–302. Wageningen: Wageningen Pers.
18. Mao, Z. 1996. Environmental impact of water-saving irrigation for rice. *Irrigation Scheduling: From Theory to Practice*, eds. Smith, M., L. S. Pereira, J. Berenjena,

B. Itier, J. Goussard, R. Ragab, L. Tollefson, and P. van Hoffwegen, pp. 141–145. FAO Water Reports 8. Rome: FAO.

19. American Society of Agricultural Engineers. 1998. *ASAE Standards 1998: Standards, Engineering Practices, Data*. St. Joseph, MI: ASAE.
20. Walker, W. R., and Skogerboe, G. V. 1987. *Surface Irrigation, Theory and Practice*. Englewood Cliffs, NJ: Prentice-Hall.
21. Vogel, T., and Hopmans, J. W. 1992. Two-dimensional analysis of furrow infiltration. *J. Irrig. Drain. Eng.* 118(5):791–806.
22. Tabuada, M. A., Z. C. Rego, G. Vachaud, and L. S. Pereira.1995. Two-dimensional infiltration under furrow irrigation: Modelling, its validation and applications. *Agric. Water Manage.* 27:105–123.
23. Tabuada, M. A., Z. C. Rego, G. Vachaud, and L. S. Pereira.1995. Modelling of furrow irrigation. Advance with two-dimensional infiltration. *Agric. Water Manage.* 28:201–221.
24. Irrigation Software Engineering Division. 1989. SIRMOD, Surface Irrigation Simulation Software. User's Guide. Logan: Department of Agricultural and Irrigation Engineering, Utah State University.
25. Strelkoff, T. 1993. SRFR, a Computer Program for Simulating Flow in Surface Irrigation Furrows-Basins-Borders. Phoenix, AZ: Water Conservation Laboratory, U.S. Department of Agriculture, Agricultural Research Service.
26. Strelkoff, T. 1993. Flow simulation for surface irrigation design. *Management of Irrigation and Drainage Systems*, ed. Allen, R. G., pp. 899–906. New York: American Society of Civil Engineers.
27. Strelkoff, T. 1985. BRDRFLW: A Mathematical Model of Border Irrigation. Phoenix, AZ: Water Conservation Laboratory, U.S. Department of Agriculture, Agricultural Research Service.
28. Clemmens, A. J., A. R. Dedrick, and R. J. Strand. 1993. BASIN 2.0 for the design of level-basin irrigation systems. *Management of Irrigation and Drainage Systems*, ed. Allen, R. G., pp. 875–882. New York: ASCE.
29. Pereira, L. S. 1996. Surface irrigation systems. *Sustainability of Irrigated Agriculture*, eds. Pereira, L. S., R. A. Feddes, J. R. Gilley, and B. Lesaffre, pp. 269–289. Dordrecht: Kluwer.
30. Latimer, E. A., and Reddel, D. L. 1990. A volume balance model for real time automated furrow irrigation system. *Visions of the Future*, pp. 13–20. St. Joseph, MI: American Society of Agricultural Engineers.
31. Mailhol, J. C. 1992. Un modèle pour améliorer la conduite de l'irrigation à la raie. *ICID Bull.* 41(1):43–60.
32. Eisenhauer, D. E., C. D. Yonts, J. E. Cahoon, and B. Brown. 1993. Reactive irrigation scheduling for sloping furrow irrigation. *Management of Irrigation and Drainage Systems*, ed. Allen, R. G., pp. 198–205. New York: American Society of Civil Engineers.
33. Trout, T. J. 1996. Furrow irrigation erosion and sedimentation: on-field distribution. *Trans. ASAE* 39(5):1717–1723.

34. Elliot, R. L., W. R. Walker, and G. V. Skogerboe. 1983. Infiltration parameters from furrow irrigation advance data. *Trans. ASAE* 26:1726–1730.
35. Smerdon, E. T., A. W. Blair, and D. L. Reddell. 1988. Infiltration from irrigation advance data. I. Theory. *J. Irrig. Drain. Eng.* 114(1):4–17.
36. Blair, A. W., and E. T. Smerdon. 1988. Infiltration from irrigation advance data. II. Experimental. *J. Irrig. Drain. Eng.* 114(1):18–30.
37. Katopodes, N. D. 1990. Observability of surface irrigation advance. *J. Irrig. Drain. Eng.* 116(5):656–675.
38. Katopodes, N. D., J. H. Tang, and A. J. Clemmens.1990. Estimation of surface irrigation parameters. *J. Irrig. and Drain. Eng.* 116(5):676–696.
39. Walker, W. R., and J. D. Busman, 1990. Real-time estimation of furrow infiltration *J. Irrig. Drain. Eng.* 116(3):299–318.
40. Malano, H. M., H. N. Turral, and M. L. Wood. 1996. Surface irrigation management in real time in south eastern Australia: Irrigation scheduling and field application. *Irrigation Scheduling: From Theory to Practice*, eds. Smith, M., L. S. Pereira, J. Berenjena, B. Itier, J. Goussard, R. Ragab, L. Tollefson, and P. van Hoffwegen, pp. 105–118. FAO Water Report 8, Rome: FAO.
41. Scalopi, E. J., and L. S. Willardson. 1986. Practical land grading based on least squares. *J. Irrig. Drain. Eng.* 112(2):98–109.
42. Kumar, Y., and H. S. Chauhan. 1987. Gradient search technique for land levelling design. *Trans. ASAE* 30(2):391–393.
43. Clough, M. R. 1993. Computerised irrigation project earth work design. *Management of Irrigation and Drainage Systems*, ed. Allen, R. G., pp. 141–148. New York: American Society of Civil Engineers.
44. Sousa, P. L., A. R. Dedrick, A. J. Clemmens, and L. S. Pereira. 1995. Effect of furrow elevation differences on level-basin performance. *Trans. ASAE* 38(1):153–158.
45. Sousa, P. L., A. R. Dedrick, A. J. Clemmens, and L. S. Pereira. 1993. Benefits and costs of laser-controlled levelling - a case study. *Planning and Design of Irrigation and Drainage Systems (ICID Congress, The Hague)*, eds. Storsbergen, C., *et al.*, pp. 1238–1247. New Delhi: ICID.
46. Humpherys, A. S., 1987. *Automated Farm Surface Irrigation System World-Wide.* New Delhi: ICID.
47. Kemper, W. D., T. J. Trout, and D. C. Kincaid. 1987. Cablegation: Automated supply for surface irrigation. *Advances in Irrigation*, Vol. 4, ed. Hillel, D., pp. 1–66. Orlando: Academic Press.
48. Sousa, P. L., M. R. Cameira, and A. Monteiro. 1993. Operation and management of a cablegation system. *Desenvolvimento de Equipamentos Mecanizados para Rega de Gravidade*, eds. Sousa, P. L., and M. R. Cameira, pp. 43–58. Lisbon: DER, ISA (in Portuguese).
49. Trout, T. J., and D. C. Kincaid. 1989. Border cablegation system design. *Trans. ASAE* 32(4):1185–1192.
50. Stringham, G. E. (ed.). 1988. Surge Flow Irrigation. Research Bulletin 515, Utah Agricultural Experiment Station, Logan: Utah State University.

51. Stringham, G. E., and J. Keller. 1979. Surge flow for automatic irrigation. *Irrigation and Drainage Division Specialty Conference*, pp.132–142. New York: American Society of Civil Engineers.
52. Kemper, W. D., T. J. Trout, A. S. Humpherys, and M. S. Bullock. 1988. Mechanisms by which surge irrigation reduces furrow infiltration rate in a silty loam soil. *Trans. ASAE* 31(3):821–829.
53. Trout, T. J. 1991. Surface seal influence on surge flow furrow infiltration. *Trans. ASAE* 34(1):66–72.
54. Soil Conservation Service. 1986. *Surge Flow Irrigation Field Guide*. Department of Agriculture. Washington DC: U.S.
55. Humpherys, A. S. 1989. Surge irrigation: 1. An overview. *ICID Bull.* 38(2):35–48.
56. Humpherys, A. S. 1989. Surge irrigation: 2. Management. *ICID Bull.* 38(2):49–61.
57. Trout, T. J., and D. C. Kincaid. 1993. Cablegation evaluation methodology. *Appl. Eng. Agric.* 9(6):523–528.
58. McCornick, P. G., H. R. Duke, and T. H. Podmore. 1988. Field evaluation procedure for surge irrigation. *Trans. ASAE* 31(1):168–176.
59. Musick, J. T., J. D. Walker, A. D. Schneider, and F. B. Pringle. 1987. Seasonal evaluation of surge flow irrigation for corn. *Appl. Eng. Agric.* 3(2):247–251.
60. Coupal, R. H., P. N. Wilson. 1990. Adopting water conserving irrigation technology; the case of surge irrigation in Arizona. *Agric. Water Manage.* 18:15–28.
61. Bondurant, J. A., and W. D. Kemper. 1985. Self-cleaning non-powered trash screens for small irrigation flows. *Trans. ASAE* 28(1):113–117.
62. Dedrick, A. R. 1989. Improvements in design and installation of mechanised level-basin systems. *Appl. Eng. Agric.* 5(3):372–378.
63. Sousa, P. L., and L. S. Pereira. 1989. Modeling a farm canal for automated basin irrigation. *Agricultural Engineering. I. Land and Water Use*, eds. Dodd, V. A., and P. M. Grace, pp. 691–697. Rotterdam: Balkema.
64. Humpherys, A. S. 1990. A semiautomation of basin and border irrigation. *Visions of the Future*, pp. 28–33. St. Joseph, MI: American Society of Agricultural Engineers.
65. Trout, T. J. 1990. Furrow inflow and infiltration variability impacts on irrigation management. *Trans. ASAE* 33(4):1171–1178.
66. Trout, T. J., R. E. Sojka, and L. I. Okafor. 1990. Soil management. *Management of Farm Irrigation Systems*, eds. Hoffman, G. J., *et al.*, pp. 875–896. St. Joseph, MI: American Society of Agricultural Engineers.
67. Fornstrom, K. J., J. A. Michel, J. Borrelli Jr., and G. D. Jackson. 1985. Furrow firming for control of irrigation advance rates. *Trans. ASAE* 28(2):529–531.
68. Allen, R. R., and A. D. Schneider. 1992. Furrow water intake reduction with surge irrigation or traffic compaction. *Appl. Eng. Agric.* 8(4):455–459.
69. Pereira, L. S., J. R. Gilley, R. M. Fernando, M. A. Tabuada, P. L. Sousa, A. L. Santos, J. L. Teixeira, and F. Avillez. 1994. Development of Irrigation Technologies for Southern Portugal. Final Rept., Project NATO-PO-Irrigation; Lisbon: Dept. Engenharia Rural, Instituto Superior de Agronomia.
70. Eisenhauer, D. E., and D. F. Heermann, and A. Klute. 1992. Surface sealing effect on infiltration with surface irrigation. *Trans. ASAE* 35(6):1799–1807.

71. Yonts, C. D., J. A. Smith, and J. E. Bailie. 1991. Furrow irrigation performance in reduced-tillage systems. *Trans. ASAE* 34(1):91–96.
72. Threadgill, E. D., D. E. Eisenhauer, J. R. Young, and B. Bar-Yosef. 1990. Chemigation. *Management of Farm Irrigation Systems*, eds. Hoffman, G. J., *et al.*, pp. 749–780. St. Joseph, MI: American Society of Agricultural Engineers.
73. Pereira, L. S., A. Hamdy, N. Lamaddalena, and M. Abdelmalek. 1994. Fertigation with surface irrigation: Experimental study on dispersion of nitrogen in a farm canal. *International Conference on Land and Water Management in the Mediterranean Region*, pp. 309–325. Bari: Istituto Agronomico Mediterraneo.
74. Santos, D. V., P. L. Sousa, and R. E. Smith. 1997. Model simulation of water and nitrate movement in a level-basin under fertigation treatments. *Agric. Water Manage.* 32:293–306.
75. Cameira, M. R., P. L. Sousa, H. J. Farahani, L. R. Ahuja, and L. S. Pereira. 1998. Evaluation of model RZWQM for the simulation of water and nitrates in level-basin, fertigated maize. *J. Agric. Eng. Res.* 69:331–341.
76. Bos, M. G. (ed.). 1989. *Discharge Measurement Structures*. Publ. 20, Wageningen: ILRI
77. Replogle, J. A. 1997. Practical technologies for irrigation flow control and measurement. *Irrig. Drain. Sys.* 11:241–259.
78. Smith, M., L. S. Pereira, J. Berenjena, B. Itier, J. Goussard, R. Ragab, L. Tollefson, P. van Hoffwegen (eds). 1996. *Irrigation Scheduling: From Theory to Practice*. FAO Water Report 8. Rome: FAO.
79. Hill, R. W., and R. G. Allen. 1996. Simple irrigation scheduling calendars. *J. Irrig. Drain. Eng.* 122(2):107–111.
80. Liu, Y., R. M. Fernando, Y. Li, and L. S. Pereira. 1997. Irrigation scheduling strategies for wheat-maize cropping sequence in North China Plain. *Sustainable Irrigation in Areas of Water Scarcity and Drought*, eds. de Jager, J. M., *et al.*, pp. 97–107. Oxford: British National Committee ICID.
81. Pereira, L. S., A. Perrier, M. Ait Kadi, and P. Kabat (eds). 1992. Crop water models. *ICID Bull.*(Special Issue), 41(2):1–200.
82. Pereira, L. S., B. van den Broek, P. Kabat, and R. G. Allen (eds.). 1995. *Crop-Water-Simulation Models in Practice*. Wageningen: Wageningen Pers.
83. Ragab, R., D. El-Quosy, B. van den Broek, and L. S. Pereira (eds.). 1996. *Crop-Water-Environment Models*. Cairo: Egypt National Committee. ICID.
84. Gilley, J. R. 1996. Sprinkler irrigation systems. *Sustainability of Irrigated Agriculture*, eds. Pereira, L. S., R. A. Feddes, J. R. Gilley, and B. Lesaffre, pp. 291–307. Dordrecht: Kluwer.
85. Rochester, E. W., and S. G. Hackwell. 1991. Power and energy requirements of small hard-hose travellers. *Appl. Eng. Agric.* 7(5):551–556.
86. Kincaid, D. C., I. McCann, J. R. Busch, and M. Hasherninia. 1990. Low pressure center pivot irrigation and reservoir tillage. *Visions of the Future*, pp. 54–60. St. Joseph, MI: American Society of Agricultural Engineers.
87. Kincaid, D. C., K. H. Solomon, and J. C. Oliphant. 1996. Drop size distributions for irrigation sprinklers. *Trans. ASAE* 39(3):839–845.

88. Li, J., H. Kawano, and K. Yu. 1994. Droplet size distributions from different shaped sprinkler nozzles. *Trans. ASAE* 37(6):1871–1878.
89. Li, J., Y. Li, H. Kawano, and R. E. Yoder. 1995. Effects of double-rectangular-slot design on impact sprinkler nozzle performance. *Trans. ASAE* 38(5):1435–1441.
90. Yazar, A. 1984. Evaporation and drift losses from sprinkler systems under various operating conditions. *Agric. Water Manage.* 8:439–449.
91. Han, S., R. G. Evans, and M. W. Kroeger. 1994. Sprinkler distribution patterns in windy conditions. *Trans. ASAE* 37(5):1481–1489.
92. von Bernuth, R. D., and I. Seginer. 1990. Wind considerations in sprinkler system design. *Visions of the Future*, pp. 334–339. St. Joseph, MI: American Society of Agricultural Engineers.
93. Tarjuelo, J. M., M. Valiente, and J. Lozoya. 1992. Working conditions of sprinkler to optimize application of water. *J. Irrig. Drain. Eng.* 118(6):895–913.
94. Addink, J. W., J. Keller, C. H. Pair, R. E. Sneed, and J. W. Wolfe. 1980. Design and operation of sprinkler systems. *Design and Operation of Farm Irrigation Systems*, ed. M. E. Jensen, pp. 621–660. St. Joseph, MI: American Society of Agricultural Engineers.
95. Rolland, M. 1982. *Mechanized Sprinkler Irrigation*. FAO Irrigation and Drainage, Paper 35. Rome: FAO.
96. Pair, C. H., W. H. Hinz, K. R. Frost, R. E. Sneed, and T. J. Schiltz. 1983. *Irrigation* (5th ed.), Arlington, VA: Irrigation Association.
97. Kay, M. 1983. *Sprinkler Irrigation. Equipment and Practice*. London: Batsford Academic and Educational.
98. Rieul, L. (ed.). 1992. *Irrigation: Guide Pratique*. Paris: CEMAGREF & CEP-Groupe France Agricole.
99. Tarjuelo, J. M. 1995. *El riego por aspersion y su tecnología*. Madrid: Ediciones Mundi-Prensa.
100. Allen, R. G. 1990. Applicator selection along center-pivots using soil infiltration parameters. *Visions of the Future*, pp. 549–555. St. Joseph, MI: American Society of Agricultural Engineers.
101. von Bernuth, R. D., and J. R. Gilley. 1985. Evaluation of center-pivot application packages considering droplet induced infiltration reduction. *Trans. ASAE* 28(6):1940–1946.
102. Lyle, W. M., and J. P. Bordowsky. 1981. Low energy precision application (LEPA) irrigation system. *Trans. ASAE* 26(5):1241–1245.
103. Fipps, G., and L. L. New. 1990. Six years of LEPA in Texas—less water higher yields. *Visions of the Future*, pp. 115–120. St. Joseph, MI: American Society of Agricultural Engineers.
104. Dubalen, J. 1993. Utilisation des matériels d'irrigation par aspersion. Diagnostic de fonctionnement au champ. *Houille Blanche* 2/3:183–188.
105. Pitts, D., K. Peterson, G. Gilbert, and R. Fastenau. 1996. Field assessment of irrigation system performance. *Appl. Eng. Agric.* 12(3):307–313.
106. Heermann, D. F., and R. A. Kohl. 1980. Fluid dynamics of sprinkler systems. *Design and Operation of Farm Irrigation Systems*, ed. Jensen, M. E., pp. 583–618, St. Joseph, MI: American Society of Agricultural Engineers.

107. Scalopi, E. J., R. G. Allen. 1993. Hydraulics of center-pivot laterals. *J. Irrig. Drain. Eng.* 119(3):554–567.
108. Scalopi, E. J., and R. G. Allen. 1993. Hydraulics of irrigation laterals: Comparative analysis. *J. Irrig. Drain. Eng.* 119(1):91–115.
109. Allen, R. G. 1996. Relating the Hazen-Williams and the Darcy-Weisbach friction loss equations for pressurized irrigation. *Appl. Eng. Agric.* 12(6):685–693.
110. Merriam, J. L., M. N. Shearer, and C. M. Burt. 1980. Evaluating irrigation systems and practices. *Design and Operation of Farm Irrigation Systems*, ed. Jensen, M. E., pp. 721–750. St. Joseph, MI: American Society of Agricultural Engineers.
111. Heermann, D. F. 1990. Center pivot design and evaluation.*Visions of the Future*, pp. 564–570. St. Joseph, MI: American Society of Agricultural Engineers.
112. Evans, R. G., S. Han, W. M. Kroeger. 1995. Spatial distribution and uniformity evaluations for chemigation with center pivots. *Trans. ASAE* 38(1):85–92.
113. Martin, D. 1991. Effect of frequency on center-pivot irrigation. *Irrigation and Drainage*, ed. Ritter, W. F., pp. 38–44. New York: American Society of Agricultural Engineers.
114. Phene, C. J., R. J. Reginato, B. Itier, and B. R. Tanner. 1990. Sensing irrigation needs. *Management of Farm Irrigation Systems*, eds. Hoffman, G. J., T. A. Howell, and K. H. Solomon, pp. 207–261. St. Joseph, MI: American Society of Agricultural Engineers.
115. Tizaoui, C., L. S. Pereira, and N. Lamaddalena. 1997. Irrigation par aspersion et localisée: Conditions hydrauliques, performances et production. *Proceedings of Workshop on Collective Irrigation Systems*, pp. 55–75, Bari: CIHEAM CIS-Net, Istituto Agronomico Mediterraneo.
116. Duke, H. R., L. E. Stetson, and N. C. Ciancaglini. 1990. Irrigation system controls. *Management of Farm Irrigation Systems*, eds. Hoffman, G. J., *et al.*, pp. 265–312. St. Joseph, MI: American Society of Agricultural Engineers.
117. Giannerini, G. 1995. RENANA: A model for irrigation scheduling, employed on a large scale. *Crop-Water-Simulation Models in Practice*, eds. L. S., Pereira, B. J. van den Broek, P. Kabat, and R. G. Allen, pp. 17–27. Wageningen: Wageningen Pers.
118. Buchleiter, G. W. 1995. Improved irrigation management under center pivots with SCHED. *Crop-Water-Simulation Models in Practice*, eds. Pereira, L. S., B. J. van den Broek, P. Kabat, and R. G. Allen, pp. 27–47. Wageningen: Wageningen Pers.
119. Threadgill, E. D. 1985. Chemigation via sprinkler irrigation: Current status and future development. *Appl. Eng. Agric.* 1(1):16–23.
120. Rawlins, S. L. 1977. Uniform irrigation with a low head bubbler system. *Agric. Water Manage.* 1:166–178.
121. Reynolds, C., M. Yitayew, and M. Petersen. 1995. Low-head bubbler irrigation systems, Part I: Design. *Agric. Water Manage.* 29:1–24.
122. Papadopoulos, I. 1996. Micro-irrigation and fertigation. *Sustainability of Irrigated Agriculture*, eds. Pereira, L. S., R. A. Feddes, J. R. Gilley, and B. Lesaffre, pp. 309–322. Dordrecht: Kluwer.
123. Vermeiren, L., and G. A. Jobling. 1980. *Localized irrigation*. FAO Irrigation and Drainage Paper 36, Rome: FAO.

124. Burt, C. M., and S. W. Styles. 1994. Drip and Microirrigation for Trees, Vines, and Rows Crops. San Luis Obispo, CA: Irrigation Training and Research Center, Cal Poly.
125. Ozekici, B., and R. E. Sneed. 1995. Manufacturing variation for various trickle irrigation on-line emitters. *Appl. Eng. Agric.* 11(2):235–240.
126. Boman, B. J. 1995. Effects of orifice size on microsprinkler clogging rates. *Appl. Eng. Agric.* 11(6):839–843.
127. Howell, T. A., D. S. Stevenson, F. K. Aljibury, H. M. Gittlin, I-P. Wu, A. W. Warrick, P. A. C. Raats. 1980. *Design and Operation of Farm Irrigation Systems*, ed. Jensen, M. E., pp. 663–717. St. Joseph, MI: American Society of Agricultural Engineers.
128. Rodrigo, J., J. M. Hernandez, A. Perez, J. F. Gonzalez. 1992. *Riego Localizado.* Madrid: YRIDA and Ed. Mundi-Prensa.
129. Hanson, B., L. Schwankl, S. R. Grattan, and T. Prichard. 1994. Drip Irrigation for Row Crops. Davis: University of California, Department of Land, Air, and Water Resources.
130. Hanson, B. 1994. Micro-Irrigation of Trees and Vines. Davis: University of California, Department of Land, Air, and Water Reources.
131. Burt, C., K. O'Connor, and T. Ruehr. 1995. Fertigation. San Luis Obispo, CA: Irrigation Training and Research Center, Cal Poly.
132. Angelakis, A. N., D. E. Rolston, T. N. Kadir, and V. H. Scott. 1991. Soil-water distribution under trickle source. *J. Irrig. Drain. Eng.* 119(3):484–500.
133. Bralts, V. F., W. H. Shayya, and M. A. Driscoll. 1990. An expert system for the hydraulic design of microirrigation systems. *Visions of the Future*, pp. 340–347. St. Joseph, MI: American Society of Agricultural Engineers.
134. Kang, Y., and S. Nishiyama. 1995. Hydraulic analysis of microirrigation submain units. *Trans. ASAE* 38(5):1377–1384.
135. Kang, Y., and S. Nishiyama. 1996. Analysis and design of microirrigation laterals. *J. Irrig. Drain. Eng.* 122(6):75–81.
136. Kang, Y., and S. Nishiyama. 1996. Design of microirrigation submains. *J. Irrig. Drain. Eng.* 122(2):83–89.
137. Hanson, B., W. Bowers, B. Davidoff, D. Kasapligil, A. Carvajal, and W. Bendixen. 1995. Field performance of microirrigation systems. *Microirrigation for a Changing World*. Orlando, 5th Drip Irrigation Congress.
138. DeTar, W. R., G. T. Browne, C. J. Phene, and B. L. Sanden. 1996. Real-time scheduling of potatoes with sprinkler and subsurface drip systems. *Evapotranspiration and Irrigation Scheduling*, eds. Camp, C. R., E. J. Sadler, and R. E. Yoder, pp. 812–824. St. Joseph; MI: American Society of Agricultural Engineers.
139. Wu, I-P. 1995. Optimal scheduling and minimizing deep seepage in microirrigation. *Trans. ASAE* 38(3):1385–1392.
140. Hlavek, R. 1992. *Selection Criteria for Irrigation Systems*. New Delhi: ICID.
141. Keller, J. 1992. Irrigation scheme design for sustainability. *Advances in Planning, Design and Management of Irrigation Systems as Related to Sustainable Land Use*, eds. Feyen, J., *et al.*, pp. 217–234. Leuven: Center for Irrigation Engineering.

5.5 Crop Water Management

J. M. Tarjuelo and J. A. de Juan

5.5.1 Introduction

Water is a finite resource essential for all forms of life; without water, life as currently known could not exist. Most of the hydrosphere is salt water (97.20%); freshwater represents only 0.65% (8.5×10^6 km^3). This quantity of water is not uniformly distributed, in either space or time. In many countries and continents the freshwater supply is becoming scarce or has become an actual problem. Drought has aggravated the situation. During the past 20 years, the population has increased significantly. As a result, water requirements also have increased.

Over the next 40 years, the world's population will increase from 5.3×10^9 to about 9×10^9 inhabitants. Thus, the agricultural sector will be encouraged to duplicate the world's food production and to improve average yields and the proportion of irrigated lands. More than 82% of the world's crop land is not irrigated, depending only on rain. In many potentially arable lands of semiarid areas, water is a limiting factor and high costs often make irrigation schemes unattractive. A large part of rainfed agriculture requires soil and water conservation measures and techniques aimed at preventing degradation and maintaining yield potentials. Sustainable rainfed agriculture requires management of the land so that production and productivity are enhanced while a healthy ecological balance is sustained within the agricultural ecosystem.

Irrigated lands include 18% of cultivated land, but produce 33% of the food that is necessary for the world population. Technologies and management techniques are available to improve the use of irrigation water, including off-farm and on-farm measures for improving efficiency and conservation. Irrigation scheduling for example, is a management technique that helps to determine the timing and the amounts of water to be applied.

In summary, water resources play a major role in agriculture in general, and in irrigated agriculture in particular. Unfortunately, our ability to substantially increase crop production in dryland farming has not been given as much attention as it has been in irrigated agriculture. Efficient water use in irrigated agriculture can achieve several goals, such as conserving freshwater, maintaining high standards of public safety, and improving environmental quality. Water availability can be controlled through irrigation and also through a range of other management techniques. To maximize water-use efficiency when water supplies are limited, the farm manager has a variety of options available. This chapter discusses water management in rainfed and irrigated agriculture.

5.5.2 Effects of Water Deficits

All of the processes of carbon gain and loss can be influenced by stress. In fact, stress can be defined as any factor, in excess or in deficit, that detrimentally affects the normal functioning of a plant, altering any aspect of a plant's carbon balance [1].

The ecological influence of water is the result of its physiological importance. The major role of water can be summarized by listing its most important functions under four general headings: constituent, solvent, reactant, and maintenance of turgidity [2].

A water deficit therefore can affect the majority of aspects related to growth: anatomy, morphology, and physiology [3].

The system that describes the behavior of water and water movement in soils and plants is based on a potential-energy relationship. Water has the capacity to do work; it will move from an area of high potential energy to one of low potential energy. The potential energy in aqueous systems is expressed by comparing it with the potential energy of pure water. The water in plants and soils usually is not chemically pure because of solute and it is physically constrained by forces such as polar attractions, gravity, and pressure. As a result, the potential energy is less than that of pure water. In plants and soils, potential energy of water is called water potential. It is usually negative; the more negative the value, the lower is the water potential.

Water potential (kPa) is the sum of several components of potential:

1. Ψ_m, matrix potential, the force with which water is held to plant and soil constituents by forces of adsorption and capillarity;
2. Ψ_s, solute potential (osmotic potential), the potential energy of water as influenced by solute concentration;
3. Ψ_p, pressure potential (turgor pressure), the force caused by hydrostatic pressure, and it usually has a positive value; and
4. Ψ_g, gravitational potential, which is always present but usually is insignificant in short plants, compared with the other three.

Because gravity can be omitted in most cases, water potential remains $\Psi = -\Psi_s + \Psi_p$. In a transpiring plant, the absolute value of Ψ_s will exceed that of Ψ_p, yielding a negative value.

The concept of water potential fulfills two main functions: It governs the direction of water flow across cell membranes and is the driving force for water movement from the soil into the roots, and it is a measure of the water status of a plant. Water potential is the most commonly measured parameter which is closely connected to plant functions. Thus, a decrease in Ψ under given conditions relative to Ψ of well-watered plants can be correlated with yield and productivity. The water movement through the Soil-Plant-Atmosphere Continuum (SPAC) is explained as occurring along a decreasing water-potential gradient, Ψ. The potential gradients among the components of the system constitute the driving force for the flow within the system. Processes might be noted as analogous to the Ohm's Law:

$$\text{Water flow} = \frac{\text{Difference in water potential } (\Delta\Psi)}{\text{Resistance}} \tag{5.169}$$

In other words, the water flow through each component of the system is determined by the existing potential gradient and by the resistance in the own segment.

It is assumed that forces causing movement are tensions originated by the large mass flux generated during transpiration and, occasionally, by the root pressure when a plant transpirates very slowly. Thus, the direction and, partly, the water movement rate in the SPAC depend on the gradient of water potential. The higher water potential in the soil, the higher the gradient. Irrigation increases the soil's water potential and decreases the resistance to flow. Hence, root water absorption is favored. Water balance and the water movement within the plant are determined by water relationships at cell level. The hydric potential in a plant cell is the same throughout when the cell has hydric equilibrium. However the potential components might vary. When the cell undergoes changes in its hydric potential then water comes in or goes out of the cell. These changes affect its turgor, volume and solute concentration. During the daytime the behavior of stomata depends on the guard cell's Ψ_p. This potential depends on capability of the plant to absorb water and to replace water losses provoked by transpiration. When Ψ_p comes close to zero, the stomata begins to close. Then the resistance to the water vapor transport goes up and transpiration is consequently limited. This function of the stomata lessens the development of severe water stress that would damage tissues. A plant becomes water stressed when hydric and pressure potentials are sufficiently lowered so as to alter normal performance [3]. These low values of hydric potential at the leaf level can be reached due to various causes: low values of hydric potential in the soil, high transpiration flows, or elevated resistance to flow:

$$\Psi_l = \Psi_s + \Delta\Psi_g - Tr \cdot R_{sl}, \tag{5.170}$$

where Ψ_l is the hydric potential of the leaf, Ψ_s is the hydric potential of the soil, $\Delta\Psi_g$ is the difference in gravitational potential between the soil and the leaf, Tr is the transpiration flux, and R_{sl} is flow resistance from soil to leaf.

The hydric status of a plant is determined by an entire series of environmental and physiological factors represented in diagram form in Fig. 5.37.

Four main factors control the plant's water balance [4]:

1. soil water potential and factors involved at its level, such as rain, irrigation water, crop water extraction, soil texture, structure, and depth, and root distribution and dynamics.
2. evapotranspiration rate, which implies several environmental factors (e.g., radiation, air humidity, temperature, and wind) and physiological factors (leaf area and exposition, plant canopy, stomatal conductance).
3. water conductance in roots, stems, and leaves, which depends on the physicochemical characteristics of the plant tissues and affects the water-movement velocity through the plant and the water equilibrium among the plant organs.

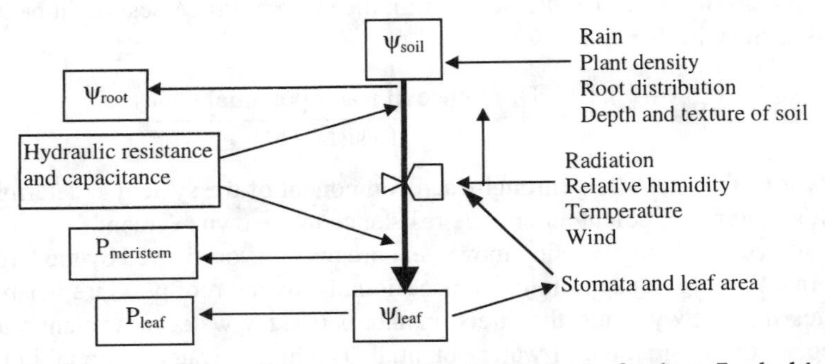

Figure 5.37. Diagram of factors controlling hydric status of a plant: Ψ = hydric potential; P = turgidity; and E = evaporation *Source:* Adapted from [4].

4. relation between the hydric and other water-status measures (e.g., turgor pressure, osmotic potential, or water content), which might be affected by factors such as osmotic regulation, cell-size alterations, and cell-wall elasticity.

The parameters commonly used as indicators of the plant's hydric status are the relative water content (RWC); the hydric potential of a tissue or organ, Ψ; and the hydric potential components, Ψ_s and Ψ_p. A comprehensive review [5] of methods for measuring water deficit will provide newcomers to the field with an array of direct and indirect techniques for measuring water status of a plant.

An indicator of water status is tissue water content, which can be measured easily with high accuracy, on either a fresh-weight or a dry-weight basis. However, earlier works [6] have shown that the water-holding capacity of different organs or tissues of a plant can vary widely, because of variation in the dry-matter content of the tissue. Consequently, the parameter used is normalized RWC, which is the water content relative to that when the tissue is saturated with water and Ψ is zero. As implied in its definition, RWC is a measurement of tissue volume maintenance and water holding and is linked closely with plant functions. This parameter and its possible errors have been reviewed [7].

Many methods have been developed to determine Ψ within tissues. These methods have been widely reviewed [5, 8]. At present, methods based on thermocouple psychrometry and the pressure chamber method are the most commonly used and are considered to be accurate.

Technical bases and management of psycrometers and hygrometers have been discussed [7–11]. These methods determine Ψ by measuring air humidity in equilibrium with the tissue sample. They are based on the fact that Ψ is the same throughout the water phases when there is equilibrium. Then, Ψ is determined in the liquid phase from Ψ of the vapor phase. With the psychrometer technique, thermocouples function like wet and dry bulbs to measure humidity. With the hygrometric technique, humidity is determined by using thermocouples to measure the dew point of the air.

Measures of Ψ using psycrometric techniques are subject to several errors [7], for example, from heat produced by respiration of the tissue sample, incomplete equilibrium because of resistance of leaf tissues to the vapor transfer, or calibration errors.

The pressure chamber is the device more commonly employed for estimating water potential of leaves and shoots. The development of this technique is essentially due to Scholander *et al.* [12], although other authors [13–14] worked on the same bases before them. There is experimental evidence that leaf Ψ measured with the pressure chamber is quite similar (numerically) to its potential measured by using psycrometry techniques.

The solute or osmotic, Ψ_s, and the pressure, Ψ_p, components of Ψ can be measured separately, but the most common practice is to measure Ψ and Ψ_s, and then to calculate Ψ_p as the difference between them. The solute potential is measured directly from the cell liquid extracted from the tissue after destroying the cell walls by freezing (immersion in liquid nitrogen). The measuring methods are refractometry, based on the refraction change; cryoscopy, based on determination of the decline in the freezing point of water in a sample of cell liquid; and thermocouple psycrometry in which Ψ_s is measured after eliminating the turgor pressure [7].

Several nondestructive methods have been developed for the study of the water content changes within plant organs [7]. The change of leaf water content is detected from β particles absorbed by a leaf irradiated with these particles. In trunks and shoots, the diameter size allows estimation of changes in water content. Dendrometers allow detection of both stem contractions and dilations associated with the day cycle of the plant's water content [15].

Water stress affects stomata behavior, and so, the stomatal conductance can be considered as a plant-stress indicator. In addition, stomatal conductance is well correlated with the rate of photosynthesis, which also depends on water status. Stomatal conductance usually is measured with a diffusion porometer, which allows precise measurement *in situ*. There are two kinds of porometers [11]: carrier porometers, where the leaf is in contact with a chamber equipped with sensors capable of responding to changes in humidity, thus changing electrical resistance; and steady-state diffusion porometers, where measurement of the dry-air flow that is necessary to balance water transpired by the leaf within the chamber offers an estimate of stomatal resistance, by keeping the vapor density constant within the chamber.

When stomata close partially or fully under water stress, the energy balance of the crop canopy is altered. Hence, increasing its values also alters the canopy temperature. The transpiration process leads to a cooling effect on the leaf in relation to air temperature. Differences between air and leaf temperature are related to the leaf's hydric potential [16].

In stressed plants, Ψ is low and the water tension within the xylem is very high, which suddenly induces discontinuities in the liquid continuum. As a result, spaces are formed and filled with air or water vapor and the water flow is disabled. This process is known as cavitation. The greater the water deficit, the greater is the number of cavitated spaces [17]. The explosion frequency might be considered as a specific index of the hydric status for a crop species [7].

The growth of leaves and branches is undoubtedly the process most susceptible to water stress [7]. Several studies have shown that leaf Ψ depletion under a threshold (obtained with nonstressed plants) reduces organ growth [11]. This suggests that the water status of plants might be inferred from measures of leaf and branch growth.

The most obvious general effects of water stress are the reduction in plant size, leaf area, and crop yield. Water deficits affect the major growth and metabolic processes of plants. Plants can maintain their turgor potential by osmotic balance. However, species and cultivars vary in amount of osmotic balance. Organs of the same plant differ in the amount of osmotic balance, which may be important for short-term survival under drought, but not for long-term survival. Betaine and proline are two osmolytes that accumulate under drought and permit a plant to lower its osmotic potential without damaging metabolic functions. Accumulation of osmolytes differs among species and cultivars and may be subject to genetic control.

The quantity and quality of plant growth depend on cell division, enlargement, and differentiation, and all are affected by water deficits, but not necessarily to the same extent. In addition to affecting leaf expansion, water deficits can affect leaf area by senescence and death of leaves during all phases of growth. Water deficits also reduce tilling and increase the death rate of tillers in multitilled species, thereby influencing the leaf area of a plant canopy. The effect of stress during the vegetative stage is the development of smaller leaves, which reduces the leaf area index (LAI) at maturity and results in less light interception by the crop.

The effects of water stress on reproductive differentiation are even greater. Spike elongation and spikelet formation of cereals are inhibited by water stress. If this occurs at early anthesis, flowers are injured and the number and size of seeds are reduced.

Water deficits are sufficient to close stomata and reduce photosynthesis and also to decrease dark respiration. The decrease in the rate of dark respiration is less than the decrease in photosynthesis. A review of the literature [18] concluded that photorespiration was unaffected by short-term stress in crop species but that, ultimately, photorespiration decreased as the substrates for photorespiration were depleted.

Generally, mild water stress for brief periods results in reductions in protein synthesis, hydrolysis of proteins, and accumulation of amino acids, especially proline. Severe prolonged stress is necessary for major damage to occur. Under moderate to severe stress conditions, the amino acid proline increases in larger concentration than any other amino acid. Proline seems to aid in drought tolerance, acting as a storage pool for nitrogen and/or as a solute molecule reducing the Ψ_s of the cytoplasm [19]. At extreme levels of stress, respiration, CO_2 assimilation, assimilate translocation, and xylem transport rapidly decrease to lower levels whereas the activity of hydrolytic enzymes increases. Chlorophyll synthesis is inhibited at higher water deficits.

The effects of water deficits on the actual distribution of assimilates to the various plant organs depends on the stage of development of the plant, the degree of and previous periods under stress, and the degree of sensitivity to stress of the various plant organs.

With reduced water potentials, plant hormones also change in concentration. For example, abscisic acid (ABA) undoubtedly increases in stressed leaves and fruits, leading to stomatal closure. Several studies indicate that drought-resistant cultivars accumulate more ABA than do drought-sensitive cultivars. Increased ethylene production is a general response of plant tissues to environmental stress, including water stress. Cytokinin (CK) content decreases in tissues under conditions of water stress. Leaf senescence associated with water stress could be due to a reduced supply of CK. Both CK and ABA can modify the rate of ethylene synthesis in water-stressed leaves, emphasizing the importance of considering hormonal interaction. Evidence exists that water stress causes a moderate reduction in indoleacetic acid.

The effect of water stress on production has been the most relevant aspect from an agronomic point of view. The amount of injury caused by water stress depends to a considerable extent on the stage of plant development at which it occurs. Reproductive stages of growth are often the most sensitive to drought. The effect of soil water deficit on flowering and fruit formation depends on the timing and severity of the water deficit.

Flowering is affected most severely by water stress at or just before peak flowering. Possible causes include reduced photosynthate supply, reduced turgor, and low relative humidity. The effect of water deficit during fruit formation is primarily to reduce the number of fruits formed while scarcely affecting weight per fruit. The effect of drought depends on the timing of water deficit relative to achieving a sufficient number of fruits of that minimum size and on availability of photoassimilates to fill the existing fruits. Grain yield is determined more by the total photosynthesis taking place during the entire season than by photosynthesis occurring during the grain-filling period alone.

Water deficits during seed filling have been reported to reduce weight per seed and weight per fruit. Even different stages of grain filling are differentially sensitive to drought. Kernel growth is decreased at lower water potential than photosynthesis.

For seed yield, the timing of water stress may be as important as the degree of stress. In the case of several species, such as maize, a severe four-day stress at certain stages of the reproductive cycle might be critical [20]. Pollination (silking) and the two weeks following are the periods most sensitive to water stress; the number of kernels per ear is the yield component most drastically affected. Three weeks after pollination, water stress no longer affects kernel number but does decrease kernel weight. A similar pattern also exists for wheat, another determinate species [21].

Indeterminate crop species that have the potential to flower over a longer period of time may be less sensitive to water stress. Short-term severe water stress during early flowering of soybeans causes little reduction in seed yield because the plant has time to germinate more flowers after stress is removed [22]. However, flowers produced late in the flowering period are less likely to produce mature pods. The yield component most influenced by water stress at flowering is the number of pods per plant. The stage most sensitive to water stress is late pod development and midbean filling.

Root elongation and dry weight are not affected by water deficit as much as leaf area, stem elongation, and dry weight of tops. Roots extend into areas where available water is not depleted, resulting in less reduction of cell elongation. Thus, root-shoot ratios generally are increased by water stress.

Water stress is not always injurious. Although it reduces vegetative growth, it sometimes improves the quality of plant products. For example, moderate water stress is said to improve the quality of apples, peaches, pears, and prunes, and to increase the protein content of hard wheat [3].

5.5.3 Effects of Water Excess on Plant Physiology

In waterlogged soils, the volume occupied by air is reduced to less than 10% of the total volume of the soil. A low availability of oxygen in the root zone also can be caused by high temperatures, organic matter, or a combination of these factors. The presence of excess water greatly modifies the soil profile, altering its chemical and biological properties. This same physical change in the rooting medium, characterized by the complete disappearance of a gaseous stage, could prove to be the most damaging element for plant survival. In this situation the absorption of water by roots is limited, producing stress symptoms similar to those of drought, although the causes and mechanisms are distinct [23]. Figure 5.38 outlines the processes leading to reductions in plant growth and crop yields due to excessive soil water.

The wilting of aerial organs is one of the first signs of the effects of excess water. In anaerobiosis, even during a short period of time, roots become less permeable to water, therefore increasing resistance to absorption. Water lost through transpiration exceeds the water absorbed by the roots, causing changes in the water status of plants, drop in

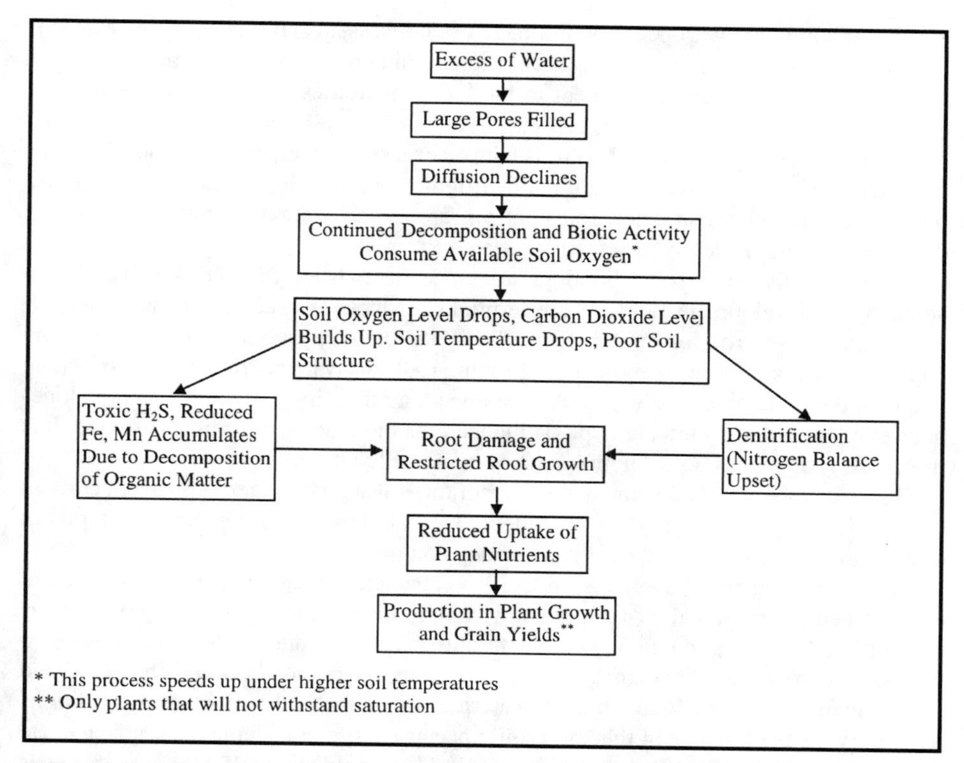

Figure 5.38. Processes leading to reductions in plant growth and grain yields due to excessive soil water.

leaf water potential, stomatal closing, reduction in stomatal conductance, and decrease in the rate of photosynthesis [24]. Nevertheless, many studies show that flooding provokes stomatal closure while maintaining leaf water potential [25].

Several biochemical reactions that are required for plant nutrition and growth depend directly on respiratory metabolism. The absorption of nutrients is relatively constant as long as the concentration of oxygen does not descend below 10% of the total volume of the soil [26]. When oxygen is lacking, a differential rate of absorption of elements is observed. The order of sensitivity is $K > N > P > Ca > Mg$ [27]. Sodium, on the other hand, accumulates in plants when aeration is insufficient.

Other effects are chlorosis and lower enzymatic activity (carboxylase), epinasty, abscission of leaves, hypertrophy of shoots, adventitious root formation, lenticel formation, decreased dry-weight accumulation, reduced root and root hair formation, root death, greater susceptibility to pathogen attacks, and finally plant death. In some species, flooding causes formation of aerenchyma in stems and roots or, less commonly, promotes petiole or shoot elongation. These symptoms are very similar to the effects of ethylene. Oxygen deficiency or hypoxia in the soil also may influence root and shoot hormone links, which affects plant response to waterlogging. In general, prolonged or intense periods of rainfall combined with poor soil drainage often increases ABA content and ethylene and indoleacetic acid in the aerial organs, whereas gibberellins and cytokinins at the root apex diminish. Also, proline content in the leaves increases and the nitrate-reductase enzyme is reduced.

Some species are able to adapt to soils with an excess of water, thanks to modifications in structure and functioning. The roots can lignify and develop tissues containing air pockets. On the other hand, new roots appear rapidly in resistant species, whereas they develop much more slowly in sensitive species.

Until recently, studies of flooding tolerance have been focused specifically on aerenchyma development [28] and the terminal products of anaerobic respiration [29] rather than on the role of hormones, such as ethylene production. However ethylene apparently plays an important role in improving root survival and thus survival of whole plants under anaerobic conditions of prolonged waterlogging [30]. Enhanced endogenous ethylene levels following hypoxia induced the breakdown of critical cells and thus promoted lysigeneous aerenchyma in roots of maize [31].

However, the role of ethylene in the promotion of lysigeneous aerenchyma in cortical cells of rice is apparently cultivar dependent [32]. Ethylene also stimulates development of adventitious roots and influences root extension rate.

The amount of injury depends on the crop species, the cultivar, the stage of plant development, the soil and air temperatures, and on the duration of waterlogging. Intermittent flooding is less damaging than continuous flooding, and the longer the duration of flooding, the greater the damage. There is a strong interaction between crop yield and soil temperature during inundation or waterlogging. High soil temperature conditions lead to oxygen deficiency in the soil profile because oxygen is depleted at a faster rate than it can be replenished from the atmosphere. Flooding during different growth stages has different effects on crop growth and final yield. Flooding during the early growth stage has the greatest impact. In grain crops, oxygen deficits provoke a yellowing of the plant and premature aging of the oldest leaves. Tilling, flowering, and grain filling are the growth stages most sensitive to water excess in cereals.

5.5.4 Water-Use Efficiency (WUE)

The term "transpiration ratio" (TR) was introduced [33] to quantify the relation between crop transpiration and crop dry-matter production. TR is defined as the mass ratio of crop transpiration to crop dry-matter production. This term was later called the transpiration coefficient. The TR concept has been expanded and improved by normalizing the transpiration by the mean daily potential evaporation during the growing season [34].

The analysis of photosynthesis and transpiration based on physiology allows a more clear comprehension of those processes [35]. Water evaporates from the inside of the stomatal cavity and passes to air through the stomatal opening. During photosynthesis, CO_2 transfers in the opposite direction. The transpiration rate, T_L, per unit of leaf area is determined by the vapor-pressure difference between the evaporative surface within the leaf (e_o) and air (e_a), and by the resistance to the water-vapor transfer:

$$T_L = \frac{\rho\varepsilon(e_o - e_a)}{p(r_s - r_a)}, \tag{5.171}$$

where r_s and r_a are stomatal resistance and layer (aerodynamic) resistance to the diffusion of water vapor, ρ is the humid-air density, ε is the ratio of molecular weight of water vapor in relation to that of the air, and p is the air pressure. The photosynthesis rate, N_L, per unit of leaf area, is determined by the difference between the concentration of CO_2 within the leaf (c_o) and the outside air (c_a) and the resistance to the CO_2 transfer:

$$N_L = \frac{c_a - c_o}{r'_{st} + r'_a} \tag{5.172}$$

where r'_{st} and r'_a are the stomatal resistance and the boundary-layer resistance, respectively, to the CO_2 transfer. Thus, the photosynthesis transpiration ratio is expressed as

$$\frac{N_L}{T_L} = \frac{(c_a - c_o)(r_{st} + r_a)p}{(e_o - e_a)(r'_{st} + r'_a)\rho\varepsilon}. \tag{5.173}$$

It has been noticed that, because the daily average temperature of a leaf is normally very close to the air temperature, $(e_o - e_a)$ will be almost equal to the vapor-pressure deficit $(e_s - e_a)$, where e_s is the saturation vapor pressure and e_a is the actual vapor pressure [36]. Equation (5.174) can be inferred by using the former estimates, which are derived by considering the behavior of an isolated leaf, as well as assuming that crops may be similarly described:

$$\frac{N}{T} = \frac{k}{e_s - e_a}, \tag{5.174}$$

where k is a crop-specific constant. The review by Tanner and Sinclair [37] suggests that k is a very stable term within the crop system and that transpiration efficiency, N/T, depends on $(e_s - e_a)$.

WUE is defined differently. The photosynthetic WUE is defined as the ratio of leaf net assimilation to leaf transpiration, A/T [38]. When WUE is enhanced, more carbon is gained per unit of water used by plants [39].

The biomass WUE of field crops is defined as follows:

$$\text{biomass WUE} = \frac{\int_{t_0}^{t_f} \text{aboveground biomass}}{\int_{t_0}^{t_f} ET} \tag{5.175}$$

where ET is plant or canopy evapotranspiration and t_0 and t_f are initial and final time limits for the integration. The timescales are medium (week) and long term (season). When soil evaporation in ET is assumed to be negligible, biomass WUE equals biomass water ratio (BWR) [38].

Agronomists have preferred the ratio of usable or marketable yield (which is expressed as fresh weight) to unit mass of ET. The yield WUE is defined as the product of aboveground consumptive biomass WUE (biomass WUE) times the harvest index (HI). In this case, the timescale is the whole crop cycle. HI is the ratio of usable reproductive-organ yield to total dry-matter production.

The two main factors that directly affect the WUE of a crop are crop species and genotypes (through regulation of internal CO_2 concentration by the photosynthesis pathway) and environmental conditions (mainly through the atmospheric humidity, radiative flux, temperature, precipitation, and evaporative demand).

Biomass WUE may differ between species. Large differences in biomass WUE occur when species are categorized by CO_2 fixation pathways. It is now accepted that the biomass WUE of C_4 species, so-called because of the foul-carbon malic or aspartic acid present in the basic photosynthesis, is generally higher than that of C_3 species, so-called because of the three-carbon phosphoglyceric acid (PGA) present in the basic photosynthesis reaction. Earlier field data for biomass WUE [33], when regrouped into C_3 and C_4 species, illustrate a two-fold increase for C_4 species. Stanhill [40] reported values of BWR of 2.9 to 3.7 $kg\,m^{-3}$ and 1.2 to 2.7 $kg\,m^{-3}$ for the C_4 and C_3 species, respectively. Differences between C_3 and C_4 species increase as the temperature rises from 20°C to 35°C. The factors contributing to the higher biomass WUE of C_4 species include higher photosynthesis and growth rates under high light and temperature, and more stomatal resistance [18].

The major environmental factor influencing the WUE of a crop is vapor-pressure deficit (VPD). Monteith [41] questioned whether transpiration or photosynthesis was the independent process in the TR. Because the internal CO_2 concentration is regulated by the plant species through its photosynthesis pathway, he argued that plants dynamically balance their leaf conductance to maintain this CO_2 concentration, thereby indirectly regulating transpiration.

Increasing atmospheric CO_2 concentrations generally escalates photosynthetic rates and the plant biomass production. A doubling of atmospheric CO_2 concentrations increases biomass production by an average of 33% in several vegetal species [42]. This buildup in biomass production, coupled with a reduction in ET, results in a significant increase in plant WUE. Both forest and agricultural species have been shown to double WUE under a doubling of atmospheric CO_2 concentration [43]. More than 500 studies analyzing the effects of increasing atmospheric CO_2 concentration have reported an acceleration in crop yield, biomass production, leaf area, and photosynthetic rates, as well as decrease in plant water-use requirements [44]. An increase in biomass and leaf area implies that plants can transpire more water. However, the reduction in transpiration caused by an increase in stomatal resistance may result in a cumulative decrease in evapotranspiration [45]. The photosynthetic reactions of C_3 plants are more sensitive than those of C_4 plants to increased CO_2 concentrations, resulting in a larger escalation in biomass production in C_3 plants.

Air temperature usually operates through its effect on VPD, which is the major environmental factor influencing WUE of a crop species.

Other environmental factors that influence WUE are radiative flux and soil water availability. Diurnal changes in biomass WUE occur in response to radiative flux and VPD. Seasonal changes may be related to different temperature or growth stages. For maximum biomass WUE, there is an optimum radiation that is usually lower than the radiation incident on a leaf oriented normal to the sun [46].

Biomass WUE generally is increased by drought because of reduced evaporation from the soil surface, whereas yield WUE is generally less because of adverse effects on reproductive growth. These effects depend on intensity and duration of the stress, along with the phenological stage of the plant at which stress is occurring. In dryland cropping, where water deficits cannot be controlled by irrigation, a severe deficit can considerably lower yield WUE.

In subhumid regions, because of the low VPD, crop production and BWR may be substantially higher than in arid regions [37].

Other factors interact with these two main factors, including soil infiltrability, water-holding capacity and water depth, soil salinity, soil fertility, weed competition, pests and diseases, cultivation techniques (seeding rates, seeding date, row spacing, and tillage), and institutional and socioeconomic factors.

Improved crop management and plant breeding have led to substantial gains in WUE. Most of these gains are derived from increased leaf area production, larger water availability due to deeper roots and/or better water extraction, and an increase in HI.

WUE can be increased [47] by increasing the HI; reducing TR; reducing the root dry matter; and increasing the transpiration component relative to the other water-balance components, in particular reducing the evaporation of soil surface, the drainage below the root zone, and the runoff.

Three options exist for improving the TR for dryland crop production systems [48], which also apply to irrigated crop production: plant manipulation to increase the photosynthesis rate; utilization of the growing season with the lowest atmospheric VPD; and manipulations of crop nutrition so that less dry matter is partitioned into the roots without limiting their water extraction capacity.

WUE can be enhanced by modifying cultural practices, such as land preparation, irrigation, cultivation, and crop selection or genetic modification of the plant [49]. The latter provides short-term solutions, whereas the former is more of a long-term solution. Five options for improving the WUE are biochemical modifications, control of stomatal physiology, HI improvement; crop microenvironment changes, and increase in the fraction of transpired water out of the total amount of water used during the growing season [50]. Among these options, the first two are in the domain of plant breeding and genetics and the third is partly so.

WUE is not the same as drought resistance. Most of the research on WUE has been oriented toward attaining high yield WUE while maintaining high productivity. In drought resistance research, emphasis often is placed on survival during periods of high atmospheric demand and low water availability. In many cases, the ability to withstand severe moisture stress is negatively correlated with productivity. Many species that can tolerate severe water deficits do not make efficient use of water in the absence of stress. Some species, well-adapted to severe water deficits, have moderate efficiency even in the presence of stress.

Drought resistance can take the form of tolerance or avoidance. Avoidance is accomplished by reducing the length of the season or by increasing rooting depth and water extraction from the lower profile. Drought tolerance refers to plant adaptation to deficits through physiological and morphological processes that limit transpiration losses and/or condition plants to withstand lowered water potentials. The seasonal progression of temperature, the distribution and intensity of rainfall, and the availability of soil moisture largely govern which attributes of the plant might be beneficially altered to enhance WUE. These attributes might include deep rooting to exploit stored soil moisture during long droughts, stomata that close quickly at threshold water potentials, waxy leaves to increase reflection of radiation, reduced stomatal density to limit transpiration, leaf rolling to reduce light interception, survival of protracted drought, premature leaf

Table 5.17. Drought resistance mechanism and traits for plant breeding strategies

Mechanism	Characteristics/Traits	Benefits
Drought escape		
Rapid phenological development	Short biological cycle	Lower total water demand
Development plasticity	Branching/tillering and variation in flower, floret, and panicle	Lower reduction in seed numbers
Drought avoidance		
Reduction of water losses		
stomatal resistance(+)	Size, number, and opening of stomata	Less transpirate
evaporative surface(-)	Leaf rolling, smaller and fewer leaves, senescence	Smalter loss surface and less radiation absorbed
radiation interception(-)	Leaf pubescence and leaf orientation	Higher reflectivity and less radiation
cuticular resistance(+)	Thicker and tighter cuticules	Lower transpiration, higher resistance to desiccation
epicuticular wax(+)	Waxiness	Lower transpiration, higher resistance to dissecation
Maintenance of water extraction		
root depth and density(+)	More extensive and intensive rooting	Lower root and soil resistances
liquid- phase conductance(+)	More or larger xylems in roots and stems	Lower resistance to water fluxes
Drought tolerance		
Maintenance of turgor		
osmotic adjustment	Water potential kinetics	Decrease osmotic potential in response to stress
cellular elasticity(-)	Cell membranes	Larger changes in volume
cell size(-)	Cell size	Increased bound water fraction (in cell wall)
Tissue water capacitance	Favorable water potential kinetics	Ability to maintain the daily water balance
Dissecation tolerance	Photoplasmic and chloroplast conditions	Maintenance of photosynthesis
Accumulation of solutes	Abscisic acid, ethylene, proline, betaine	Regulation of senescence and abscission

(+)Increment.
(-)Depletion.

senescence, or maintenance of extension growth under severe stress. These processes have been reviewed [9]. Table 5.17 summarizes this information from the literature from which plant breeders can better plan their strategies.

Genetic variation in WUE has been detected in crop species [33, 51–53], but most crop improvement programs do not emphasize WUE, even though this should be an important trait in water-limited environments. The lack of simple, rapid, and reliable screening criteria and measurement techniques for WUE has greatly restricted progress in this critical area of crop improvement.

Plant characteristics are needed for breeding programs that are associated with WUE and are easier to measure, and thus, more suited to selection [52].

Highly significant genotype and drought effects on specific leaf weight (SLW = dry weight/unit projected leaf area) have been observed [53]. However, genotype differences in SLW were not associated with differences in WUE. Under drought, the root/shoot ratio was 32% higher than under wet conditions, but genotype differences were not also associated in this ratio with WUE classes.

WUE increases under dry conditions because of decreased water use due to reduced stomatal conductance [54], whereas the ratio of total plant biomass to leaf area (l_a) increases because of drought-induced reductions in the rate of leaf expansion and an increase in allocation of carbohydrates to roots. Virgona *et al.* [55] reported a similar association between l_a and WUE for sunflower genotypes under well-watered conditions, and suggested that they both might be positively associated with photosynthetic capacity.

Techniques involving stable C isotopes may provide an efficient method for estimating integrated WUE in tissue of C_3 plants [56]. During photosynthetic C assimilation, plants discriminate against ^{13}C, fixing a relatively larger portion of CO_2 composed of ^{12}C. Variation in ^{13}C discrimination in C_3 plants depends on leaf intercellular CO_2 concentration (c_o) [57]. Data supporting this theoretical relationship have been provided for a broad range of C_3 plants and have been recently reviewed [56]. That variation has been shown to be negatively related to WUE in several C_3 crop species and has been proposed as a criterion to select for improved WUE in plant breeding programs [58].

Nonetheless, although such manipulation of plant characteristics remains a goal for plant breeders, the most immediate and dramatic increase in WUE will be achieved through improved crop management.

Comparisons between dryland and irrigated conditions indicate that the yield reduction under dryland conditions can be attributed to a decrease in aboveground biomass production and to a reduction in HI. During the past 100 years, HI of new cultivars has been increased through plant breeding. Essentially, the same total dry matter is produced by new cultivars and, if the growing season has not been shortened, ET remains about the same. The resulting increases in WUE with new cultivars have been mainly from increases in HI. For many crops, HI is not totally controlled genetically but is influenced by environmental factors. This allows the farm manager some responsibility for efficient water use.

5.5.5 Water Yield Functions

Yield ET Relationships

Crop-water productions relate total dry matter (DM) or crop yield (Y) to either soil moisture content or soil water tension, evaporation, transpiration (T), evapotranspiration (ET_c), or applied irrigation water (IR).

It has been shown, in container experiments [34], that DM production is linearly related to the ratio of transpiration to pan evaporation in climates with mostly clear skies during the growing season. The relationship is expressed as

$$\mathrm{DM} = mT/E, \tag{5.176}$$

where DM is crop dry-matter production (kg), m is a crop-specific proportionality rate (mm day^{-1}), T is transpiration (kg), and E is pan evaporation (mm day^{-1}) averaged over the growing period. However, in several experiments in The Netherlands, production was

$$\mathrm{DM} = nT, \tag{5.177}$$

where n is a crop-specific coefficient (kg kg^{-1}).

Arkley [59] showed that the ratio between T and crop production may be distorted by advection; therefore, he suggested a modified equation for crop production:

$$\mathrm{DM} = k_a T/(100 - \mathrm{RH}), \tag{5.178}$$

where k_a is a crop-specific coefficient and RH is the mean daily relative air humidity (in percentages) during the growing season.

Arkley [59] also studied the relationship between crop dry-matter production and the ratio of transpiration to VPD arranged in the following form:

$$\mathrm{DM} = k'_a T/\mathrm{VPD}, \tag{5.179}$$

where k'_a is a crop-specific coefficient. Because $H = 100e_a/e_s$, the coefficients are related as follows:

$$k'_a = 0.01 e_s k_a \tag{5.180}$$

Because e_s is temperature dependent, the coefficient k'_a would be temperature dependent also.

For studying only the effects of limited water on crop production, Eqs. (5.176) and (5.179) could be simplified because m and E would be constant for a given crop in a given year. The following expression was proposed [60]:

$$\mathrm{DM}/\mathrm{DM}_m = T/T_m, \tag{5.181}$$

where DM_m is maximum or potential dry-matter production (kg ha^{-1}), T is transpiration for the growing season (mm), and T_m is the maximum or potential transpiration when soil water does not limit transpiration or yield. T/T_m is an indicator of the relative amount of CO_2 exchange and water exchange taking place at the leaf surface through stomata.

A generalized production function is described [61] as follows:

$$1 - Ya/Ym = \beta(1 - ET_a/ET_m), \tag{5.182}$$

where Y_a is the actual yield (kg), ET_a is the actual seasonal ET (mm), Y_m is the maximum yield (kg), ET_m is the seasonal ET for maximum yield (mm), and β is the water-yield sensitivity coefficient. The normalization of production functions (i.e., its graphic representation) considering the grain yield and the seasonal ET in relative terms (as percentage of its maximum values) is very useful for its comparison with the results attained in other works.

Figure 5.39 represents the normalized production function for maize crops from California [61], North Dakota [62], and Zaragoza, Spain [63].

FAO adopted this model [Eq. (5.182)] to develop a method of quantifying the relationship between yield and water [64]. Y_m is calculated by de Wit's [34] approach

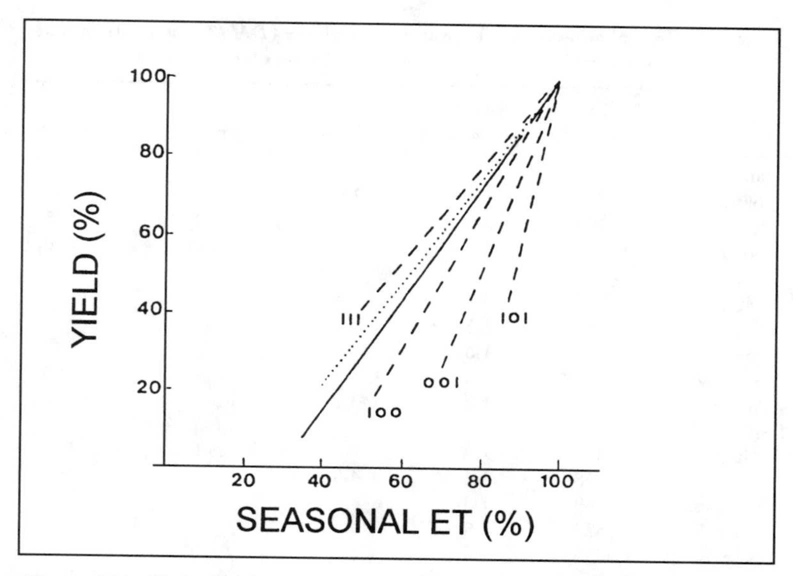

Figure 5.39. Normalized corn production functions: (- - -) [61], (........) [62], and (—) [63].

modified with several correction factors, and ET is determined by FAO methods [65]. Table 5.18 gives some selected values of β for a few crops [64]. For further reading see [66–68].

Equation (5.182) is used widely because of its simplicity. It is not difficult in most instances to estimate a reasonable value for β for prediction purposes. In many situations, β is close to the same from year to year for the same crop.

An equation accounting for variable water supply at different crop growth stages has been developed. Each stage is weighted by a sensitivity factor. It is suggested that the productivity related to the end of a growth stage be determined in order to estimate the reduction of yield in the subsequent growth stages. Such a hypothesis generally is accepted for the majority of herbaceous crop species. The following multiplicative equation applies:

$$Y_a/Y_m = \prod_{i=1}^{n}\{1 - \beta_i[1 - (ET_a)_i/(ET_m)_i]\} \qquad (5.183)$$

where i is the generic growth stage; and n is the number of growth stages considered. This equation was found to give improved predictions in only a few instances compared to the simpler Eq. (5.182).

Hanks [68] and Howell *et al.* [66] reviewed many types of yield ET models. Advanced yield ET models mainly address the effects of severe water deficits during critical crop growth periods and are called multiperiod or dated models. The effects of water deficits at specific crop growth stages are most likely related to stress effects on HI and not necessarily on TR.

Table 5.18. Transpiration ratio (TR), water use efficiency (DM/T), normalized water use efficiency (k_d), and yield sensitivity (β) values for selected crops

Crop	TR[a] (kg kg^{-1})	DM/ T[b] (kg m^{-3})	k_d[c] (kg k Pa m^{-3})	k_d[d] (kg k Pa m^{-3})	β[e]
Grains					
Barley	518	1.9	2.9		
Corn	350	2.9		9.5 (11.8)	1.25
Sorghum	304	3.3	8.3	13.8 (11.8)	0.90
Oats	583	1.7			
Millet	267	3.7	9.4		
Rice	682	1.5			1.1
Rye	634	1.6			
Sunflower	557	1.8			0.95
Wheat	557	1.8	3.1		1.10–1.15
Legumes					
Alfalfa	844	1.2		4.3 (5.0)	0.7–1.1
Chickpeas	638	1.6	4.8		
Cowpeas	569	1.8			
Guar	523	1.9			
Navy beans	656	1.5			
Peanuts			3.9		0.7
Soybeans	715	1.4		4.0 (4.1)	0.9
Root					
Potatoes	575	1.7		6.5 (5.5)	1.1
Sugar beets	377	2.7			0.7–1.1
Fiber					
Cotton	568	1.8			0.85
Flax	783	1.3			

[a] [40].
[b] [40].
[c] [67].
[d] [37]. Numbers in parentheses are theoretical estimates.
[e] [65].
Source: [66.]

Crop management (cultivar selection, planting date, sowing date, spatial distribution of the crop stand, fertilizer applications, etc.) directly affect crop yields through effects on radiation interception, the partitioning of transpiration from total ET, and the partitioning of economic yield from total dry matter.

Crop selection is an important management decision for maximizing profits. The cultivars or hybrid chosen may affect yield in several ways [66]: crop cycle length; HI (partitioning of dry matter into economic yield); disease and pest insect resistance; harvesting influences (avoidance of lodging, fruit, grain, or lint retention, etc.); and harvesting quality (marketing and/or storage properties). Although small differences may exist for TR within a species, major differences in other yield factors exist.

Early sowing can allow the crop to grow under conditions of lower evaporative demand, resulting in increased water-use efficiency as analyzed by Fereres *et al.* [69].

HI is affected by sowing rate and row spacing of the crop. Sowing density and spatial distribution of the crop determine the resulting crop geometry and affect crop yield and water influencing the solar radiation interception and the partitioning of T from ET.

Crop-production evapotranspiration relations assume that crop nutrition is adequate and nonlimiting to production. Crop-fertility management can have both positive and negative effects on crop productivity when the water supply is fixed and/or limited. Fertilizer applications to a crop with limited available water could result in early depletion of the limited soil water and the development of severe water deficits during later critical crop development stages. With sufficient available soil water and a nutrient-deficient soil, nutrient additions can result in improved root development and soil water extraction from deeper soil layers, resulting in yield increases. Inadequate crop nutrition under irrigation most often affects crop yield through reductions in leaf area, dry matter, and HI.

Removal of crop residues likewise affects soil nutrient availability and water relations. Such losses of nitrogen over a period of years eventually would reduce soil, fertility levels and also reduce WUE.

Weeds compete with crop plants for light, water, and nutrients, and especially under conditions of water deficits, this leads to yield losses. Weeds can be controlled by tillage and herbicides, as well as by single or combination crop rotations. The increased WUE resulting from tillage and herbicide applications manipulates the field water balance to differentially increase the partitioning of T from ET. Reducing tillage reduces the exposure of wet soil to evaporation, thereby conserving soil water but requiring chemical weed control to minimize transpiration from weeds. Too many tillage operations may increase the risk of soil crusting, runoff, and erosion.

Yield Salinity Relationships

Many arid or semiarid soils contain concentrations of soluble salts that have a negative impact on the efficiency of water use. In irrigated agriculture, salinity is probably the second most important yield constraint to irrigation. In addition to a direct osmotic effect and a possible toxicity of specific ions, soil salinity and salts present in the irrigation water may have a deleterious impact on physical properties such as infiltrability, water holding, and aeration, especially if the soil or the water are rich in exchangeable sodium.

The most desirable characteristics in selecting a crop for irrigation with saline water are high marketability, high economic value, ease of handling, tolerance to salts and specific ions, ability to maintain quality under saline conditions, and compatibility in a crop rotation [70]. However, no crop is outstanding in all of these categories. For example, the economic value per crop area is negatively correlated with crop salt tolerance [71], and many high-value crops are sensitive to specific ions.

Because saline conditions reduce both plant growth and seasonal ET, it is important to develop information on crop water-production functions under saline conditions.

Van Genuchten [71] established four crop response models or functions of salinity. SALT software offers several options for each model, whose application depends on information experimentally generated. Most salinity response studies have reported only the economic-yield component. Few studies have reported any differential yield effects between the economic yield and dry-matter yield as a result of salinity, although crop quality may be differentially affected.

A linear response model was developed [72]:

$$Y = Y_m \qquad 0 \leq \text{EC} \leq \text{EC}_t \tag{5.184}$$

$$Y = Y_m - Y_m s(\text{EC} - \text{EC}_t) \qquad \text{EC}_t < \text{EC} \leq \text{EC}_0 \tag{5.185}$$

$$Y = 0 \qquad \text{EC} > \text{EC}_0, \tag{5.186}$$

where EC_t is the salinity threshold (dS m^{-1}) (salinity from Y starts to decrease), s is the sensitivity of the crop to salinity above the threshold level and EC is the electrical conductivity of the soil solution on the basis of saturation extracts (dS m^{-1}). These relationships are applicable where C is the main ion that affects yield.

Even though the model in Eq. (5.184) is the most widespread due to its conceptual comprehension and the relative simplicity of fits, it is evident that the concept of threshold salinity is a simplification of the crop response, which is actually curvilinear in shape.

Yield Irrigation-Depth Relationships

Irrigation is the primary management practice that is utilized to increase crop yields (or to permit crop production) in semiarid and arid climates. In addition, irrigation is used to stabilize crop yields when rainfall is unreliable in subhumid and humid climates.

Including irrigation depths (IR) in the ratios of DM production or marketable yield (Y) to transpiration (T) or evapotranspiration (ET) is necessary when making both engineering and economic decisions, to determine the optimal irrigation depths. The difference between both kinds of ratios is due first, to the fact that all of the water evapotranspirated by a crop does not come from irrigation and, second, that all of the water applied by irrigation is not evapotranspirated by the soil-plant set.

Far fewer studies [73] have expressed the relationship of yield to applied irrigation amount. Water-production functions relating seasonal irrigation depth and crop yield have been derived for several crops by polynomial regression analyses [74]. The relation of irrigation to crop production is essentially site specific [75].

Evidence exists that the function of yield applied IR is convex in shape, in contrast to the linear shape of the yield ET function [76]. The linear [$Y = f(\text{ET})$] and the convex [$Y = f(\text{IR})$] functions coincide up to a point and then diverge as the amount of applied water increases [77]. The difference between the two curves is the non-ET portion of applied water. The relationship between DM or Y and IR depends mainly on the ET function (DM/ET or Y/ET), the irrigation salinity, and the irrigation hydrology determined by the partitioning of ET and T from IR as influenced by irrigation uniformity.

Crop yields typically are related to seasonal ET and seasonal IR as shown in Fig. 5.40 [77]. In a given season, the available soil water at planting (ASWP) and the effective rainfall (P_e) supply the crop with water to reach its dryland yield level. The applied-water yield relationship may be roughly linear up to approximately 50% of full irrigation [64].

For larger depths of applied water, the function begins to curve. Deep percolation increases with additional applied water. If the increase in applied water is associated with higher irrigation frequencies, larger evaporation may occur. The irrigation system will become less efficient as water use approaches full irrigation.

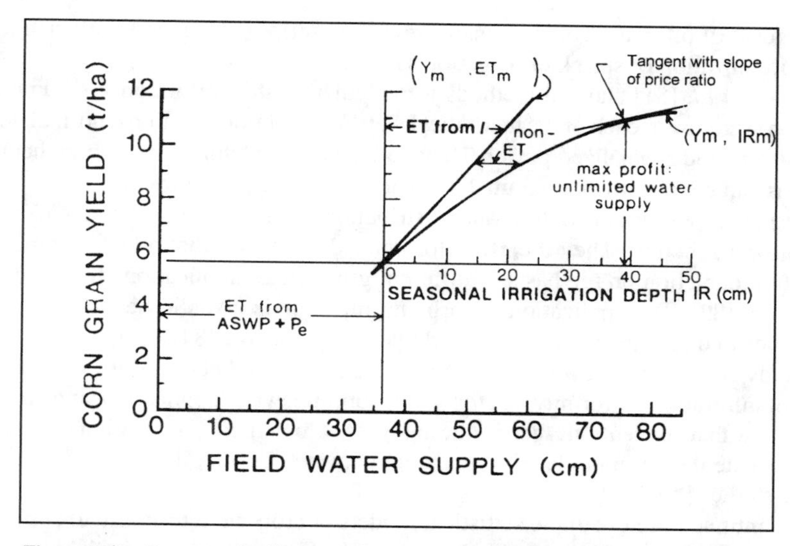

Figure 5.40. Processes leading to reductions in plant growth and grain yields due to excessive soil water. ***Source:*** **Adapted from [77].**

Because *Y* response to applied IR is a diminishing-return function, it follows that, at some point on this curve, further water application cannot be justified economically [77].

In irrigated agriculture, the uneveness (nonuniformity) with which water infiltrates the root zone is determined mainly by the spatial variability of irrigation water application and on the variability of hydrologic soil properties [78]. The yield of a given crop, grown during a specific season in a certain field and under certain management and cultivation conditions, is also spatially variable because it is assumed to be directly dependent on the spatially variable water infiltration [79]. Unfortunately, none of the known irrigation systems are able to apply water with perfect uniformity. Because of this lack of uniformity, part of the surface is adequately irrigated but other parts are not. In the areas receiving an excess of water, the price of pumping can increase, yields may decline and mineral nutrients may be leached from the soil. This leaching of fertilizers represents an economic loss of applied resources, lowering of yields, and pollution of groundwater. In areas receiving insufficient irrigation water, a drop in yield can be noticed along with a poor use of fertilizers, which are not used by crops even though the whole water volume is retained in the rhizosphere.

Uneven irrigation or soil infiltration variability can produce nonlinear yield ET functions even though the basic yield function is exactly linear.

The economic implications of nonuniform irrigation application on crop yield and profit maximization have been analyzed by various researchers. Results are critically influenced by the nature of the water/yield relations postulated for the crop species. Results show that conventional economic analyses ignoring infiltration uniformities underestimate optimal levels of applied water, often substantially.

Seginer [80] offered a comprehensive review describing a general approach to economic optimization of sprinkler irrigation systems.

Warrick *et al.* [81] showed methods for calculating the performance parameters of irrigation systems for each of the theoretical water distribution functions: normal, lognormal, uniform and specialized power. They noticed that matching raw data to theoretical functions can better estimate parameters than the direct use of raw data.

Orgaz *et al.* [82] assumed that water distribution in a sprinkler system could be fitted to a uniform function. They therefore formulated a relation that allows estimation of production depletion on the basis of different gross water applications from the irrigation system [83]. The application to sprinkler injection is given in Section 5.4. Similar approaches to drip injection are proposed by Anyoji and Wu [84].

A study by de Juan *et al.* [85] provides a general methodology for analyzing the effects of infiltration uniformity on crop yield, optimum water application, and profit. The results show that conventional economic analyses, which ignore infiltration uniformities, underestimate the optimum level for the applied water. Further developments are given in other studies [86, 87].

The continued use of simple statistical models of crop-yield response to applied irrigation water could be replaced by process-orientated models in which individual components can be evaluated and analyzed. As the development of comprehensive crop-growth models increases, the irrigation economic analyses and real-time irrigation decisions can be accomplished with expert systems that rely on crop simulation.

5.5.6 Crop Water Models

Agricultural planning is a wide management field [88].

Irrigation management consists of determining when to irrigate, the amount to apply at each irrigation and during each stage of plant growth, and the operation and maintenance of the irrigation system. The primary objective is to manage the production system for profit without compromising the environment. The major management activity involves irrigation scheduling. Most irrigation water management concepts include salinity control and the improvement of soil–air–water environment. The management strategy for an irrigated farm or region normally is dictated by economics. The benefits include increased and predictable yields, enhanced crop quality, and reduced farming risks. The principal economic variables include the price of irrigation water, the cost of applying the water, and the price received for the crop. The optimal level of applied water depends on the crop-water production function. Generally speaking, the management of irrigation systems aims to achieve optimal crop production and efficient water use or, in other terms, a reliable, predictable, and equitable irrigation water supply to farmers.

The management of irrigation systems for highest benefit requires an understanding of the many physical, biological, technical, and socioeconomic factors involved in crop production. The more information that water managers have, and the better that information is, the better the decisions they can make. During the past 25 years, the development and application of computers and computer models to describe main aspects of processes and links relative to crop growth and yield as influenced by water availability have been relevant. This development has resulted from an increased understanding of soil–plant–atmospheric processes and the rapid expansion of personal computers. However, only a few of the models have been implemented, with varying success. Limitations of most models are that they aim to support only one aspect of irrigation management, or that they are programmed for one irrigation system alone, or that their user-friendliness does not fit the skills of the users.

The development of computer simulation tools leads to several applications of modeling in irrigation management. Crop-water models focusing on the crop and the farm scales probably have been the most important development in the area of irrigation management, as evidenced in models presented by Pereira *et al.* [89, 90].

Irrigation modeling can be focused on three levels according the technical detail and application scope [90]: irrigation scheduling models, crop growth simulation models, and models for irrigation management at the system level.

The first group of models places primary emphasis on irrigation scheduling. Irrigation scheduling provides information that can be used to develop strategies for different crops under varying soil and climatic conditions. These strategies can be determined using long-term data representing average conditions or in-season factors based on real-time information and short-term predictions. Several computer-assisted irrigation scheduling techniques relying on the water-balance method are available [89–94]. The computer programs usually generate an irrigation schedule such that the irrigation event takes place when the estimated soil moisture of the root zone is depleted to a certain level. Crop water relations are described by simple yield-response functions to predict relative yield loss based on simulated evapotranspiration relative to simulated potential evapotranspiration. Some of these models are currently utilized in field practice, with interactive communication with users in real or near-real time. In general, these models apply to the field scale for irrigated crops; however, procedures usually can be expanded to enlarge the scale of application. In addition, the irrigation scheduling models are usually not crop specific and can apply to different types of irrigation methods.

A second group of models includes models that aim at a more detailed description of soil-water fluxes and/or a more sophisticated simulation of crop growth and yield formation. Thus, these models have more state variables than the former models have, but they vary considerably in level of detail. These models require specific soil and crop characteristics for a more detailed description of each physical and biological process. Consequently, the models developed at this level are used mostly in research and special applications [89, 90].

Crop systems are highly complex. The crop in the field is affected by weather, soil physicochemical factors, pests, diseases, weeds, and interactions of these many factors. A classification of systems of crop production, based on growth-limiting factors, has been proposed by de Wit [95].

Most crop simulation models are a mixture of empiricism and mechanism, and even the most mechanistic models make use of empiricisms at some hierarchical level. A mechanistic crop model generally is considered to be based on physiological and physical processes and considers cause and effect at the process level. Material (carbon, nitrogen, water) and energy balance usually are included. However, the most useful models for studying irrigation management of crops under several weather and soil conditions have been largely functional models. The term *dynamic* is used to mean that the crop model responds to daily (or more frequent) changes in the environment.

In the past decades, much research effort has been devoted to the development of crop-growth models [96]. The CERES [97, 98] and GRO [99] families of models are among the most widely used. These models simulate crop growth, development, and yield for specific genotypes, taking into account weather, soil water, and nitrogen dynamics in the soil and crop in a mechanistic manner. Because these models are based on physiological and biological concepts along with experimental data, it is believed that the simulation provides a reasonable estimation of the relationships between management practices, weather conditions, and yield.

A number of crop modeling groups have attempted to optimize water and nitrogen management over long-term historical weather data [100–104]. Models used for this purpose must have reliable soil-water and soil-nitrogen-balance components. These optimization efforts can suggest the best long-term strategy for water or nitrogen-applications (often growth-stage dependent) and also show that enhanced decisions can be made if additional information, such as weather forecast, is available. In addition to phenology, dry-matter, and final-yield data, SOYGRO [105, 106], CERES-maize [107, 108], and CERES-wheat [109] models simulate daily values of leaf area index; root-length density; biomass of leaves, stems, grain, and roots; number of leaves; soil water content; evapotranspiration; potential evapotranspiration; transpiration; yield components; and water stress. They include processes that describe the development of a reproductive structure, photosynthesis, respiration, and tissue senescence. The irrigation management components allow users to specify different strategies for managing the crop such as specific dates and amounts of irrigation for comparison and selection of the best strategies.

Crop growth models have an advantage over the response functions in that they are designed to be more robust for use in other weather conditions and they include the possibilities for studying irrigation decisions in combination with other management decisions, such as planting date, row spacing, and nitrogen fertilizer use. However, all models have certain limitations because they do not comprise all possible parameters and influences that represent the biophysical environment. Moreover, they require local calibration and validation. These models are not a panacea for all agricultural problems, and care must be used in incorporating them into research studies, but many efforts have been made to improve the general application and accuracy of the crop-growth models, particularly for design or management of irrigation systems [90].

Crop models also can be integrated with optimization procedures to allow the computer to automatically search for the irrigation strategy that maximizes profit or satisfies other optimization objective functions. Other objective functions may include considerations for energy use, WUE, nitrogen leaching, and constraints on water availability.

Management practices have to be selected so that the levels of salinity in the soil are not harmful for the crop. The selection of appropriate practices for salinity control requires quantification of the movement of salts and water in the soil, the response of the crop to soil water and salinity, and how the environment and management conditions affect these interactions. There are a larger number of research and management models reported in scientific literature that deal with water and solute movement. Seasonal models assume steady-state conditions for crop response [110]. However, these models are not suitable for irrigation management in saline conditions [111].

Transient models compute water and solute flux in the soil and include a water-uptake term. Available transient models differ in their conceptual approach, degree of complexity, and in their application for research or management purposes.

Current models for quantifying the stress imposed on crops by excessive soil water conditions are based on water-table depth [112, 113].

Several computer models have been proposed to simulate the behavior of drainage and irrigation systems [114]. These include numerical solutions of Richards' equation for combined saturated-unsaturated flow in two or three dimensions.

A useful approach is provided by SWATRE, a model for transient vertical water flow in soil [115, 116]. The model includes a sink term for water uptake by roots and two functions for flow to drains. A modern version [117] incorporates solute transport. The advantage of this approach is that it is based on sound theory for vertical water movement in the unsaturated zone. Because most of the unsaturated water movement tends to be in the vertical direction, even in drained soil, this approach should provide reliable predictions of the soil water conditions above the water table. Another advantage is that it also can be applied for soils without water tables. Because SWATRE combines concepts for soil water flow, soil salinity, and drainage, it can be used to analyze problems of waterlogging and salinity in irrigated agriculture.

A third approach to modeling drainage systems is based on numerical solutions to Boussinesq's equation. This approach normally is applied for watershed-scale systems or where the horizontal water-table variation is critical [118, 119]. These models may be used in large nonuniform areas where lateral differences in soils, surface evaluations and water-table depths are important [119].

DRAINMOD [120, 121] is a drainage and subirrigation design and operation model for drainage-subirrigation systems and it is probably the best-known and most accepted water management model in humid areas. It includes methods to simulate subsurface drainage, surface drainage, subirrigation, controlled drainage, and surface irrigation. Input data include soil properties, crop parameters, drainage-system parameters, and climatological and irrigation information. The model can be used to simulate the performance of a water management system over a long period of climatological record.

The SWARD model [122] predicts the effects of drainage on grass yield. The model estimates daily soil water budget, including drainage and irrigation, and then links with a physiological model to predict a daily grass budget.

The recently developed drainage module of model MUST [123] is capable of simulating drainage and subirrigation in relation to ET and vertical flow in the unsaturated zone. The latest version of MUST has been used to design a subsurface irrigation system and to establish operational rules.

More complex models, such as the leaching estimation and chemistry model (LEACHM) [124], PRZM [125], the groundwater-loading effects of agricultural management systems (GLEAMS) model [126], and ADAPT [127], include leaching of agrochemical pesticides and soil fertilizer elements such as nitrate. The OPUS model [128] treats the water, plant, nutrients, and pesticide systems in the agricultural ecosystem.

It includes surface-water flow and transport, processes of water flow, nitrogen cycling, pesticide movement and decay within the soil profile, and crop usage of nitrogen, phosphorus, and water. Soil water flow is simulated by using Richards' equation adapted to a multiple-horizon soil profile.

Crop models also can be used to help schedule irrigation or control pests. Use of the entire crop-growth models in this mode is not very common, except for the experience of the GOSSYM cotton-modeling group [129]. GOSSYM predicts the response of crops to variations in both environment and cultural management. The model estimates the ratio of appearance of several plant organs and takes into account the placement of every organ, the content of nitrogen and carbon, the water content, and the thermal exposition. (Crop Management Expert) COMAX is a framework of expert systems explicitly developed for working in crop-simulation models. COMAX is based on rules that imply a reasoning system, a file-storage system for the simulation-model requirements, a database system for the knowledge base, and a system for interaction with the user. It is based on menus. COMAX was developed to allow farmers and technicians using the GOSSYM model to incorporate water, nitrogen, and chemical management.

The goal of the International Benchmark Sites Network for Agrotechnology Transfer Project (IBSNAT) [130] is to accelerate the flow of agrotechnology and to increase the success rate of technology transfer from agricultural research centers to farmers' fields. To do this, IBSNAT has developed computer software that helps to match crop requirements to land characteristics using crop simulation models, databases, and strategy evaluation programs. The resulting system is called Decision Support System for Agrotechnology Transfer (DSSAT). In the DSSAT, users select as many combinations of alternative irrigation (or other) management practices as they wish; then, simulations are performed for all that were selected to allow users to select the best one.

Cropping Systems Simulation Model (CropSyst) [131] is an existing multiyear, multicrop, daily time-step model with a mechanistic approach, including a variety of agronomic management options (irrigation, fertilization, tillage, residue management, cultivar selection, and rotation selection) and environmental impact analysis capabilities (erosion and chemical leaching). CropSyst simulates the soil water budget, soil-plant nitrogen budget, crop phenology, crop canopy and root growth, biomass production, crop yield, residue production and decomposition, and soil erosion by water. CropSyst was modified for assessing crop response and water management under saline conditions [132].

The ever-increasing developments in geographical information systems (GIS) and decision support systems (DSS) may contribute to the set of input data for simulation studies in irrigation management. The package HYDRA [133] is an example. It was designed to be a versatile software package that can be employed for both strategic and operational irrigation and management purposes.

To utilize the water resources fully, to match water supply and requirements, and to reach the maximum economic benefit, the crop's response to applied water, the technical efficiency of the system, the farm's resource constraints, economic considerations at both the micro- and sectoral level, institutional factors along with the decision-maker's objectives and risk attitudes need to be considered [85]. There are many techniques that can be used to evaluate alternative irrigation equipment and strategies [134]. Crop models can be used to evaluate economic risk [135], as considered with prices and complete enterprise production costs. Bogges and Ritchie [135] also showed that irrigation reduces weather-related variability in yield, and thus may be preferred by risk-averse producers.

Crop models that predict yield and irrigation demand can be used in planning regional and watershed-level strategies for water withdrawals for irrigation [136]. In regions where water resources are limited, it is particularly important to plan the permitted area of irrigated crops and water demand for drought years.

5.5.7 Strategies for Water Conservation in Rainfed Agriculture

General Considerations

Rainfed agriculture is used as synonymous with dryland farming and deals with crop production under the constraints typical of conditions in a semiarid zone, with mainly inadequate and unpredictable precipitation. The importance of optimizing the utilization of water resources in dryland farming is disputed by many. Unfortunately, our ability to achieve substantial increases in crop production under these conditions has not been as great as it has been in irrigated agriculture. However, over the next few decades, agriculture will not be able to rely on vast increases in irrigated area to maintain the required growth in output necessary to match increases in population. This means that much of the world's future food increase will have to come from the dryland regions, where productivity is currently low. In these areas, crop production often is directed toward the reliability of yields rather than their maximization.

Dryland farming can be highly productive even though the system is, by definition, extensive rather than intensive, and naturally a riskier undertaking for the farmers. Because the dominant feature of dryland agriculture is erratic and often scarce rainfall, it is implausible to expect any great degree of intensification in the dryland farming system.

Successful rainfed farming systems require efficient, low-risk management of the soil water regime and improved WUE by crops. Genetic engineering may provide great opportunities for adapting crops to these harsh environments, although this will not be in the near future. However, further development and improvement of current and new practices of soil and water conservation could offer greatly improved exploitation of dryland regions while reducing the risks of soil degradation and the costs and risks for the farmer.

Soil Management

Water storage in soil is essential because plants use water from soil between precipitation or irrigation events. Adequate water storage in soil is important not only in dry regions (arid, semiarid, and subhumid) but also in humid regions where short-term droughts at critical growth stages can greatly reduce crop yields. Because of its importance, water storage in soil and subsequent efficient use of that water for crop production have long been studied by researchers.

In hot, dry climates, water storage is improved by minimizing soil evaporation, decreasing runoff losses, limiting deep percolation, and reducing transpiration by weeds.

Direct evaporation from the soil is a net loss to crop production and is a substantial component of the total water use of crops grown in arid and semiarid environments. In Mediterranean climatic regions, water loss by direct evaporation from the soil during the cropping season may represent 30% to 50% of the annual rainfall. Recent estimates of soil water evaporation under a lupin crop was 1 to 1.6 mm day^{-1} when total crop evapotranspiration was 1.8 to 3.4 mm day^{-1} [137]. Evaporation from soil is reduced by weed control, tillage, and leaving the stubble or other crop residue in the fields during the dry or fallow season.

Materials such as crop residues, plastic films, petroleum-based products, gravel, and soil itself have been studied widely as potential mulches for decreasing evaporation. However, the effect of mulches on evaporation is difficult to establish because of the interacting influences of mulches on soil water infiltration, distribution, and subsequent evaporation.

Adding external materials on the soil surface is generally, but not always, restricted to ornamentals, vegetables, or other high-value crops. The most economical mulch for large-area application is plant residue (standing or flattened). Residues reduce surface temperatures and shade the soil from solar radiation, decreasing the rate of soil evaporation.

Mulching is most effective early in the growing season, before a crop canopy forms and evaporation exceeds transpiration. Theoretical analyses and field experiments show that surface residues are of greatest value for water conservation during the winter rainy season and early spring when the soil surface is often wet.

Mulch effectiveness for decreasing evaporation increases as mulch thickness increases. Because material density largely influences mulch thickness, low-density materials, such as wheat straw, more effectively decrease evaporation than sorghum stubble or cotton stalks, which are more dense. Because the water-content difference between mulched and bare soil is mainly near the soil surface, especially soon after water additions, mulches are very useful for seedling establishment.

Tillage is needed primarily on heavy soils that may crack during fallow and lose water by evaporation through the cracks. Sands and other light-textured soils that do not crack are self-mulching and do not need tillage.

Surfactants may decrease evaporation by decreasing capillary rise of water to the surface. Surfactants probably cause decreased surface tension at the solid–liquid interface, thereby decreasing capillary flow of water and causing the formation of a dry diffusion barrier. Seemingly, the use of surfactants for making more efficient use of water has limited potential. Although evaporation is decreased, total water use by crops is not affected and yields are decreased [138].

Water infiltration and runoff control are virtually inseparable. Reducing surface runoff and increasing infiltration of rainfall is probably the most important way in which the agronomic practices affect the total water supply of the soil. Runoff will occur whenever the surface detention is full and rainfall rate exceeds soil infiltrability. This situation is favored by: poor soil tilth and structure management, small surface holding, and unstable surface soil structure with a tendency to crust.

Infiltration is influenced by soil conditions at the surface, in the tillage layer, and within the profile. Adequate surface residues or increased surface roughness (e.g., microbasins, tied ridges) and aggregation to increase surface retention and infiltration, and stabilization of surface soil structure to rainfall impact may reduce runoff losses out of a field and also help in retaining a uniform soil water distribution.

A very effective alternative solution is to mechanically create a soil surface microrelief so as to increase surface storage after the soil has been brought to the tilth desired for crop production. The effectiveness of a roughened surface depends, among other factors, on the intensity and amount of precipitation, and on the stability of the surface soil. In ridge-and-furrow cultivated row crops, this technique is variously called basin tillage, furrow damming, or tied ridges, and consists of constructing small dams in the furrow at intervals of approximately 1–5 m.

The use of tillage to increase the infiltration of rainfall into the soil in the short term may have long-term negative effects. If loosening of the upper soil to increase infiltration comes at the expense of compaction below the tillage zone, this can disable infiltration to lower storage zones and induce a perched water table above the tillage pan. This, in turn, can increase the chance of waterlogging, thereby increasing runoff and decreasing infiltration.

Tillage near the contour associated with ridge terraces has long been used to control runoff on relatively steep land in humid regions. Under contour tillage, surface roughness is least along the contours, and largest along the slope of the land. In water-deficit farming areas, contour furrowing for row crops is an effective runoff control and water conservation practice.

Conservation tillage is one of the best means to conserve water under adverse supply conditions by using tillage practices as the principal management tool. The term *conservation tillage* means any tillage sequence that minimizes soil and water losses; thus, every region, crop, and climate will require a somewhat different set of practices. The objectives are to achieve a soil surface with high infiltrability and adequate holding storage that will retain advantageous properties over extended time periods and will provide a favorable seedbed and rooting medium for agricultural crops.

Modern conservation tillage and herbicidal practices conserve crop residue on the surface, thereby leading to better soil water conservation and cooler soil temperatures during springtime compared with more soil-disturbing, conventional tillage practices. In humid or subhumid regions, cooler, wetter soils can lead to slower crop development, and reduced crop growth under some circumstances. However, in the water-limited environments of dryland cropping regions, no tillage or other residue-conserving practices will increase small-grain growth and yield as well as conventional tillage practices.

Residues intercept rainfall, absorbing and dissipating impact energy. This transfer of energy reduces degradation of soil aggregates at the surface, which in turn reduces the sealing of the surface against infiltration.

Increased residue on the soil surface also has been associated with increased organic-matter content of soils or altered distribution in the profile, thereby affecting infiltration. Little information exists to separate the effects of surface residues on rainfall capture from the effects of organic matter on infiltration.

A commonly used method is stubble mulch (SM) tillage, in which a sweep or blade undercuts the soil surface to control weeds and prepare a seedbed, yet retains most residues on the surface. The depth of tillage usually ranges between 7 and 10 cm. Each operation with a properly operated SM implement reduces surface residues only about 10% to 15%. Hence, adequate surface residues usually can be retained by using SM tillage.

One traditional method to increase storage capacity has been to increase the volume of soil suitable for rooting. When impeding layers or horizons (fragipans, handpans, and plowpans) are present, subsoiling or deep tillage can improve infiltration and crop rooting, although its effectiveness in conserving soil water for a succeeding crop is limited. The effects of deep tillage are often transitory because of the inherently poor structure of the soil or the dispersive nature of the subsoil. In such cases, deep tillage needs to be supplemented with organic matter or chemical treatments for aggregates.

A second method acts by increasing the capacity to store water in a unit of soil volume. Addition of organic matter or other soil amendments should increase the available water-holding capacity. The effects of organic-matter management on soil properties encompass almost all parameters of the soil environment. Microbial activity and growth are stimulated, resulting not only in larger pools of labile nitrogen, but also improved soil physical conditions that maintain porosity and promote aeration and favorable water regimes. Organic substances that contribute to soil aggregation are derived from plant materials, either after alteration by soil animals, bacteria, and fungi, or directly from the plants themselves.

Earthworms improve soil structure, which increases infiltration. While feeding on organic materials and burrowing in soils, earthworms secrete gelatinous substances that coat and stabilize soil aggregates. Water-stable aggregates result also from water-insoluble gummy substances secreted by bacteria, fungi, and actinomycetes. Earthworm activity and intensive soil tillage are not very compatible. Hence, little earthworm activity occurs in many intensively cultivated soils [138]. For maximum earthworm activity, no tillage is desirable.

Plants directly influence soil aggregation through exudates from roots, leaves, and stems. Additional influences are a result of weathering and decaying plant materials, which bind soil particles together; plant canopies, which protect surface aggregates against breakdown due to raindrop impact; and root action in soil, which promotes aggregate formation.

In addition to the direct osmotic effect and possible toxicity of specific ions, soil salinity may have a deleterious effect on physical properties such as infiltrability, water holding and aeration, especially if the soil is rich in exchangeable sodium. The surface soil is more dispersible by raindrops, and crusting and runoff will follow. Reclamation of salt-affected soils under these water-deficient dryland conditions is very difficult because removal of the excess salts by leaching is an essential part of the reclamation process. Gypsum, which often is applied to displace adsorbed sodium from the soil exchange complex, is itself a contributor of soluble salts. Unless it can be leached from the soil in the last stage of reclamation, it thus may further depress yields. Even if temporary relief is obtained, it may not be economical because of the cost of the amendments and their application.

A soil conditioner's effectiveness often is related to its ability to promote flocculation. Polymers induce flocculation (or coagulation) of dispersed clay particles by electrostatic absorption of polymer molecules on clay particles, which helps to compensate for the clay surface charge, bridging soil particles together. Anionic polymers are effective flocculants, especially in the presence of polyvalent cations. When applied to soil, these substances result in larger aggregates. Although not economical, except possibly for some high-value crops, the substances improve aggregation, which increases water holding and infiltration in the soil profile and decrease runoff.

Mixing fine-textured materials from outside sources or from larger depths within the profile with sandy surface horizons can increase the water-holding capacity of sandy surface soils. Adding materials from outside sources may be practical in limited areas, but not in large areas because of the large amount of material needed and the expenses involved in transporting the materials. In contrast, mixing finer materials from deeper in the profile with coarser materials near the surface is within the reality of practice. The water-holding capacity of the entire profile may not be increased, but soil near the surface should hold more water, thus improving seedling establishment and early growth [138].

Compaction of the subsoil and placement of relatively impervious materials at depth are methods used to reduce percolation. In sandy soils, the problem of deep drainage is especially acute because of the limited water-storage capacity. Thus, incorporation of clay or organic colloids or residues into the profile may increase its water-storage volume. However, the cost of these agronomic practices would be prohibitive in the context of dryland farming.

Land-farming practices are used widely to control runoff from excess rainfall. Precipitation amount and distribution strongly influence the type of terrace used in a particular region. In higher-precipitation regions, excess water is channeled from fields by graded terraces. Such terraces drain the water at non-erosive velocities into grassy waterways or other grassy areas. When designed for water conservation, terraces often are leveled and may have closed ends (level terraces). The conservation-bench terrace is a practice that has been used successfully in dryland and irrigated farming to reduce runoff and retain precipitation.

Although microwatersheds and vertical mulches serve two distinct purposes, they often are used in a combination system. Microwatersheds increase runoff from a portion of the field and concentrate the water on a relatively small area to increase depth of water penetration. Vertical mulching, by providing a residue-filled soil slot open to the surface, results in rapid channeling of water into soil.

In areas where natural rainfall is insufficient for dryland farming, nonarable areas or hillsides may be used for water harvesting to support an economically valuable crop. Water harvesting conveys runoff to cultivated fields.

Microcatchment water harvesting is a low-cost method of collecting surface runoff in the area (A) and storing it in the root zone of an adjacent infiltration basin (B) to cover the crop water requirement. The crop may be a single tree, bush, or annual crops. This method is also applicable in areas with a high rainfall but low soil permeability. To design an optimum ratio between A and B, the impact of climatic conditions and soil physical properties on the water dynamics must be considered. The design should aim at sufficient available water in an average year: Deep percolation losses during a wet year then must be accepted as well as some shortage during a dry year [139].

In a runoff-farming system, crops are grown in widely spaced strips or rows on the contour, where the areas between the strips or rows are treated to enhance runoff from rainfall. Depending on rainfall distribution, the crops in a runoff-farming system should be able to survive extended periods without rain.

Crop Management

WUE depends on the relative rates of assimilation and evapotranspiration as influenced by crop genotype and environment. Those traits are subject to optimization through variations in cropping practice and through special treatment that might be applied to the crop-plant community. The diversity of crop-management techniques involving water management is great (Table 5.19), but, in this section, the cropping practices of

Table 5.19. Crop Management Techniques for drought and water-stress conditions

Crop Management Techniques	Benefits	Effectiveness
Drought risk management		
Change of crop patterns, replacing sensitive crops with tolerant ones (eventually decreasing irrigation surface)	Limits the effects of droughts	High
Choice of drought-tolerant crops over highly productive varieties	Limits drought requirements	High
Use of short-cycle varieties	Low water requirements	High
Early seeding	Avoids terminal stress	High
Early cutting of forage crops	Avoids degradation of stressed crops	High
Grazing drought-damaged fields	Alternative use, livestock support	High
Supplemental irrigation of rainfed crops	Avoids stress at critical stages	High
Management for controlling effects of water stress		
Use of appropriate soil management techniques	Increases available soil water	High
Adaptation of crop patterns to environmental constraints and resource conservation	Coping with water-stressed environment	High
Use of fallow cropping in rainfed systems	Increases soil moisture	Controversial
Use of mixed cropping and intercropping, namely for forages	Better use of resources	Low
Increased plant spacing of perennials and some row crops	High individual explorable soil volume	Limited/high
Cultivation techniques		
Minimizing tillage	Avoids E_s	High
Adequate seed placement	Prevents rapid drying of soil layers around the seed	High
Preemergence weed control	Alleviates competition for water, avoids herbicide effects on stressed crop plants	High
Reduced and delayed fertilization	Favors deep rooting, adaptation to crop responses under water stress	Variable
Dry-soil land preparation and seeding of paddy rice	Saves water	High
Antitranspirants	Reduces plant transpiration	Controversial
Reflectants (increasing albedo)	Decreases energy available for transpiration	Limited/high
Windbreaks	Decreases energy available for evaporation	Limited
Growth regulators	Improves responses of physiological processes to water stress	Promising

Source: [140].

greatest interest are discussed. The review of crop-management techniques in Table 5.19 distinguishes the techniques or decisions that are associated with a given risk from those adapted to water-stressed environments that can be adopted easily for drought control [140].

Decreasing water loss by soil water evaporation provides the potential for improved crop production on the same rainfall input. Selection of cultivars with early growth and/or earlier dates of planting to increase early growth are mechanisms for reducing soil water evaporation and increasing crop water use.

Crop rotations, including grasses and legumes, reduce the impact of raindrops and surface aggregate breakdown, crusting, runoff, and erosion, and soil water evaporation while improving soil water storage. Improved soil structural stability is attributed to the positive influence of root systems in increasing infiltration. If legumes are used as cover crops, they can provide nitrogen through fixation for subsequent crops. Cover crops prevent leaching of nitrogen, potassium and possibly other nutrients by incorporating them into their biomass. Cover crops add organic matter to the soil and they can cause an increase in microbial activity by providing a readily available carbon source, which subsequently increases aggregation.

Advances in cropping systems productivity can be accomplished by developing and using crop rotations that either are more timely in utilizing available water or are able to conserve water. However, few reports exist regarding the effect of crop rotation on efficient water use. The one exception is the cereal-fallow rotation used in dry areas.

In cropping systems, the term fallow is used to describe land that is resting, that is, not being cropped. The inclusion of a weed-free fallow is used widely in semiarid regions to conserve soil water storage and increase the water available for the next crop. Fallow also maintains a change in crop succession to minimize carryover of pests, diseases, and weeds, and plays an important role in the mineralization of nitrogen.

As with many other techniques, the use of a fallow period has been questioned. Because fallows inevitably lose water by evaporation from the surface and drainage below the root zone, their efficiency may be low and varies from year to year.

It is obvious than each crop rotation and fallow combination needs to be tested over a long period to evaluate its sustainability. Information pertinent to a given location does not seem to be simply transferable to other locations without additional prolonged testing before adoption.

Weeds pose severe problems in many fields in dry areas. They can dramatically decrease the WUE of crop production, particularly in crops that compete less vigorously with weeds. During late spring and early summer, weeds compete vigorously for available moisture and, under conditions of water deficit, this leads to yield losses. Weeds can be controlled by tillage, herbicides, and crop rotations.

Antitranspirants have been used to control transpiration at the leaf-air interface. These materials may induce stomatal closure, cover the mesophyll surface with a thin, monomolecular film, or cover the leaf surface with a water-impervious film. However, antitranspirants have potential detrimental effects on net assimilation of photosynthates and evapotranspirational cooling. Moreover, the cost of applying antitranspirants at present disqualifies their use in dryland farming.

5.5.8 Water Management Under Specific Environmental Conditions

Arid Areas

A generally accepted definition of arid or semiarid lands are those lands where the crop water requirements exceed the plant water availability (growing season precipitation plus soil water stored in the root zone) by a significant amount [141]. At least one-third of the world's land surface is taken up by arid or semiarid lands.

The ratio between precipitation (P) and reference evapotranspiration roughly delineates the severity of water deficit in various zones. Various aridity indices are defined as follows [142]:

- *Hyperarid zones*, $P/ET_0 \leq 0.03$, yearly rainfall under 100 mm. The zones include true deserts with virtually no rainfall over periods of one to several years.
- *Arid zones*, $0.03 \leq P/ET_0 \leq 0.20$, annual rainfall of 100 to 200 mm. Zones of pastoral nomadism and irrigated farming include sparce and scanty vegetation of perennials and annuals.
- *Semiarid zones*, $0.2 \leq P/ET_0 \leq 0.5$, annual rainfall of 200 to 400 mm. Zones of pastoral nomadism and irrigated and rainfed agriculture include semidesert or tropical shrublands with an intermittent grass cover.
- *Subhumid zones*, $0.5 \leq P/ET_0$, yearly total rainfall of 400 to 800 mm. Zones of traditional rainfed agriculture include Mediterranean communities of the maquis and chaparral types.

With precipitation being such a major limitation to crop production, its efficient use is an important consideration in any sustainable system. Dry farming can be an excellent means to cultivate regions of low productivity. A small amount of rainfall can be sufficient to grow drought-resistant crops. With extremely low rainfall rates, irrigation water is required to create possibilities for agricultural practices. However, in many arid and semiarid regions, the availability of either surface or groundwater is limited. Water in deep aquifers often has collected over many years or even centuries. If not replenished at the rate of utilization, groundwater ultimately will be a finite resource. Runoff farming in dry environments is still practiced in modern agriculture, albeit on a different scale. Systems that supplement rainfall can be sustainable in the long term. Irrigation management to prevent overwatering, soils impermeable to deep percolation, and crops that utilize water efficiently all contribute to long-term sustainability. Likewise, runoff farming systems need to be balanced carefully to utilize all incoming rainfall without creating excesses during wet years or shortages and crop failures during dry years.

Dryland farming should be designed to cope with the negative impact of highly variable rainfall timing and amounts on crop production. Dryland farming has to increase crop-available soil water by enhancing water infiltration into the profile, by increasing its storage, by reducing evaporation, and by minimizing water losses.

Dryland farming techniques to improve infiltration include tillage operations that leave a coarse tilth on the soil surface, create layers that cause the soil to self-mulch; and leave a mat of crop residues on the surface to protect the surface tilth and to reduce evaporation. Conservation of water stored in the root zone from one rainy season to the next under fallow requires weed control and minimizing surface evaporation and deep percolation.

Disturbing crusts during periods between rains to increase infiltration probably constitutes an early example of soil surface modifications. Other approaches to increase infiltration retain residue on the surface by means of reduced-tillage or conservation-tillage practices. The average rainfall holding from six conservation-tillage systems was 9% higher than that from four conventional tillage systems [143]. Improvement in rainfall holding was attributed to a buildup of surface residue and improved soil tilth in the surface horizon, especially protection of the surface from raindrop impact. This reduces the sealing of the surface against infiltration. Residues also shade the soil surface, which reduces the surface temperature and soil evaporation. Increased residue on the soil surface also has been associated with an increased organic-matter content of soils or altered distribution in the profile, thereby affecting infiltration.

Because infiltration is to be kept as high as possible, any restriction caused by shallow plow pans or a clay layer requires correction by plowing, chiseling, or sweep plows. Although the primary objective of subsoiling is to create zones acceptable for rooting in lower horizons, clearly the operations also increase infiltration. Burrows of earthworms and other soil fauna provide channels for water infiltration.

Subsoiling, deep tillage, the use of gypsum to produce stable aggregates, and breeding programs that select for stronger rooting cultivars are traditional methods to improve storage capacity. Where such methods increase water penetration, root growth, and water use by the crop, they will have a major impact on sustainability.

Cover crops provide an effective method for improving water quality because they accumulate and retain plant nutrients as well as reduce soil erosion.

High levels of fertility in soil lead to both optimal crop growth and enhancement of root-system spread, thus helping the crop to endure water deficits [144].

Tillage systems designed to conserve water and eradicate weeds also must contribute to pest and disease control. Crop rotations maintain a change in crop succession to minimize carryover of pests, diseases, and weeds, and attempt to maintain soil fertility. Many factors are involved in selecting a crop rotation, including soil water-storage capacity; amount and variability of rainfall; crops and their profitability; pressure of weeds, pests, and diseases; and requirements and cost of tillage operations [145].

Every arid, semiarid, and subhumid region has developed appropriate water conservation techniques, some of them quite particular. In North Africa, among others are biological stabilization of gullies, small earth dams, meskat, loose-rock belts, and broad-base level terraces [146].

Water harvesting can significantly increase plant production in drought-prone areas by concentrating the rainfall/runoff in parts of the total area [147]. The goals of water harvesting are restoring the productivity of land that suffers from inadequate rainfall; increasing yield of rainfed farming; minimizing the risk in drought-prone areas; combating desertification by tree cultivation; and supplying drinking water for animals. Of the great number of forms in existence with various names, six forms generally are recognized [147]: roof top harvesting; water harvesting for animal consumption; inter-low water harvesting; microcatchment water harvesting; medium-sized-catchment water harvesting; and large-catchment water harvesting.

Bucks [148] discussed several technologies for improved water management and conservation of irrigated agriculture in arid and semiarid regions. Although no single

Table 5.20. Demand management for water conservation in irrigated agriculture

Objective	Technology
Reduce water delivery	Increase irrigation efficiency and water application uniformity.
	Irrigation scheduling and control based on monitoring the soil, the plants, and/or the microclimate.
	Reduce water evaporation from lakes, reservoirs, or other water surfaces.
	Reduce evaporation of water from soil surfaces.
	Reduce water use by non-economic and phreatophyte vegetation.
Reduce evapotranspiration	Limit irrigation by applying less water than maximum ET demand.
	Limit irrigated cropland acreages by converting irrigated cropland in water-short areas to dryland farming.
	Change crops by introducing those with lower water requirements.
	Crop selection and modification for drought-resistant strains that can withstand dry periods.
	Decision-making models and systems for irrigation scheduling and crop simulation, and using expert systems.

Source: Adapted from [141].

technology can solve all water quantity and quality problems confronting irrigated agriculture, he did indicate that advanced irrigation scheduling, increased irrigation efficiency, limited irrigation, soil moisture management, and wastewater irrigation can be used more effectively in the future.

Table 5.20 lists the primary technologies available for demand management for water conservation in irrigated agriculture. Demand management objectives include reducing water delivery (nonbeneficial ET) and reducing water requirements (beneficial ET) [141].

Supply management objectives include storing runoff water, increasing water yield, capturing precipitation, and adding to available water supply. Table 5.21 lists the major technologies available for supply management for water conservation in irrigated agriculture for arid areas.

Because of a combination of low application efficiencies and inadequate drainage, irrigation has led to large-scale waterlogging and salinization of irrigated land. In semiarid and arid regions, prevention of irrigation-induced salinization is the main concern. Drainage requirements can be reduced by improving the efficiency and management of irrigation. Thus drainage for irrigated land must be treated as a component of the water management system and its design should depend on the design and management of other components. The primary design and operational objectives of water-table management systems in the arid, semiarid, and subhumid zones are to provide trafficable or workable conditions for farming operations, to reduce crop stresses caused by waterlogging to control salinity and alkalinity, to minimize harmful offsite environmental impacts, and to conserve water supplied by rainfall, thus minimizing irrigation water requirements. Modern research emphasizes the importance of controlling the water table by using a combination of drainage and irrigation.

As the demand for water increases, wastewater reclamation and reuse have become an increasingly important source for meeting some of this demand. The level of wastewater treatment required for agricultural and landscape irrigation uses depends on the soil

Table 5.21. Supply management for water conservation in irrigated agriculture

Objective	Technology
Increase storage runoff water	Small reservoirs to catch and retain floodwater for release during droughts
	Groundwater recharge by conveying or confining surplus runoff to recharge areas to increase water storage.
Increase water yield	Water harvesting by constructing an impermeable surface to reduce infiltration and store runoff.
	Vegetative management by manipulating vegetative cover to increase or decrease runoff for improved groundwater recharge storage.
Capture and retain precipitation	Snow management.
	Soil moisture management by cultural and mechanical practices to decrease runoff and evaporation, thereby increasing soil moisture storage.
	Increased crop rooting depths by breaking up hardpans and selecting crop species and cultivars that root more deeply to expand soil moisture extraction.
Add to available water supply	Inter- and intrabasin transfers.
	Wastewater irrigation, using moderately saline drainage waters and renovated wastewater effluents for irrigation.

Source: Adapted from [141].

characteristics, the crop irrigated, the type of distribution and application systems, and the degree of worker and public exposure.

Humid Areas

Humid regions are characterized by the fact that rainfall equals, or even exceeds, ET. Many humid areas have sufficient annual precipitation for crop production but do not have the proper seasonal distribution. The spatial distribution and intensity of rainfall can vary greatly within a season and from year to year.

Water in these areas is plentiful, generally of good quality, without inducing salinity problems.

In humid regions, amount and distribution of precipitation are largely beyond the control of people, but soil management may have a major influence on how effectively the available precipitation, or even irrigation, is used for crop production. One of the foremost requirements for more efficiently using precipitation is retaining the water-storage reservoir.

Irrigation provides predictability and stability of water supplies, enables farmers to obtain higher yields by using modern agricultural technology and optimal levels of production inputs, and increases intensity of cropping. Most irrigation in humid areas is supplemental to reduce crop stress caused by short- duration rainfall deficits. The purpose of irrigation in these areas thus is to increase ET but with minimal losses from runoff, poor drainage, and leaching of fertilizers and pesticides. Scheduling under conditions of rainfall thus is the major goal of irrigation in humid areas. Irrigation systems should be operated mostly to partially refill the profile. Problems arise when irrigating soils with low water-holding capacities, which do not maximize the effects of rainfall.

In humid areas, irrigation systems frequently are designed with a smaller capacity than required for extended dry periods, or the systems are moved from one field to another. These factors complicate the water management when rainfall is low and evaporative demand is high.

Irrigation scheduling used in arid, semiarid, and subhumid areas should not be used in humid areas. One of the most common methods for irrigation scheduling is based on field soil water balance. The effective rainfall must be considering in computations, but its determination is very difficult. Rainfall of high intensities or in large amounts may produce significant runoff, and only part of the rainwater can be considered to be effective. Similarly, rainfall on a wet soil profile will produce losses through drainage.

Field observation of soil water status or plant water potentials or other plant parameters, and the combination of these, can be used to determine irrigation timing.

The water balance can be based on real-time meteorological data or average climatic data. Various forms of computer-based, water-balance scheduling methods have been developed and evaluated for humid-area conditions. Several of these include a short-term forecast of rainfall.

The most widespread system is gravity irrigation, especially in areas where water availability is not a limiting factor and the water price is low. Furrow or flood irrigation is used in humid areas for sugar cane or rice, respectively. Frequent low-volume irrigation often is practiced in humid areas, which permits chemigation with little change in irrigation schedules. Most frost protection in humid areas is accomplished with overhead sprinkler irrigation. Although rainfall is often sufficient for crop establishment, irrigation can ensure uniform germination and emergence. Most sprinkler systems are well adapted for this purpose.

Water Scarcity and Drought

In many areas of the world, available irrigation water supplies are either very low or are being depleted rapidly. When water supplies diminish and population increases and municipal and industrial water demands increase, the optimization of irrigation water use becomes increasingly imperative. Droughts in many parts of the world have raised this question.

According to WMO drought is defined as a sustained, extended deficiency in precipitation [149]. Aridity is a permanent climatic feature of a region, resulting from low average rainfall [150]. Drought, by contrast, is a short-term lack of precipitation, a temporary feature of climate [151]. It is necessary to differentiate drought, aridity, and water shortage [152]:

- Drought is a natural temporary imbalance of water availability, consisting of persistent lower- than-average precipitation of uncertain frequency, duration, and severity, of unpredictable occurrence, with overall diminished water resources and carrying capacity of the ecosystem.
- Aridity is also a natural phenomenon but it is a permanent imbalance in the availability of water: The average annual precipitation is low. The spatial and temporal variation of precipitation is high, with overall low moisture, extreme temperature variations, and a low carrying capacity.

- Water shortage is a human-induced temporary water imbalance and results in groundwater overdraft, reduced reservoir capacities, disturbed and reduced land use, and an altered carrying capacity.

Droughts create temporary water-stressed environments similar to aridity. Thus, responses to aridity are partly applicable to droughts. Agricultural water shortages induced by droughts have much in common with human-induced shortages [140].

Drought is generally extensive in space and time. For irrigated agriculture, drought does not necessarily begin with the cessation of rain, but rather when available irrigation water falls below normal for a large period of time, when water supply is not sufficient to meet normal demands. Drought may require both managerial and technological adjustments in farming operations, depending upon its length and severity. Because of knowledge in prediction and forecasting of droughts, which could help in the development and application of nonstructural water management measures, is insufficient, structural measures to improve supply are preferred [152].

Drought periods can produce important negative impacts on agriculture. The severity of these impacts depends on the farmer's ability to adapt production systems to water availability. Drought can affect the irrigation-water demand by farmers. In these cases, farmers in rainfed agriculture resort to irrigation for supplementing natural rainfall for crop production.

In establishing an agricultural production system, several factors must be carefully studied (e.g., crop species, management strategies, irrigation technologies). All of these factors are based mainly on expected regional weather patterns. Depending on drought strength and duration, farmers must react rapidly to the reduced supplies and increased demands for water.

The increase in hydrometeorological information makes it possible to better use stochastic analysis for prevention of reservoir inflow and for reservoir management, to use demand forecasting models for planning releases, as well as to use optimization and simulation models to help in decision making.

Drought also affects irrigation water quality in several ways. There is less leaching of mineral nutrients and pesticides toward aquifers and, on the other hand, less transport of sediments and certain chemicals to surface streams.

Haas [153] described several general strategies for managing water under drought, although they are not specific to irrigation. Under limited water availability, it is necessary to look for new, specific solutions for the following [151]:

- Farmers must select and optimize cropping patterns as a function of water availability. The application of techniques of dynamic, linear programming, as well as the use irrigation-scheduling simulation models are tools of farmers and experts for achieving these goals.
- When designing irrigation-scheduling techniques, farmers and experts must take into account not only the water-availability limitation, but also the irrigation system used for applying water to the crops. Irrigation-scheduling simulation models and the implementation of irrigation-scheduling programs can be useful tools.
- A factor that interacts with irrigation adequacy and affects irrigation efficiency is irrigation uniformity. With any irrigation system, uniformity is strongly affected

by irrigation management. Programs for field evaluation of irrigation practices and technical assistance for adoption of improved water application techniques are necessary.

- Farmers should be advised on the limitations of the irrigation system. They should be trained for collaboration in the operation and management of the system, which improves the interaction between off-farm and on-farm demand management.
- Methods of demand forecasting, aiming at rational planning of reservoir releases and deliveries, should be enhanced.
- Water pricing policies should be used as an incentive to farmers for using water-saving techniques related to both irrigation management and water application, and penalties should be imposed on farmers for wasting water or not using appropriate crop patterns.

The process of technology transfer must be accelerated to achieve this goal, so that farmers can benefit from the efforts of research to enhance WUE.

The study of the deficit irrigation strategies is highly relevant to the preceding discussion. These strategies are capable of reducing the amount of water applied with minimum impact on production. Deficit irrigation is an optimizing strategy under which crops are deliberately allowed to sustain some degree of water deficit and yield depletion [154]. Deficit irrigation may provide higher economic return per unit of surface than returns attained with irrigation for achieving maximum production [155].

Much information on crop response to deficit irrigation is available, which concludes that water deficit reduces yields. However, these irrigation strategies may have a favorable effect on yields when reducing the incidence of diseases and on crop quality. Late-season deficits that develop as temperature declines also can enhance the conditioning of tree crops for winter dormancy.

In a strategy known as high-frequency deficit irrigation, water is supplied throughout the crop cycle below demand levels, but at a frequency high enough to avoid situations of significant stress.

Although high-frequency deficit irrigation represents a clear alternative in several circumstances, it has important limitations, such as the significance of the water deficit to the phenological stage. Thus, a new concept has been introduced, called regulated deficit irrigation [156], which is based on reducing water supplies in the phenological stages where the water deficit does not affect either production and harvest quality. At the same time, plant requirements during the sensitive periods of the crop cycle are addressed.

Irrigation during noncritical periods is one of the essential aspects of regulated deficit irrigation. The water application must be estimated by means of tests to select the most suitable conditions as a function of the impact on harvest and on environment. When there is insufficient water availability, the possibility of fully covering plant water requirements during the critical stages must be considered.

Deficit irrigation in the Columbia Basin is typically practiced on silt loam and fine sandy loam soils. With the introduction of center-pivot systems, deficit irrigation also has been used on sandy soils [154]. Nevertheless, the nearly continuous irrigation required even for deficit irrigation of such soils reduces because of WUE the nearly constant surface evaporation that results.

Deficit irrigation can involve modification of some cultural practices [154], including use of moderate plant densities, reduced application of fertilizers and other farm chemicals, use of fallowing when crops are grown in rotations and when a fallow interval is desired for precipitation storage, and flexible planting dates and associated use of shorter-maturity-length cultivars.

Scheduling for deficit irrigation is potentially more challenging than for full irrigation. Ideally, the decision maker should evaluate not only the amount of water remaining in the soil profile but also the level of stress that the crop is experiencing and how that stress will affect yields.

Salinity Management in Irrigated Agriculture

Introduction

Salinity is defined as the salt concentration present in soil or in water per unit of volume or weight. Salinity usually involves two different kinds of problems: the negative effect on plant growth caused by a soluble salt excess (salinity) that causes an increase in osmotic pressure; and the effect on soil-structure rupture and the depletion of both soil-infiltration and water-holding capability produced by the excess of changeable sodium (sodicity). Some times, both effects (salinity and sodicity) may appear together. In addition, high values of trace elements (e.g. boron, cadmium, selenium) may alter plant growth.

The increase in salinity levels, either in the root zone or in runoff and seepage irrigation water, appears as a result of the alteration of the hydrologic balance of the surface water, influenced by soil and water management practices carried out during a certain period of time. Thus, both salinity prevention and control can be performed by appropriate water management.

Formation of saline soils in nature usually results from the combination of geologic, meteorologic, and hydrologic factors. The main processes involved in salinization are evaporation, capillary rise when a shallow water table layer is present, weathering, and the input of salts with the irrigation water.

Soil salinization is quite common when irrigating arid or semiarid regions. Salinization problems occur in about 23% of cultivated land in the world [157].

Salinity Management

Salinity management must be considered at two levels [158]: in the root zone and at a regional level or irrigable-zone level. Several possible actions are recommended for addressing the first level [158]:

- Using chemical amendments with the goal of improving soil physicochemical properties, adding chemicals with soluble calcium to replace the exchangeable sodium.
- Applying irrigation scheduling appropriate to maintain a specific moisture content in soil and to provoke a periodic leaching of salts, always keeping an adequate drainage.
- Alternating the use of saline and freshwater or mixing them, depending on the soil water conditions and crop conditions.

Salinity management at a regional level is developed around the objectives of environmental quality and economic development of a region or a country. Water requirements must be considered in a sustainable way by every user, as related to both quantity and quality.

The need to improve irrigation systems and their management and to increase uniformity and water application efficiency is also relevant.

References

1. Amthor, J. S., and K. J. McGree. 1990. Carbon balance of stressed plants: A conceptual model for integrating research results. *Stress Responses in Plants: Adaptation and Acclimation Mechanisms*, Vol. 12. Plant Biology, eds. Alscher, R. G., and J. R. Cumming, pp. 1–15. New York: Wiley.
2. Kramer, P. J., and T. T. Kozlwski. 1979. *Physiology of Woody Plants*. New York: Academic Press.
3. Kramer, P. J. 1983. *Water Relations of Plants*. New York: Academic Press.
4. Jones, H. G. 1990. Physiological aspects of the control of water status in horticultural crops. *Hortscience* 25:19–26.
5. Slavik, B. 1974. *Methods of Studying Plant Water Relations*. Berlin: Springer-Verlag and Prague: Academic Publ. House.
6. Barrs, H. D., and P. E. Weatherley. 1962. A re-examination of the relative turgide technique for estimating water deficits in leaves. *Austr. J. Biol. Sci.* 15:413–428.
7. Hsiao, T. C. 1990. Measurements of plant water status. *Irrigation of Agricultural Crops. Agronomy 30*, eds. Stewart, B. A., and D. R. Nielsen, pp. 243–279. Madison, WI: American Society of Agronomy, Crop Science Society of America, and Soil Science Society of America.
8. Barrs, H. D. 1968. Determination of water deficits in plant tissues. *Water Deficits and Plant Growth. Development, Control and Measurement*, ed. Kozlowski, T. T., pp. 235–368. New York: Academic Press.
9. Boyer, J. S. 1969. Measurement of the water status of plants. *Annu. Rev. Plant Physiol.* 20:351–364.
10. Brown, P. W., and C. B. Tanner. 1981. Alfalfa water potential measurement: A comparison of the pressure chamber and leaf dew–point hygrometers. *Crop Sci.* 21:240–244.
11. Turner, N. C. 1986. Crop water deficits: A decade of progress. *Adv. Agron.* 39:1–151.
12. Scholander, P. F., H. T. Hammel, E. D. Bradstreet, and E. A. Hemmingsen. 1963. Sap pressure in vascular plants. *Science* 148:339–346.
13. Dixon, H. H. 1914. *Transpiration and the Ascent of Sap in Plants*. London: Macmillan.
14. Richards, L. A., and C. N. Wadleigh. 1952. Soil water and plant growth. *Soil Physical Conditions and Plant Growth*, ed. Shaw, B. T., pp. 73–251. New York: Academic Press.
15. Huguet, J. G., V. Benoit, and P. Orlando. 1987. Aplications de la micromorphometrie sur tige au pilotage de l'Irrigation du maïs. *Proc. Le Maïs et l'Eau. Colloque Alimentation Hydrique du Maïs*, Agen, France, 8–9 Décembre: AGPM, ITCF, INRA et CEMAGREF, pp. 1–16. Paris, France.
16. Puech, J. 1987. Indicateurs de sechresse externes de la plante: Cas de la thermometrie infra-rouge. *Proc. Le Maïs et l'Eau. Colloque Alimentation Hydrique du Maïs*,

Agen, France, 8–9 Décembre: AGPM, ITCF, INRA et CEMAGREF, pp. 1–13. Paris, France.
17. Tyree, M. T., and M. A. Dixon. 1983. Cavitation events in *Thuja occidentalis* L. *Plant Physiol.* 72:1094–1099.
18. Begg, J. E., and N. C. Turner. 1976. Crop water deficits. *Adv. Agron.* 28:161–217.
19. Gardner, F. P., R. Brent Pearce, and R. G. Mitchell. 1990. Water relations. *Physiology of Crop Plants*, pp. 76–97. Ames, IA: Iowa State Univ. Press.
20. Claasen, M. M., and R. H. Shaw. 1970. Water deficit effects on corn. I. Vegetative components. *Agron. J.* 62:649–652.
21. Hooker, M. L., S. H. Mohiuddin, and E. T. Kanemasu. 1983. The effect of irrigation timing on yield and yield components of winter wheat. *Can. J. Plant Sci.* 63:815–823.
22. Shaw, R. H., and D. R. Laing. 1966. Moisture stress and plant response. *Plant Environment and Efficient Water Use*, eds. Pierre, W. H., *et al.*, pp. 73–94. Madison, WI: American Society of Agronomy, Crop Science Society of America, and Soil Science Society of America.
23. Wample, R. L., and R. K. Thorton. 1984. Differences in the response of sunflower (*Helianthus annuus* L.) subjected to flooding and drought stress. *Physiol. Plant* 61:611–616.
24. Domingo Miguel, R. 1994. Respuestas del limonero Fino al riego deficitario controlado. Aspectos fisiológicos. Ph.D. Thesis. Murcia, Spain: Universidad de Murcia y Centro de Edafología y Biología Aplicada del Segura.
25. Kozlowsky, T. T., and J. G. Pallardy. 1979. Stomatal responses of *Fraxinus pennsylvanica* seedlings during and after flooding. *Physiol. Plant* 16:155–158.
26. Duthion, C., and M. Mingeau. 1976. Les reactions des plantes aux excés d'eau et leurs conséquences. *Ann. Agron.* 27:221–246.
27. Lo Giudice, V. 1985. Eccesi idrici negli agrumeti. *Terra Vita* 13:95–97.
28. Laan, P., M. J. Berrevoets, S. Lytye, W. Armstrong, and C. W. P. M. Blom. 1989. Root morphology and aerenchyma formation as indicators for the flood-tolerance of *Rumex* species. *J. Ecol.* 77:693–703.
29. Konings, H., and H. Lambers. 1991. Respiratory metabolism, oxygen transport and the induction of aerenchyma in roots. *Plant Life under Oxygen Deprivation*, eds. Jackson, M. B., *et al.*, pp. 247–265. The Hague: Academic.
30. Jackson, M. B. 1985. Ethylene and the responses of plants to soil waterlogging and submergence. *Annu. Rev. Plant Physiol.* 36:145–174.
31. Konings, H. 1982. Ethylene–promoted formation of aerenchyma in seedlings of *Zea mays* L. under aerated and non-aerated conditions. *Physiol. Plant* 54:119–124.
32. Justin, S. H. F. W., and W. Armstrong. 1991. Evidence for the involvement of ethylene in aerenchyma formation in adventitious roots of rice (*Oryza sativa* L.). *New Phytol.* 118:49–62.
33. Briggs, L. J., and H. L. Shantz. 1913. The water requirement of plants. I. Investigations on the Great Plains in 1910 and 1911. Bureau Plant. Ind. Bull. No. 284. Washington, DC: U.S. Department of Agriculture.
34. de Wit, C. T. 1958. Transpiration and crop yields. *Versl. Landbouwk. Onderz* 64:69–88.

35. Fischer, R. A., and N. C. Turner. 1978. Plant productivity in the arid and semiarid zones. *Ann. Rev. Plant Physiol.* 29:277–317.
36. Bierhuizen, J. F., and R. O. Slatyer. 1965. Effect of atmospheric concentration of water vapour and CO_2 in determining transpiration-photosynthesis relationships of cotton leaves. *Agric. Meterol.* 2:259–270.
37. Tanner, C. B., and T. R. Sinclair. 1983. Efficient use of water in crop production: Research or pre-search. *Limitations to Efficient Water Use in Crop Production*, eds. Taylor, H. M., W. R. Jordan, and T. R. Sinclair, pp. 1–27. Madison, WI: American Society of Agronomists, Crop Science Society of America, and Soil Science Society of America.
38. Steduto, P. 1994. Water use efficiency. *Sustainability of Irrigated Agriculture*, eds. Pereira, L. S., R. A. Feddes, J. R. Gilley, and B. Lesaffre, pp. 193–209. Dordrecht, The Netherlands: Kluwer Academic.
39. Ludlow, M. M. 1976. Ecophysiology of C_4 grasses. *Water and Plant Life: Problems and Modern Approaches*, eds. Lange, O. L., L. Kappen, and E. D. Shulze, pp. 364–386. New York: Springer-Verlag.
40. Stanhill, G. 1986. Water use efficiency. *Adv. Agron.* 39:53–85.
41. Monteith, J. L. 1988. Does transpiration limit the growth of crops or vice versa? *J. Hydrol.* 100:57–68.
42. Kimball, B. A. 1983. Carbon dioxide and agricultural yield. An assemblage and analysis of 430 prior observations. *Agron. J.* 75:779–788.
43. Rogers, H. H., J. F. Thomas, and B. E. Bingham. 1983. Response of agronomic and forest species to elevated atmospheric carbon dioxide. *Science* 220:428–429.
44. Allen, R. G., F. N. Gichuki, and C. Rosenzweig. 1991. CO_2-induced climate changes and irrigation-water requirements. J. Water Resour. Plan. Manage. 117:157–178.
45. Idso, S. B., K. L. Clawson, and M. B. Anderson. 1986. Foliage temperature: Effects on environmental factors with implications for plant water stress assessment and CO_2 climate connection. *Water Resour. Res.* 22:1702–1716.
46. Turner, N. C. 1979. Drought resistance and adaptation to water deficits in crop plants. *Staple Stress Physiology in Crop Plants*, eds. Mussell, H., and R. C. Staples, pp. 341–372. New York: Wiley-Interscience.
47. Howell, T. A., R. H. Cuenca, and K. H. Solomon. 1990. Crop yield response. *Management of Farm Irrigation Systems*. eds. Hoffman, G. J., *et al.*, pp. 93–122. ASAE Monograph No. 9. St Joseph, MI: ASAE.
48. Cooper, P. J. M., P. J. Gregory, D. Tully, and H. G. Harris. 1987. Improving water use efficiency of annual crops in rainfed farming systems of West Africa and North Africa. *Exp. Agric.* 23:113–158.
49. Zobel, R. W. 1983. Crop manipulation for efficient use of water: Constraints and potentional techniques in breeding for efficient water use. *Limitations to Efficient Water Use in Crop Production*, eds. Taylor, H. M., W. R. Jordan, and T. R. Sinclair, pp. 381–392. Madison, WI: American Society of Agronomists, Crop Science Society of America, and Soil Science Society of America.
50. Sinclair, T. R., C. B. Tanner, and J. M. Bennett. 1984. Water use efficiency in crop production. *Biol. Sci.* 34:36–40.

51. Garrity, D. P., D. D. Watts, C. Y. Sullivan, and J. R. Gilley. 1982. Moisture deficits and grain sorghum performance: Evapotranspiration-yield relationships. *Agron. J.* 74:815–820.
52. Fischer, R. A. 1979. Growth and water limitation to dryland wheat yield in Autralia: A physiological framework. *J. Aust. Inst. Agric. Sci.* 45:83–94.
53. Abdelbagi, M., and A. E. Hall. 1992. Correlation between water-use efficiency and carbon isotope discrimination in diverse cowpea genotypes and isogenic lines. *Crop Sci.* 32:7–12.
54. Hall, A. E., R. G. Muters, and G. D. Farguhar. 1992. Genotypic and drought-induced differences in carbon isotope discrimant and gas exchange of cowpea. *Crop Sci.* 32:1–6.
55. Virgona, J. M., K. T. Hubick, H. M. Rawson, G. D. Fargujar, and R. W. Downes. 1990. Genotypic variation in transpiration efficiency, carbon-isotope discrimination and carbon allocation during early growth in sunflowers. *Aust. J. Plant Physiol.* 17:207–214.
56. Farguhar, G. D., J. R. Ehleringer, and K. T. Hubick. 1989. Carbon isotope discrimination and photosynthesis. *Ann. Rev. Plant Physiol.* 40:503–537.
57. Farguhar, G. D., M. H. O'Leary, and J. A. Berry. 1982. On the relationship between carbon isotope discrimation and the intercellular carbon dioxide concentration in leaves. *Aust. J. Plant Physiol.* 9:121–137.
58. Peng, S., and R. Krieg. 1992. Gas exchange traits and their relationship to water use efficiency of grain sorghum. *Crop Sci.* 32:386–391.
59. Arkley, R. J. 1963. Relationships between plant growth and transpiration. *Hilgardia* 34:559–584.
60. Hanks, R. J. 1974. Model for predicting plant yield as influenced by water use. *Agron. J.* 66:660–665.
61. Stewart, J. T., R. H. Cuenca, W. O. Pruitt, R. N. Hagan, and J. Tosso. 1977. Determination and Utilization of Water Production Functions for Principal California Crops. Calif. Contrib. Proj. Rept. No. 67. Davis: University of California.
62. Stegman, E. C. 1982. Corn grain yield as influenced by timing of evapotranspiration deficits. *Irrig. Sci.* 3:75–87.
63. Cosculluela, F., and J. M. Faci. 1992. Obtencion de la función de producción del maíz (*Zea mays* L.) respecto al agua mediante una fuente lineal de aspersión. *Invest. Agric. Prod. Prot. Veg.* 7:169–194.
64. Doorenbos, J., and A. H. Kassam. 1979. Yield Response to Water. Irrigation and Drainage Paper No. 33. Rome: FAO.
65. Doorenbos, J., and W. O. Pruitt. 1977. Crop Water Requirements. Irrigation and Drainage Paper No. 24. Rome: FAO.
66. Howell, T. R., R. H. Cuenca, and K. H. Solomon. 1992. Crop yield response. *Farm Irrigation Systems*, eds. Hoffman, C. J., T. A. Howell, and K. H. Solomon, pp. 93–122. ASAE Monograph No. 9. St. Joseph, MI: ASAE.
67. Monteith, J. L. 1990. Steps in crop climatology. *Proceedings of the International Conference on Dryland Farming*, eds. Unger, P. W. *et al.*, pp. 273–282. College Station: Texas A&M Univ.

68. Hanks, R. J. 1983. Yield and water-use relationships: An overview. *Limitations to Efficient Water Use in Crop Production*, eds. Taylor, H. M., W. R. Jordan, and T. R. Sinclair, pp. 393–411. Madison, WI: American Society of Agronomy, Crop Science Society of America, and Soil Science Society of American.
69. Fereres, E., J. M. Fernández, C. Giménez, F. Orgaz, and M. Pastor. 1991. Dryland cropping systems of southwestern Spain: Avenues for improvement. *Improvement and Management of Winter Cereals Under Temperature, Drought and Salinity Stresses*, eds. Acevedo, E., *et al., Proceedings of ICARDA-INIA Symposium*, pp. 467–477. Córdoba, Spain: MAPA-INIA.
70. Grattan, S. R., and J. D. Rhoades. 1990. Irrigation with saline ground water and drainage water. *Agricultural Salinity Assessment and Management*, ed. Tanji, K. K., pp. 432–449. Manual Rept. on Engineering Practices No. 71. New York: American Society of Civil Engineers.
71. Van Genuchten, M. T. 1983. Analyzing crop salt tolerance data: Model description and user's manual. U.S. Salinity Lab. Res. Rep. No. 120. Washington, D.C.: U.S. Department of Agriculture, Agricultural Research Service.
72. Maas, E. V., and G. J. Hoffman. 1977. Crop salt tolerance: Current assessment. *J. Irrig. Drain. Div.* 103:115–134.
73. Vaux, H. J., Jr., and W. O. Pruitt. 1983. Crop water production functions. *Adv. Irrig.* 2:61–97.
74. Hexem, R. W., and E. O. Heady. 1978. *Water-Production Functions for Irrigated Agriculture*. Ames, IA: Iowa State Univ. Press.
75. Yazon, D., and E. Bresler. 1983. Economic analysis of on-farm irrigation using response functions of crops. *Advances in Irrigation*, ed. Hillel, D., pp. 223–255. New York: Academic Press.
76. Stewart, J. I., and R. M. Hagan. 1973. Functions to predict effects of crop water deficits. *J. Irrig. Drain. Div. ASCE* 93(IR4):421–439.
77. Skogerboe, G. V., J. W. Hugh Barrett, B. J. Treat, and D. B. McWhorter. 1979. Potential effects of irrigation practices on crop yield in Grand Valley. U.S. Environ. Protection Agency. Tech. Ser. Rep. No. EPA 600/2-79-149. Washington, DC: U.S. Gorvernment Printing Office.
78. Dagan, G., and E. Bresler. 1988. Variability of yield of an integrated crop and its causes. III. Numerical simulation and field results. *Water Resour. Res.* 24: 396–401.
79. Bresler, E., and A. Laufer. 1988. Statistical inferences of soil properties and crop yields as spatial random functions. *Soil Sci. Soc. Am. J.* 52:1234–1244.
80. Seginer, I. 1987. Spatial water distribution in sprinkle irrigation. *Adv. Irrig.* 4:119–168.
81. Warrick, A. W., N. E. Hart, and M. Yitayew. 1989. Calculation of distribution and efficiency for nonuniform irrigation. *J. Irrig. Drain. Div. ASCE* 115:674–686.
82. Orgaz, F., L. Mateos, and E. Fereres. 1992. Season length and cultivar determine the optimum evapotranspiration deficit in cotton. *Agron. J.* 84:700–706.
83. Mantovani, E. C. 1993. Desarrollo y evaluación de modelos para el manejo del riego: Estimación de la evapotranspiración y efectos de la uniformidad de apliación

del riego sobre la producción de los cultivos. Ph.D. dissertation Córdoba, Spain: Universidad de Córdoba.
84. Anyoji, H., and L. P. Wu. 1993. Normal distribution water application for drip irrigation schedules. *Trans. ASAE* 37:159–164.
85. de Juan, J. A., J. M. Tarjuelo, M. Valiente, and P. García. 1996. Model for optimal cropping patterns within the farm based on crop water production functions and irrigation uniformity. I: Development of a decision model. *Agric. Water Manage.* 31:115–143.
86. Feinerman, E., Y. Shani, and E. Bresler. 1989. Economic optimization of sprinkle irrigation considering uncertainty of spatial water distribution. *Aust. J. Agric. Econ.* 33:88–107.
87. Feinerman, E., E. Bresler, and H. Achrish. 1989. Economic of irrigation technology under conditions of spatially variable soils and non-uniform water distribution. *Agronomie* 9:819–826.
88. Glen, J. J. 1987. Mathematical models in farm planning: A survey. *Operations Res.* 35:641–666.
89. Pereira, L. S., A. Perrier, M. Ait Kadi, and P. Kabat (eds.). 1992. Crop-water models. ICID Bull. 41 (Special Issue):1–200.
90. Pereira, L. S., G. van den Broek, P. Kabat, and R. G. Allen. 1995. *Crop-Water-Simulation Models in Practice*. Wageningen, The Netherlands: Wageningen Press.
91. Couch, C. E., W. E. Hart, G. D. Jardinea, and R. J. Brase. 1981. Desktop data system for irrigation scheduling. *Proceedings of the Irrigation Scheduling Conference* ASAE Special Publ. 23–81. St. Joseph, MI: ASAE.
92. Miyamoto, S. 1984. A model for scheduling pecan irrigation with microcomputer. *Trans. ASAE* 27:456–463.
93. Camp, C. R., G. D. Christenbury, and D. W. Doty. 1988. Scheduling irrigation for corn and soybean in southeastern coastal plain. *Trans. ASAE* 31: 513–518.
94. Clarke, N., C. S. Tan, and J. A. Stone. 1992. Expert system for scheduling supplemental irrigation for fruit and vegetable crops in Ontario. *Can. J. Agric. Eng.* 34:27–31.
95. Penning de Vries, F. W. T., and H. H. Van Laar (eds.). 1982. *Simulation of Plant Growth and Crop Production*. Simulation Monographs. Wageningen, The Netherlands: Pudoc.
96. Jones, J. W., and J. T. Ritchie. 1990. Crop growth models. *Management of Farm Irrigation Systems*, eds. Hoffman, G. J., T. A. Howell, and K. H. Solomon, pp. 63–89. ASAE Monograph No. 9. St Joseph, MI: ASAE.
97. Singh, U., D. C. Godwin, J. T. Ritchie, G. Alagarswamy, S. Otter-Näcke, C. A. Jones, and J. R. Kiniry. 1988. Version 2 of the CERES models for wheat, maize, sorghum, barley and millet. *Agronomy Abstracts*. Madison, WI: American Society of Agronomy?
98. Singh, U., J. T. Ritchie, and D. C. Godwin. 1993. A user's guide to CERES-Rice. Ver. 2.10. Int. Fert. Dev. Center, Muscle Shoals, AL. pp. 86.

99. Hoogenboom, G., J. W. Jones, and K. J. Boote. 1992. Modeling growth, development, and yield of grain legumes using SOYGRO, PNUTGRO, and BEANGRO: A review. *Trans. ASAE* 35:2043–2056.
100. Hood, C. P., R. W. Mclendon, and J. E. Hook. 1987. Computer analysis of soybean irrigation management strategies. *Trans. ASAE* 30:417–423.
101. Keating, B. A., R. L. McCown, and B. M. Wafula. 1993. Adjustment of nitrogen inputs in response to seasonal forecasts in a region with high climatic risk. *Systems Approaches for Agricultural Development*, Vol. 2, eds. Penning de Vries, F., P. Teng, and K. Metselaar, pp. 233–252. Dordrecht, The Netherlands: Kluwer Academic.
102. Singh, U., P. K. Thornton, A. R. Saka, and J. B. Dent. 1993. Maize modeling in Malawi: A tool for soil fertility research and development. *Systems Approaches for Agricultural Development*, Vol. 2, eds. Penning de Vries, F., P. Teng, and K. Metselaar, pp. 253–276. Dordrecht, The Netherlands: Kluwer Academic.
103. Thornton, P. K., and J. F. MacRobert. 1994. The value of information concerning near-optimal nitrogen fertilizer scheduling. *Agric. Syst.* 45:315–330.
104. Aggarval, P. K., and N. Kalza. 1994. Analyzing the limitations set by climatic factors, genotype, and water and nitrogen availability on productivity of wheat: II. Climatically potential yields and management strategies. *Field Crop Res.* 38: 93–103.
105. Wilkerson, G. G., J. W. Jones, K. J. Boote, K. T. Ingram, and J. W. Mishoe. 1983. Technical Documentation. Gainesville, FL: Agricultural Engineering Dept., University of Florida.
106. Jones, J. W., K. J. Boote, S. S. Jagtap, G. Hoogenboom, and G. G. Wilkerson. 1989. SOYGRO V5.42. Soybean crop growth simulation model. User's guide. Ganiesville, FL: Agric. Exp. Station Journal No. 8304. University of Florida.
107. Jones, C. A., and J. R. Kinizy. 1986. CERES-Maize: Assimilation Model of Maize Growth and Development. College Station: Texas A&M Univ. Press.
108. Ritchie, J. T., B. S. Jonhson, S. Otter-Nacke, and D. G. Godwin. 1989. Development of Barley Yield Simulation Model. Proc. Final Progress Rept. USDA No. 86-CRSR-2–2867. East Lansing, MI: Michigan State Univ.
109. Ritchie, J. T., and S. Otter. 1984. CERES-WHEAT: A user–oriented wheat yield model. Preliminary documentation. Agristar Publ. No. YM-U3-04442-JSC-18892.
110. Letey, J., A. Dinar, and K. C. Knapp. 1985. Crop-water production model for saline irrigation waters. *Soil Sci. Soc. Am. J.* 49:1005–1009.
111. Bresler, E., and G. J. Hoffman. 1986. Irrigation management for soil salinity control: Theories and tests. *Soil Sci. Soc. Am. J.* 5:1552–1560.
112. Sieben, W. H. 1964. Het verband tussen ontwatering en regen bij de jonge zavelgronden in de Noordoost polder. Van Zee tot Land, 40. Tjeenk Willink V. Zwolle, The Netherlands.
113. Hiler, E. A. 1969. Quantative evaluation of crop drainage requeriments. *Trans. ASAE* 12:499–505.
114. Feddes, R. A., P. Kabat, P. J. T. van Bavel, J. J. B. Bronswijk, and J. Haberstsma. 1989. Modelling soil water dynamics in the unsaturated zone—state of the art. *J. Hydrol.* 100:69–111.

115. Feddes. R. A., P. J. Kowalik, and H. Zaradny. 1978. Simualtion of field water use and crop yield. Simulation Monographs, Wageningen, The Netherlands: PUDOC.
116. Belmans, C., J. G. Wesselinga, and R. A. Feddes. 1983. Simulation of the water balance of a cropped soil: SWATRE. *J. Hydrol.* 63:271–286.
117. Van Dam. J. C. 1991. Additional input instruction for solute in SWACROP. Tech. Note. Wageningen, The Netherlands: Dept. of Water Resources, Wageningen Agricultural University.
118. de Laat, P. J. M., R. H. C. M. Atwater, and P. J. T. van Bavel. 1981. GELGAM—a model for regional water management. In: Mede Comm. Hydrol. Onderz-TN027. Proc. Tech. Meet 37 Versl., 25–53. The Hague: the Netherlands.
119. Parsons, J. E., R. W. Skaggs, and C. W. Doty. 1990. Simulation of controlled drainage in open ditch drainage systems. Agric. Water Manage. 18:301–306.
120. Skaggs, R. W. 1978. A water management model for shallow water table soils. Water Resources Res. Inst. Report No. 34, Raleigh, NC: Univ. of North Carolina.
121. Skaggs, R. W. 1980. DRAINMODL Reference Report: Methods for Design and Evaluation of Drainage Water Management Systems for Soils with High Water Tables. Ft. Worth, TX: U.S. Department of Agriculture, Soil Conservation Service, South Tech. Center.
122. Armstrong, A. C., D. A. Castle, and K. C. Tyson. 1995. SWARD: A model of grass growth and the economic utilization of grass land. *Crop-Water-Simulation Models in Practice*, eds. Pereira, L. S., B. J. van deen Broek, P. Kabat, and R. G. Allen, pp. 189–197. Wageningen, The Netherlands: Wageningen Press.
123. de Laat, P. J. M. 1995. Design and operation of a subsurface irrigation scheme with MUST. *Crop-Water-Simulation Models in Practice*, eds. Pereira, L. S., B. J. van deen Broek, P. Kabat, and R. G. Allen, pp. 123–140. Wageningen, The Netherlands: Wageningen Press.
124. Wagenet, R. J., and J. L. Hutson. 1989. Leaching Estimation and Chemistry Model: A Process Based Model for Water and Solute Movement, Transformation, Plant Uptake and Chemical Reactions in the Unsaturated Zone Continuum. New York: Water Resources Inst. Center for Environmental Research, Cornell Univ.
125. Carsel, R. F., C. N. Smith, L. A. Mulkey, J. D. Dean, and P. P. Jowise. 1984. *User's Manual for the Pesticide Root Zone Model (PRZM)*, Release 1. EPA-600/3-84-109. Athens, GA: U.S. Environmental Protection Agency.
126. Leonard, R. A., W. G. Knisel, and D. A. Still. 1987. GLEAMS: Groundwater loading effects of agricultural management systems. *Trans. ASAE* 30:1403–1418.
127. Chung. S. O., A. D. Ward, and C. W. Schalk. 1992. Evaluation of hydrologic component of the ADAPT water table management model. *Trans. ASAE* 35:571–579.
128. Smith, R. E. 1995. Simulation of crop water balance. With OPUS. *Crop-Water-Simulation Models in Practice*, eds. Pereira, L. S., B. J. van deen Broek, P. Kabat, and R. G. Allen, pp. 215–227. Wageningen, The Netherlands: Wageningen Press.
129. Whisler, F. D., B. Acok, D. N. Baker, R. E. Fye, H. F. Hodges, J. R. Lambert, H. E. Lemmon, J. M. Mckinion, and V. R. Reddy. 1986. Crop simulation models in agronomic systems. *Adv. Agron.* 40:141–208.
130. IBSNAT 1989. Decision support system for agrotechnology transfer. Ver. 2.1. *User's Guide*. Honolulu, HI: Dept. of Soil Science, Univ. of Hawaii.

131. Stockle, C. D., F. K. van Evert, R. Nelson, and G. S. Campbell. 1991. CropSyst: A cropping system simulation model with GIS link, pp. 113–119. *Proc. Agronomy Abstracts*. Madison, WI: American Society of Agronomy.
132. Ferrer, F., and C. D. Stockle. 1995. A model for assessing crop response and water management in saline conditions. *Proc. ICID/FAO Workshop on Irrigation Scheduling: From Theory to Practice*. Rome: FAO/ICID.
133. Jacucci, G., P. Kabat, P. J. Verrier, J. L. Teixeira, P. Steduto, G. Bertanzon, G. Giannerini, J. Huygen, R. M. Fernando, A. A. Hooijer, W. Simons. G. Toller, G. Tziallas, C. Uhrik, B. J. van den Broek, J. Vera Moñoz, and P. Youchev, 1995. HYDRA: A decision support model for irrigation water management. *Crop-Water-Simulation Models in Practice*, eds. Pereira, L. S., B. J. van deen Broek, P. Kabat, and R. G. Allen, pp. 315–332. Wageningen, The Netherlands: Wageningen Press.
134. Mjelde, J. W., R. D. Lacewell, H. Jalpaz, and R. C. Taylor. 1990. Economics of Irrigation Management. *Management of Farm Irrigation Systems*, eds. Hoffman, G. J. *et al.*, pp. 461– 493. ASAE Monograph No. 9. St Joseph, MI: ASAE.
135. Bogges, W. G., and J. T. Ritchie. 1988. Economic and risk analysis of irrigation decisions in humid regions. *J. Prod. Agric.* 1:116–122.
136. Hook, J. E. 1994. Using crop models to plan water withdrawals for irrigation in drought years. *Agric. Syst.* 45:271–289.
137. Greenwood, E. A. N., N. C. Turner, E. D. Schulze, G. D. Watson, and N. R. Venn. 1992. Groundwater management through increased water use by lupin crops. *J. Hydrol.* 134:1–11.
138. Unger, P. W., and B. A. Stewart. 1983. Soil management for efficient water use: An overview. *Limitations to Efficient Water Use in Crop Production*, eds. Taylor, H. M., W. R. Jordan, and T. R. Sinclair, pp. 419–460. Madison, WI: American Society of Agronomists, Crop Science Society of American, and Soil Science Society of America.
139. Beers, T. M., T. de Graff, R. A. Feddes, and J. Ben-Asher. 1986. A linear regression model combined with a soil balance model to design micro-catchments for water harvesting in arid zones. Agric. Water Manage. 11:187–206.
140. Pereira, L. S. 1990. Mitigation of droughts: 1. Agricultural. ICID Bull. 39:62–79.
141. Bucks, D. A., T. W. Sammis, and G. L. Dickey. 1990. Irrigation of arid areas. *Management of Farm Irrigation Systems*, eds. Hoffman, G. J., *et al.*, pp. 499–548. ASAE Monograph No. 9. St Joseph, MI: ASAE.
142. United Nations Educational, Scientific, and Cultural Organization (UNESCO). 1977. World map of desertification. United Nations Conference on Desertification. Conf. 74/2. Rome: FAO.
143. Mills, W. C., A. W. Thomas, and G. W. Langdale. 1988. Rainfall retention probabilities computed for different cropping-tillage systems. *Agric. Water Manage.* 15:61–71.
144. Stewart, B. A., and J. L. Steiner. 1990. Water use efficiency. *Advances in Soil Science*, Vol. 13, ed. Stewart, B. A., pp. 151–174. Berlin: Springer.

145. Steiner, J. L., J. C. Day, R. I. Papendick, R. E. Meyer, and A. R. Bertrand. 1988. Improving and sustaining productivity in dryland regions of developing countries. *Advances in Soil Science*, Vol. 5, ed. Stewart, B. A., pp. 111–138. Berlin: Springer.
146. Missaoui, H. 1996. Soil and water conservation in Tunisia. *Sustainability of Irrigated Agriculture*, eds. Pereira, L. S., R. A. Feddes, J. R. Gilley, and B. Lesaffre, pp. 121–135. Dordrecht, The Netherlands: Kluwer Academic.
147. Prinz, D. 1996. Water harvesting-past and future. In *Sustainability of Irrigated Agriculture*, eds. Pereira, L. S., Feddes, R. A., Gilley, J. R., and Lesaffre, B., pp. 137–168. Dordrecht, The Netherlands: Kluwer Academic.
148. Bucks, D. A. 1990. Water conservation for drought preparedness. *The Role of Irrigation in Mitigating the Effects of Drought*, Pereira, L. S., *et al.*, pp. 83–98. New Delhi: ICID.
149. Farago, T., E. Kozma, and C. Nemes. 1989. Drought indices in meteorology. Idojaras 93:9–17.
150. Felch, R. E. 1978. Drought: Characteristics and assessment. *North American Droughts*, ed. Rosemberg, N. J., AAAS Select. Symp. 15. Washington, DC: American Association for the Advancement of Science.
151. Pereira, L. S., and J. L. Teixeira. 1992. Irrigation under limited water availability: Water saving techniques. *Water Saving Techniques for Plant Growth*, eds. Verplancke, H. J. W., E. B. A. de Strooper, and M. F. L. de Boodt, pp. 33–54. Dordrecht, The Netherlands: Kluwer Academic.
152. Pereira, L. S. 1990. The role of irrigation in mitigation of the effects of drought. General report. In *Proceedings of the* 14*th* *Congress on Irrigation and Drainage*, pp. 1–27. New Delhi: ICID.
153. Haas, J. E. 1990. Strategies in the event of drought. *North American Droughts*, ed. Rosemberg, N. J., AAAS Select. Symp. 15. Washington, DC: American Association for the Advancement of Science.
154. English, M. J., J. T. Musick, and Murty, V. V. N. 1992. Deficit irrigation. *Farm Irrigation Systems*, eds. Hoffman, C. J., T. A. Howell, and K. H. Solomon, pp. 631–663. ASAE Monograph No. 9. St. Joseph, MI: ASAE.
155. Hargreaves, G. H., and Z. A. Samani. 1984. Economic consideration of deficit irrigation. *J. Irrig. Drain. Eng.* 110:343–358.
156. Mitchell, P. D., P. H. Jerie, and D. J. Chalmers. 1984. Effects of regulated water deficits on pear tree growth, flowering, fruit growth and yield. *J. Am. Soc. Hort. Sci.* 109:604–606.
157. Brinkman, R. 1980. Saline and sodic soils. *Land Reclamation and Water Management*, pp. 62–72. Wageningen, The Netherlands: International Institute for Land Reclamation and Improvement.
158. Tyagi, N. K. 1996. Salinity management in irrigated agriculture. *Sustainability of Irrigated Agriculture*, eds. Pereira, L. S., R. A. Feddes, J. R. Gilley, and B. Lesaffre, pp. 345–358. Dordrecht, The Netherlands: Kluwer Academic.

5.6 Land Drainage

J. Martinez Beltran

5.6.1 Introduction

Land drainage deals with the control of waterlogging and soil salinization in agricultural lands. In flatlands, a first problem emerges if soil infiltration rates are low and rainfall or irrigation water stands on the ground surface, generally in small depressions or at the edges of the irrigation basin. This problem can be solved by leveling and smoothing the land and providing it with a uniform slope for excess water to flow through furrows or shallow ditches toward the surface drainage outlet. Surface water is discharged into a collector drain through pipes to prevent the erosion of the open ditch bank.

Waterlogging of the root zone due to the presence of perched water tables also occurs if the percolation rate is lower than the amount of water infiltrated, because of poor internal drainage of the soil. In flatlands, percolation can be improved by means of subsoiling to break hardpans and other types of less pervious layers of the topsoil. Thus, the soil structure, porosity, and hydraulic conductivity are enhanced and thus so is water percolation.

In areas with insufficient natural drainage, subsurface drainage systems are required to maintain the groundwater table below the root zone, thus keeping it free from waterlogging and salinization. These systems consist of parallel, horizontal, generally perforated plastic pipes, although deep open field drains also are used as a first step of drainage development. Groundwater flows toward the lateral drains and through them to the outlet, where it is discharged into the collector drain by means of a rigid unperforated pipe.

The drainage water, removed from the field through the surface and subsurface drainage systems, is conveyed through a network of main drains to the outlet, where it is discharged into a water body by gravity flow, by pumping, or through tidal gates.

Land drainage has contributed to agricultural development in rainfed areas of the temperate regions, in irrigated lands of the arid and semiarid regions, and in the humid tropics. In the temperate humid regions, land drainage promotes good aeration of the root zone and provides moisture appropriate in the topsoil to ensure workability. Therefore, in these regions, drainage has been an efficient means to increase crop production and to decrease farming costs. In irrigated lands, drainage is indispensable to prevent the permanent hazard of waterlogging and salinization: Of 260 million ha of irrigated lands, approximately 60 million ha currently suffer from the effects of salinization [1] and FAO estimates that salt buildup has severely damaged about 30 million ha. In the humid tropics, particularly in lowland areas where, traditionally, only paddy is grown, land drainage is required to increase rice production and to diversify agriculture by growing dry-foot crops.

However, drainage may have adverse environmental effects such as conversion of wetlands to agricultural lands, deterioration of water quality, and landscape destruction. Therefore, new drainage systems must be designed, constructed, and managed, taking into account not only agricultural objectives but environmental factors as well. These factors are essential in the rehabilitation projects that are currently under way.

In this section, the following phases of a drainage project are described: identification and characterization of the problem areas, planning and design of the drainage systems, implementation and control of the quality of the drainage works, and operation and maintenance of the systems installed. Books that provide basic concepts, applications, and guidelines on land drainage can be found in the literature [2–8].

5.6.2 Drainage Investigations

The first phase of a drainage project is the identification of the waterlogged or salt-affected lands and their further characterization for planning the reclamation procedure. This phase is followed by design of the drainage system. During these two phases, some drainage investigations must be carried out by means of two fundamental studies: a soil survey and a hydrologic study, both based on a sound topographic survey. The climatic, soil, and hydrological data can be compiled in geographic information systems (GIS), by means of thematic maps, which can be continuously updated.

Additional investigations should be carried out to assess the environmental impact of the drainage project and to follow up its performance in order to address the protective measures that may be necessary (see Section 5.6.6). Investigations are also necessary to deal with the socioeconomic issues related to the drainage project.

Soil Surveys

To identify the problem areas, soil mapping, focused on assessing land drainability, is required to differentiate those areas affected by waterlogging and salinity from those lands with natural drainage and that are salt free. The intensity of the survey required depends on several factors, such as the size of the project area, the complexity of the problem, and the time and funds available for the project. The intensity varies from reconnaissance surveys at a scale of 1:50,000 to semidetailed studies at a scale of 1:25,000. The observation densities depend on the intensity of the survey, but they can vary from one to four observations per 100 ha.

The physiographic approach to mapping soils [9] is particularly suitable for drainage purposes because a close relationship exists between geomorphology and drainage conditions. Following this approach, once the landforms of the studied area have been mapped through a photointerpretation study, field observations are made. Their main purpose is to determine those soil characteristics relevant to assess land drainability, namely, the position of the mapping unit in the landscape; its salt content; the internal drainage of the soil and the characteristics of the pans impeding the water percolation, if present; and the transmissivity of the layers down to the impervious barrier. Sometimes the hydraulic resistance of a semipervious layer must be known if vertical seepage from a semiconfined aquifer toward the phreatic aquifer is expected or if there is deep percolation through that layer.

For drainage design, a detailed soil survey is needed to delineate more accurately the soil units mapped in the previous survey and to measure the hydrologic soil qualities relevant to the design of the drainage system, such as the infiltration rate, the thickness and the hydraulic conductivity of the soil layers down to the impervious barrier, and the drainable pore space of the layer where the groundwater level oscillates. These soil characteristics are rather constant in homogeneous soils, but in stratified soils they vary in each layer. Moreover, the hydraulic conductivity gets the same value in all directions in isotropic soils but not in anisotropic soils. For example, if laminar soil structure prevails,

Table 5.22. **K and μ values according to soil texture and structure**

Texture[a]	Structure	μ	K (m day^{-1})
C, heavy CL	Massive; columnar	0.01–0.02	0.05
	With permanent wide cracks	0.10–0.20	>10
C, CL, SC, sCL	Prismatic, angular blocks and laminar	0.01–0.03	0.05–0.1
C, SC, sC, CL, sCL, SL, S	Prismatic, angular blocks and laminar	0.03–0.08	0.1–0.5
CL, S, SL, very fine sL, L	Prismatic, medium subangular blocks	0.06–0.12	0.5–1.5
Fine sL, sL	Subangular blocks, coarse granular	0.12–0.18	1.5–3.0
Ls	Medium granular, sand grain	0.15–0.22	3.0–6.0
Fine s, medium s		0.22–0.26	>6
Coarse s, gravel	Sand grains, gravel	0.26–0.35	>6

[a] C-clay; L-loam; S-silt; s-sand.
Source: Adapted from [6].

then the horizontal conductivity is higher than the vertical one, whereas in cracking clay soils the opposite holds. Mapping scales can vary from 1:10,000 to 1:5,000 and observation densities from one per 5 ha to one per hectare.

Methods to Determine Soil Hydraulic Characteristics

Note that main concepts in soil physics are given in Section 5.2. The saturated hydraulic conductivity K (m day^{-1}), the transmissivity KD (m^2 day^{-1}) of a soil layer with a thickness D (m), and the drainable pore space μ (dimensionless) can be estimated, measured by means of field observations, and derived from investigations on drained lands. In the following paragraphs, the methods most commonly used to measure those soil hydraulic characteristics are described. More detailed descriptions are presented elsewhere [6, 10–14].

Estimation from Soil Properties

Estimated values of K and μ can be obtained from observations of the soil profile because these soil hydraulic qualities depend on soil texture and structure. In Table 5.22, average values of K and μ are presented.

Field Methods for Measurement of Hydraulic Conductivity

The transmissivity of a certain area can be estimated by applying Darcy's law to the flow region if, previously, the hydraulic gradient s (dimensionless) is determined on the isohypses map and the discharge is measured. The K value can be measured directly in the soil layers situated below the groundwater level. Above this level, the K value can be estimated from the capillary hydraulic conductivity measured in the unsaturated zone.

The field methods for determining K are based on the following principle: Water flows through a volume of soil, whose boundary conditions are known, and the discharge is measured; the K value is calculated by applying an equation derived from Darcy's law for the specific characteristics of the method. It is not necessary to obtain in each measurement a very accurate result, but to obtain a series of data for each type of soil from which the mean value can be derived. In this way, the effect of the spatial variability of the soil is prevented. It is always recommended that the calculated mean value be compared with the estimated value from the soil texture and structure.

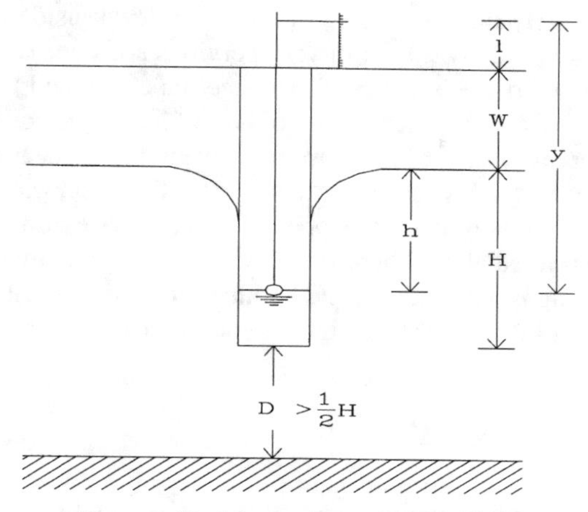

Figure 5.41. Measurements for auger hole.

The auger hole method [15] is the most suitable to measure the K value of homogeneous soils down to a depth of approximately 3 m. This method is based on the relationship between the K value of the soil surrounding a hole and the rate of the rise of the water level inside the hole after pumping water. The hole is augered below the groundwater level and its geometric characteristics are known (Fig. 5.41). With this method the hydraulic conductivity of a soil column, with a radius of 30 cm and a thickness equal to the distance between the groundwater level and 20 cm below the hole bottom, is measured.

The K value can be calculated from

$$K = Cv, \tag{5.187}$$

where K is the hydraulic conductivity (m/day); $C = f(H, \bar{h}, r, D)$ is a geometric factor; H is the thickness of the hole below the groundwater level (cm); r is the radius of the hole (cm); D is the distance from the bottom of the hole to the impervious layer (cm); $v = -dh/dt$ is the velocity of the rise of the water level inside the hole (cm s^{-1}); $h = y - (W + 1)$ is the distance from the water level inside the hole to the groundwater level (cm); W is the depth of the groundwater level (cm); l is the height of the reference level on the ground surface (cm); y are the measurements of the water level inside the hole from the reference level (cm); $\bar{h}$ is the mean value of the h measurements for $h > 3/4h_0$; h_0 is the initial value of h (cm); and t is the time interval (s).

The C value can be calculated with one of the following equations:

When

$$D > 1/2H, \qquad C = \frac{4000\, r^2}{(H + 20r)(2 - \frac{\bar{h}}{H})\bar{h}}. \tag{5.188}$$

When

$$D = 0, \qquad C = \frac{3600\, r^2}{(H + 10r)(2\frac{\bar{h}}{H})\bar{h}}. \tag{5.189}$$

To assess appropriately the rate of rise of the water level inside the hole, this level must be lowered approximately 20 cm in very pervious soils and about 80 cm in soils with low K. The rate of rise is measured by a steel tape fastened to a float. The time intervals vary between 5 and 30 s according to the soil hydraulic conductivity.

The method of the piezometer [10] is more convenient for measurements of the K value in stratified soils and in layers deeper than 3 m. In this case, water is pumped out of a piezometer of which the bottom only is open. The rate of rise of the water level inside the tube is measured immediately. Therefore, the K value of only that portion of the soil surrounding the opening is determined. The Kirkham formula describes the relationship among the rate of rise of the water level, the geometric factors of the measurement, and the K value:

$$K = A' \frac{r^2}{t_i - t_f} \ln \frac{h_f}{h_i}, \tag{5.190}$$

where K is the hydraulic conductivity (m day^{-1}); $A' = 864\pi/A$ (cm^{-1}), A being a geometric factor that depends on r_0 and L (cm) (A values were determined by Youngs by means of an electric analog [5, 10]); r_0 is the radius of the cavity (cm); L is the length of the cavity (cm); r is the internal radius of the tube (cm); $(t_i - t_f)$ is the time interval between the first and the last measurements (s); $h_i = y_i - (W + l)$ is the distance from the water table to the water level inside the tube (cm); y_i is the measurement from the reference level in time t_i (cm); W is the depth of the groundwater level (cm); and l is the height of the entrance of the piezometer above the ground surface (cm).

The inverse auger-hole method [11] is applied to determine the capillary hydraulic conductivity above the groundwater level. In this case, water is poured into an augered hole and the rate of the lowering of the water level inside the hole is measured from the reference level by means of a tape fastened to a float. The measurements are performed after water has been infiltrating for a long time, to diminish the effect of the matric potential in the rate of drawdown. The equation used to calculate the K value has been derived from the balance between the water flowing through the side walls and bottom of the hole and the rate of lowering of the water level in the hole:

$$K = 864 \frac{r}{2t} \ln \frac{h_0 + (r/2)}{h_f + (r/2)}, \tag{5.191}$$

where K is the hydraulic conductivity (m day^{-1}); r is the radius of the hole (cm); h_0 is the initial height of the water column in the hole (cm); h_f is the final height of the water column in the hole (cm); and t is the time elapsed (s) between both measurements.

The infiltrometer method [11] used for measurements of the infiltration rate also can be applied to determine the K value of the top layers of the soil profile, provided that the soil surrounding the infiltrometer is sufficiently moist to avoid the effects of the soil suction in the lowering of the water level.

Determining Transmissivity

Once the hydraulic conductivity has been measured, to calculate the transmissivity it is necessary to determine the thickness of each layer down to the impervious barrier. The field measurements can be done by manual boring down to a depth of 3 to 5 m. If the thickness of the aquifer is greater, pumping tests in drilled wells are required. Methods for pumping tests are described by Boonstra and De Ridder [16] and Kruseman and De Ridder [17].

The impervious layer usually can be identified by observations in an auger hole. For example, when a net change of the soil texture or a sharp increase of the soil compactness are observed and, specifically, if a dry material is found below a layer saturated with water. However, it is not always easy to distinguish an impervious layer; in this case, a layer can be considered less pervious if its K value is less than $1/10$ of the K value of the overlying layer or less than one-fifth of the average value of the pervious soil profile. A layer also can be considered absolutely impervious if the hydraulic resistance for vertical flow ($c = D/K_v$) is higher than 250 days, D being the thickness of the layer (m) and K_v the vertical hydraulic conductivity (m day^{-1}) [19].

The distance from the drain level to the impervious layer is considered the average thickness of the aquifer, if the barrier is deep. However, an average value of the hydraulic head should be added if the water flow above the drain level is relevant in comparison with the flow through the region below the drain level.

Considerations of Soil Anisotropy and Stratification

Alluvial soils frequently present a soil profile with several layers of different materials. Occasionally, the hydraulic conductivity is also anisotropic and then, an isotropic K value equivalent to the actual anisotropic one must be estimated. The model developed by Boumans [20] can be applied to this transformation. This model takes into account the coefficient of anisotropy, which is defined by the relation between the horizontal and vertical conductivity of a soil layer ($R_K = K_h/K_v$). According to this model, the equivalent value of the isotropic conductivity of a soil layer (K') is the geometric mean of K_h and K_v:

$$K' = \sqrt{K_h K_v}. \tag{5.192}$$

In addition, the thickness of each layer is modified by multiplying the actual value (D) by the square root of the coefficient of anisotropy:

$$D' = D\sqrt{R_K}. \tag{5.193}$$

In a stratified soil with n layers, average values of K_h and K_v can be obtained by applying the following formulas:

$$K_h = \frac{\sum_{i=1}^{n} K_{hi} D_i}{\sum_{i=1}^{n} D_i}, \tag{5.194}$$

$$K_v = \frac{\sum_{i=1}^{n} D_i}{\sum_{i=1}^{n} \frac{D_i}{K_{vi}}}. \tag{5.195}$$

Anisotropy is difficult to measure in the field because the usual methods provide K_h only. However, neglecting it involves a greater error. For clearly stratified soils, a ratio $K_h/K_v = 16$ is a better guess than neglecting the anisotropy, which means assuming a ratio of *one*.

Determination of Drainable Soil Porosity

The μ value can be determined by field measurements, for example, by applying the method developed by Guyon, as discussed by Chossat and Saugnac [21]. This method is based on the relationship between the volume of water pumped from an auger-hole and the drawdown of the water observed in four piezometers installed close to the hole. However, sometimes it is not possible to apply this direct method of measurement and an estimation of the μ value is required.

One option is to estimate the μ value on a pF curve as the difference of the water content by volume at saturation and the water content at field capacity, because it is equivalent to the air content at field capacity. This procedure has an important drawback because of the low similarity between a small undisturbed soil sample and the actual field conditions. However, if several laboratory measurements are performed, then a mean value can be obtained that shows the tendency of the μ value.

The μ value also can be estimated from the K value. There are several empirical relations between both hydraulic characteristics, such as those developed by Van Beers [19] and the USBR [6], which are shown in Fig. 5.42. Chossat and Saugnac [21] found that water only flows through a fraction of the total macroporosity; therefore, they obtained lower values of μ (Fig. 5.42).

Investigations on Drained Lands

The results obtained by applying the conventional field methods described above must be checked with those derived from investigations on drained lands, where a large volume

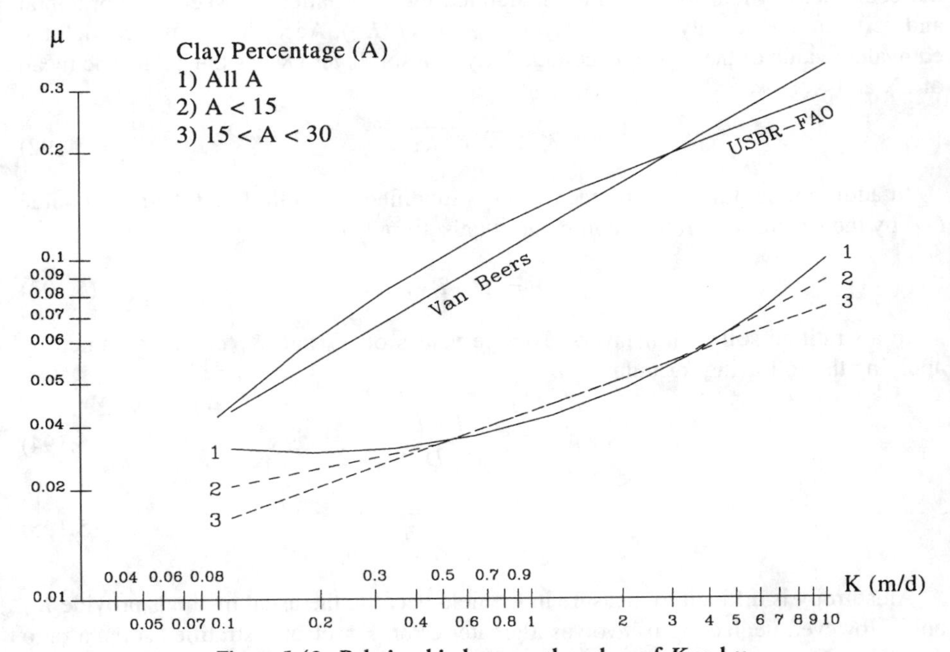

Figure 5.42. Relationship between the values of K and μ.

of soil is considered. These investigations are based on observations of the drawdown of the water table and the specific drainage discharge.

First of all, the best-fitting relationship between the hydraulic head h_h at the mid-point between two drains and the specific drain discharge q must be established. For example, if water flows under steady-state conditions toward drains that reach the impervious barrier, through a layer with good permeability, the following relation can be obtained:

$$q = \frac{8KDh_h}{L^2}, \tag{5.196}$$

where K is the hydraulic conductivity (m day^{-1}), h_h is the hydraulic head for horizontal flow (m), D is the average thickness of the region of flow (m), and L is the drain spacing (m).

In Eq. (5.196) the KD value is the unknown term because L is a design parameter, and the values of q and h_h are measured through field observations from piezometers and at the drainage outlet. Once the KD value is known, the hydraulic conductivity can be calculated if the thickness of the aquifer is measured, by taking the difference between the depth of the impervious layer and the drain level and adding the average hydraulic head above the drain level.

The drainable pore space also can be measured through investigations on a drained field if the water balance in the saturated zone is established:

$$R + S = G + D_r - \mu\Delta h, \tag{5.197}$$

where R is the amount of percolation that recharges the water table (mm), S is the amount of seepage (mm), G is the capillary rise (mm), D_r is the amount of drainage water (mm), Δh is the average drawdown of the water table during the time considered (mm), and μ is the drainable pore space.

If an area free of seepage is chosen during a dry period with low evaporation, outside of the irrigation season when no recharge to the water table occurs, the water balance can be simplified. Then Eq. (5.197) reads

$$D_r = \mu\Delta h. \tag{5.198}$$

Then, to determine the average μ value, it is only necessary to measure the average drawdown of the water table during the interval of time selected and the amount of water drained during that same period.

Hydrological Study

The main purpose of the hydrological study is to determine the natural drainage conditions of the project area by means of a groundwater study and to estimate the magnitude of the recharge of the groundwater table. The mapping scale depends on the intensity level of the study, but it is recommended that work be at the same scale used in the soil survey. The contour lines of the topographic map must be at least 50 cm apart and 10 cm if the lands are extremely flat.

Groundwater Study

The groundwater map is drawn once the hydraulic head data from the aquifer have been obtained by means of piezometric recording. Its main purpose is to determine the groundwater flow through the project and surrounding areas and the recharge and discharge zones.

The *first draft* of the isohypses map can be obtained with data from field observations carried out during the soil survey. On each observation site, an approximate value of the hydraulic head can be obtained if the depth to the groundwater table is recorded and the height above the mean sea level is estimated on the topographic map. Then the isohypses map can be drawn by interpolating these data, but always bearing in mind the continuity of the section of the aquifer considered. On this draft of the isohypses map, the direction of the flow lines, which are perpendicular to the isopiezometric lines, and the recharge and discharge areas can be observed, as shown in Fig. 5.43. This draft map should be checked with the topographic and geomorphologic maps.

The *next phase* is to design and install the piezometer network for permanent recording. In this phase the following criteria should be considered: the piezometer rows must

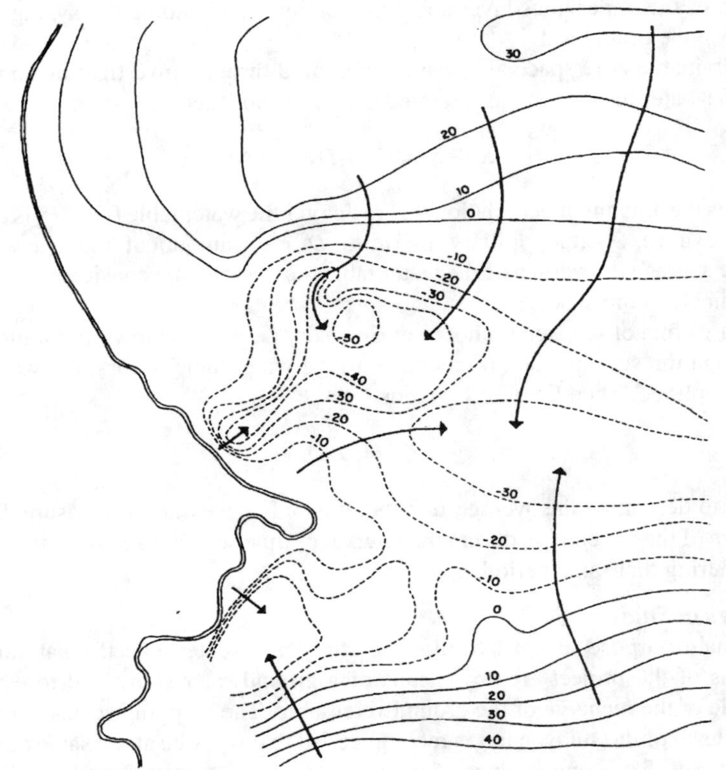

Figure 5.43. Isohypses map with groundwater flow lines.

be installed in the same direction as the flow lines; at least two piezometers and a recorder of the free water level must be installed in a section transverse to the water courses, to know if they drain or recharge the groundwater table; in addition, those transversals also should be installed throughout borders with adjacent areas to estimate the lateral seepage, if any. The piezometer density depends on the complexity of the project area and the resources available for drainage investigations. Densities of 10 to 20 piezometers per 100 ha are recommended for detailed studies for drainage design.

The observations of the piezometer network should be done during the shortest time interval in order to describe the groundwater flow conditions during the same period. At least two annual records should be obtained: for example, in monsoon climates, one for the rainy season and the other for the irrigation season. Both records should focus on the most critical periods of recharge: the period of maximum irrigation requirements when the interval between irrigation applications is the shortest, or at the end of the irrigation season; and the period of heavy rainfall, particularly if it coincides with critical periods of the cropping season, such as sowing or harvesting when there is the need to ensure soil workability.

A geohydrological study is required if a semiconfined aquifer is present below the unconfined aquifer, to determine whether there is vertical seepage from the semiconfined aquifer or leakage from the phreatic aquifer. Deep piezometers must be installed below the semipervious layer to determine the vertical hydraulic gradient. Pumping tests are also needed to measure the hydraulic resistance of the semipervious layer. If these data are known, the amount of seepage or leakage can be calculated by applying Darcy's law. Detailed information on groundwater investigations can be found elsewhere [18].

Determination of Recharge

To design a subsurface drainage system, it is necessary to assess the magnitude of the recharge, which can have different components: percolation of irrigation water, percolation of excess rainfall, and seepage.

The water losses from the network for conveyance and distribution of the irrigation water are not considered as recharge of a groundwater table controlled by a system of parallel horizontal drains; in some instances, they should be controlled by interceptor drainage (see Section 5.6.3). Therefore, only the percolation losses at the field level are taken into account, and they can be estimated from the water balance of an irrigated soil:

$$R = I\left(1 - \frac{e_a}{100}\right) - (E + Q_r), \tag{5.199}$$

where R is the recharge (mm), I is the gross irrigation depth (mm) applied at the field level; e_a is the application efficiency, which represents the ratio (%) between the net I_n and the gross application depths (see Sections 5.1 and 5.4); E is the evaporation loss (mm), and Q_r is the amount of surface runoff (mm).

The options to determine these factors depend on the data immediately available and the information that can be collected over time and at reasonable costs. If the drainage project is located in lands currently irrigated, those agrohydrological factors can be measured through irrigation evaluations. For example, the amount of water effectively applied and the surface runoff can be calculated if the respective flows are measured with

Table 5.23. FAO guidelines to estimate the values of e_a and R for fine or coarse soil

Irrigation Method	Water Management	e_a (%)		R(% I)	
		Fine	Coarse	Fine	Coarse
Sprinkler	Day application; moderate wind	60	60	30	30
	Night application	70	70	25	25
Drip		80	80	15	15
Basin	Poor leveling and design	60	45	30	40
	Good leveling and design	75	60	20	30
Furrows and borders	Poor design	55	40	30	40
	Good design	65	50	25	35

Source: [19].

Table 5.24. Mean values of drainage coefficient for irrigated lands

Soil Conditions	Water Management	q (mm day^{-1})
Less pervious soils	Internal drainage restricted	<1.5
Pervious soils	According to the internal drainage and crop intensity	1.5–3.0
Pervious soils	Poor irrigation or leaching for salinity control	3.0–4.5
Very pervious soils	Irrigation of paddy fields	>4.5

Source: Adapted from [19].

flumes. The I_n value can be estimated by determining the water content of soil samples taken before and after the irrigation application; the calculated value should be checked with ET_c and P data from the previous period.

If the irrigation and drainage systems are being designed jointly, the irrigation requirement is calculated with the water retention data:

$$I = 100\frac{I_n}{e_a} = 100\frac{(\theta_{cc} - \theta_i)1{,}000Z_r}{e_a}, \tag{5.200}$$

where I is the gross application depth (mm), θ_{cc} is the soil water retained at field capacity (m^3 m^{-3}), θ_i is the minimum soil water fraction that does not stress the crop (m^3 m^{-3}), and Z_r is the average thickness of the root zone (m). If the θ_i value is unknown, the amount of water easily available to the crops can be estimated as approximately half the interval between field capacity and the permanent wilting point.

If no data are available, tentative values of e_a and R can be estimated from the Food and Agriculture Organization (FAO) guidelines [19] which are shown in Table 5.23 (Other information on e_a is given in Section 5.4.6). To use these guidelines, only data that can be obtained easily are required, such as the soil texture, the irrigation method, and some qualitative information on water management at the field level.

FAO also has provided data for the average drainage discharge in irrigated lands, which are shown in Table 5.24.

To check if the average amount of percolation water satisfies the minimum leaching requirement to avoid soil salinization, the leaching requirements must be calculated. In the long term, for example, the irrigation season, this could be done by using the salt equilibrium equation from Dieleman [22] and later modified by Van Hoorn [23]:

$$R^* = (ET_c - P_e)\frac{1 - f_i(1 - \mathrm{LF})}{f_i(1 - \mathrm{LF})}, \quad (5.201)$$

where R^* is the long-term leaching requirement (mm); ET_c is the actual crop evapotranspiration (mm); P is the effective precipitation (mm); f_i is the leaching efficiency coefficient (fraction) as a function of the irrigation water applied; $\mathrm{LF} = f_r R / f_i I$ is the leaching fraction, which depends on the soil characteristics, the irrigation method, and the water management; and f_r is the leaching efficiency coefficient as a function of the percolation water.

The f_i coefficient takes into account the fraction of the infiltrated water that flows through cracks and macropores without mixing with the soil solution. This coefficient depends on soil texture and structure and on the irrigation method: It is higher (from 0.95 to 1) in well-structured loamy soils than in heavy clay cracking soils (<0.85), and it is also higher with sprinkler than with surface irrigation.

If the value of the electrical conductivity of the irrigation water (EC_i) is known, the LF value can be calculated by means of the approach developed by Van Hoorn [23] (Fig. 5.44), once the threshold value of soil salinity, expressed as electrical conductivity of the saturated soil paste (EC_e) that on average must not be exceeded in the root zone, has been established from crop salt-tolerance data [24]. This method is based on the water

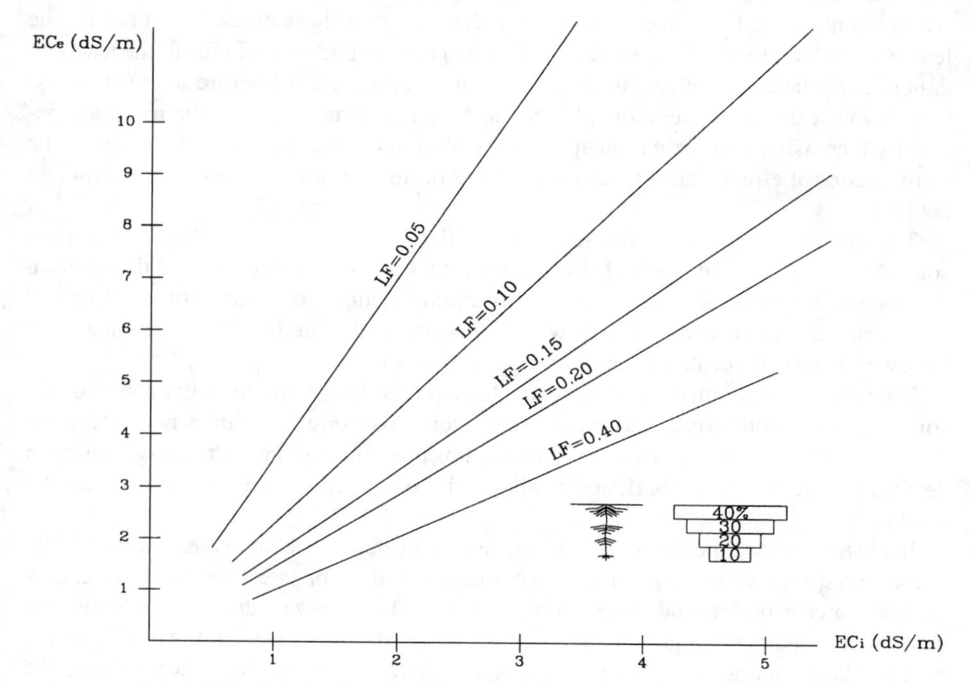

Figure 5.44. Relation between EC_e and EC_i for different leaching fractions.

and salt balances in equilibrium status for the model of water intake represented in the same figure: Water extraction decreases within the root zone from 40% of the total in the top quarter to 10% in the deepest quarter.

In the event of the presence of slightly soluble salts in the irrigation water, such as gypsum, magnesium, and calcium carbonates, the leaching requirement is calculated first for the soluble salts. Afterward, the contribution of the slightly soluble salts to the total soil salinity is added. For average salt content, the total solubility of gypsum and carbonates is approximately equal to 40 meq L^{-1}, equivalent to an EC of some 3.3 dS m^{-1}. If bicarbonates predominate in the irrigation water, it is advisable to decrease the sodium adsorption ratio by increasing the calcium content of the soil solution by applying gypsum. Total amounts of gypsum from 5 to 20 t ha^{-1} are recommended but applications should be applied as fractionated.

If the excess rainwater is sufficient to cover the leaching deficit, the salt content in the root zone can increase at the end of the irrigation season, but the rainfall will supply enough leaching to start the next irrigation season with a low salt content. However, in arid zones, no effective precipitation is usually available for leaching. Therefore, the deficit must be covered by increasing the annual allocation of irrigation water:

$$I = (ET_c - P_e) + R^*. \tag{5.202}$$

A security margin ($R > 1.3R^*$) is advisable [19] because the irrigation is not always uniformly applied in the field. If, under the actual irrigation management, the leaching requirements are not satisfied ($R < 1.3R^*$), there are two alternatives: Either reduce the leaching fraction for growing more salt-tolerant crops or find a way to fulfill the leaching deficit. In the latter case, there are two possibilities: either irrigate before sowing the next crop to lower the salt content or split the leaching requirement during the irrigation period by increasing each irrigation application. With the second possibility, however, the salinity control affects the subsurface drainage design because the drainage coefficient is higher.

The increase of the irrigation requirement depends on the availability of water resources during or at the end of the growing season. It also depends on the internal drainage of the soils: Coarse-textured soils admit leaching fractions between 0.15 and 0.25, whereas in fine-textured soils with low permeability the leaching fraction should be lower than 0.10 because of their limited internal drainage.

When soil salinity increase is not expected over the long term, the salt content of the soil solution should be checked over the short term to ensure that it does not exceed the threshold value for the crop's salt tolerance. For this purpose, the salt storage equation derived for predicting the buildup of soil salinity on a weekly or monthly basis can be used [23].

In the temperate zones the major component of the design recharge is excess rainfall in winter and spring, which is generally lower than the infiltration rate. Consequently, excess surface water is only found in less pervious soils. The average value of the percolation water that recharges the groundwater table is generally less than 10 mm day^{-1}. In the irrigated lands of arid and semiarid regions, the design recharge is commonly due to the percolation of excess irrigation water and to salinity control. However, in some areas, excess precipitation is also present during the rainy season. In the humid tropics, heavy rainfall over short time periods causes surface drainage problems.

The amount of percolation water that recharges the groundwater table during a time interval can be estimated from the water balance of the unsaturated zone:

$$R = F - \Delta W = (P - Q_r - E) - \Delta W, \tag{5.203}$$

where R is the recharge (mm), F is the depth of water infiltrated (mm), ΔW is the variation of the soil water retained in the root zone during the same time interval (mm); P is the rainfall (mm); Q_r is the excess of surface water to be removed through surface drainage (mm); E is composed by interception and surface water evaporation (mm).

The magnitude of the terms of this balance depends on the climatic characteristics of the project area, specifically the rainfall conditions and the evaporation demand, and some characteristics of the unsaturated zone, such as the infiltration rate and the water retention capacity.

First, it is necessary to calculate the design rainfall. The Gumbel distribution generally is used to predict it for certain frequency of exceedance, from data series of daily rainfall recorded over at least 15 to 20 years. Generally, a return period of 5 years is considered for the design of subsurface drainage systems; a period of 10 years can be considered if high-value crops, such as horticultural crops, are grown. To design the major structures of the main drain network, a return period of 25 years generally is adopted. Detailed information on the methods used to determine the rainfall depth–duration–frequency relations are described by Oosterbaan [25].

Once a depth–duration curve has been selected for the design return period, the fraction of excess rainfall that infiltrates must be estimated because part of the water is evaporated directly to the atmosphere and part is discharged through the surface drainage system as surface runoff. It is difficult to measure the infiltration rate because the volume of soil involved in measurements done with the infiltrometer is so small, and normally the soil conditions at the moment of heavy rainfall are quite different from those at the moment of the measurement. For irrigated lands, it is more reliable to obtain average data of the infiltration from field observations after irrigation applications. Also, field measurements of surface runoff after heavy rainfall can be compared with the amount of precipitation recorded by pluviometers to estimate the actual infiltration. If no data are available, mean reference values for different soils and land uses can be utilized.

The recharge of the water table is only a fraction of the water infiltrated because part is absorbed in the unsaturated zone. It is difficult to determine an accurate value of ΔW because it depends on the soil conditions, the precipitation deficit, the land use, and moisture conditions before the period of design rainfall. An average value can be obtained through simulation models for daily water balances in the root zone, taking into account the soil water retention capacity and daily data on evapotranspiration and precipitation. Anyhow, the results obtained should be checked with field observations and direct measurements of the soil water content at the beginning of the rainy season.

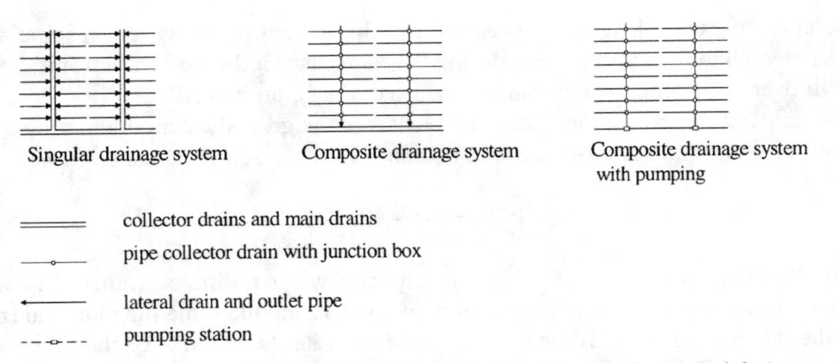

Figure 5.45. Layout of a subsurface drainage system of parallel horizontal drains.

5.6.3 Subsurface Drainage Systems

Main Concepts

Subsurface drainage systems are installed in flatlands to control the groundwater level in order to achieve water and salt balances favorable for crop growth. The groundwater level can be controlled with a system of horizontal parallel drains or by pumping groundwater in wells that penetrate into the aquifer. The first method is known as horizontal drainage; the second is called vertical drainage.

A horizontal subsurface drainage system consists of a network of parallel drains. The groundwater flows toward the field lateral drains, which discharge into collector drains through an outlet pipe. The drainage water is conveyed through the network of collector drains and main canals toward the drainage outlet, where water is discharged into a water body by gravity through a tidal gate or by pumping. In Fig. 5.45, several systems of horizontal drains are shown. Singular drainage systems consist of pipe lateral drains and collector drains that are open ditches, whereas pipe collector drains are used in composite drainage systems. In some areas, especially where maintenance of deep open drains is difficult, groups of collector drains discharge into sumps, from which the water is pumped into a shallow open system (Fig. 5.45).

Several factors must be considered to select the appropriate system, such as the need to discharge surface runoff, the slope of the land to be drained, the depth of the lateral outlets, and the maintenance requirements. Singular drainage systems are appropriate if, in addition to the subsurface flow, it is necessary to discharge excess rainfall through a surface drainage system; a certain amount of water can be stored in the open ditch system and, consequently, the peak flow in the outlet can be reduced. These systems also are recommended for very flat lands because pipe collector drains with large diameters are required if the drainage flow is high and the slope available is low. However, in the irrigated lands of arid regions, composite drainage systems are recommended because the lateral depth is usually higher than in the temperate zones and, consequently, large excavations are required if open ditches are used as collector drains; generally, the excess rainfall is negligible; and weed proliferation increases the maintenance costs of open ditches. Composite systems also are recommended in sloping areas, where soil conservation is required.

Layout

Two basic maps are necessary to draw the layout of a subsurface drainage system: the isohypses map (Fig. 5.43), where the flow lines can be observed, and the map containing the contour lines and the layout of the actual infrastructures, namely, rural roads and irrigation and drainage canals.

Some design principles must be borne in mind. No outside water should be allowed to go into the project area, which must be protected by embankments; surface water must be diverted and lateral seepage must be intercepted. The lateral drains should be drawn perpendicularly to the flow lines to intercept the lateral seepage if any fraction of it has not been discharged through the interceptor drainage; in this way, the collector drains benefit from the natural slope of the ground. Most of the existing drainage canals should be included in the new network of main drains after reshaping their section to obtain the appropriate one. The outlet structures must be situated in the discharge areas, which generally coincide with the lowest topographic zones.

Design Parameters

A subsurface drainage system is defined by the following design parameters: the average depth of the laterals Z, which is equal to the sum of the depth to the water table z and the hydraulic head h, both midway between two drains; the drain spacing L; and the diameter of the pipe d, and under specific circumstances the hydraulic section of the open ditches, which are rarely used as laterals. These parameters can be observed in Fig. 5.46, where the average thickness of the aquifer D also is shown.

In addition, the drainage materials to be used, namely, the type of pipe and envelopes (if necessary), must be chosen.

Other design parameters can be observed in a longitudinal section of a lateral drain (Fig. 5.47): the drain length B and the drain slope s as well as the outlet structures for the drainage water to be discharged into the collector drains.

The drain slope must be as high as possible to discharge the drainage flow with the minimum drain section. Drain slopes in the range between 0.1% and 0.3% are recommended.

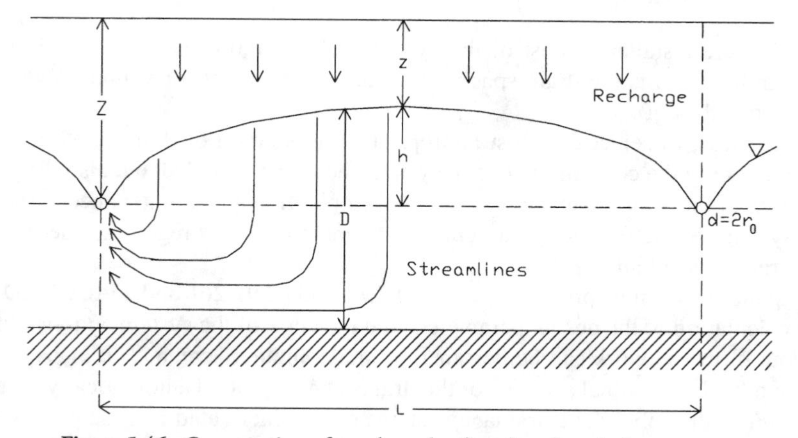

Figure 5.46. Cross section of two laterals of a subsurface drainage system.

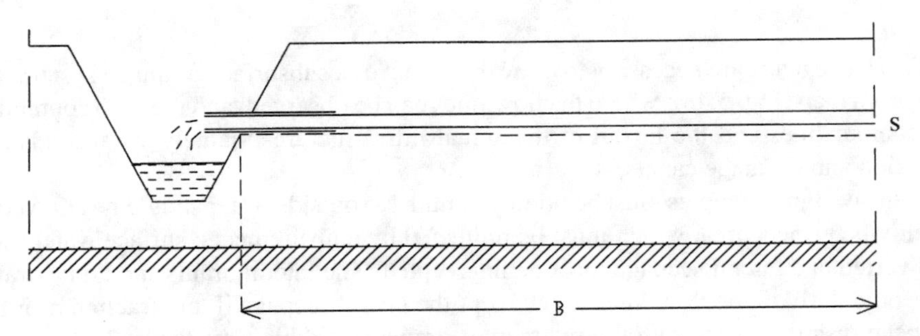

Figure 5.47. Longitudinal section of a lateral drain.

The pipe can be installed with the same slope as the ground slope, especially if it is homogeneous after land-leveling operations; in this way the drain depth is maintained along the lateral. However, in very flat lands, it is often necessary to have a lower depth at the upstream end of the lateral in order to get the design slope. This fact limits the maximum drain length obtainable.

Moreover, the design drain length depends on the geometric characteristics of the agricultural fields and the maintenance requirements of the drain. In flatlands drained with singular systems, lengths up to 250–300 m are common, especially if the need to discharge surface runoff requires open collector drains set 500–600 m apart. However, laterals as long as 750–1,000 m can be designed if there is enough slope. In this case, junction boxes must be installed along the lateral, approximately 250 m apart, to facilitate the connection between pipes of different diameters, as well as the inspection and maintenance of the drain (Fig. 5.45).

The drain spacing and the drain depth are mutually interrelated: The deeper the level of the drains, the wider the drain spacing and, consequently, the lower the drainage intensity and the lower the cost of the subsurface drainage works. That is,

$$C = \frac{10{,}000}{L} C_u, \tag{5.204}$$

where C is the installation cost of the system ($/ha), $10{,}000/L$ is the density of the system (m ha^{-1}), L is the drain spacing (m), and C_u is the cost per unit length of the lateral installed ($/m).

Nevertheless, to select the most appropriate combination of drain depth and drain spacing, not only the costs of the lateral system should be included, but also the proportional costs of the main drainage system and the operation and maintenance costs of the whole system. In the following paragraphs, the main concepts regarding the design of these parameters are analyzed.

Computer-aid design programs, such as LANDRAIN [26] and DACCORD [27], enable calculation of the design parameters and the costs of the system and, in addition, the layout of the network can be designed on the computer screen. The maps with the layout and the longitudinal sections of the drains can be plotted automatically. Planning and design of the subsurface drainage systems can be facilitated if these programs are integrated with GIS [28].

Design Criteria

To calculate the drain spacing of a subsurface drainage system, once a design depth has been selected, there are two options: one is to design the system under conditions of steady-state groundwater flow toward the drains; the other is to design it under unsteady-state flow conditions.

If the recharge to the water table is low and approximately constant during a period of time, one possibility is to design the system in such a way that the drain discharge is approximately equal to the recharge. In this case, the water balance in the saturated zone is in equilibrium and the groundwater level remains permanently at a depth more or less constant. In practice, these conditions occur: In the temperate zones, where low-intensity rainfall over several consecutive days is common in winter; in areas recharged by deep vertical seepage from a semiconfined aquifer; in polder areas, where there is lateral seepage from the outside water bodies; and in irrigated lands, where water is applied continuously through high-frequency irrigation methods, such as drip irrigation and central-pivot systems. Under steady-state conditions, the first design criterion is to assume a drainage coefficient equal to the design recharge:

$$q = r, \tag{5.205}$$

where q is the specific discharge rate (mm day^{-1}) and r is the design recharge rate (mm day^{-1}). The second criterion is to select the hydraulic head midway between two drains, which depends on the minimum depth to a permanent water table and the drain depth:

$$h = Z - z, \tag{5.206}$$

where Z is the drain depth (m); z is the depth to the water table in the midpoint between two drains (m); and h is the hydraulic head in this point (m) as represented in Fig. 5.46.

The z value depends on the thickness of the crop root zone and on the hazard of capillary rise of saline groundwater, mainly if seepage permanently recharges the groundwater table. For field crops grown on clay soils of the temperate zones, a minimum depth to a permanent groundwater table of 0.8–0.9 m is recommended, and 0.6–0.75 m if soils are sandy loam; a depth of 0.5–0.6 m is sufficient if vegetables are grown on sandy loam soils [29]. In irrigated lands of arid regions, an average depth of 1.0–1.2 m is recommended for field crops and 1.2–1.6 m for fruit trees, according to the soil texture [19].

The steady-state approach is not economically feasible if the recharge is high and it occurs over a short period of time. Nevertheless, it is possible to design a drainage system that enables the discharge of the amount of recharge in a period of time equal to the interval between two consecutive recharge events. In this case, the water balance is not in equilibrium: When the recharge is higher than the discharge, the water level rises; when the recharge ceases and the system is still draining, the water level declines. These unsteady conditions are frequent in areas with heavy rainfall of short duration, which is common in the humid tropics and in irrigated lands where irrigation applications from 50 to 100 mm are common if surface or conventional sprinkler irrigation is used.

Under unsteady conditions, the hydraulic head depends not only on the distance to the drain but also on the time. Midway between two drains it varies from a maximum value h_0 just after the end of the recharge to a minimum value h_t just before a new recharge event. Therefore, it is necessary to define the minimum depth of a rising water table z_0 and the period of time Δt for the water table to decline to a maximum depth z_f. The water table can rise close or even up to the ground surface for a short period of time if the irrigation water is good quality, provided that the water level drops below the root zone during a period of approximately three days [30] and the average depth of the water table is 0.8–1 m [30, 31]. The corresponding hydraulic heads are obtained with the following equations:

$$h_0 = Z - z_0, \tag{5.207}$$

$$h_f = Z - z_f, \tag{5.208}$$

where Z is the drain depth (m); h_0 and h_f are, respectively, the maximum and the minimum hydraulic head midway between drains (m).

The relationship between both design parameters is as follows:

$$h_0 = h_f + \Delta h = h_f + R/\mu, \tag{5.209}$$

where Δh is the rise of the water table (m), R is the recharge due to the percolation of excess irrigation water (m), and μ is the mean value of the drainable pore space above the drain level.

The design criteria for the irrigation season should be established for the period of peak water requirements, when the interval between irrigation applications is shortest and, consequently, the recharge is higher. The first approach is based on the dynamic equilibrium concept for a decreasing groundwater level: During the interval Δt between two consecutive irrigation applications, the drainage system should facilitate a drawdown of the water level $-\Delta h$ similar to the rise $+\Delta h$ due to the percolation of the irrigation water losses, as shown in Fig. 5.48.

The Δt value can be determined from irrigation evaluations or from the irrigation schedule if the project area is not yet irrigated:

$$\Delta t = \frac{e_a I}{ET_c}, \tag{5.210}$$

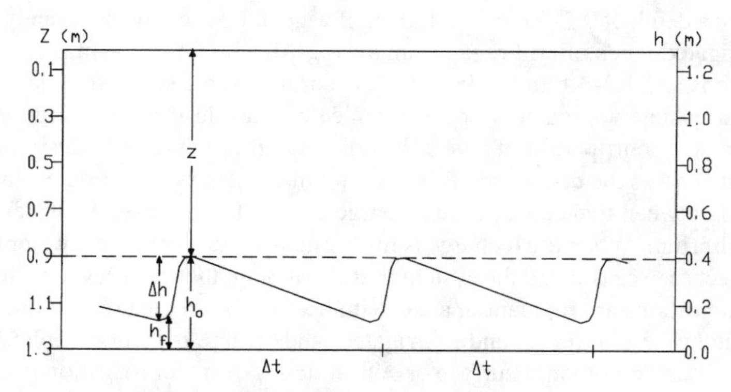

Figure 5.48. A fluctuating water table is controlled below a design depth during the irrigation season.

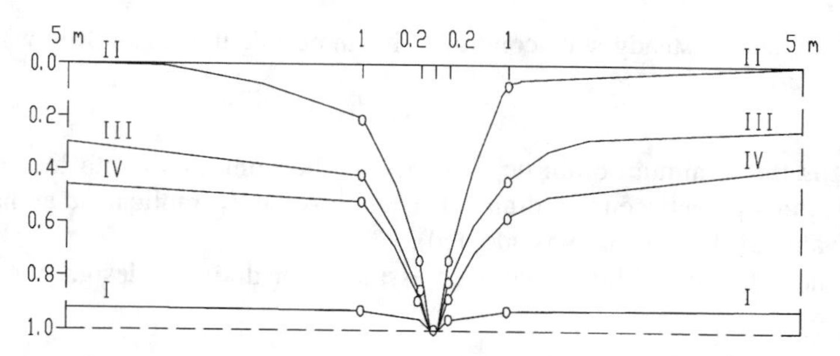

Figure 5.49. Drawdown of water table after an irrigation application in a heavy clay soil.

where Δt is the interval of time between two irrigation applications (day), e_a is the application efficiency (fraction); I is the gross depth of water applied (mm), and ET_c is the average daily crop evapotranspiration (mm day^{-1}).

If, after the irrigation application, the groundwater level rises to near the ground surface, in addition to the dynamic equilibrium concept, the rate of drawdown of the water level should be considered to formulate appropriate h_0/h_t criteria. This situation is frequent in draining heavy clay soils with a shallow impervious layer and low drainable pore space, as is the case of the example shown in Fig. 5.49. Design criteria for draining such soils can be found in the literature [30, 32].

Sometimes steady-state drainage criteria must be applied during the irrigation season. Then geometric mean values of the hydraulic head and the specific discharge during the critical drainage period can be considered:

$$\bar{h} = \sqrt{h_0 h_t}, \tag{5.211}$$

$$q = r/\Delta t, \tag{5.212}$$

where q is the average specific discharge (mm day^{-1}), r is the recharge due to the percolation of excess irrigation water (mm); and Δt is the interval of time between two irrigation applications (day).

Criteria for a falling water level after a sudden rise of the water table due to the percolation of excess water from short-duration heavy rainfall are also based on crop tolerance to temporal waterlogging of the root zone and water content of the arable layer required to ensure the soil workability and the transit of agricultural machines. A drawdown of the water level over a week to a depth of 1.0–1.5 m is recommended or to a depth of 0.2 in 1–2 days. Horticultural crops in warm climates may require this drawdown within 6 h.

Drain Depth

Once the layout and the design criteria of the drainage system have been established, it is necessary to define the possible range of drain depth from a minimum to a maximum value. The minimum drain depth depends on crop requirements. If the drainage system is being designed for steady-state conditions it can be calculated as it follows:

$$Z_m = z + h, \tag{5.213}$$

where Z_m is the minimum drain depth (m); z is the minimum depth to a permanent water table midway between two drains (m), and h is the hydraulic head at that point (a minimun value of 0.1 m usually is adopted).

If unsteady-state conditions have been assumed for drainage design, the minimum depth is

$$Z_m = z_0 + h_0 = z_0 + (h_f + \Delta h), \tag{5.214}$$

where z_0 is the minimum depth to the water level after its rise due to the recharge (m), h_0 is the maximum hydraulic head (m), h_f is the hydraulic head at the end of the water table drawdown (a minimum value of 0.1 m also is adopted), and Δh is the elevation of the groundwater level due to the recharge (m).

A minimum drain depth of 0.8 m generally is considered because pipes installed at a shallower depth may be clogged if crop roots penetrate into the drain through the pipe slots. In addition, shallow drains can interfere with subsoiling practices, which are common in the management of clay soils with low permeability. However, if the conditions of the drainage project enable a design depth higher than the minimum one, the h_0/h_t values can be increased and, consequently, the drain spacing also can be increased.

The major factors that determine the maximum drain depth are the hydraulic conductivity and the soil stability of the layers situated above the impervious barrier because drains should be installed in pervious but stable layers free of silt and quicksand; the drainage level at the outlet of the lateral into the collector drain, which depends on the water level at the outlet of the drainage system and the available slope, bearing in mind that discharge should be by gravity flow if possible; and the depth obtainable by the drainage machines available in the project area.

In the temperate regions, drain depths range from 0.9–1.2 m in rainfed areas to 1.0–1.5 m in irrigated lands. In arid zones, where salinity control is a priority, laterals usually are installed at depths of 1.5–2.0 m, and even down to 2.5–3.0 m in silt-loam soils where seepage of saline water is a major component of the acting recharge.

Drain Spacing

The distance between two consecutive laterals may vary between 50 and 150 m in permeable soils. In pervious clay soils, 20 to 50 m spacings are common and, in heavy clay soils, spacings of 10–20 m frequently are required. The drain spacing of a drainage system can be determined for each land unit, by applying to a drainage equation the mean hydrological constants and the drainage design criteria, formulated for the specific land use and water management technique. The drainage equations are mathematical expressions derived for the specific characteristics of the flow region and for the type of flow considered in the design of a subsurface drainage system. In the following paragraphs, some equations frequently used to calculate drain spacings are described. More detailed descriptions can be found in the literature [33].

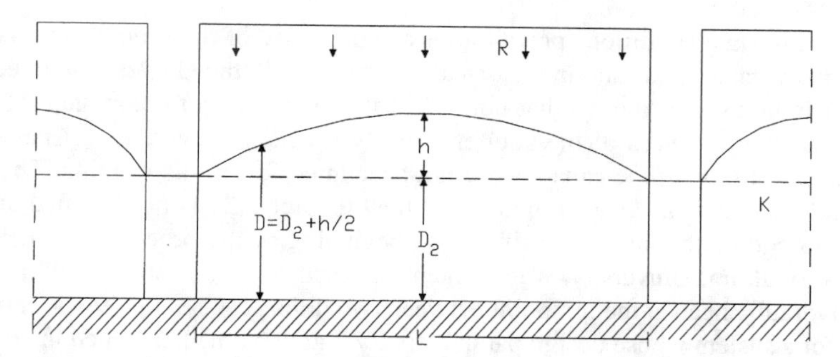

Figure 5.50. Characteristics of the flow region of a homogeneous soil with shallow impervious layer.

Homogeneous Soils

If a steady-state approach has been chosen to design a drainage system for a homogeneous soil, where the impervious layer is not so deep that the laterals can be installed on it (Fig. 5.50), the Hooghoudt equation [34] can be used to calculate the drain spacing for the design drain depth:

$$L^2 = \frac{8KD_2h + 4Kh^2}{q} = \frac{8KDh}{q}, \tag{5.215}$$

where L is the drain spacing (m); $D = D_2 + h/2$ is the average thickness of the flow region (m); D_2 is the thickness of the aquifer below the drain level (m); h is the hydraulic head midway between two drains (m); K is the hydraulic conductivity (m day^{-1}); $q = Q/LB$ is the specific discharge, namely, the water flow per surface area (m day^{-1}); Q is the drainage flow rate (m^3 day^{-1}); and B is the drain length (m).

With Eq. (5.215) the L value can be calculated for a drain depth Z once: The K value and the depth to the impervious layer $(Z + D_2)$ have been measured and drainage design criteria for the discharge $(q = r)$ and for the hydraulic head $(h = Z - z)$ corresponding to a design depth to the water level z have been established.

In some cases, the design drain level can be situated above the impervious layer because the thickness of the phreatic aquifer allows it. It is then necessary to have not only hydraulic head for horizontal flow (h_h) but an extra head (h_r) to overcome the hydraulic resistance due to the radial flow in proximity to the drain trench. In this case the Hooghoudt equation reads

$$L^2 = \frac{8Kdh}{q}, \tag{5.216}$$

where

$$d = \frac{D}{1 + \frac{8D}{\pi L} \ln \frac{D_2}{u}}$$

is the thickness of the equivalent layer (m) for $D_2 < \frac{1}{4}L$, $u = \pi r_0$ is the wetted perimeter of the drain (m), and r_0 is the radius of the drain pipe (m).

If unsteady-state conditions prevail after a sudden rise of the groundwater table, as happens in irrigated lands and in areas with heavy rainfall, the Glover-Dumm equation [35, 36] can be used, provided that horizontal flow is the major component of the total flow and the thickness of the aquifer is fairly constant. However, resistances near the drains and limited pipe capacity are not accounted for in this formula. The use of Hooghoudt's *equivalent layer* in nonsteady flow theoretically is not justified and may give rise to contradictions, especially at the beginning of the process (W. H. Van der Molen, Agricultural University, Wageningen, personal communication, 1998.).

The hydraulic head at the midpoint between two drains can be foreseen as a function of time, for a system whose design parameters Z/L are assumed, installed in an aquifer whose hydraulic characteristics KD and μ have been measured, after a sudden rise of the water level:

$$h_t = 1.16 h_0 e^{-\alpha t}, \tag{5.217}$$

where $\alpha = \pi^2 KD/\mu L^2$ is the reaction factor of the system (day^{-1}), $D = D_2 + \bar{h}/2 = D_2 + \sqrt{h_0 h_t}/2$ is the average thickness of the aquifer (m), h_0 is the maximum hydraulic head midway between two drains (m) at the beginning of the drawdown process, and h_t is the hydraulic head (m) as a function of time (day^{-1}).

The α coefficient describes the response capacity of the system to a sudden rise of the water table. It can be determined in operating systems and it varies from 0.1 to 0.3 (days) and from 2 to 5 (days) in systems with low and high response, respectively.

The Glover-Dumm equation also can be used to calculate the drain spacing if a design drawdown criteria (h_0/h_f; Δt) has been formulated:

$$L^2 = \frac{\pi^2 KD\Delta t}{\mu \ln\left(1.16 \frac{h_0}{h_f}\right)}. \tag{5.218}$$

In some instances the recharge is variable with a fluctuating water level that rises through the effect of the percolation of water and declines through the effect of the drainage system, and a sudden rise of the water table is not expected. In this case the theory developed by Krayenhoff van der Leur [37] can be applied, provided that the flow also is predominantly horizontal.

Stratified Soils

Frequently in drainage projects, alluvial soils are found in which a fine-textured, less pervious layer overlies a coarse-textured, more pervious layer ($K_2 > K_1$). In draining clay soils, sometimes the well-structured topsoil has a higher conductivity than the massive subsoil ($K_1 > K_2$). In both cases it is usually possible to design the drain level so that it is just at the limit between the layers of distinct permeability. Then, the Hooghoudt equation can be applied if the system is designed for steady-state conditions and the respective K value is assigned to the components of the equation:

$$L^2 = \frac{8K_2 D_2 h + 4K_1 h^2}{q}, \tag{5.219}$$

where L is the drain spacing (m), K_1 is the hydraulic conductivity above the drain level (m day^{-1}), and K_2 is the hydraulic conductivity below the drain level (m day^{-1}). D_2 is the thickness of the aquifer below the drain level (m), h is the hydraulic head midway between two drains (m), and q is the specific discharge.

If the drain level is above the impervious layer the Hooghoudt equation still can be used in a stratified soil providing that the drain level is on the boundary between the two layers:

$$h = \frac{qL^2}{8KD} + \frac{qL}{\pi K_2} \ln \frac{D_2}{u}, \tag{5.220}$$

where $KD = K_1 h/2 + K_2 D_2$ is the average transmissivity (m^2 day^{-1}) and u is the drain wetted perimeter (m).

If a marked difference of permeability is found ($K_2 > K_1$), most of the water flows below the drain level ($KD \approx K_2 D_2$). Then the Hooghoudt equation is simplified to

$$L^2 = \frac{8K_2 d_2 h}{q}, \tag{5.221}$$

where

$$d_2 = \frac{D_2}{1 + \frac{8D_2}{\pi L} \ln \frac{D_2}{u}}. \tag{5.222}$$

Sometimes an alluvial soil consists of several layers and occasionally the drain level cannot be designed on the boundary between the layers of different permeability because it is found either deep or very shallow, as can be observed in Fig. 5.51.

In these cases, it is also possible to apply a steady-state approach to calculate the drain spacing by means of the Ernst equation [38] or the Toksöz-Kirkham method [7],

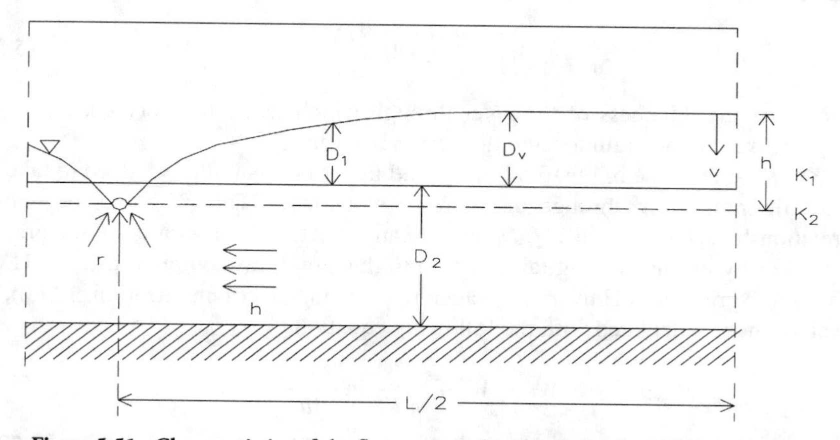

Figure 5.51. Characteristics of the flow region of a stratified soil with drains laid in the pervious layer.

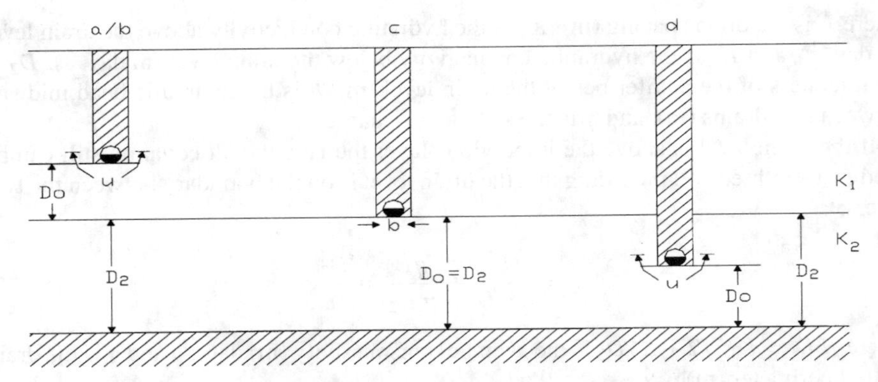

Figure 5.52. Diagram to calculate radial resistance according to Ernst equation.

of which only the former is discussed:

$$h = q\frac{D_v}{K_1} + \frac{qL^2}{8KD} + qLW_r, \tag{5.223}$$

where D_v is the thickness of the layer (m) where vertical flow can exist if there is marked difference of permeability; K_1 is the K value of the less pervious top layer (m day^{-1}); $KD = \sum_{i=1}^{n} K_i D_i$ is the mean transmissivity of n layers (m^2 day^{-1}); and W_r is the radial resistance (day m^{-1}), which depends on the situation of the drain level relative to the position of the boundary between the top layer and the most permeable layer (Fig. 5.52).

If the drain level is in the less pervious top layer (Fig. 5.52), the W_r value depends on the relationship between the K values of both layers. If $K_2/K_1 > 20$ (case a, Fig. 5.52), radial flow only exists through a layer of the topsoil of thickness D_0. Then, the W_r value can be calculated with the following equation:

$$W_r = \frac{1}{\pi K_1} \ln \frac{4D_0}{u}, \tag{5.224}$$

where D_0 is the thickness of the layer through which radial flow occurs, that is, the distance (m) from the drain level to the pervious layer.

If $K_2/K_1 < 20$ (case b, Fig. 5.52), a second factor (W') should be added to take into account the radial flow through the more pervious layer. This W' factor depends on the relationships K_2/K_1 and D_2/D_0 and it can be determined with the nomograph of Fig. 5.53 or by means of a digital approach to this graph for computer use (W. H. Van der Molen, Agricultural University, Wageningen, personal communication, 1998.). The W_r value can be calculated with the following equation:

$$W_r = W' + \frac{1}{\pi K_1} \ln \frac{\pi D_0}{4u}. \tag{5.225}$$

If the drain level is situated at the boundary between both layers (case c, Fig. 5.52), in addition to the Hooghoudt equation (5.219), the Ernst equation (5.223) can be applied

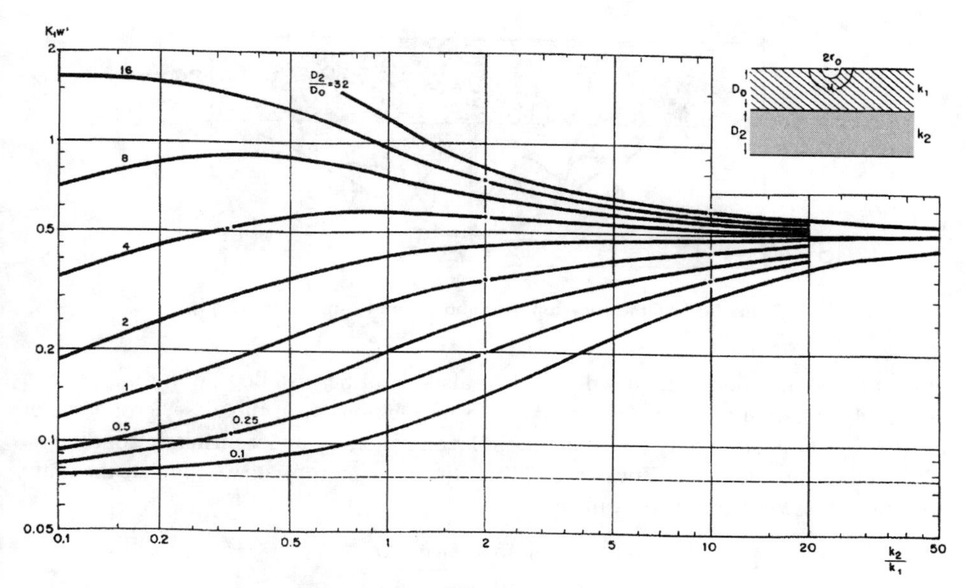

Figure 5.53. Nomograph to determine the W′ value.

by using the following formula:

$$W_r = \frac{1}{\pi K_2} \ln \frac{4D_0}{\pi b}, \tag{5.226}$$

where b is the width (m) of the trench where the drain pipe has been laid; in this case, D_0 is equal to D_2, which is the thickness (m) of the most permeable layer.

If the drain level is situated in the more pervious layer (case d, Fig. 5.52), the W_r value can be calculated with the following equation:

$$W_r = \frac{1}{\pi K_2} \ln \frac{D_0}{u}, \tag{5.227}$$

where D_0 is the distance (m) from the drain level to the impervious layer.

No drainage equations are available for stratified soils if an unsteady-state approach is envisaged. In this case simulation models can be used or fictitious design criteria for steady-state conditions equivalent to the actual unsteady-state conditions must be established. It therefore is recommended to design for steady state and to check for nonsteady behavior if necessary.

Heavy Clay Soils

Waterlogging problems are common in heavy clay soils, in which the hydraulic conductivity depends on soil structure. The K value generally decreases as the soil depth increases, the subsoil being almost impervious.

The permeability of the topsoil layers can be improved by means of deep ploughing and subsoiling down to a depth of 60–80 cm but, for draining such soils, a subsurface

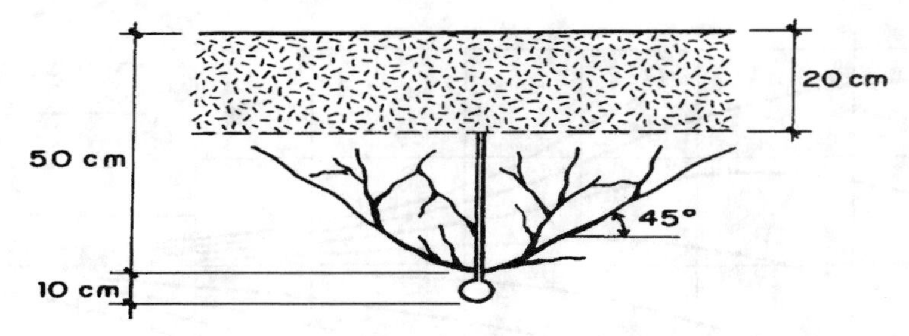

Figure 5.54. Crack development above a mole drain. ***Source:*** **[5].**

drainage system often is required. The laterals should be installed on the impervious layer, which usually lies at a depth of 1–1.5 m. Obviously, in this case, groundwater flows toward the drain only above the drain level, generally under unsteady conditions. The equation developed by Boussinesq [39] is the most appropriate to calculate the drain spacing if these conditions prevail:

$$L^2 = \frac{4{,}46\, K\, h_0\, h_f\, \Delta t}{\mu(h_0 - h_f)}, \tag{5.228}$$

where h_0 is the initial hydraulic head midway between two drains (m); h_f is the h value at the end of the tail recession period (m); and Δt is the time (days) for the groundwater level to decline from h_0 to h_f.

However, to control a perched water table in heavy soils with very low K values ($K < 0.1\,\mathrm{m\,day^{-1}}$) mole drainage is sometimes more suitable than subsurface drainage. For moling, soils should have a clay content of at least 45% and a sand content below 30%. In addition, to obtain a stable mole channel the soil must be wet at the mole depth and dry enough above the critical depth for crack development, as is shown in Fig. 5.54.

In flatlands ($s < 1\%$), mole drains should be laid in the direction of the ground slope with lengths from 20 to 50 m; if the slope is higher than 2%, mole drains 80 m long can be designed. Mole depth depends on the critical depth but it often varies between 40 and 70 cm. Spacings from 1.5 to 3 m are common and mole slopes between 0.5% and 3% are appropriate.

Composite systems of moles, used as lateral drains, and pipe drains as collector also are recommended. In Fig. 5.55, the layout of a composite system can be observed. The spacing between the pipe collectors depends on the mole length and their depth usually varies from 0.7 to 1 m. The mole outlet is connected with the pipe through permeable material, usually gravel filling the pipe trench up to the more pervious top layer.

If soils are suitable for moling and the mole operation is adequate, a system of mole drains can function properly for 3 to 5 years. More details on mole drainage [40] and a comprehensive review of drainage of clay soils made by FAO can be consulted [32].

Interceptor Drainage

To intercept the lateral seepage flowing from sloping lands and water courses, a system of horizontal drains is not often necessary because waterlogging generally occurs

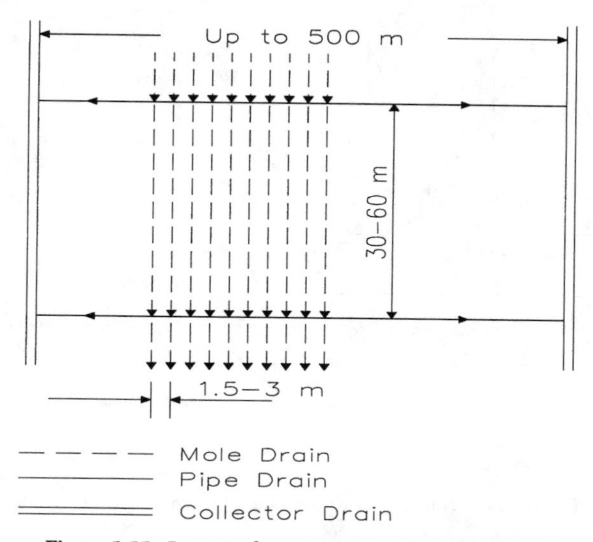

Figure 5.55. Layout of a composite drainage system of mole and pipe drains.

locally. The water level can be controlled by means of interceptor drains, which discharge the lateral seepage into the main drainage system. If the interceptor drain does not reach the impervious layer, a fraction of seepage may flow downstream from the drain and the excess must be discharged through a system of parallel drains. To design an interceptor drain, some parameters must be defined: the distance to the limit between problem and no-problem areas; the drain depth; and the outer diameter of the pipe, which depends on the amount of flow to be intercepted and the available gradient. Furthermore, the drainage materials must be selected. The interceptor drain should be installed very close to the impervious layer to diminish the radial resistance and intercept as much seepage as possible; in sloping lands, there is usually sufficient gradient to drain by gravity. The design depth depends mainly on the hydraulic characteristics and stability of the soil.

Sloping lands

The effective distance at which the drain can be installed can be calculated with the equation developed by Glover and Donnan [41] for a homogeneous soil:

$$x_e = \frac{1}{tg\alpha}\left(D_1 \ln \frac{D_1 - D_0}{(D_1 - h_1) - D_0} - h_1\right), \tag{5.229}$$

where x_e is the effective distance (m) from the drain to the upstream boundary of the sloping land, $tg\alpha$ is the hydraulic gradient upstream from the drain, D_1 is the average thickness (m) of the aquifer upstream from the drain, D_0 is the distance (m) from the drain level to the impervious layer, $h_1 = Z - z$ is the hydraulic head at x_e (m), Z is the drain depth (m), and z is the design depth (m) of the water level. The intercepted flow can be

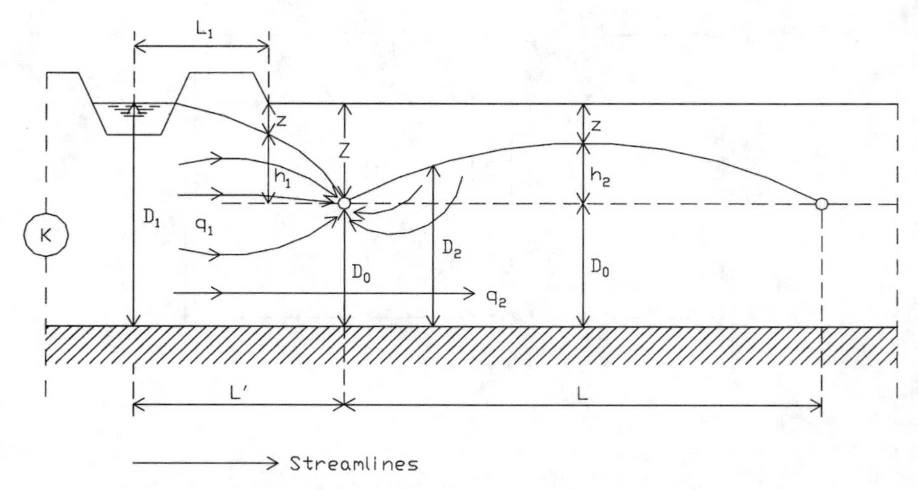

Figure 5.56. Interceptor drain parallel to an irrigation canal.

calculated with the following equation, if the radial resistance is low:

$$q = q_1\left(1 - \frac{D_0}{D_1}\right), \tag{5.230}$$

where q is the flow intercepted by unit length of the drain (m^2 day^{-1}), $q_1 = KD_1 tg\alpha$ is the flow upstream from the drain (m^2 day^{-1}), $tg\alpha$ is the hydraulic gradient upstream from the drain, and K is the hydraulic conductivity (m day^{-1}). If the water table is recharged by excess rainfall or irrigation in addition to seepage, the flow due to percolation should be added [$q' = rx_e$], with r being the specific recharge (m day^{-1}).

Solutions to calculate drain spacings between parallel drains in sloping lands have been developed by Schmid and Luthin [42] and Wooding and Chapman [43]. A review was done by Van Hoorn and Van der Molen [44].

Water Courses and Irrigation Canals

In this case, seepage also can be intercepted through a drain that runs parallel to the water course, as shown in Fig. 5.56. The distance of the interceptor drain from the axis of the canal can be calculated with the Glover-Donnan equation [6, 41] for a homogeneous soil:

$$L' = L_1 + \frac{h_1^2 + 2D_0h_1}{2s(h_1 + D_0)}, \tag{5.231}$$

where L' is the distance (m) from the axis of the canal to the drain; L_1 is the distance (m) between the axis of the canal and the cropped land; $s = [D_1 - (h_1 + D_0)]/L_1$ is the hydraulic gradient of the water table; D_1 is the distance (m) between the water level at the canal and the impervious layer; D_0 is the distance (m) from the drain level to the impervious layer, $h_1 = Z - z$ is the hydraulic head in the limit of the cropped area, defined as the drainage criterion (m); Z is the drain depth (m); and z is the design depth to the water table (m).

If the radial resistance is low, the intercepted flow can be calculated using the following equation:

$$q = q_1 = KD\frac{D_1 - D}{L_1}, \tag{5.232}$$

where q is the flow intercepted by the unit length of drain (m^2 day^{-1}); q_1 is the flow upstream from the drain (m^2 day^{-1}); K is the hydraulic conductivity (m day^{-1}); and $D = h_1 + D_0$ is the thickness (m) of the saturated layer at the boundary of the cropped area.

Drainage Materials

Pipes

Corrugated polyvinyl chloride (PVC), polyethylene (PE), and polypropylene (PP) pipes are the drainage materials most frequently used because of their flexibility, low weight, and suitability for trenchless drainage machines, even if the drain depth reaches 2.5 m. Nevertheless, clay and concrete pipes are still utilized, the latter mostly for collector drains. Pipes 80–100 mm in outer diameter are common for wide drain spacings; 60- to 80-mm pipes frequently are used in systems in the temperate regions and 50- to 60-mm pipes are required in tight drainage systems for draining clay soils.

The diameter of the lateral required to convey the subsurface drainage flow at full capacity can be calculated with the following formulas, which are based on the drainage principle with nonuniform flow, respectively, for smooth (clay, concrete, and plastic), corrugated PVC, and PE-PP pipes [45, 50]:

$$Q = 89d^{2.714}s^{0.572}, \tag{5.233}$$

$$Q = 38d^{2.667}s^{0.5}, \tag{5.234}$$

$$Q = 27d^{2.667}s^{0.5}, \tag{5.235}$$

where Q is the maximum drainage flow (m^3 s^{-1}) at the lateral outlet into the collector drain, d is the internal pipe diameter (m), and s is the drain slope. First, the maximum flow at the outlet of the lateral must be calculated with the following formula:

$$Q_M = \frac{q_M A}{86{,}400}, \tag{5.236}$$

where Q_M is the maximum flow (m^3 s^{-1}), q_M is the maximum specific discharge (m day^{-1}), $A = L \times B$ is the maximum area drained by a lateral (m^2), L is the drain spacing (m), and B is the maximum drain length (m).

If the drainage system has been designed for steady-state flow, the maximum discharge is equal to the recharge selected as drainage criterion ($q_M = r$). If unsteady-state conditions prevail, the maximum discharge usually corresponds to the highest position of the groundwater table; in this case, the design discharge must be calculated with the appropriate equation; For example, if the drains are situated on the impervious layer with the Boussinesq equation [39],

$$q_M = \frac{3.46K}{L^2}h_0^2, \tag{5.237}$$

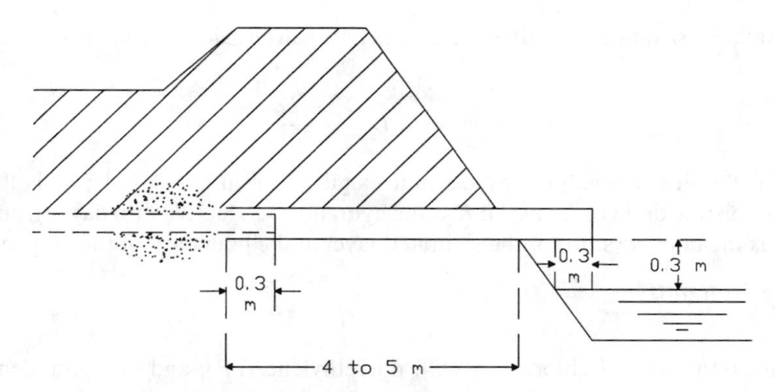

Figure 5.57. Outlet of a lateral drain into a collector drain.

With Eqs. (5.233–5.235), the diameter for clean pipes to discharge at full capacity is calculated. However, if siltation of the drain pipe is a hazard, a higher hydraulic section could be required the diameter of which can be calculated with the following formula:

$$d' = d/\sqrt{e}, \tag{5.238}$$

where d' is the corrected diameter (m); d is the diameter required for flow at full capacity (m); and e is an efficiency coefficient. ILRI recommends an e value of 0.6 for lateral drains [45]. The diameter available on the market that is proximate to the calculated one should be adopted, taking into account the risk of siltation and the availability of envelopes to prevent it.

If singular drainage systems are used, a rigid pipe, long enough for water to flow directly onto the ditch water, as shown in Fig. 5.57, is necessary to prevent soil erosion from water flowing directly through the ditch bank. Rigid pipes also are used where a drain crosses unstable soil or where there is a row of trees or other obstacles. Other minor ancillary structures are pipe fittings, junction boxes, and cleaning facilities.

Drainage pipes are not ideal drains because water enters into the lateral only through the unions of clay tiles and through the perforations of plastic pipes. Therefore, in addition to the radial resistance through the drain trench, the water flow must overcome an extra resistance, called the entrance resistance. The total resistance is then

$$W_t = W_r + W_e = \frac{r_0}{uK_r}\ln\frac{R}{r_0} + \frac{2\pi r_0}{uK_r}\alpha, \tag{5.239}$$

where W_t is the total resistance through the drain trench (day m^{-1}); W_r is the radial resistance through the drain trench (day m^{-1}); W_e is the entrance resistance (day m^{-1}); r_0 is the pipe radius (m); u is the wet perimeter (m); K_r is the hydraulic conductivity of the soil surrounding the drain (m day^{-1}); R is the distance for radial flow (m); and α is a dimensionless coefficient for the entrance resistance which depends on the pipe characterictics. The α coefficient varies from 0.3 to 0.6 for corrugated plastic pipes and from 1.0 to 3.0 for clay and concrete tiles; smooth plastic pipes have intermediate values from 0.6 to 1.0 [46].

To diminish the entrance resistance, a permeable material must be laid around the drain. This can be achieved by means of a pervious backfill with well-structured soil or by installing an envelope with high K_r value.

In addition to the entrance resistance restriction, drainage pipes have to face other problems such as erosion of the soil close to the drain and the subsequent entrance of soil particles into the pipe; clogging of the pipe openings by biochemical compounds, such as ochre, and less soluble salts, such as gypsum and carbonates; and penetration of roots into the pipe. Envelopes also can contribute to solve those problems.

Envelopes

These are drainage materials installed around the drain pipe, which have three main functions: to increase the hydraulic conductivity around the drain and subsequently to reduce the entrance resistance of the water flow into the drain pipe; to prevent the clogging of the pipe openings and the entrance of soil particles into the drain and subsequent siltation; and to stabilize the drain trench surrounding the pipe to ensure the correctness of the drain slope and the alignment of the pipe.

There are different types of envelopes: granular mineral material, such as a 5- to 10-cm cover of gravel or coarse sand laid around the drain, which has excellent performance but high cost; thin sheets of mineral fibers; prewrapped 1-cm-thick coconut fiber, peat, and other organic fibers, which are useful but decompose over time; and synthetic materials such as granular polystyrene and fibrous polypropylene, which are being used currently. Detailed information on envelopes is provided in the literature [46–49].

It is difficult to predict the need for an envelope. Clay content and the size distribution of soil particles are factors that have been taken into account, but they seem to be insufficient because soil structure also plays a role. However, some tentative criteria can be used as guidelines for prediction: Soils with a clay content of more than about 25%–30% show enough structural stability that no envelope is necessary; unstable coarse-textured soils with high hydraulic conductivity need only a thin envelope for filtering purposes; and the most problematic soils are unstable soils with high content of silt and fine sand.

The SCS and the USBR have developed procedures to design gravel envelopes based on the principle that the particle size distribution of the soil should be related to the particle size distribution of the gravel. Guidelines for prediction and design of the appropriate envelope are being developed by FAO [50].

Implementation of the Drainage Works

During the past decade, much progress has been achieved in the implementation of drainage systems because of improvements in drainage machines and the introduction of laser technology to control the drain depth for obtaining an appropriate design slope. These improvements have enabled an increase of the drain depth; an increase in the installation velocity up to 600 m h^{-1}, with three laborers attending one machine and, subsequently, a decrease in the drainage costs.

There are light trench machines, with a power of 100 kW a weight of 12 t, and the capacity to install pipes at a depth of 1.5 m; more powerful trenchers (200 kW and 20 t), are able to install drains with a gravel envelope at a depth of 3 m. Big trenchers with a power of 300 kW and a weight of 40 t are used to excavate drain trenches 60 cm wide and install collector drains with a high diameter down to a depth of 3.5 m [51].

If drain depth does not exceed 1.4 m, the maximum efficiency and the lowest costs are achieved with delta trenchless drainage machines. However, these machines are not suitable to install pipes with diameters larger than 100 mm and granular envelopes, as often are required in draining irrigated lands [52].

Control of the Quality of the Drainage Works and Maintenance Requirements

If a subsurface drainage system has been installed adequately, the maintenance requirements are very low. However, deficiencies in performance sometimes are observed if defective drainage materials have been used or the works have been improperly implemented, specific defects in alignment of the pipe, in drain depth, and consequently in the slope. These deficiencies limit the entrance of water into the lateral, reduce its discharge capacity, and make the maintenance difficult.

One measure to ensure the quality of the drainage materials is to use pipes and envelopes certified by an official institution. Deviations in the drain depth and alignment are easy to check if the laterals have been installed by means of trench machines because they can be observed before the drain trench has been closed. However, if trenchless machines have been used, the only possibility is to use a metal bar 0.9 m long, which can be inserted into the pipe by means of a flexible rod of glass fiber. However, rodding is a difficult operation in composite drainage systems. Detailed description of this equipment can be found in the literature [53, 54].

Injection of water at low pressure is used to remove sediments, roots, and crusts from the pipe walls and to clean the clogged openings. The material removed is flushed by the water flow and discharged into the collector drain. Pressures at the pump from 200 to 300 kPa and water flows of approximately 1.2 L s^{-1} are utilized to clean pipes with diameters from 50 to 90 mm. Higher pressures are recommended only if the drains are completely silted up, because there is the risk of destabilizing the soil around the pipe [55].

5.6.4 Surface Drainage Systems

In flatlands, a drainage system is required to discharge standing water, caused by excessive rainfall or irrigation, over the soil surface. Two major components are essential: a graded and smoothed ground surface free of small depressions with an appropriate slope to enable the surface runoff to flow without producing soil erosion; and surface drains and shallow ditches to convey the drainage water toward the outlet into the main drainage system.

In sloping and undulating lands, the main issue is to prevent soil erosion due to overland flow; this can be achieved by land grading. In this way, surface runoff decreases, infiltration is enhanced, and the soil water availability is thus increased.

Types of Systems

Different systems are used in irrigated lands and rainfed areas; the type depends on the hydraulic characteristics of the soil, the slope, and the land use. In the following paragraphs, some of the most commonly used surface drainage systems are described. Detailed information on the design and construction of surface drainage systems and procedures for land grading has been provided by Sevenhuijsen [56].

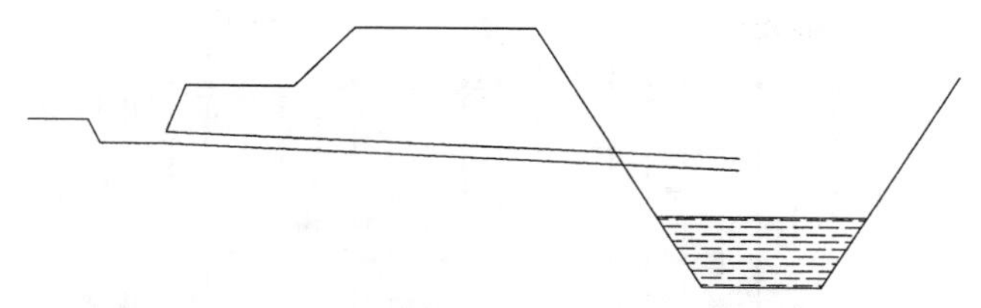

Figure 5.58. Outlet of surface runoff from a shallow ditch to a collector drain.

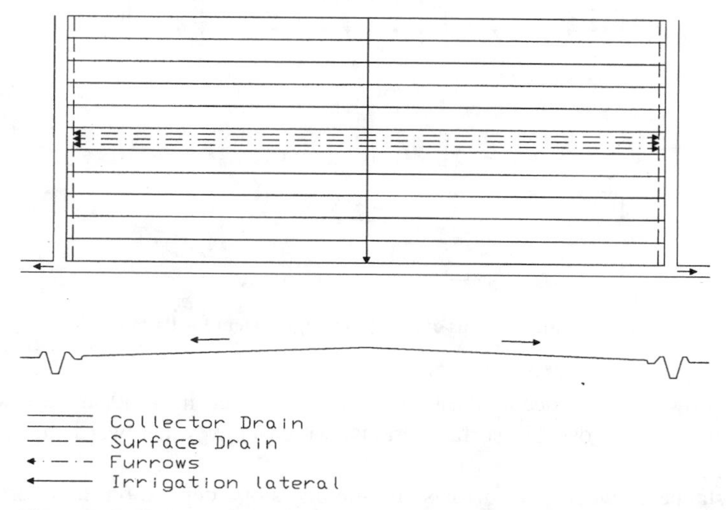

Figure 5.59. Layout of a surface drainage system for flatlands with furrows and shallow ditches.

Furrows and Shallow Ditches

In flatlands, the overland flow can be discharged directly into a shallow ditch running parallel to the collector drain if crops cover the ground surface and there is sufficient slope. To protect the bank from erosion, water should be discharged from this ditch into the collector drain through several short pipes (Fig. 5.58).

In lands irrigated by surface irrigation, better drainage conditions are obtained if furrows are used because, in addition to maintaining the plants in a relatively high position, they transport the excess water toward the outlet ditch, as can be observed in the layout shown in Fig. 5.59.

The length and slope of the furrows can vary according to the type of soil. Lengths from 150 to 200 m and slopes from 0.05% to 0.5% are common, bearing in mind that flow velocities in the furrow should not exceed 0.5 m s^{-1}.

Beds and Dead Furrows

This system is suitable to drain the excess rainfall from clay soils with low infiltration rate, especially if grassland is the major land use. The system consists of beds provided with transversal slope, which are constructed by ploughing, and dead furrows running with the ground longitudinal grade. Excess water flows through the bed surface although

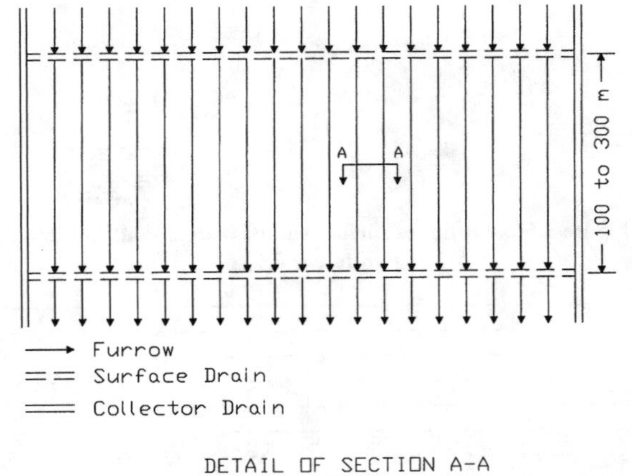

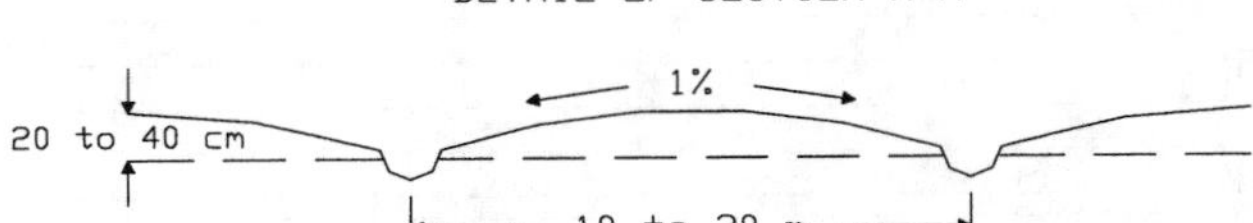

Figure 5.60. Layout of a bed drainage system for flatlands.

some interflow also can occur; the total flow is collected in the dead furrows, which transport it toward crossable surface drains, which convey it toward the field outlet (Fig. 5.60).

The distance between two furrows and the bed slope depend on the hydraulic conductivity and the land use: Spacings can vary between 10 and 30 m and the difference in height from the bed top to the bottom of the dead furrow can vary from 40 cm in grassland to 20 cm in arable land. The furrow length varies from 100 to 300 m and the slope is usually 0.1%.

Parallel Surface Drains

For field crops, a system of parallel surface drains is generally more appropriate. Once the land has been graded and smoothed with the appropriate slope, surface runoff flows through furrows and is discharged into surface drains, which transport the drainage water toward a shallow ditch from which it is discharged into an open collector drain. Usually, the surface drains have lengths up to 250 m and are spaced from 100 to 200 m, according to the ground slope, land use, and land consolidation of the project area.

These surface drains should be designed bearing in mind not only their hydraulic capacity but also the farming operations required for the specific land use. They are usually 20–30 cm deep, with banks from 1:6 to 1:10 and slope between 0.1% and 0.3% (Fig. 5.61).

Parallel Ditches

If, in addition to surface drainage, some amount of subsurface drainage is required, instead of surface drains, a system of parallel shallow ditches would be the most appropriate.

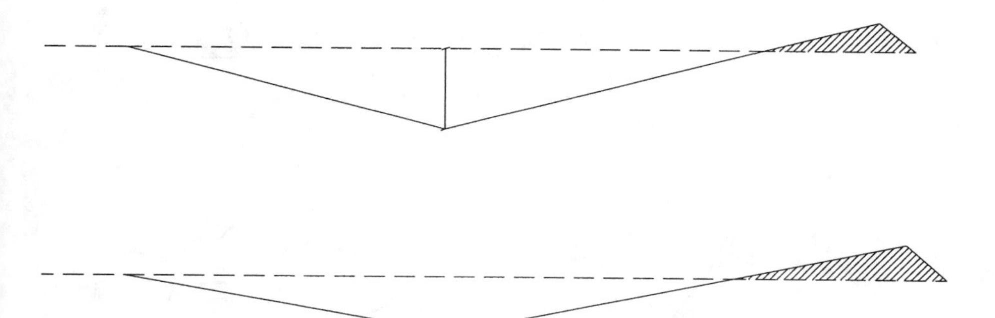

Figure 5.61. Cross sections of surface drains.

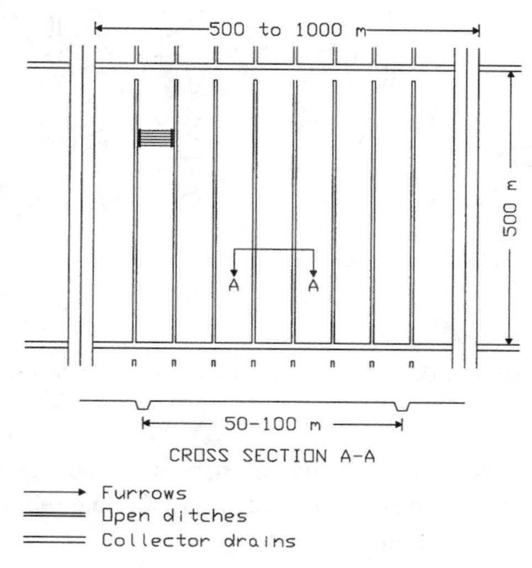

Figure 5.62. Layout of a drainage system with parallel ditches.

In Fig. 5.62, the layout of such a system can be observed. Drain spacing varies from 50 to 100 m, drain depth from 0.6 to 1 m, and the bank slope from 1:1 to 1:1.3.

Surface Drains in Undulating and Sloping Lands

In undulating lands, surface runoff can be collected through a surface drain running along the low-lying valley, as shown in Fig. 5.63. The drain depth depends on the amount of subsurface drainage required: If it is not relevant, shallow drains are appropriate, otherwise, deeper open ditches are preferred. If the valley is broad enough and there is the need to protect the bottomlands from the surface runoff flowing from the adjacent highlands, interceptor drains must be constructed in the valley at the foot of the slope.

In sloping lands with gentle and uniform slope up to 4%, surface runoff can be intercepted by surface drains designed perpendicularly to the ground slope with grades

Figure 5.63. Drainage system for undulating lands.

between 0.1% and 1%. The water collected discharges into collector drains running with the natural slope (Fig. 5.64). In some instances, to facilitate the agricultural practices, the last section of the surface drain is a pipe.

Design Discharge

There are different methods to calculate the contribution of surface drainage to the design discharge of the main drainage system, according to the data available. A first distinction should be done between flat and sloping lands.

Flatlands

Generally, the design discharge is due to excess rainfall. The design discharge can be calculated if the amount of excess rainfall has been estimated from the water balance at the soil surface, and the tolerance of crops to ponding is known. The period of discharge should be lower than the critical period for crops to tolerate inundation, which depends on the crop itself and its growing phase. Empirical data are available on crop tolerance to flooding [57, 58]. Vegetables are sensitive to periods of flooding over 6 to 8 h; fruit trees, however, can tolerate inundation up to 4 days; field crops can be inundated for 1 day and grassland up to 3 days. In addition, a function to calculate the relative crop yield in relation with the submergence time has been developed by Gupta *et al.* [58]. An

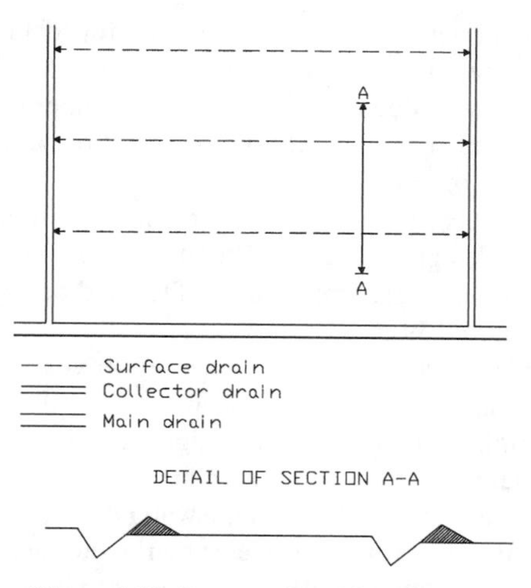

Figure 5.64. Layout of a surface drainage system in a slope.

empirical approach could be to discharge during the first 6 h half of the excess rainfall accumulated during a period of 12 h and the rest during the following 6 h [5].

In irrigated lands, surface runoff can vary from 5% to 15% of the amount of irrigation water applied, according to land use and water management. To calculate the design discharge to the main drain system, the percentage of the area simultaneously irrigated must be known (see Section 5.6.5).

Sloping Lands

In addition to the precipitation intensity, the design flow depends on the size and characteristics of the catchment area. To calculate it, first the fraction of rainfall that becomes surface runoff must be estimated, once the amounts of water evaporated, intercepted, and infiltrated are discounted. Then, a hydrograph showing the basin response must be obtained to determine the maximum flow in each section of the main drainage system.

There are several methods to estimate the surface runoff according to the available hydrological data. If flow data can be measured in a gauge station localized at the outlet of the catchment area, the design discharge can be calculated for a certain return period by means of a statistical analysis of the available data. Unfortunately, this case is not common in small agricultural basins. However, sometimes a relationship between the surface runoff produced by a certain amount of rainfall can be obtained; then, a model can be designed to predict the surface runoff due by using precipitation, data from a longer period. The unit hydrograph method developed by Sherman is based on this principle.

If some data of the discharge produced in small areas are available, a relationship among the surface area, the hydrological characteristics of the catchment area, the precipitation, and the surface runoff can be obtained. This relationship can be applied to other areas in the basin with similar characteristics, by means of the rational method.

Frequently, only rainfall data are available and only few data are known on the characteristics of the project area. In these cases, the amount of surface runoff produced by the design rainfall can be estimated only by the curve number method developed by the SCS [8]. The distribution of the runoff during a period of time can be determined by applying standard unit hydrographs.

The results obtained by the rational and the curve number methods can be considered only an estimation of the design discharge. Therefore, they must be verified later when measured values are obtained at gauging stations. Detailed description of these methods can be found in the literature [59].

Sometimes, a reasonable estimate of the "once in 50 years" flood may be made by means of field observations and enquiries to old farmers. The peak discharge can be estimated by using Manning's equation, if the highest water level observed during the farmer's life is known, taking a cross section and the longitudinal slope and estimating the roughness from the size of the boulders. If the water course floods a large lowland and the water depth is known (e.g., from high-water marks on the trees), the volume of the pool also gives an estimate. Often the outflow is only a small fraction of the accumulated volume, but it can be estimated as well and used as a correction (W. H. Van der Molen, Agricultural University, Wageningen, personal communication, 1998.).

Construction, Operation, and Maintenance

Surface drains usually are constructed during the grading operations with the same earthmoving and smoothing machinery. However, the shallow ditches are constructed by excavators. Both types of drains should be maintained with conventional mowers.

5.6.5 Main Drainage Systems

The main drainage system consists of collector drains, which can be either pipes or open ditches; main drains, which generally are open canals; and auxiliary structures. Drainage disposal can be made effective through gravity outlets, tidal gates, or, in polder areas, through pumping stations. In areas receiving seepage from adjacent lands, interceptor drains are necessary to divert the outside flow. In polder areas, the agricultural area must be protected from flooding by means of embankments. In this way the project area forms a hydrologically independent unit.

The major functions of the main drainage system are to convey the drainage water toward the drainage outlet by the shortest route, to store excess rainfall during the rainy season to reduce the discharge at the outlet, and to maintain appropriate water levels during the dry season for controlled drainage management. This topic is covered in depth elsewhere [5, 60].

Layout

To plan the main drainage system, in addition to the design principles described in Section 5.6.3, another design criterion should be borne in mind: The water flow in the main network should be adapted to the natural flow as much as possible; therefore, the outlets should be located at the lowest sites and the main drains should run along the depressions; most of the existing water courses should be used as drainage canals, with only the adaptations that are necessary to perform their hydraulic function.

To reduce the adverse effect of a network of straight lines on the landscape, water courses must be conserved in their natural configuration and the riparian vegetation should not be removed. Smooth rectangular bends may be used in tertiary-order collector drains (water flows up to 1–2 $m^3 s^{-1}$) but, for larger canals curves are recommended. The radii of curvature may vary from 5 m for medium canals (flow less than 5 $m^3 s^{-1}$) to about 5 to 10 times the bed width for large canals (flow greater than 10 $m^3 s^{-1}$) [5].

The spacing between two consecutive collector drains equals twice the lateral length; therefore, spacings up to 500–600 m are common. The length of the collector drains depends on several factors, namely, the arrangement of the hydraulic and rural road infrastructure and the field plot configuration.

In Fig. 5.65 the layout and the different components of the irrigation and drainage system constructed in a polder, which covers an area of about 14,000 ha, is shown. It can be observed that the whole area is protected from flooding by a perimetric embankment and from lateral seepage by a perimetric interceptor drain. The area is drained by two main drains, which discharge into the Guadalquivir River through tidal gates. During winter, supplementary pumping is used to discharge excess rainfall. Secondary collector drains are spaced at 500 m and primary collector drains at 2,000 m. The irrigation and drainage systems and the rural road network are arranged in such a way that field plots of 12.5 ha are formed, in which surface and subsurface drainage systems have been constructed.

Design Criteria for Drainage Canals

In this section, only open drainage canals are considered; for pipe collectors, reference is made to the paragraph on drainage materials (Section 5.6.3). Collector and main drains are designed with a unique section in each canal reach between two auxiliary structures. In each reach, the discharge, the flow velocity, and the water depth are constant.

First, the average hydraulic gradient available for gravity discharge must be estimated by means of the following equation:

$$s = \frac{h_f - (h_0 + \Delta h)}{B}, \tag{5.240}$$

where h_f is the hydraulic head (m) of the average water level of the collector drain, which collects the drainage discharge of the most remote field plot of the project area from the outlet; h_0 is the hydraulic head (m) of the outer water level at the drainage outlet; Δh is the total head loss (m) due to the auxiliary hydraulic structures of the canal reach; and B is the length of the canal reach (m).

The average slope of the canal network should be equal to the average hydraulic gradient: s values between 0.00005 and 0.0001 are common in flat areas. If the hydraulic gradient is insufficient, discharge by pumping should be considered. A similar procedure can be applied to determine the average slope of each reach of the main drainage network.

Generally, a trapezial hydraulic section is adopted for open ditches (Fig. 5.66).

Although banks should be designed with the highest slope to save agricultural land, and the maximum permissible slope should be selected to design the canal with its minimum cross section, the slope of the bank, the hydraulic gradient, and the maximum

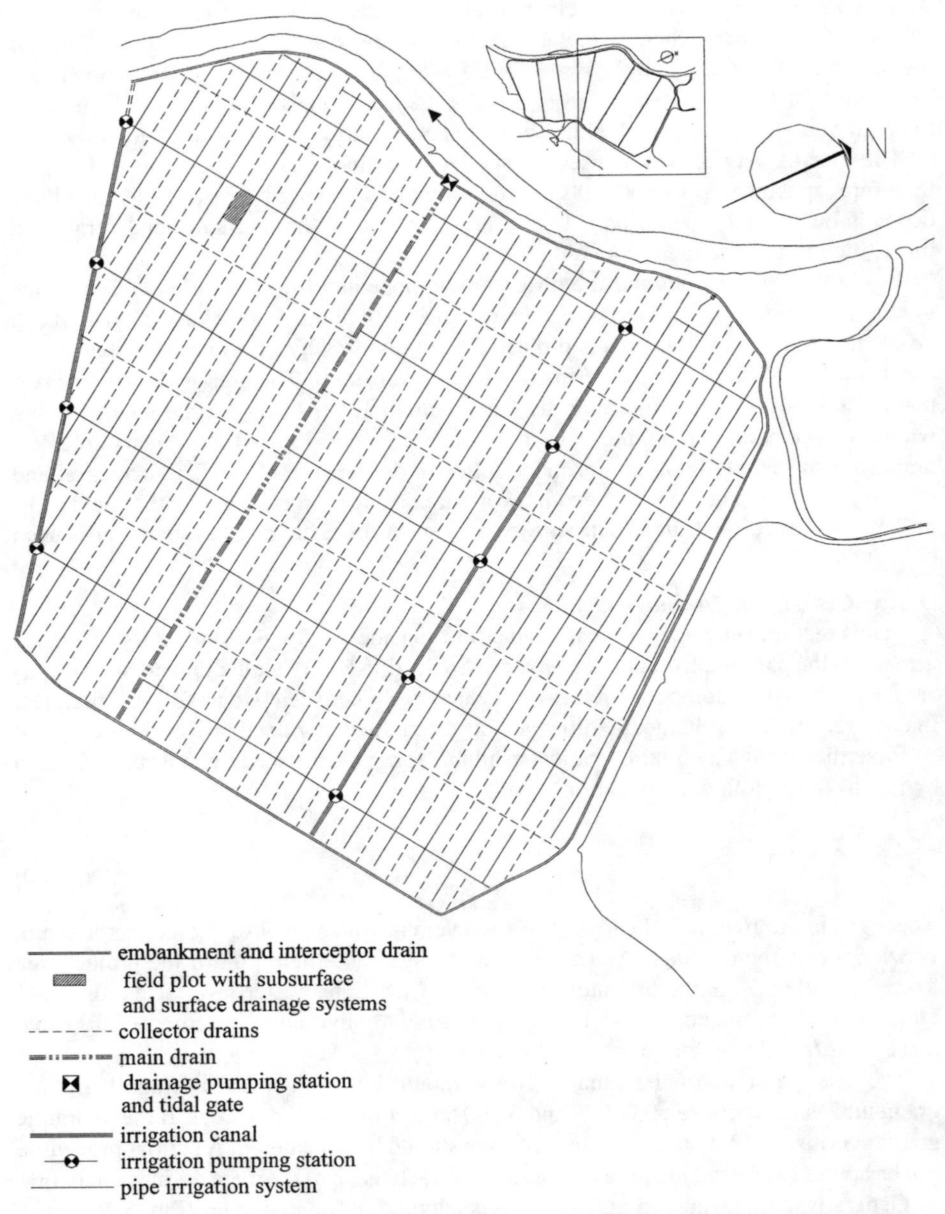

Figure 5.65. Layout of irrigation and drainage systems of a polder area (B-XII Irrigation District, Bajo Guadalquivir Irrigation Project, Spain).

Table 5.25. Bank slopes and maximum water velocity for open ditches (adapted from [5])

Soil Type	v_M (m s^{-1})	Bank (1:α)
Heavy clay	0.60–0.80	1:3/4 to 1:2
Loam	0.30–0.60	1:1.5 to 1:2.5
Fine sand	0.15–0.30	1:2 to 1:3
Coarse sandy	0.20–0.50	1:1.5 to 1.3
Tight peat	0.30–0.60	1:1 to 1:2
Loose peat	0.15–0.30	1:2 to 1:4

Source Adapted from [5].

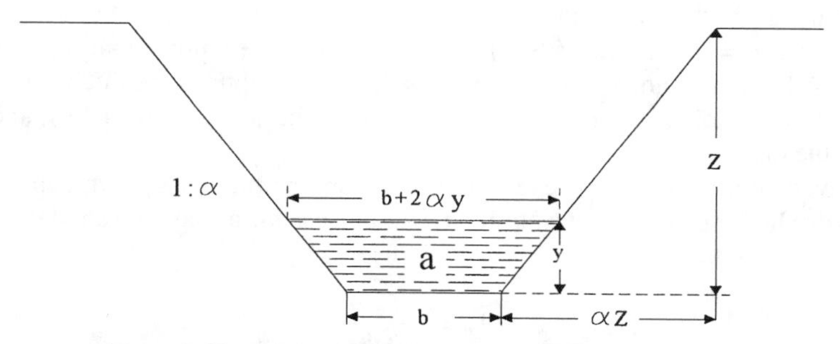

Figure 5.66. Cross section of an open-ditch collector drain.

allowed flow velocity should be determined in accordance with the future stability of the canal, which depends on the type of soil and on the existence of a grass cover. These design parameters should be determined empirically but, in Table 5.25, some values related to the soil texture that can tentatively be adopted are shown.

The depth of a collector drain should be equal to

$$Z = Z_d + F + y, \tag{5.241}$$

where Z is the collector depth (m), Z_d is the depth of the lateral drain at its outlet (m), F is the freeboard (m), and y the water depth (m). F values may vary from 30 cm for small ditches (see Fig. 5.57) to 50 cm in main drains with flows over 5 m^3 s^{-1}.

In addition to these design parameters, it is necessary to select an adequate relationship between the bottom width of the ditch and the water depth; this relationship depends on the canal size (Table 5.26).

Once these design parameters have been selected, it is necessary to check if the hydraulic section as designed has enough capacity to convey the maximum drainage flow with the available slope. The Manning formula can be used for this calculation, assuming steady uniform flow:

$$Q = K_m a r^{2/3} s^{1/2} = \frac{1}{n} a r^{2/3} s^{1/2}, \tag{5.242}$$

Table 5.26. Design parameters for open drains

Size of Drain	y (m)	b/y	Soil Texture	K_m	n
Small	<0.75	1–2	Sandy	20	0.050
			Clayey	15	0.067
Medium	0.75–1.5	2–3	Sandy	30	0.033
			Clayey	20	0.050
Large	>1.5	3–4		40–50	0.020–0.025

Adapted from [5].

where Q is the design flow ($m^3\,s^{-1}$); K_m is the Manning roughness coefficient ($m^{1/3}\,s^{-1}$), which depends on the size of the ditch, the type of soil, and the grass cover of the bank (Table 5.26); $n = 1/K_m$ ($m^{-1/3}\,s^{-1}$); a is the hydraulic section (m^2); b is the bottom width (m); 1:α is the slope of the bank; y is the water depth in the ditch (m); $r = a/u$ is the hydraulic radius (m); $u = b + 2y\sqrt{1+\alpha^2}$ is the wet perimeter (m); and s is the slope of the drain.

The average flow to be conveyed through the main drainage system depends on the average discharge produced during the irrigation season; it may be calculated with the following equation:

$$Q_m = \frac{fqA}{86{,}400}, \tag{5.243}$$

where Q_m is the average flow ($m^3\,s^{-1}$); q is the drainage coefficient due to irrigation losses (m day^{-1}); A is the area drained by a collector drain (m^2); and f is a coefficient that depends on the ratio of the area irrigated simultaneously and the area A. The f coefficient varies from 0.7 to 1; specific f values can be found in the literature [19]. An additional reduction factor can be used to consider the ratio between the area actually cropped and the total area drained [19].

If, in addition to the irrigation return flow, the drainage network conveys water from other sources, for example, water from tail scapes, seepage, or wastewater, the amount of flow calculated with Eq. 5.243 should be increased.

In the temperate regions, the maximum flow is produced generally during the rainy season and can be determined with the following equation:

$$Q_M = \frac{q_M A}{86{,}400}, \tag{5.244}$$

where Q_M is the maximum flow ($m^3\,s^{-1}$); q_M is the drainage coefficient due to excess rainfall (m day^{-1}) (see Section 5.6.4); and A is the area drained by the collector drain (m^2).

Finally, the flow velocity ($v = Q/a$) must be checked with the maximum permissible value, to prevent the erosion of the ditch banks and bottom (Table 5.25). Because of construction and maintenance requirements, tertiary ditches with a hydraulic capacity up to 0.5 $m^3\,s^{-1}$ should have at least a minimum cross section, with a bottom width of 0.5–1 m and a water depth of 0.30–0.50 m.

Implementation, Operation, and Maintenance of Open Drains

In large drainage projects, the implementation of the works usually starts with the construction of the outlet structure to enable water discharge; then the main canal and the primary and lower-order collector drains may be excavated. Open canals up to a width of about 20 m may be constructed by means of excavators: Axial excavation enables the construction of ditches with a width of 8–10 m; larger canals are excavated with the machine working from one side only. Draglines often are used for large canals with a width of 20–35 m.

Erosion of the ditch banks and subsequent sedimentation in the canal bottom and vegetation growth reduce the hydraulic capacity of the drainage canals; consequently, their water levels rise. Therefore, to achieve a stable canal the banks must be protected with grass, which should be mown periodically. In large canals, maintenance operations can be performed by means of a boat equipped with a cutterbar or by dredging. Clearance and reshaping of the cross section of small and medium canals can be made using backhoe excavators; grass-cutting operations usually are performed with a mowing bucket. To facilitate the mobility of maintenance machinery a rural road alongside the drainage canal is necessary. If trees are planted along the road, in addition to the positive impact on the landscape, weed proliferation in the canal is restricted.

In most countries, natural water courses and major hydraulic works of a drainage scheme are generally under public control. However, with the exception of rivers, which should remain under public domain, there is a general trend to transfer operation and maintenance of most of the components of the system to water-user associations. The technical staff of these organizations is in charge of the operation and maintenance of the outlet structures and of inspection and clearance of the main and primary collector drains, which should be performed at least once a year. Farmers usually maintain the tertiary canals and the field drainage systems. To train the above technical staff and promote farmers' participation in drainage projects, FAO and ILRI have published an excellent training manual [61].

Auxiliary Structures

The major structures of a main drainage system are culverts, drops, and weirs. Culverts are necessary where a drainage canal crosses a farm entrance or a rural road. One metal pipe, buried at least 50 cm deep, is commonly used if the water flow is less than 0.5 $m^3\,s^{-1}$ and two pipes for discharges up to 1 $m^3\,s^{-1}$. If the flow is higher, box-type culverts and bridges are preferred [5]. The hydraulic section of the culvert can be calculated with the following equation:

$$Q = \alpha a \sqrt{2g\Delta h}, \tag{5.245}$$

where Q is the discharge ($m^3\,s^{-1}$), α is a coefficient that depends on the conditions at the entrance and at the exit of the structure, a is the area of the hydraulic section (m^2), $g = 9.8$ m s^{-2} is the acceleration of gravity, and Δh is the head loss along the culvert.

The design discharge is often taken some 25% to 50% higher than those for drainage canals. Values of α can vary from 0.8 for short culverts (<10 m long) to 0.7 for long ones; common values for bridges are 0.8–0.9. Head losses of 5 cm for small structures and

10 cm for large ones generally are taken. To calculate the cross section of the structure, in addition to the wet section, a minimum 10-cm clearance is considered [5].

In flatlands, second-order collectors usually join primary collectors and main drains at the same level; however, in sloping lands, where the canal gradient is lower than the ground slope, drop structures are necessary to maintain the permissible flow velocity and to dissipate the excess head. If the energy drop exceeds 1.5 m, inclined drops should be constructed and, if it is not straight, drop structures are preferred.

Rectangular weirs may be used as drop structures and to maintain the desired water levels in the main drains. To calculate the crest width of a weir, the following equation can be used for free-flow discharge [5]:

$$Q = 1.7\alpha b h_1^{3/2}, \tag{5.246}$$

where Q is the discharge (m^3 s^{-1}); α is a coefficient, which in semisharp crested weirs varies from 1.0 to 1.1 [5]; b is the crest width (m); and h_1 is the upstream water level above the crest level (m).

For submerged discharge the following equation may be used [5]:

$$Q = \alpha b h_2 \sqrt{2g(h_1 - h_2)}, \tag{5.247}$$

where Q is the discharge rate (m^3 s^{-1}), α is a coefficient (1.0–1.1), b is the crest width (m), $g = 9.8$ m s^{-2} is the gravitational acceleration, h_1 is the upstream water level above the crest level (m), and h_2 is the downtream water level above the crest level (m).

Other details to design these auxiliary structures can be found in the literature [60].

Outlet Structures

If the outer water level of the drainage outlet is permanently below the inner water level, then gravity discharge is possible. The discharge rate through a sluice can be calculated with Eq. 5.247, being in this case the width of the sluice b and the coefficient α from 0.9 to 1.1 (m). The water depth h_2 should be increased by 3.5% if the sluice discharges directly into the sea [5].

However, in tidal rivers the outer water level varies over time. In this case a tidal gate is required to prevent the entrance of water into the drainage system at the time of high water levels. During the period in which the gate is closed, gravity discharge is interrupted and water is stored in the main canal. At low water levels the gate is opened and gravity discharge is possible. Flap gates placed in a small outlet sluice or in a culvert outlet through a dike are commonly used for low discharges. Manually operated gates and automatic gates are used if the discharge is higher. Details on the design, construction, operation, and maintenance of sluices and gates can be found in the literature [62].

If the outer water level is always above the inner water level, a pumping station is necessary. Often, drainage by pumping is only necessary during the rainy season; then, a gate and a pumping station are joined in the same outlet. A pumping station consists of an approach canal, which enables uniform flow; a sump where drainage water is collected; a suction pipe; the group of pumps; and a delivery pipe with an outlet to the receiving water body.

To select the most appropriate type of pump, some design parameters should be calculated: the design discharges, the lift and the dynamic head, and the power requirement.

Two discharge rates should be considered, following the same criteria as for designing gravity outlet structures: the most frequently occurring discharge, which usually is due to groundwater flow; and the peak discharge, which is mainly due to surface drainage during the rainy season. Generally, the peak flow is discharged through two pumps and the base flow through only one. An additional nonoperational pump usually is installed.

The dynamic head can be calculated with the following equation:

$$h = h_s + \Delta h + \frac{v_d^2}{2g}, \tag{5.248}$$

where h is the total head (m); h_s is the lift or static head, that is, the difference between the inner and outer levels (m); Δh is the total head loss in the suction and delivery pipes; $g = 9.8$ m s^{-2}; and v_d is the flow velocity at the oulet of the delivery pipe (m s^{-1}).

The power requirement can be calculated with the following equation:

$$P = \frac{\rho g Q h}{\eta_t \eta_p}, \tag{5.249}$$

where P is the power required (kW); ρ is the density of water ($\approx$1 kg L^{-1}); $g \approx 9.8$ (m s^{-2}); Q is the discharge rate (m^3 s^{-1}); h is the total head (m); and η_t and η_p are, respectively, the transmission (0.90–0.95) and pump efficiencies. Some correction factors also may be considered in Eq. 5.249 to take into account the elevation of the site and the safe load [5].

Three types of pumps are commonly used according to the drainage flow to be elevated and the lift: if the flow is less than 200 L s^{-1} even if the lift is high, radial pumps are recommended because they have greater flexibility in relation to flow variations; axial pumps are the most suitable for water flows up to 1 m^3 s^{-1} and low lift (2–4 m); and if high flows up to 6 m^3 s^{-1} are to be elevated, the outer level is almost constant, and the water transports.vegetation, the Archimedean screw is appropriate, even if its efficiency is not always the highest.

Once the design parameters are known, pump selection should be made by analyzing the pump characteristics, which describe the relationship among the discharge, the total head, the power consumed, and the efficiency. Pump efficiency η_p values can vary for axial pumps from 0.65 for 1-m lift to 0.80 for 2.5- to 3.0-m lift; for radial pumps from 0.6 for 1-m lift to 0.80–0.85 for lifts higher than 4.0 m [5]; a good Archimedean screw may have 80%–85% efficiency, although if there is too large a gap between the screw and its bed, the efficiency will be lower. The design of pump and pumping stations is broadly described in the literature [63].

Pumping stations may have a negative impact on the landscape, especially those equipped with Archimedean screws or radial flow pumps. In these instances, large buildings are required to house the pumps and additional electrical equipment. The civil works required to house axial flow pumps are simpler and can be integrated into the landscape more easily.

5.6.6 Drainage and the Environment: Water-Quality Aspects

Wetlands are not currently being drained, but sometimes natural areas are situated downstream of drainage projects, where fish and wildlife habitats, mangroves, and other natural resources are to be preserved. Protection of the water environment may be required and frequently a hydrological buffer zone of extensive grassland between agricultural lands and wetlands must be designed.

The negative impacts of drainage works on soil conservation and on the landscape must be mitigated in the phases of design and implementation of the drainage systems (see Sections 5.6.3–5.6.5). Therefore, in this section, only water quality aspects of environmental protection are considered.

Deterioration of the Quality of the Drainage Water

Increasing drainage intensity on agricultural lands, by the installation of subsurface drainage systems, may have positive impacts on the water quality, such as decreasing the load of sediments and chemicals in surface drainage water. In fact, lowering the groundwater table and increasing the drainable pore space reduces the proportion of the total outflow occurring as surface runoff and increases the fraction that is removed slowly by subsurface drainage over longer periods of time.

However, intensive agriculture related to irrigation and drainage may be a source of contamination of surface-water bodies and aquifers, through the deterioration of the drainage water quality due to salinization, pollution by agrochemicals, or mobilization of toxic trace elements in the soil. This is particularly significant in irrigation return flows because, in irrigated lands, salinity control is essential to ensure sustainable agriculture, but drainage-water disposal can contribute to water salinization.

Percolation water also can dissolve soil microelements such us boron, selenium, chromium, molybdenum, and arsenic. Boron, which is common in coastal areas of volcanic origin, may be toxic for plants in small concentrations. Selenium, which is present in certain shales, has been detected in the drainage water of irrigation projects in the San Joaquin Valley of California, in concentrations toxic for the wildlife around the lakes into which the drainage systems discharge [64]. Chromium was found in Egypt because the Blue Nile comes from a basalt area in Ethiopia (W. H. Van der Molen, Agricultural University, Wageningen, personal communication, 1998.).

Another source of water quality degradation is chemicals used in intensive agriculture, such as fertilizers (mainly nitrates), pesticides, fungicides, and herbicides, which can be conveyed by surface runoff, deep percolation to aquifers, and groundwater flow to surface water bodies. Because subsurface drainage generally delays the drainage discharge, increasing the intensity of subsurface drainage generally reduces losses of phosphorus and organic nitrogen while increasing losses of nitrates and soluble salts. Conversely, improved surface drainage tends to increase phosphorus losses and reduce nitrate outflows [65]. A detailed overview on the issue of agricultural drainage water quality can be found elsewhere [66].

Desalinization of agricultural drainage water is considered a promising technique, but it is still expensive [67]. However, to reduce the pollution of surface water bodies, several options are available: to reduce the pollutant load at the field level through sound irrigation water management and farming practices, to reuse the drainage water inside the project area, and to design a safe disposal. To protect the quality of groundwater, measures at the field level must be taken. A detailed review of these options can be found in the literature [68].

Improvement of the Quality of the Drainage Water at the Field Level

In addition to an efficient application of pesticides and fertilizers adapted to crop requirements, the most reasonable strategy to minimize the above adverse environmental impacts of drainage projects is to improve on-farm water management, by reducing surface runoff and deep percolation.

An advisable practice is to reduce surface drainage discharge, because the water excess discharged by surface flow is the main source of chemical pollution of drainage water. Subsurface drainage flows seem to have a less harmful impact on the environment than surface flows because chemicals diluted in infiltrated water are partly fixed in the soil, which has a limited buffer effect.

If salinity control is not a major need, the volume of drainage water to be disposed of can be reduced by means of controlled drainage. Water-table management is therefore an option in drainage projects of the temperate regions to significantly reduce the nitrate concentration of drainage water [69].

To achieve a reasonable equilibrium between crop production and environmental protection, if salinity control is necessary, the most advisable practice is to use a moderate leaching fraction, to retain some salts in the subsoil below the root zone, and to grow suitable crops according to their salt tolerance (main concepts are given in Section 5.6.2).

To reduce water pollution attributable to the mobilization of soil toxic microelements, irrigation projects first must be planned carefully by using sound soil surveys, leaving areas with potential hazards out of the irrigation command. If problematic soils are irrigated, water control by reducing the leaching fraction is also beneficial in reducing the concentration of microelements in drainage water.

Reuse of the Drainage Water

An additional practice to reduce the volume of water to be disposed of is to reuse the drainage water to irrigate salt-tolerant crops and forest trees. The long-term feasibility of reuse increases with regional scale as opposed to on-farm-scale reuse because a system for the collection and redistribution of drainage water is required to reduce the need for storage reservoirs [70].

Direct reuse by pumping from the open drains is a common practice in areas where water resources are scarce at some period during the crop season. Salt buildup is then a risk that must be controlled when fresh water is available for irrigation and leaching or during the rainy season if there is excess rainfall. Successive irrigation with drainage water starting with less salt-tolerant crops and using the last secondary drainage water to irrigate halophytes is another reuse option. The final reduced flow is conveyed to an evaporation pond where salts precipitate.

At the irrigation-scheme level, two options often are applied: to blend the drainage water with freshwater to obtain water of medium quality; and to alternate use of saline water and freshwater according to a strategic approach.

The practice of mixing saline return flows with water of good quality to obtain a blended water of moderate salt content has been applied for a long time in the Nile delta, where 4,000 million m^3 of drainage water are reused annually [71]. Drainage water is pumped from the main drains to the irrigation canals through drainage pumping stations. To determine the amount of drainage water to be blended with fresh water, it is necessary to know the quantity and quality of both water resources and the salt-tolerance requirements of the crops to be irrigated. To predict the quantity and quality of drainage water available for reuse, simulation models such as the SIWARE model [72], developed for the Nile delta, can be applied.

The option of alternating irrigation water with water of good quality, particularly in the germination period when seedlings are more salt sensitive, and full irrigation of salt-tolerant crops with drainage water, seems feasible in areas with sufficient rainfall for postharvest leaching or with alternative sources of irrigation water to control soil salinization due to drainage water reuse. However, on clay soils, irrigation with fresh water after brackish may give problems with soil structure (crusting, slaking). The feasibility of this strategy depends on additional requirements such as the possibility of interception in specific areas; and the availability of storage facilities and delivery systems. Details of the application of this option in the Imperial Valley of California are described [73, 74].

Disposal of Drainage Water

Drainage water usually is discharged directly to surface freshwater bodies such as rivers and less frequently into lakes, or directly into the sea, which under natural conditions is the permanent receptor of salts. The water flow of the river, the interest of downstream users, and the concentration of salts and pollutants in the river and drainage waters, which may vary during the year, determine the volume of drainage water to be discharged.

In coastal areas, where irrigation return flows discharge into tidal rivers, salt contamination is not a severe problem but, if direct disposal into rivers is to be made, strict control of chemicals is necessary. In inland areas, several options exist to minimize the degradation of the quality of water bodies as a result of drainage water disposal: drainage canals to convey the water directly to the sea, evaporation ponds, constructed wetlands, and injection of drainage water into deep wells. A general review of these alternatives can be found in the literature [75].

The discharge of highly concentrated drainage water can be regulated by diverting it into evaporation ponds, when the flow of the river where the system discharges is low. The stored salts can be flushed away during the rainy season when the flow in the disposal river is sufficient to dilute the salt load. In this way the volume of water is reduced and the water discharge is regulated to reduce its pollution effect. These ponds can be natural depressions where the drainage water is disposed of by gravity flow or artificial basins where it is pumped.

However, this option constitutes a risk to groundwater aquifers if leakage occurs and to the river system if there is seepage. Therefore, the ponds must be excavated in fine-textured soils and be closed with embankments. To avoid undesired effects on wildlife, the water level in the pond also must be reduced to a minimum, particularly if toxic trace elements are present in drainage waters. Therefore, the entrance of drainage flow must be limited to the daily evaporation rate if possible. Some examples from the evaporation ponds of the San Joaquin Valley of California have been reviewed by Tanji [76].

The construction of wetlands, where vegetation of grass and reeds grows, also may be an appropriate option to dispose of drainage water, specifically if sediments, nitrogen, and phosphorus have to be removed from the drainage water. Construction details and hydrological requirements of such wetlands can be found elsewhere [77].

The injection of drainage waters into aquifers through deep wells may be another cost-effective option for discharge disposal. If this option is chosen, fresh groundwater must be protected by isolating fresh aquifers; the aquifer therefore must be confined to prevent the upward flow of polluted water. In addition, it must be deep and have sufficient porosity to store the drainage water and its transmissivity high enough to match the injection rate. To prevent salt precipitation and subsequent clogging of the injection well, chemical compatibility of the drainage water and the groundwater is also necessary. To avoid adverse impacts, geohydrological and other environmental conditions must be carefully considered, and continuous monitoring should be implemented to detect leaks. A detailed case study of injection into aquifers in the San Joaquin Valley is presented in the literature [78].

Environmental Assessment and Followup

Environmental assessment of the drainage project and monitoring of the performance of the drainage systems implemented are necessary to detect potential and actual negative impacts, respectively. Investigations related to these environmental issues are noted in the following paragraphs (W. Ochs, World Bank, personal communication 1998.).

If hydrological changes are noted in the assessment as potential for environmental problems in downstream areas, a hydrological evaluation on timing of peak flows or maintaining base flows would be necessary for alternative designs. Some additional storage in the outlet area may be needed, open drains may require development or improvement, changing of outlet locations may be necessary, and the land area to be developed may need to be changed to meet downstream flow requirements.

Pollution of downstream waters normally is indicated in environmental assessments as a concern, and special studies on how to control the pollution are usually necessary. The investigations require collection of information and development of scenarios to address these concerns, which usually relate to erosion and sediment control, water quality control, toxic substances, and health-related items such measures to control waterborne disease and vector problems (snails and mosquitoes).

Sometimes, investigations to provide socioeconomic benefits, such as drainage requirements for rural villages, also are needed. A lot of benefits to justify some projects can be gained by providing a village with drainage outlets and, in many cases, drains within the village.

References

1. Jensen, M. E. 1993. The impacts of irrigation and drainage on the environment. 5th N.D. Gulhatí Memorial Lecture. The Hague: International Commission on Irrigation and Drainage.
2. Ritzema, H. P. (ed.). 1994. Drainage Principles and Applications (2nd ed.), ILRI Publ. 16, Wageningen: International Instute for Land Reclamation and Improvement.
3. Ochs, W. J., and B. G. Bishay. 1992. Drainage guidelines, Tech. Paper No. 195. Washington, DC: World Bank.

4. Eggelsman, R. 1987. Subsurface drainage instructions. *Bull. National Committee of Germany* (2nd ed.), No. 6, Hamburg: International Commission on Irrigation and Drainage.
5. Smedema, L. K., and D. Rycroft. 1983. *Land Drainage: Planning and Design of Agricultural Drainage Systems*. London: Batsford Academic.
6. US Bureau of Reclamation. 1978. *Drainage Manual*. Washington, DC: US Department of Interior.
7. Van Schilfgaarde, J. (ed.). 1974. *Drainage for Agriculture*. Publ. No. 17. Madison, WI: American Society of Agronomy.
8. Soil Conservation Service. 1973. *Drainage of Agricultural Land*. Washington, DC: Water Information Center.
9. Veenenbos, J. S. 1972. Soil survey. Lecture notes of M.Sc. course on Soil and Water Management. Wageningen: Agricultural University.
10. Bouwer, H., and R. D. Jackson. 1974. Determining soil properties. *Drainage for Agriculture*, ed. Van Schilfgaarde, J., Publ. No. 17., pp. 611–666. Madison, WI: American Society of Agronomy.
11. Van Hoorn, J. W. 1979. Determining hydraulic conductivity with the inversed auger hole and infiltrometer methods. *Proceedings of the International Drainage Workshop*, ed. Wesseling, J., pp. 150–154. ILRI Publ. 25. Wageningen: International Institute for Land Reclamation and Improvement.
12. Martínez Beltrán, J. 1993. Soil survey and land evaluation for planning, design and management of irrigation districts. *Cahiers Options Méditerranéennes*, Vol. 1, No. 2. Soils in the Mediterranean Region: Use, Management and Future Trends. Zaragoza, Spain: Centre International de Hautes Etudes Agronomiques Méditerranéennes (CIHEAM).
13. Oosterbaan, R. J., and H. J. Nijland. 1994. Determining the saturated hydraulic conductivity. *Drainage Principles and Applications* (2nd ed.), ed. Ritzema, H. P., pp. 435–475. ILRI Publ. 16. Wageningen: International Institute for Land Reclamation and Improvement.
14. Van Aart, R., and J. G. Van Alphen. 1994. Procedures in drainage surveys. *Drainage Principles and Applications* (2nd ed.), ed. Ritzema, H. P., pp. 691–724. ILRI Publ. No.16. Wageningen: International Institute for Land Reclamation and Improvement.
15. Van Beers, W. F. J. 1970. The auger-hole method: A field measurement of the hydraulic conductivity of soil below the watertable. ILRI Bulletin No.1, rev. ed. Wageningen: International Institute for Land Reclamation and Improvement.
16. Boonstra, J., and N. A. De Ridder. 1994. Single-well and aquifer tests. *Drainage Principles and Applications* (2nd ed.), ed. Ritzema, H. P., pp. 341–375. ILRI Publ. 16. Wageningen: International Institute for Land Reclamation and Improvement.
17. Kruseman, G. P., and N. A. De Ridder. 1994. *Analysis and Evaluation of Pumping Tests Data*. (2nd ed.). ILRI Publ. No. 47. Wageningen: International Institute for Land Reclamation and Improvement.
18. De Ridder, N. A. 1994. Groundwater investigations. *Drainage Principles and Applications* (2nd ed.), ed. Ritzema, H. P., pp. 33–74. ILRI Publ. No.16. Wageningen: International Institute for Land Reclamation and Improvement.

19. Food and Agriculture Organization. 1980. Drainage design factors. Irrigation and Drainage Paper No. 38, Rome: FAO.
20. Boumans, J. H. 1979. Drainage calculations in stratified soils using the anisotropic soil model to simulate hydraulic conductivity conditions. *Proceedings of the International Drainage Workshop*, ed. Wesseling, J., pp. 108–123. ILRI Publ. No. 25. Wageningen: International Institute for Land Reclamation and Improvement.
21. Chossat, J. C. and A. M. Saugnac. 1985. Relation entre conductivite hydraulique et porosite de drainage mesurees par la methode du puits et des piezometres. *Sci. Sol* n° 1985/3, Plaisir-France.
22. Dieleman, P. (ed.). 1963. *Reclamation of Salt Affected Soils in Iraq*. ILRI Publ. No. 11, Wageningen: International Institute for Land Reclamation and Improvement.
23. Van Hoorn, J. W., and J. G. Van Alphen. 1994. Salinity control. *Drainage Principles and Applications* (2nd ed.), ed. Ritzema, H. P., pp. 533–600 ILRI Publ. No. 16. Wageningen: International Institute for Land Reclamation and Improvement.
24. Mass, E. V., and G. L. Hoffman. 1977. Crop salt tolerance current assessment. *J. Irrig. Drain. Div.* 103:115–134.
25. Oosterbaan, R. J. 1994. Frequency and regression analysis. *Drainage Principles and Applications* (2nd ed.), ed. Ritzema, H. P., pp. 175–223. ILRI Publ. No. 16. Wageningen: International Institute for Land Reclamation and Improvement.
26. Sands, G. R., P. Peterson, and R. J. Gaddis. 1985. LANDRAIN: A computer-aided design (CAD) program for subsurface drainage systems. ASAE Paper 85-2556. American Society of Agricultural Engineers, St. Joseph, MI.
27. Penel, M. 1989. DACCORD: A computer-aided design (CAD) program for subsurface drainage. ASAE/CSAE Paper 89-2142. American Society of Agricultural Engineers, St. Joseph, MI.
28. Chieng, S. T. 1990. Integrating GIS technology and CADD techniques for agricultural drainage planning system, system design and drafting. In: Land Drainage, 4th International Workshop, ed. Lesaffre, B., Cairo: International Commission on Irrigation and Drainage.
29. Williamson, R. E., and G. J. Kriz. 1970. Response of agricultural crops to flooding, depth of watertable and soil gaseous composition. *Am. Soc. Agric. Eng. Trans.* 13:216–220.
30. Martínez Beltrán, J. 1988. Drainage criteria for heavy clay soils with a shallow impervious layer. *Agric. Water Manage.* 14:91–96.
31. Abdel-Dayem, M. S. 1987. Development of land drainage in Egypt. *Proceedings of the Symposium of the 25th International Course on Land Drainage*, ed. Vos, J., pp. 195–204. ILRI Publ. No. 42. Wageningen: International Institute for Land Reclamation and Improvement.
32. Rycroft, D. W., and M. H. Amer. 1995. Prospects for the drainage of clay soils. Irrigation and Drainage Paper No. 51, Rome: FAO.
33. Ritzema, H. P. 1994. Subsurface flow to drains. *Drainage Principles and Applications* (2nd ed.), ed. Ritzema, H. P., pp. 263–303. ILRI Publ. No. 16. Wageningen: International Institute for Land Reclamation and Improvement.
34. Hooghoudt, S. B. 1940. Bidragen fot de kennis van enige natuurkandige grootheden van de ground. *Versl. Landbounwkd. Onderz.* 46:517–707.

35. Dumm, L. D. 1954. Drain spacing formula. *Agric. Eng*. 35:726–730.
36. Dumm, L. D. 1960. Validity and use of the transient flow concept in subsurface drainage. ASAE meeting Dec. 4–7. Memphis, TN: American Society of Agricultural Engineers.
37. Krayenhoff van de Leur, D. A. 1958. A study of non-steady groundwater flow with special reference to a reservoir-coefficient. *De Ingenieur*. 40:87–94.
38. Ernst, L. F. 1962. Groundwaterstroningen in de verzading de zone en hun berekening big aanwezighind van horizontale evenwijdage open leidingen. *Versl. Landbounwkd. Onderz*. 189:67–15.
39. Boussinesq, M. J. 1904. Recherches théoriques sur l'écoulement des nappes d'eau infiltrées dans le sol et sur le débit des sources. *J. Math. Pures Appl*. 10:1–78.
40. Spoor, G. 1994. Mole drainage. *Drainage Principles and Applications* (2nd ed.), ed, Ritzema, H. P., pp. 913–927. ILRI Publ. No.16, Wageningen: International Institute for Land Reclamation and Improvement.
41. Donnan, W. W. 1959. Drainage of agricultural lands using interceptor lines. *J. Irrig. Drain. Div.* Proceedings ASAE IR 85:13–23.
42. Schmid, P., and J. Luthin. 1964. The drainage of sloping lands. *J. Geophys. Res*. 69(8):1525–1529.
43. Wooding, R. A., and T. G. Chapman. 1966. Groundwater flow over a sloping impermeable layer. *J. Geophys. Res*. 71(12):2895–2902.
44. Van Hoorn, J. W., and W. H. Van der Molen. 1974. Drainage of sloping lands. *Drainage Principles and Applications*, Vol. IV, pp. 328–339. ILRI Publ. No.16. Wageningen: International Institute for Land Reclamation and Improvement.
45. Cavelaars, W. F., and W. F. Vlotman. 1994. Subsurface drainage systems. *Drainage Principles and Applications* (2nd ed.), ed. Ritzema, H. P., pp. 827–712. ILRI Publ. No.16, Wageningen: International Institute for Land Reclamation and Improvement.
46. Dierickx, W. 1993. Research and developments in selecting subsurface drainage materials. *Irrig. Drain. Sys*. 6:291–310.
47. Willardson, L. S. 1979. Synthetic drain envelope materials. *Proceedings of the International Drainage Workshop*, ed. Wesseling, J., pp. 297–305, ILRI Publ. No. 25. Wageningen: International Institute for Land Reclamation and Improvement.
48. Stuyt, L. C. P. M. 1987. Developments in land drainage envelope materials. *Proceedings of the Symposium of the 25th International Course on Land Drainage*, ed. Vos, J., pp. 82–93. ILRI Publ. No. 42. Wageningen: International Institute for Land Reclamation and Improvement.
49. Vlotman, W. F., L. S. Willardson, and W. Dierickx. 1998. Envelope Design for Subsurface Drains. Publ. No., Wageningen
50. Stuyt, L. C. P. M., and W. Dierickx. 1998. Subsurface drainage materials. Irrigation and Drainage Paper, Rome: In preparation, FAO.
51. Zijlstra, G. 1987. Drainage machines. *Proceedings of the Symposium of the 25th International Course on Land Drainage*, ed. Vos, J., pp. 74–81, ILRI Publ. No. 42. Wageningen: International Institute for Land Reclamation and Improvement.
52. Van Zeijts, T. E. J., and W. H. Naarding. 1990. Possibilities and limitations of trenchless pipe drain installation in irrigated areas. *Proceedings of the Symposium*

on Land Drainage for Salinity Control in Arid and Semi-Arid Regions. Cairo, Egypt: Drainage Research Institute.

53. Van Zeijts, T. E. J. 1987. Quality control of subsurface drainage works in the Netherlands, pp. 117–124. *Proceedings of the Third International Workshop on Land Drainage*. Ohio State University.
54. Van Zeijts, T. E. J., and L. Zijlstra. 1990. Rodding, a simple method for checking mistakes in drain installation. *Proccedings of the Symposium on Land Drainage for Salinity Control in Arid and Semi-Arid Regions*, Vol. III: pp. 84–93. Cairo: Drainage Research Institute.
55. Van Zeijts, T. E. J., and A. Bons. 1991. Jet flushing, a method for cleaning subsurface drainage systems. Government Service for Land and Water Use, Utrecht.
56. Sevenhuijsen, R. J. 1994. Surface drainage systems. *Drainage Principles and Applications* (2nd ed.), ed. Ritzema, H. P., pp. 799–826. ILRI Publ. No. 16, Wageningen: International Institute for Land Reclamation and Improvement.
57. Rojas, R. M., and L. S. Willardson. 1984. Estimation of the allowable flooding time for surface drainage design. *Proceedings of the ICID 12th Congress*, Q-390, R 31, pp. 493–509. New Delhi: International Commission on Irrigation and Drainage.
58. Gupta, S. K., R. K. Singh, and R. S. Pandey. 1992. Surface drainage requirement of crops: Application of a piecewise linear model for evaluating submergence tolerance. *Irrig. Drain. Syst.* 6:249–261.
59. Boonstra, J. 1994. Estimating peak runoff rates. *Drainage Principles and Applications* (2nd ed.), ed. Ritzema, H. P., pp. 111–143. ILRI Publ. No.16, Wageningen: International Institute for Land Reclamation and Improvement.
60. Bos, M. G. 1994. Drainage canals and related structures. *Drainage principles and applications*, ed. Ritzema, H. P., pp. 111–143. ILRI Publ. No. 16, Wageningen: International Institute for Land Reclamation and Improvement.
61. Ritzema, H. P., R. A. L. Kseiik, and F. Chanduvi. 1996. Drainage of irrigated lands. FAO Irrigation Water Management, Training Manual No. 9, Rome: FAO.
62. De Vries, W. S., and E. J. Huyskens. 1994. Gravity outlet structures. *Drainage Principles and Applications* (2nd ed.), ed. Ritzema, H. P., pp. 1001–1040. ILRI Publ. No. 16. Wageningen: International Institute for Land Reclamation and Improvement.
63. Wijdieks, J., and M. G. Bos. 1994. Pumps and pumping stations. *Drainage Principles and Applications* (2nd ed.), ed. Ritzema, H. P., pp. 965–998. ILRI Publ. No. 16. Wageningen: International Institute for Land Reclamation and Improvement.
64. Hoffman, G. J. 1990. Environmental impacts of subsurface drainage. 4th International Drainage Workshop, pp. 71–78. International Commission on Irrigation and Drainage, Cairo, Egypt.
65. Skaggs, R. W., *et al.* 1992. Environmental impacts of agricultural drainage. *Irrigation and Drainage, Saving a Threatened Resource—In Search of Solutions. Proceedings of the Irrigation and Drainage Sessions at Water Forum'92*, ed. Engman, T., American Society of Civil Engineers, New York.
66. Westcot, D. 1997. Drainage water quality. *Management of Agricultural Drainage Water*, ed. Madramootoo, C. A., W. R. Johnston, and L. S. Willardson, pp. 11–20. FAO Water Reports No. 13, Rome: FAO.

67. Lee, E. W. 1990. Drainage water treatment and disposal options. *Agricultural Salinity Assessment and Management*, ed. Tanji, K. K., pp. 450–468. American Society of Civil Engineers, New York.
68. Madramootoo, C. A. 1998. Options for the management of agricultural drainage water. *Proceedings of the Workshop on Drainage Water Reuse in Irrigation*, Sharm El-Sheik, Egypt. (not yet published).
69. Zimmer, D., and C. A. Madramootoo. 1997. Water table management. *Management of Agricultural Drainage Water*, ed. Madramootoo, C. A., W. R. Johnston, and L. S. Willardson, pp. 21–28. FAO Water Reports No. 13. Rome: FAO.
70. Grattan, S. R., and J. D. Rhoades. 1990. Irrigation with saline groundwater and drainage water. *Agricultural Salinity Assessment and Management*, ed. Tanji, K. K., pp. 432–449, ASCE, New York.
71. Abdel-Dayem, S. 1998. Guidelines for drainage water irrigation: Possibilities and constraints. *Proceedings of the Workshop on Drainage Water Reuse in Irrigation*, Sharm El-Sheik, Egypt.
72. Drainage Research Institute and DLO Winand Staring Centre. 1995. SIWARE: Reuse of drainage water in the Nile delta; monitoring, modelling and analysis. Final Report on Reuse of Drainage Water Project. Reuse Report No. 50. Cairo, Egypt: DRI and Wageningen: DLO-SC.
73. Rhoades, J. D. 1989. Intercepting, isolating and reusing drainage waters for irrigation to conserve water and protect quality. *Agric. Water Manage*. 16:37–52.
74. Rhoades, J. D., A. Kandiah, and A. M. Mashali. 1992. The use of saline water for crop production. FAO Irrigation and Drainage Paper No. 48. Rome: FAO.
75. Westcot, D. W. 1988. Reuse and disposal of higher salinity subsurface drainage water: A review. *Agric. Water Manage.* 14:483.
76. Tanji, K. K. 1997. Evaporation ponds. *Management of Agricultural Drainage Water*, ed. Madramootoo, C. A., W. R. Johnston, and L. S. Willardson, pp. 53–57. FAO Water Reports No. 13. Rome: FAO.
77. Ochs, W. 1997. Drainage water treatment in constructed wetlands. *Management of Agricultural Drainage Water*, ed. Madramootoo, C. A., W. R. Johnston, and L. S. Willardson, pp. 45–48. FAO Water Reports No. 13. Rome: FAO.
78. Burns, R. T. 1997. Deep well injection. *Management of Agricultural Drainage Water*, ed. Madramootoo, C. A., W. R. Johnston, and L. S. Willardson, pp. 57–60. FAO Water Reports No. 13. Rome: FAO.

5.7 Off-Farm Conveyance and Distribution Systems

H Depeweg

5.7.1 Irrigation Systems Definitions

Irrigation is the artificial supply of water for agriculture, the controlled distribution of this water, and the removal of excess water to natural or artificial drains after the water has been used in an optimal way. The aim of the irrigation facilities is to divert water from a source and to bring it to the fields via a network of open or closed canals and/or pipes, and to evacuate excess water to outside the area via a drainage network to ensure or to improve agricultural production. A main aspect of irrigation is that the facilities are made, managed, and used by people. In view of its aim, an irrigation system has to be planned, constructed, operated, and managed in such a way that all of the farm fields in the commanded area will receive and discharge water in an appropriate, conveniently arranged, and adjustable manner.

An irrigation scheme consists of the irrigation network (drains, roads, etc.) and the organization(s) in charge of managing the physical infrastructure according to preset objectives. These objectives are diversion, conveyance, distribution, and application of irrigation water and removal of excess water.

The irrigation network is the physical part of the irrigation system and includes all of the structural components required for the acquisition, conveyance and protection, regulation and division, measurement, distribution, and application of the irrigation water. At the head of the system, the diversion point being a weir, reservoir, pumping station, dam, or free intake, the water enters the conveyance system that transports the water to a delivery point. This point might be a turnout or offtake structure, where the water is diverted and delivered via an off-farm distribution and on-farm application network to the fields within the farm plots. A group of farmers usually operates the off-farm distribution system. If there is only one group of farmers in a unit, the quaternary unit coincides with the tertiary one. The responsibility for the management of the on-farm water distribution and the water application belongs to an individual farmer. The water is applied to the fields by one of the field irrigation methods described in Section 5.4.

The drainage system is a complementary component at any level of an irrigation system. No system is complete without a well-designed and well-functioning drainage system. Excess water from heavy rainfall or excessive irrigation can be very harmful to crops when not removed as quickly as possible. The drainage network is a system of open channels or pipes and structures, which conveys excess water from the individual fields to the main drains and then outside the area. Sometimes the drainage water is reused for irrigation. Drainage systems are discussed further elsewhere [1].

The management is responsible for the operation and maintenance of the irrigation and drainage system. Generally, three management levels can be distinguished:

- *conveyance or main level* by the government or an irrigation authority;
- *off-farm distribution or tertiary level*, by a group of formally or informally organized farmers or water users, e.g., in a water users' organization;
- *field level or on-farm distribution and application system* managed by the farmer.

5.7.2 Irrigation Areas

Each irrigation scheme is characterized by its specific land features and differs from other areas by topography, water availability, soil types, land use, and specific objectives of the irrigation and drainage system. The design of an irrigation network is based on engineering, agricultural, hydraulic, social, legal, and economic considerations. A large number of physical and economic parameters, such as crops, topography, hydrology, climate, soils, hydraulics, and cost-benefit analysis, are required for an optimal design [2]. However, an optimal design is difficult to find and it will have to integrate all of the criteria to find the best solution for the specific conditions of the area to be irrigated.

An additional difficulty is that today's best solution may not necessarily be the optimal solution 10 years from now: An (educated) guess on future development also is needed. Evaluation of these specific features will result in the most appropriate method for the conveyance and distribution of the irrigation water by the off-farm (main and tertiary) irrigation system.

For most irrigation areas, the following aspects have to be considered, not separately but in an integrated way [3]:

- irrigation and drainage systems;
- irrigation units;
- irrigation practices or methods;
- management, operation, and maintenance;
- access to the area and the facilities;
- environmental [4] as well as legal and socioeconomic aspects.

Irrigation areas are divided into units at various levels. Depending on the size of the irrigation scheme, not all of the following levels are always present:

- *Field units* are the smallest areas irrigated by one of the several irrigation methods [5]. The unit might be an individual field or farm plot that receives water at one place from the irrigation system, e.g., by a field inlet. Usually a farm consists of more than one field. Water is supplied by one inlet and then distributed over the fields in the on-farm distribution system.
- A *block or quaternary unit* is an area where water is supplied to one group of farmers. The group receives water via a group inlet. When several farms are irrigated in one group, it is important that each farmer have his or her own farm inlet for the control and management of the water that he or she is entitled to.
- An *end or tertiary unit* is an area where two or more blocks are grouped and receive water from the off-farm conveyance (main) system through one structure: the tertiary offtake or turnout structure. The management responsibility within the tertiary unit lies with the water users—the farmers or a water users' association. The tertiary offtake structure is the link between the water users responsible for the off-farm distribution system (tertiary-unit level) and the irrigation authority responsible for the conveyance system (main-system level). Usually, the water flow is regulated and measured at this offtake structure. In small schemes, a tertiary unit might be only one field or a farm. In large schemes, a unit may be one large farm (sugar or tobacco plantation) or several small farms together. The size of tertiary units can differ greatly, from a few hectares to 70 ha or more. Small channels or pipes within each tertiary unit distribute the water and small hydraulic structures control the flow to the fields [6].
- A *secondary unit* is an area in which two or more tertiary units are grouped and which receives water from a main canal through a diversion structure. A secondary canal (branch canal or lateral) conveys the water to two or more tertiary units. When water is received from another secondary canal, it is called a subsecondary, sub-branch, or sublateral.
- The *irrigation area* is the area in which two or more secondary units are grouped and which receives water from the diversion point (headworks, reservoir, pumping station). The main canal conveys irrigation water to two or more secondary canals.

Small irrigation schemes may have a single irrigation canal; large schemes may have an intricate system of canals or pipes. Hydraulic structures in the canals or pipes measure and control the water flows and its level or pressure.

5.7.3 Objectives of a Conveyance and Distribution System

As follows from the preceding sections the objectives of any irrigation system are to deliver irrigation water in the right amount (size, frequency, and duration), at the required head or pressure, at the right place, at the right moment, equitably (fairly and objectively) and in a reliable and assured way. Therefore, management activity is a major component of success for any irrigation scheme. The activities to operate and to maintain the system can be divided into two general components: functional and service. The functional activities are related to the operation and maintenance and encompass the storage, conveyance, and delivery of irrigation water to the agricultural water users and an adequate maintenance of the physical infrastructure to keep it ready for use. In general, the service activities include administration, financing, engineering, and other support, which are essential to meet the objectives of the irrigation system.

The required amount and timing of the irrigation supplies are governed by climate, topography, crops, available labor, social aspects, available water quality and quantity, and soil and soil moisture characteristics. The design and management of an irrigation system and the scheduling of the irrigation supplies are based on these characteristics. The timing and location for delivery of water to meet the crop requirements mainly determine the design capacities of canals, pipelines, structures and pipe appurtenances, and storage and pumping capacities of the system. The climate governs the evaporation of the water and the evapotranspiration of the crops; the type of crops and their growing stage determine their water use. One of the main factors that controls the frequency of the irrigation applications is the capacity of the soil to retain water and to make it available to the crops at the right moment. Another significant aspect in the water delivery is the efficiency at the different irrigation levels: field application (surface, sprinkler, or drip systems); conveyance, including operation losses (supply from the source to the turnout); and on- and off-farm distribution (from the turnout to the fields) [7]. An important aspect of the management component is the implementation of water delivery schedules that specify the time, duration and amount of the water delivery.

5.7.4 Types of Conveyance and Distribution Systems

Irrigation water can be conveyed, distributed, and applied by either a gravity system, a pressurized system, or a combination of both methods.

Gravity Irrigation Systems

Gravity irrigation systems convey the water in open canals from the headworks to the offtake and from there to the farm inlet and the fields. Generally, the canals are provided with a series of check structures (regulators) that control the water level. This control is needed to divert the water by gravity to canals with lower water levels. For large canals (radial) gates are installed at these points. Also, various types of weirs and flumes are available to regulate and measure the flow in open canal systems [8]. For outlet structures vertical sliding gates and movable or adjustable weirs are used.

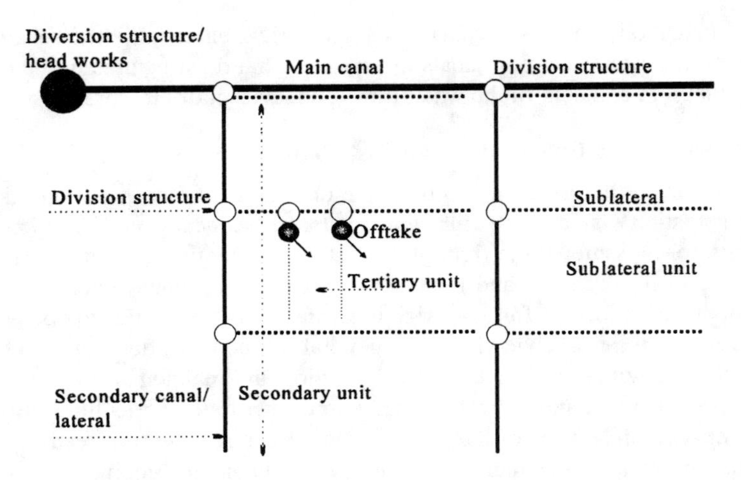

Figure 5.67. Irrigation scheme with gravity canals.

For a gravity irrigation system, the infrastructure can be divided as follows:

- Irrigation canals, that is, main (primary), secondary (lateral), tertiary, and field canals, convey and distribute the water over the area to be irrigated. Main and secondary canals form the main system.
- Flow control structures, for example, weirs, culverts, siphons, flumes, aqueducts, gates, and diversion boxes, are required to divert, convey, regulate, measure, distribute, and protect the water flow.

Whether canals are lined or unlined can play an important role in canal regulation and in the conveyance efficiency of the system (infiltration and seepage losses). In some canal systems, sedimentation can be a serious problem and will negatively influence the performance of the canal network and the structures [9]. Figure 5.67 shows a schematization of the irrigation canals and main flow-control structures in an irrigation network.

Pressurized Irrigation Systems

Pressurized or pipeline systems frequently are used for either the whole irrigation system (at main, tertiary, and field level) or for a part of the system. A pressurized irrigation system carries the water through a pipe network from the source to the point where the water will be applied to the fields (see Fig. 5.68).

Pipelines sometimes are preferred to open canals to reduce seepage and evaporation losses, to minimize maintenance costs, to improve water control, to irrigate hilly ground, to minimize the land taken up by the irrigation system, or to improve public health. Pipelines have several advantages over (lined) canals, but they can be expensive and skilled labor is needed to manufacture, install, and repair them. A pipe system might be designed for high, medium, or low pressure, depending among other things on the requirements of the field irrigation method. A classification of pipelines (derived from [10]) according to the hydraulic grade line follows:

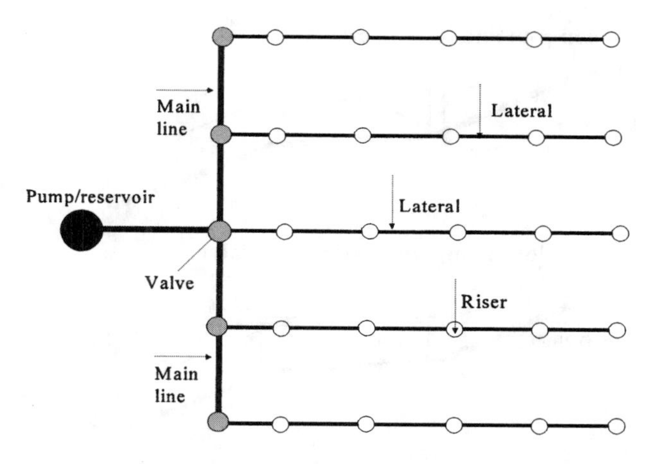

Figure 5.68. Irrigation scheme with pipes.

- *Open pipelines* are generally low-pressure pipelines with in-line risers (stand pipes). The risers are open to the atmosphere to reduce the maximum possible pressures in the pipeline. An overflow weir in the riser regulates the pressure upstream of the riser. Turnouts upstream of the riser are typically outflow points. The open pipeline is comparable to open canals but does not always provide advantages over open canals. In many cases the flow is more difficult to measure and divide properly. Any flow that is not taken by a turnout is passed downstream.
- *Semiclosed pipelines* have several advantages over open pipelines. Semiclosed pipelines give a relatively constant downstream pressure instead of a constant upstream pressure as in open pipelines. A float-controlled valve within the riser regulates this downstream pressure. The riser is open to the downstream pipe section and allows an operation without hydraulic transients (surges, water hammer). The main advantage of the semiclosed pipe system is that water not used by the turnouts will remain in the pipe. Variable flows are automatically adjusted and the system is adaptable to flexible water delivery schedules at stable heads.
- *Closed pipelines* typically operate under high pressures. Valves are used at bifurcations and turnouts for shutting off, adjusting, or regulating the flow. The flow is controlled by the pressure or flow regulation at the outlets and bifurcations. The pressure is kept at a maximum. Sudden changes in flow rate may result in hydraulic transients, which can be reduced by surge chambers or valves. Figure 5.69 presents the different types of pipelines according to their hydraulic grade lines.

Another classification of pipe systems is based on permanent, semipermanent, and portable systems. Especially at field level, the most common system is the permanent one, which typically consists of a pump or reservoir, a main line, and lateral pipes. The main line and laterals usually are installed underground. The pumps or reservoirs provide the operating pressure, and valves divert the flow from the mainline to the laterals.

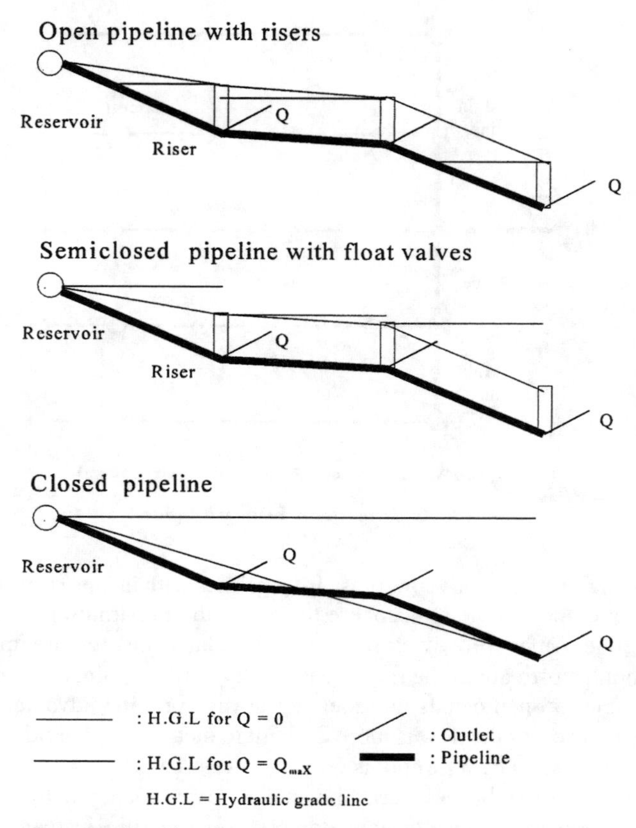

Figure 5.69. Classification of pipelines according to hydraulic grade line.

Pipe outlets (risers or hydrants) control the flow from the laterals to the field. The pipe systems can be used for all field irrigation methods.

5.7.5 Management of Conveyance and Distribution Systems

Management of conveyance and distribution systems includes some general aspects as well as some specific aspects that are related to the way in which the irrigation water will be delivered.

General Aspects

In most irrigation systems, three levels of operation can be distinguished:

- the *main or conveyance system* for water acquisition, conveyance, and delivery to the tertiary units;
- the *tertiary or off-farm distribution system* for the distribution of the irrigation water over the quaternary units and among the farmers;
- the *field system* for the on-farm distribution and water application.

Water is delivered by the irrigation authority to a group of farmers who distribute this water among themselves according to certain formal or informal rules devised by themselves or with outside assistance, which they have accepted.

The role of the government, especially in large schemes, is to provide the enabling environment for irrigation development and management that allows the irrigation authority to deliver the irrigation services in a sustainable, efficient, and effective manner beneficial to the users in particular and society in general. This enabling environment includes the legal and institutional frameworks, accountability mechanisms and procedures, regulations and criteria for service provision, and price control [11].

With a given set of objectives the irrigation authority develops a physical and management infrastructure to deliver water at the interface with a group of farmers or water users. The irrigation authority prepares and implements strategies to define and provide these services within the environment created by the government [12]. Policy choices made at this planning level concern water rights, cropping arrangements, water delivery arrangements, cost recovery, and responsibilities for system management.

The group of water users will decide within this policy setting on their water distribution arrangements, how their costs will be recovered, and how they will manage their tertiary and quaternary systems. Individual farmers will decide on crops, use of agro-inputs and labor, irrigation method, and water distribution and application system. These choices cannot be made in isolation but require an interaction between the various levels. Consultative approaches are recommended to develop sustainable solutions. Figure 5.70 presents a framework for interactions between policy choices, infrastructure, and management output at the various management levels [11].

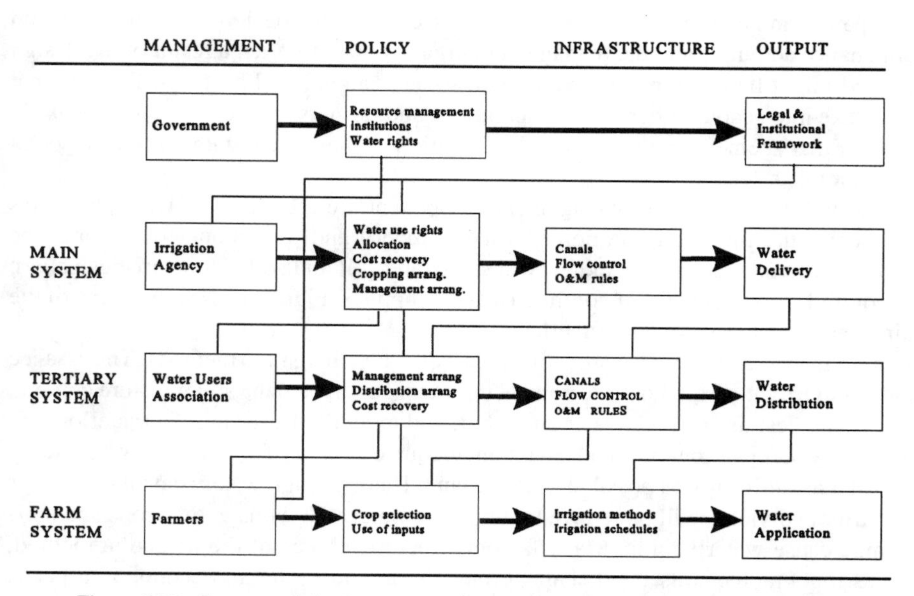

Figure 5.70. Framework for interactions between policy choices, infrastructure, and management output.

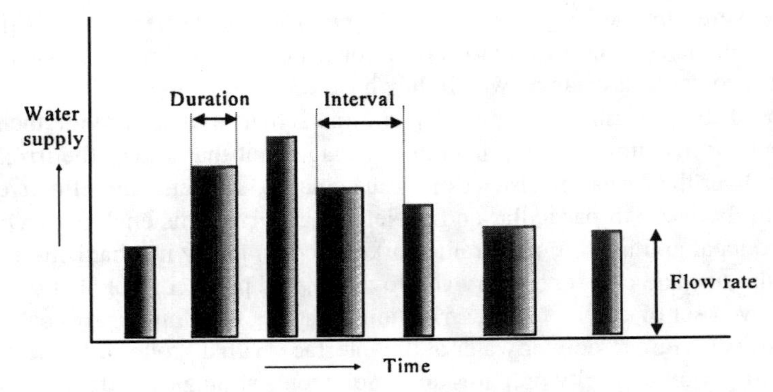

Figure 5.71. Required flow rate, duration, and frequency of irrigation relative to water supplies.

Water Delivery Aspects

The soil-water-plant relationship should be the starting point for the design of the field size and shape, but also for the water management and efficiency of each irrigation system. Knowledge of the water use by the crops and the water-retention characteristics of the soils is fundamental in the design and management of any irrigation system and in scheduling the water supplies [13]. The aim of the system is to meet the varying water requirements at field level for a specific, sometimes predetermined, cropping calendar with the highest possible efficiency, maximum operational flexibility, and optimal production [14]. However, to meet these goals, issues such as management and farmers' participation and action have to be taken into account in the design. The operation and maintenance activities and the management organization are often a result of the design only and might be incompatible with local, social, and cultural customs. Designs that do not take into account operational capabilities and farmers' participation will result in improper management, conflicts, and deterioration of the system with all related social and economic effects.

Important aspects of the management component are the water delivery schedules that specify the time and duration of water delivery and the amount of water to be delivered. Climate, crop type and growth stage, and soil and soil characteristics govern the required amount and timing of the irrigation supplies. Figure 5.71 shows some of the main characteristics in water deliveries.

The delivery schedules also are influenced by the field irrigation methods. The essence of any field irrigation method is to control the water supply and the soil moisture content, to make the best use of both water and labor, and to avoid the hazards of waterlogging and salinity. Crops grow optimally with an adequate supply of water and by spreading this water as uniformly as possible over the land. If not enough water is available, crops will suffer and yields will be reduced. Too much water will damage the crops and soils and may cause waterlogging. Many factors affect the choice of the irrigation method, such as topography, soils, field shape, crops, labor, and social and economic aspects. Several ways to deliver the water to the crops exist, for example, surface, sprinkle, or drip irrigation, but the water delivery at farm level is either continuous or intermittent.

For an individual farmer, three criteria are essential for the water delivery:

1. Is the supply sufficient (adequate)?
2. Is the supply at the scheduled time and in the right amount (reliable)?
3. Is the supply a fair share compared to what other farmers receive (equity)?

The off-farm conveyance and distribution system should deliver the water in a adequate, reliable, and equitable way to the farmers, which is a precondition for good water management by the farmers downstream of their farm inlet [15]. The irrigation authority is responsible for these water deliveries and should fulfill the requirements directly or indirectly via farmers' groups or water users' associations.

Adequacy is a measure for the delivery of the required amount of water for optimal plant growth. It is the ratio of the amount of delivered water to that required. The required amount is the quantity needed to achieve the agricultural policy and is a function of the area irrigated, crop water requirements, crop water-production factors, application losses, and cultural practices such as land preparation and salt leaching. Adequacy depends on

- ability to assess crop water requirements;
- water availability;
- specified water delivery schedules;
- capacity of the hydraulic infrastructure to deliver water according to the schedules;
- operation and maintenance of the hydraulic infrastructure.

Reliability is a measure of the confidence in an irrigation system to deliver water as promised. It is defined as temporal uniformity of the ratio of the delivered amount of water to the required or scheduled amount. A reliable water delivery allows for a proper planning by the farmers. A farmer can plan for a reliable delivery of an inadequate irrigation flow by planting less, growing different crops, or adjusting other farming inputs. However, a farmer cannot plan when the water delivery is not predictable. All farmers must have equal security on supplies and they will feel secure only if the reliability of water delivery is maintained. In continuous-flow systems, reliability refers to the expectation that a certain discharge or water level will be met or exceeded. When the flow is intermittent, in view of rotational irrigation or other water-sharing arrangements, reliability describes the predictability of the timing when the flow will start or stop.

Equity is the spatial uniformity of the ratio of the delivered amount of water to the required or scheduled amount. Equity means the delivery of a fair share of water to the water users throughout a system. Here, a share of water represents a right to use a specified amount of water. An irrigation system that covers an area with scarce water with its focus on equity is a fair application of justice and is socially more acceptable. An equitable amount of water should be ensured to all of the farmers in an irrigation area regardless of geographic location of the farm plot.

5.7.6 Water Delivery Policies

Problem Description

The fact that water users cannot use the conveyance and distribution system whenever or wherever they need it characterizes most irrigation systems. This is contrary to other infrastructure provisions such as water supply, electricity, or roads, which can be used as needed. The dependency of the end users on the irrigation management is inherent to the basic concepts of many irrigation network designs. Any design or layout of an irrigation system should be based upon organizational, operational, and farmers' participation aspects. Lack of engineering knowledge will result in faulty designs or construction, inappropriate layout, seepage and salinity problems, shortage or surplus of water, etc. Lack of knowledge of the above-mentioned aspects will result in

- problems in water distribution among the farmers at tertiary and quaternary level;
- problems in water management at main level.

Problems in water delivery at the farm level might result in unequal water application, low efficiency, or low production. At the main level, water delivery problems might be operational problems; assuming high efficiencies in the design may lead to complicated irrigation schedules and sophisticated structures, rules too strict for operation, and regulation and measurement of flows. Mismanagement, faulty operation, and wrong water deliveries might be the result of the staff's inability or preferences (not following the procedures or regulations). The design of any irrigation network should take into account the management and the water delivery aspects at all levels of the irrigation system.

General Water Delivery Policies

As discussed in the preceding sections, many conveyance systems are managed by an irrigation authority (government) and not by the farmers. Because the conveyance system supplies irrigation water to the tertiary units according to preset management objectives, it can only be well designed and operated when these objectives have been specified precisely. Otherwise, it might be questionable whether the system will perform to the requirements. This means that the planners and designers of a conveyance and distribution system and the users of this system have to agree jointly on these objectives.

The main irrigation system has to deliver water to the tertiary units with sufficient head above the terrain and with a reliable delivery in proper quantities and at proper flow rates [16]. These objectives are not very practical for the engineer who has to select the appropriate flow control structures in the conveyance and distribution system. Therefore, the water delivery practices have to be discussed in more detail. Existing classifications for water delivery, distribution, and scheduling vary and they usually are developed for water delivery at the farm level. A generally accepted terminology for the operation of main systems is not always clear. A distinction should be made between farmers and farms, water users' associations and tertiary units, and tertiary units and main systems. In some older irrigation systems, the water delivery practices were grouped according to the delivery at farm level as follows:

- *Continuous flow*. Water is delivered throughout the irrigation season;
- *Delivery on demand*. Water delivery is based on the request of the farmers to have (all of the) water during a (relatively) short period of time.
- *Rotation on schedule*. Each farmer receives his share of water based on a systematic delivery schedule, which was worked out in advance.

The FAO uses the following classification of water distribution methods and divisions in the main system:

- *On demand*. Water is available to the farmer at any time.
- *Semidemand*. Water is made available to the farmer within a few days of his request;

- *Continuous flow*. Throughout the irrigation season the farmers receive a small but continuous flow that compensates the daily crop evaporation;
- *Canal rotation and free demand*. Secondary canals receive water by turn and, once the canal has water, farmers can take the amount they need;
- *Rotational system*. Canals (secondary) receive water by turns, and the individual farmers receive the water at a preset time.

The World Bank presents four options for irrigation scheduling:

1. *continuous,*
2. *demand,*
3. *fixed-rotational with constant flow,*
4. *variable-rotational at variable flow and/or during variable periods.*

Clemmens [17] also describes various delivery schedules, with their intended frequency, rate, and duration. Another example is the classification of delivery schedules in either rigid (predetermined) or flexible (modifiable) schedules.

From the previous examples it is clear that various researchers give different descriptions for the delivery schedules. The next section uses the generally accepted descriptions of water delivery methods at the farm and main-system levels, but makes a clear distinction between water delivery to the farmers and the tertiary units and the irrigation flows through the main system.

Establishing Water Delivery Methods

The operation of conveyance and distribution systems is an important tool to improve the performance of irrigation schemes and should be based on a clear classification of water delivery at farm and at main-system levels. The operation of a main irrigation system has to meet the requirements at the tertiary offtake, the location where the two management levels meet. Therefore, the operational aspects of an irrigation system can be described as follows [18]:

1. *The actual decision making on the water delivery to the tertiary offtake*, that is, who decides on the water delivery to the tertiary unit;
2. *the method of water delivery to the off-farm distribution system (tertiary and quaternary unit)*, that is, how is the water allocated to the tertiary unit;
3. *the method of water division through the conveyance system (main system)*, that is, how the water is conveyed and divided through the main system.

Answers to these issues will determine the flow control method, the type of flow control structures, and their operation. Water users can request a change in water delivery, which may be effected immediately (the delivery is on demand) or after some time (the delivery is on request or semidemand). In many systems the users do not have any say in the water delivery (the delivery is imposed).

1. The decision making on the water delivery to the tertiary units can be divided as follows:
   - *Imposed*. The irrigation management decides about a fixed delivery schedule to a tertiary unit for longer periods (e.g., one season, one year, or seveal years). The schedule is based on their knowledge of either the crop water requirements (crop-based or productive irrigation) or the water availability at the diversion

point (supply-based or protective irrigation). The imposed delivery is extensively used in areas with water scarcity or with a monoculture.
- *On request*. Water users request the management for a certain amount of water for a certain duration at a certain time. The requests are evaluated and balanced against water availability. The processing of the water requests and the (hydraulic) response time of the conveyance system will delay the actual water delivery (1–14 days). This water delivery policy often is used in areas with constraints on water availability, canal capacity, and/or capability of the flow control system.
- *On demand*. The water users have direct access to the water. They decide about the water delivery to the tertiary unit and will receive the supply immediately from the conveyance system. On-demand systems are only applicable when water availability is ample. In general, the canal capacities for on-demand are larger than for other delivery methods. Measurement of the volume of water used and payment of fees accordingly are possibilities to limit the water use.

2. Water can be delivered to the tertiary unit according to the following flow types:
   - *Proportional (or split)*. The flow in the conveyance system is diverted at a fixed ratio of the flow to the tertiary unit. To divert the water at the offtake, an ungated diversor is all that is required.
   - *Intermittent*. An on/off flow is allocated to the tertiary unit. To divert the intermittent flow, an on-off gate in the offtake is needed.
   - *Adjustable*. A continuous or intermittent, but adjustable, flow is diverted from the conveyance system. To divert this variable flow, an adjustable regulator and a discharge measurement structure at the offtake will be required.
3. The water delivery methods in the conveyance system should be based on the water delivery methods to the distribution system in the tertiary units. Generally, the same methods are used for the main system:
   - *Proportional (or split)*. The flow is diverted at a fixed ratio at the secondary division structures, often also proportional to the commanded area.
   - *Intermittent*. The flow at the secondary division structures is divided in an intermittent, on-off way.
   - *Rotational*. The flow at the secondary division structures is supplied intermittently;
   - *Adjustable*. The flow at the secondary division structures is a continuous, adjustable flow.

Figure 5.72 presents some of the main issues in preparing water delivery schedules.

Selection Criteria for Water Delivery Methods

Selection of the objectives of the conveyance and distribution system should be based on a consensus between the planners and the users whether the decision-making process at the tertiary offtake is based on imposed, on-request, or on-demand delivery of the water. The accepted delivery policy limits the options for the water delivery to the tertiary units and for the water conveyance through the main system. The selection of delivery methods through the conveyance system is discussed according to the previously given issues:

- the decision-making process;

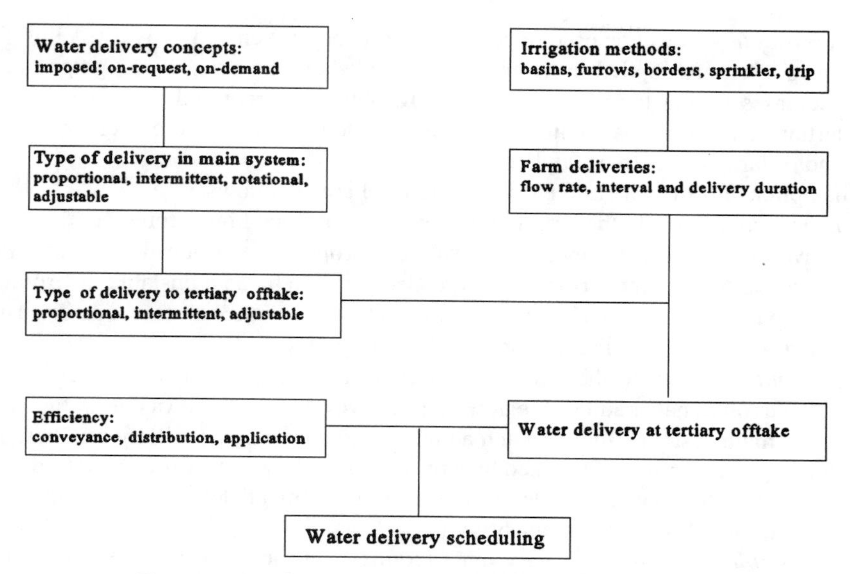

Figure 5.72. Main characteristics in water delivery scheduling.

- the method of water delivery to the tertiary unit;
- the method of water delivery through the main system.

The decision-making process on the water delivery involves consideration of the following options at various levels. The three most common water deliveries to a tertiary offtake require specific flows in the conveyance system, namely for:

- *Imposed delivery* by:
  - a *proportional flow* requires a proportional flow through the conveyance system. However, intermittent, rotational or adjustable flows in the conveyance system are also possible for a proportional flow to the tertiary unit;
  - an *intermittent flow* is possible by a rotational as well as an adjustable flow in the conveyance system. For smaller discharges a rotational flow is also possible;
  - an *adjustable flow* requires an adjustable flow through the conveyance system.
- *On-request delivery* with either an intermittent or adjustable flow requires an adjustable flow through the whole conveyance system.
- *On-demand delivery* with either an intermittent or adjustable flow requires an adjustable flow through the conveyance system.

The main options for water delivery to the tertiary units are as follows:

- *Proportional flow* is the traditional delivery method to tertiary units in many irrigation systems and is applied under imposed delivery;
- *Intermittent flow* to a tertiary unit often is applied to areas with dryland crops. The peak discharge is rotated throughout the tertiary unit and each farmer can irrigate at his turn. An intermittent flow is recommended in schemes with a limited irrigation duration, for example, only day irrigation. Intermittent flows also can be used under on-request or on-demand delivery in the main system;

- *Adjustable flow* can be used under on-request and on-demand delivery. Adjustable flows often are applied in irrigation systems where the peak discharge gradually decreases during the off-peak season. To obtain the required water level in the tertiary units for smaller flows, a rotational flow between the quaternary blocks and/or farmers has to be applied.

The options for the water delivery at main level are as follows:

- *Proportional flow* to the tertiary units by a proportional flow throughout the whole conveyance system to obtain everywhere a proportional water distribution. Proportional flows to the tertiary units are also possible by an adjustable, intermittent (e.g., pumping station), or rotational flow in the conveyance system. The discussed delivery methods require an imposed delivery policy.
- *Intermittent flows* to the tertiary units under an imposed delivery policy require either a rotational or adjustable flow in the conveyance system. An intermittent flow in the main system will not match an intermittent flow at the tertiary offtakes. The delivery can simply be obtained by a proportional flow at the offtakes. Intermittent flows to the tertiary units under on-request or on-demand delivery results always in an adjustable flow in the main system.
- *Adjustable flows* to the tertiary units require adjustable flows in the conveyance system. Proportional, intermittent, or rotational flows in the conveyance system are not possible with an adjustable flow to the tertiary units.

Figure 5.73 presents the various flow control methods in view of the policy decision-making process and the flow to the tertiary units and through the main system.

5.7.7 Flow Control Systems

Objectives of Flow Control

Flow control systems are an important component of the hydraulic infrastructure of a conveyance and distribution system. The purpose of flow control systems is to control the flows in the canal network at bifurcations to meet the level-of-service specifications. These specifications are translated in criteria and standards regarding flexibility, reliability, equity, and adequacy of delivery. The agreed level of service dictates the requirements for the flow control system in irrigation schemes. Flows can be regulated by several means through water-level control, discharge control, and volume control [19]. A combination of water-level control and discharge control is most common in irrigation systems. Flow rates at offtakes often are controlled indirectly through water-level control in the conveyance canals. Variation in discharge at the offtake is determined by the variations in upstream water levels and the sensitivity of the offtake and water-level regulator to those variations. If the rate of flow is changed at a delivery point, all upstream control structures must reflect that change. Fluctuating water depths in the canal system will change the effective head. The resulting fluctuating heads at the gates will require corresponding changes in the gate opening to maintain constant delivery rates. Because many deliveries may be affected by the changes in water surface elevation, the problem is compounded throughout the system. In case of delivery of agreed discharges, it is more desirable to maintain stable water surfaces than to adjust gate openings to try to meet delivery schedules.

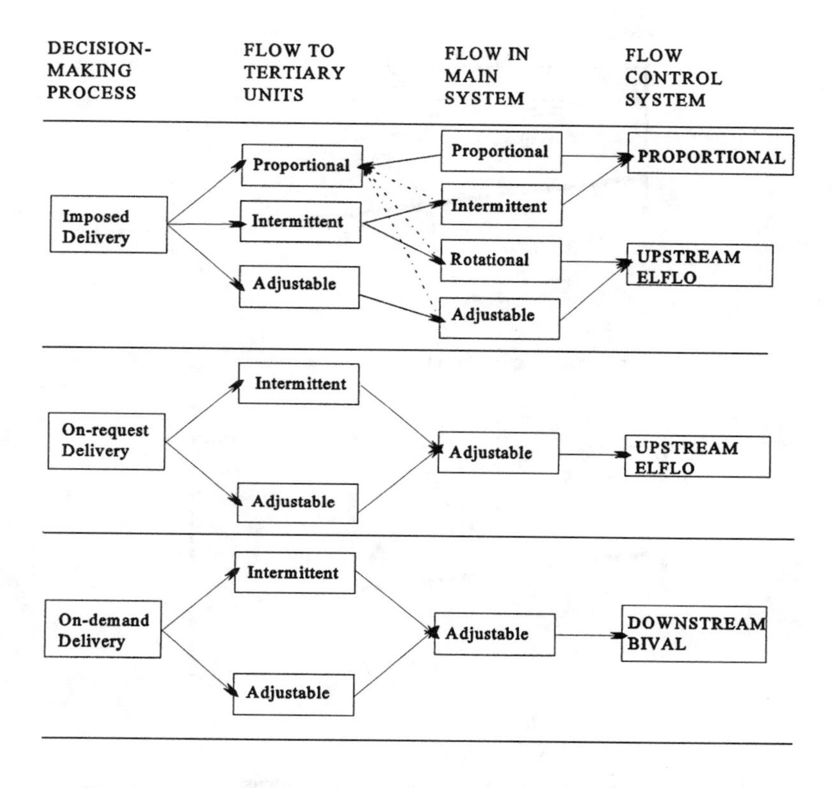

Figure 5.73. Various flow control methods in view of policy decision-making process, and the flow to the tertiary units and through the main system.

There are special modular distributors for constant discharge control (single- and double-baffle or Neyrtec distributors) that are independent of the upstream water levels and serve as on/off flow measurement and flow regulation devices. The constant discharges are obtained by the hydraulic function of the structure and not by moving parts. The baffle distributor can be used for an on/off water delivery by opening or closing a simple gate or shutter. These simple distributors are used as offtakes from one canal to another, or from one canal to a tertiary unit, or from the tertiary canal system to the fields. Additional advantages include easy flow measurement, insensitivity to upstream water-level changes, and the ability to lock them at a specific flow rate. A disadvantage of the distributor might be the required head losses.

Types of Flow Control Systems

The main available technologies (flow control systems) for the division of water and control of the water levels can be classified in three systems: according to orientation

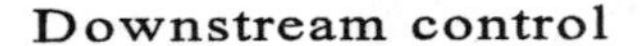

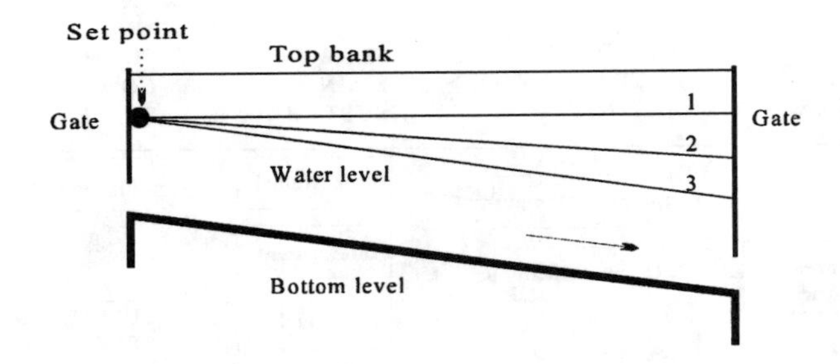

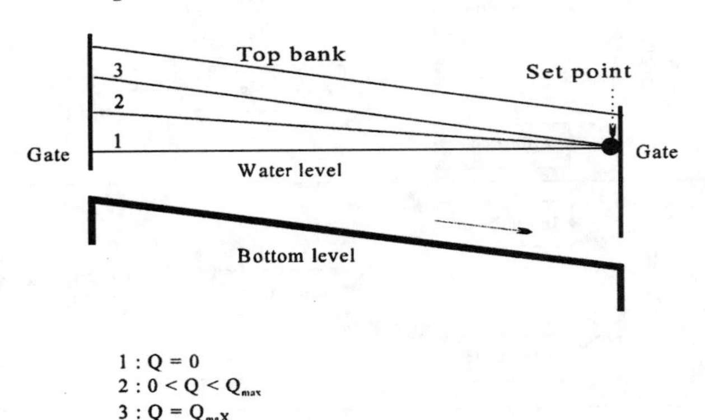

Figure 5.74. Main concept of upstream and downstream control.

of the control (upstream, downstream, and volume control), the degree of automation (manual, hydraulic, or electric automatic control) and the form of control (local or central control). Figure 5.74 shows the main concept of upstream and downstream control.

Some principal aspects of the flow control systems according to their orientation follow [20].

Upstream control is by far the most commonly used type of control around the world. Controlled flows are released to the upstream end of each canal section according to a prearranged schedule, resulting in water levels in the section in line with the released flows, the flow through the offtakes, and the control at the downstream end. Changing discharges requires changing the setting of the gates upstream of the location concerned. If the canal has to convey varying flows, an adjustable cross-regulator at the downstream end will maintain a constant water level upstream of that structure. A considerable time lag occurs before new equilibrium conditions are reached after changing the gate settings. Simple fixed weirs (such as duckbill weirs) can achieve a nearly constant level upstream of the weir without human intervention. In case of extreme backwater effects, minor raising of berms is needed. Upstream control is a serial control, is supply oriented, has limited flexibility, and requires management. It provides a high level of control in case of imposed delivery.

Downstream control systems respond to water-level changes downstream of a regulator. They are designed to permit instantaneous response to changes in demand by using water in storage in the upstream canal section (pool) in the case of an increase in demand and by storage of water in case of decreasing demand. Downstream control systems can be operated manually, but they are easily automated, either hydraulically (Neyrpic) or electrically.

Both water level and flow are controlled at the upstream end of a canal section. Changes are gradually passed on in the upstream direction untill the headworks. Most downstream control systems use balanced gates. Downstream control systems are specifically designed to maximize flexibility by minimizing system response times. The hydraulic stability of the system should be checked, preferably by a mathematical model [21]. A considerable raising of berms is needed in steep terrain; this can make the system very expensive. The application of downstream control therefore, usually, is limited to canals with a bottom slope that is smaller than 0.3%. Downstream control cannot be applied in canals with drops (structures that cater to drops in water level that are greater than the hydraulic losses over the structure). Downstream control can be applied in both on-demand and on-request delivery systems. The main design requirement is that each canal section have sufficient capacity to meet the maximum instantaneous demand. In local downstream control the set point is located just downstream of the regulator gate. In BIVAL control, the set point is located somewhere in the middle of the canal section (pool). Downstream control is also a serial process.

The *volume control method* involves the (usually simultaneous) operation of all flow control structures in the system to maintain a nearly constant volume of water in each canal section (pool). This method is used to meet operational requirements for different users of water in a flexible manner because it can provide immediate response to changing demand. With this type of operation, the water surface between control structures rotates around a point located approximately midway between the control structures that are operated to respond to changes of demand in the canal section. Implementing this type of control requires automation of the control structures coupled with centralized monitoring and control. With remote control, all control points in the system can be operated simultaneously, and not only serially. In the volume control method, the set point is located somewhere in the middle of a canal section; in ELFLOW, the set point is located at the downstream end of the canal section.

Operation Aspects of Upstream Control

In view of the operation of upstream control systems, a distinction has to be made between systems that achieve water delivery by fixed division structures and those that have gated offtakes along the canal [22].

Fixed division systems are those where water can be managed only at the head of the canal; division between subsidiary canals or offtakes is achieved through fixed division structures without gates:

- *fixed overflow weirs*, where the width of each weir section is in proportion to water rights based on a share of available water. Sometimes the width of the weirs is adjustable to cater for different flow ratios during different parts of the irrigation season;
- *submerged orifices* deliver a relatively constant discharge over a range of different operating heads.

Fixed division systems are particularly effective in meeting equity objectives based on a fixed share of available water, be it per unit area, per household, or per person. Therefore, the system is static and cannot respond easily to changes such as expansion of area or change in the number and size of the shares to be delivered. Because these systems are supply based, there is no room to adequately adapt the discharges to the demands. Individual water users adjust demand to the water availability through a careful selection of their cropping pattern. The system is generally reliable regarding its predictability. However, the variability of flows very much depends on the variability of the flow at the diversion point.

Gated division systems have gates at the offtakes along the canal to regulate the flow at any division point. Three types of regulation exist, depending on the degree of water-level control upstream of the division point:

1. *No canal cross-regulation*. The water level in the canal depends on the hydraulic relationship between discharge and water depth (head) in the canal cross section.
2. *Fixed cross regulation*. Weirs or other structures maintain stable head-discharge conditions.
3. *Gated cross regulation*. Gates control the water levels irrespective of the flow.

The majority of these systems are controlled manually but other systems utilize hydraulically or electrically operated systems. The gates in the offtakes are important tools to deliver the required amount of water in view of the management objectives. The delivery method determines the selection of the gate. If the water delivery is related to fixed discharges but with variable duration, fixed gates or on/off gates are used for a constant discharge over a wide range of upstream water levels. If the flows are variable and the duration is set, then movable gates in combination with measuring devices are required. For both cases the water level upstream of the offtake has to be maintained within a certain range. The larger the allowable range, the less the management input requirements. As mentioned before, modular distributors deliver an almost constant discharge independently of the upstream water level and serve as on/off devices to deliver the irrigation water by opening or closing a simple gate or shutter.

Gated division systems provide more flexibility than those with little or without cross regulation. They also allow for an improved flow control to meet short-term changes in demand: It is possible to manage the systems more adequately, but management (operation) requirements increase (dis)proportionally. If managed well they are also very reliable in terms of both predictability and variability.

5.7.8 Some Observations on Management Inputs

As discussed in the preceding sections the physical infrastructure of a conveyance and distribution system comprises a large number of technical works ranging from the irrigation and drainage network proper to roads, bridges, buildings etc. The irrigation network itself can be divided into two parts:

1. a *conveyance part* consisting of canals, pipes, and fixed structures, which, if well designed, constructed, and maintained, will convey and deliver the water as foreseen and operation activities will not be required;
2. an *operational part* consisting of division structures, valves, and pipe outlets at those points in the irrigation network where the irrigation water is divided, regulated, and measured.

The type of division structures largely determines the operability and manageability of the irrigation system. The structures may be simple or complicated, more or less sophisticated, fragile or sturdy, flexible or rigid, user-friendly or user-incompatible. The required level of management and whether farmers' participation is possible depends on the type and characteristics of these structures. The design of the physical part is based on the water deliveries related to the areas irrigated, cropping calendars, and crop water requirements. Division structures are the intersections, where conflicts between farmers and management and between (groups of) farmers may take place. From an operational point of view, four major types of structures exist [23]:

- fixed (weirs and orifices);
- on/off (shutter gates);
- adjustable, *stepwise* (stoplogs, modular distributors) or *gradual* (undershot gates, movable weirs);
- automatic (automatic upstream and downstream water-level control structures).

As mentioned before, the type of structure determines whether the operation is easy or difficult, the required number and skill of staff, and the understanding by the farmers. The first three types require an increasing number of skilled operating staff, whereas the last type needs fewer, but higher skilled staff. Figure 5.75 shows the relationship

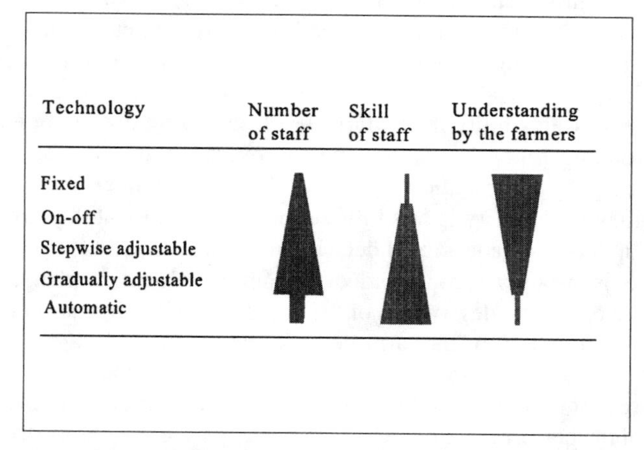

Figure 5.75. Relationship between the level of technology and the number and skill of staff and the understanding by the farmers.

between the level of technology and the required number and skill of the staff and the understanding of the operation by the farmers.

The management requirements for the operation of an irrigation system vary according to the design. A design that includes adjustable structures always requires operational inputs. They can be operated manually or automatically, locally or via remote control [24]. Irrigation systems with fixed structures require only operation at the control locations at the head of the major canal sections and therefore the operational input will be minimal. The operational requirements increase considerably for irrigation systems with gates at the division points. Gates are required at every offtake and their number will increase with the number of moveable cross regulators in the system. Automatic systems require fewer operational inputs, but they need some form of control at the head of the irrigation system to maintain hydraulically stable discharges through the main system.

Theoretically, any level of water delivery using automatic control also can be achieved with manual control, assuming that the wages are acceptable both to the irrigation authority and operators and that the available communication facilities are adequate. Although manual control may be as good as any automatic control, the latter allows for a higher level of flexibility and reduces the staff requirements in terms of number. However, it requires a higher level of expertise for maintenance and operation. Flexible water delivery in an upstream control system with local control requires suitable communication facilities; otherwise, the operation will result in rigid delivery schedules. In upstream control systems, the (hydraulic) response time between two delivery situations—from one stable state of delivery to a next one—is a long one. To increase the flexibility of water delivery and to reduce the response time of the system, remote centralized control can be considered. In automatic downstream control, the response time can be ignored and localized management is very possible.

Irrigation systems designed for high flexibility and (assumed) high efficiencies require movable and adjustable, often sophisticated, structures with measuring devices and a large number of highly qualified operating staff. Although the design of many of these systems is based on a high potential for flexibility and efficiency, this potential rarely is utilized and the flexibility is nullified by falling back on the original predetermined delivery schedules and by staff misuse of the too-complicated structures. Because of these circumstances, flexibility at the farm level, including a free choice of cropping pattern and cropping schedule, is low and the overall efficiency falls far below the assumed efficiency. These irrigation systems give a potential flexibility for the irrigation authority, but their complexity leaves little room for flexibility at the farm level and for farmers' participation in management decision making [25].

Therefore, it is strongly recommended that any development plan of an irrigation scheme be based on a clear description of the objectives of the irrigation system, aiming at the ultimate goal that all of the farm fields in the commanded area will receive and discharge water in an appropriate, conveniently arranged, and adjustable manner to obtain an optimal agricultural production. Any development plan should start with the identification of the performance objectives of the irrigation system and a clear definition of the operation plan. Planners (designers and operators) should develop a well-defined concept of the operational procedures and maintenance requirements as well as the level of performance standards. An operation plan should be based on a comprehensive view of the irrigation and drainage scheme, including the major aspects such as crops to be grown, soil type, water availability and quality, expected silt load, climate, technical and social infrastructure, farmer skills and cooperation, availability and costs of labor, and farm size. A successful plan for an irrigation system requires an understanding at all levels of the objectives and performance standards of the system. Especially the objectives and constraints at the farm level should be of overriding concern because only the farmers can support irrigation development that will be successful and sustainable. Planners and designers must understand that the farmers can increase their production only with adequate, reliable, and convenient water supply. In turn, the farmers will pay more if labor for water application is reduced and made more convenient. Therefore, the planning and design process should be based on a maximum participation by the farmers. Ultimately, they should be able to understand the operational procedures and the functioning of regulating and division structures.

There is a variety of designs concepts, structures, methods of control, and schedules, which may lead to many different combinations. Essential are the water users, that is, the farmers, who have to be satisfied with the quality of service, especially at the downstream end of the system. Planners must be aware of the resource limitations and of the implications of their design for maintenance, operation, and flexibility of water use.

As follows from the preceding considerations, a structural design is very important, but it is only the basis of a good design. A good design should be based upon the clear concept of how the final design will be operated according to understandable operational procedures. In principle, any designer should be able to follow his or her own instructions in the field once water begins to flow. The more specific design activities will include the advanced concepts in hydraulic and irrigation engineering, in agronomy and soil science, in economics, social, legal and environmental science to identify the most appropriate solutions for a manageable situation.

References

1. Ritzema, H. (ed.). 1995. Drainage Principles and Applications. Wageningen, The Netherlands: International Institute for Land Reclamation and Improvement.
2. Food and Agriculture Organization of the United Nations. 1988. Design and optimization of irrigation distribution networks. Irrigation and Drainage Paper 44. Rome: FAO.
3. Food and Agriculture Organization. 1990. Improvement of irrigation system performance for sustainable agriculture. *Proceedings of the Expert Consultation* (Bangkok). Rome: FAO.
4. Jensen, M. 1993. The impacts of irrigation and drainage on the environment. Proceedings 15th ICID Congress, The Hague Fifth N. D. Gulhati Memorial Lecture pp. 1–27. New Delhi: ICID.
5. Kay, M. 1993. Surface Irrigation. Cranfield, England: Cranfield Press.
6. Food and Agriculture Organization of the United Nations. 1993. Irrigation Water Management Training Manuals. Rome: FAO.

7. Wolters, W. 1992. Influences on the efficiency of irrigation water use. Wageningen, The Netherlands: International Institute for Land Reclamation and Improvement.
8. Bos, M. G. 1989. Discharge measurement structures. Wageningen, The Netherlands: International Institute for Land Reclamation and Improvement.
9. Mendez, V. N. 1998. Sediment Transport in Irrigation Canals. Rotterdam: Balkema.
10. American Society of Civil Engineers. 1991. Management, Operation and Maintenance of Irrigation and Drainage Systems. ASCE Manuals and Reports on Engineering Practice No. 57. New York: ASCE.
11. van Hofwegen, P. J. M., and E. Schultz. 1997. Financial Aspects of Water Management. Rotterdam: Balkema.
12. Uphoff, N., P. Ramamurthy, and R. Steiner. 1991. Managing Irrigation: Analyzing and Improving the Performance of Bureaucracies. London: Sage.
13. James, L. G. 1988. Principles of Farm Irrigation System Design. New York: Wiley.
14. Food Agriculture Organization/ICID. 1996. Irrigation Scheduling: From Theory to Practice. FAO Water Report 8. Rome: FAO.
15. Plusquellec, H., C. Burt, and H. W. Wolter. 1994. Modern water control in irrigation. Concepts, issues and applications. World Bank Technical Paper No. 246. Irrigation and Drainage Series. Washington, DC: World Bank.
16. Burt, C. 1995. Guidelines for establishing irrigation scheduling policies. *Proceedings of the ICID/FAO Workshop Irrigation Scheduling: From Theory to Practice*. Rome: FAO.
17. Clemmens, A. J. 1987. Delivery system schedules and required capacities. *Planning, Operation, Rehabilitation and Automation of Irrigation Water Delivery Systems*, ed. D. D. Zimbelman, Proceedings of Symposium Portland, USA, pp 18–34. New York: American Society of Civil Engineers.
18. Ankum, P. 1993. Operation specifications of irrigation main systems. *Proceedings 15th ICID Congress* (The Hague). New Delhi: International Commission on Irrigation and Drainage.
19. Plusquellec, H. 1988. Improving the operation of canal irrigation systems (slides/video). Washington, DC: World Bank Economic Development Institute.
20. Ankum, P. 1991. Classification of flow control systems for irrigation. J. Feijeni (ed.) *Proceedings of the International Conference on Advances in Planning, Design and Management of Irrigation Systems* (Leuven, Belgium), pp. 265–274.
21. Schuurmans, W. 1991. A Model to Study the Hydraulic Performance of Controlled Irrigation Canals. Delft, The Netherlands: Centre for Operational Watermanagement (COW).
22. Murray Rust, D. H., and W. B. Snellen. 1993. Irrigation System Performance Assessment and Diagnosis. Collaborative Research Programme on Irrigation Performance. Colombo, Sri Lanka: IIMI.
23. Horst, L. 1987. Choice of irrigation structures. *Proceedings of the Asian Regional Symposium on Irrigation Design for Management* (Kandi, Sri Lanka), pp. 45–60.
24. Schuurmans, J. 1997. Control of Water Levels in Open Channels. Delft, The Netherlands: Faculty of Civil Engineering, Group of Water Management.
25. Horst, L. 1990. Interactions between technical infrastructure and mangement. Paper 90/3b. London: ODI/IIMI Irrigation Management Network.

5.8 Water Quality in Agriculture

M. Greppi and F. Preti

5.8.1 Quality of Water for Agricultural Use: Impacts on Agriculture

Water that is intended for agricultural uses but in some way endangers agricultural activity can be defined as "unsuitable" water. Natural waters are defined as "polluted" when they become unsuitable following the addition of substances other than those normally present.

The use of water of less than optimal quality for irrigation introduces various kinds of risk (agronomic, sanitary, and environmental) and can produce damage of direct and indirect types, depending on the form of pollution involved (see Fig. 5.76; [1–3]).

Suitability of Water for Agricultural Use

Although irrigation has been practised throughout the world for several millennia, it is only in this century that the importance of irrigation water quality has been recognized. The design approach to irrigation with reclaimed water depends upon whether emphasis is placed on providing a water supply or providing wastewater treatment.

Legislation and literature provide much information on the extra-agricultural use of water (consumption, bathing, industrial, etc.). This information is not precise with respect to the suitability of irrigation water (except for salinity of natural origin). Water

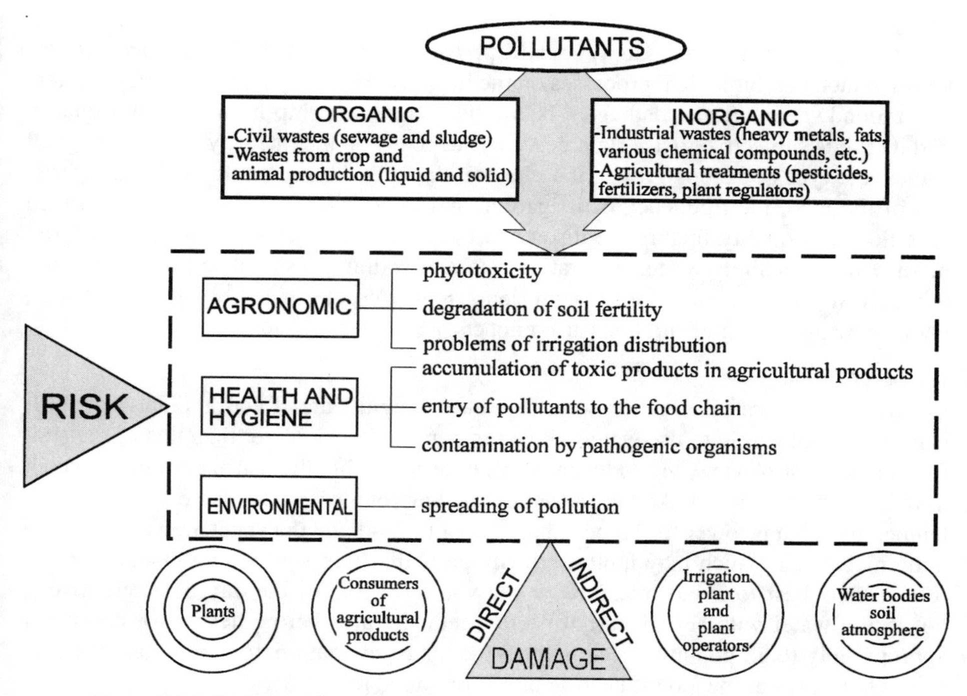

Figure 5.76. Risks and damage connected with the agricultural use of polluted water.

quality is influenced by numerous factors (analytical, agronomic, and territorial): The type and concentration of pollutant, the type and destination of the crop products, irrigation technique and period, soil characteristics, and quantity of water utilized [2–5]. Irrigation (including the use of reclaimed wastewater in recent years) now also is used to maintain recreational lands such as parks, golf courses, landscaped areas, and urban parks.

For water from conventional sources, it is usually sufficient to check the following characteristics (see Section on pH, later): content of sodium, calcium, magnesium, sulfates, chlorides, boron, the sodium adsorption ratio (SAR) and electrical conductivity (EC). In the case of wastewater, however, it is also necessary to check for the presence of organic substances, nitrogen, phosphorus, potassium, microorganisms, and heavy metals.

A number of different irrigation-water-quality guidelines have been proposed and are summarized in Table 5.27 [6].

Physical and Chemical Water Quality

The quantity and nature of solutes and suspended substances in water determines their suitability for different uses. In the definition of the qualitative requirements for irrigation water, the following risk factors must be evaluated: toxic effects on plants, pollution of water resources (groundwater in particular), and propagation of pollutants in the food chain. In addition to any immediate action, the progressive effects should be considered of any possible accumulation of a pollutant (e.g., in the soil) under continued use. Quality requirements vary with crop type and the state of the vegetation [7, 8].

All irrigation water must satisfy specific physicochemical and bacteriological characteristics such that biological processes on the irrigated land and in nearby surface waters and groundwaters are not damaged. Naturally, the relationship between water quality and its use is subjective to a certain degree. It depends on the integrity of the biological cycles, the types of crops grown, the climate, the soil, the morphological characteristics of the land, the frequency of irrigation, and the quantity of water used. Fourteen specific water-quality factors are discussed in some detail: temperature, pH, material in suspension, salinity, water-infiltration rate and sodium, residual sodium carbonate, boron, heavy metals, trace elements, nutrients, reducing substances, organic substances, pathogenic agents, and equipment-use impacts.

Temperature

The temperature of irrigation water can have a significant effect on plant growth. It is normally damaging to use water with a temperature much lower than that of the soil. From this point of view, groundwater is often less suitable than surface waters (which have a thermal regime linked with that of the environment). The use of water with a temperature that is lower (or higher) than that of the soil and the vegetation can produce a thermal shock, which may inhibit growth, particularly in very young plants. For this reason, it is best to avoid irrigation at the warmest time of the day or, alternatively, not to use water with significantly different temperatures before heating or cooling it appropriately (e.g., planning accumulation of long-flow path before administration, to paddy fields, water meadows, heating pools, or coiled pipes).

Table 5.27. Guidelines for interpretation of water quality for irrigation

Potential Irrigation Problem	Units	Degree of Restriction on Use: None	Slight to Moderate	Severe
Salinity[a]				
EC_w	dS m^{-1}	<0.7	0.7–3.0	>3.0
Total dissolved salts (TDS)	mg/L (10^{-3}kg m^{-3})	<450	450–2.000	>2.000
Permeability EC_w^b				
SAR = 0–3		≥0.7	0.7–0.2	<0.27
SAR = 3–6		≥1.2	1.2–0.3	<0.3
SAR = 6–12		≥1.9	1.9–0.5	<0.5
SAR = 12–20		≥2.9	2.9–1.3	<1.3
SAR = 20–40		≥5.0	5.0–2.9	<2.9
Specific ion toxicity [c]				
Sodium (Na)				
Surface irrigation	SAR	<3	3–9	>9
Sprinkler irrigation	mg L^{-1}	<70	>70	
Chloride (Cl)				
Surface irrigation	mg L^{-1}	<140	140–350	>350
Sprinkler irrigation	mg L^{-1}	<100	>100	
Boron (B)	mg L^{-1}	<0.7	0.7–3.0	>3.0
Trace elements		see [7]		
Miscellaneous effects [d]				
Nitrogen (total N)	mg L^{-1}	<5	5–30	>30
Bicarbonate (HCO_3) (overhead sprinkling only)	mg L^{-1}	<90	90–500	>500
pH	unit		6.5–8.4[e]	
Residual Cl (overhead sprinkling only)	mg L^{-1}	<1.0	1.0–5.0	>5.0

[a] Affects crop water availability.
[b] Affects infiltration rate of water into the soil. Evaluate using EC_w and SAR or adjSAR together. For wastewater irrigation, it is recommended that SAR be adjusted to include a more correct estimate of calcium in the soil water.
[c] Affects sensitive crops.
[d] Affects susceptible crops.
[e] Normal range.
Source: Adapted from [7].

pH

The pH is a variable that regulates all biological functions. Sometimes it has an inhibitory effect on process rates. Recommended values range from 6.0 to 8.5. Low pH also may affect conduits by corrosion. One cause of increased acidification of water sources is acid precipitation or, more appropriately, acid deposition (acid in dry dust, in rainwater, and in snow).

Material in Suspension

The presence of solids in suspension, usually made from clay or from other materials derived from drainage waters, generally does not cause significant damage to crops (although it may leave an anaesthetic covering of sediment on plant leaves). Clay and silt deposits, however, significantly affect irrigation canals in many parts of the world. Damage to irrigation equipment also can occur, giving rise to the corrosion of mechanical components (pumps) or the constriction and blocking of narrow pipes (e.g., drip-feed apparatus).

Salinity

The ionic species most commonly present in solution include both anions (chloride, sulfate, bicarbonate, and nitrate) and cations (calcium, magnesium, and sodium). Their presence, mainly due to the dissolution of minerals from the rocks and soil with which the water comes into contact, is increased with the degree of human activity in an area. The increase of salts in solution is always a factor that decreases the quality of irrigation water. A high salinity can be, in the severest cases, extremely phytotoxic or can create conditions for the progressive loss of soil fertility [6, 8].

The salinity of irrigation water is determined by measuring its EC, which is the most important parameter contributing to the suitability of a water source for irrigation. The EC of water is used as a surrogate measure of total dissolved salts (TDS) concentration. For most agricultural irrigation purposes, the values for EC and TDS are directly related and are convertible within an accuracy of about 10% according to the formula TDS (mg L^{-1}) $\approx$ EC (dS m^{-1}) $\times$ 640.

The presence of salts affects plant growth in three ways: (1) osmotic effects, caused by the total dissolved salt concentration in the soil water; (2) specific ion toxicity, caused by the concentration of an individual ion; and (3) soil particle dispersion, caused by high sodium and low salinity. With increasing soil salinity in the root zone, plants expend more of their available energy on adjusting the salt concentration within their tissues (osmotic adjustment) in order to obtain water from the soil. Consequently, less energy is available for plant growth.

Salts originate from local groundwater or from other sources in applied irrigation water. In the case of sprinkler irrigation, damage can derive from the concentration of salts on plant foliage by evaporation, resulting in leaf "burns" during the hottest periods. In this respect, waters derived from surface systems characterized by a lower salt content are preferable to those from groundwater. It is reasonable to say that the use of water with TDS levels of less than 500 mg L^{-1} does not have significant drawbacks. Nevertheless, it is nearly impossible to identify an absolute danger limit because of differences in the sensitivities of various crops and the influence of the lithological characteristics of the soil. Water can be used with a higher salt content in sandy, well-drained soils than in clay soils with poor drainage.

Salts tend to concentrate in the root zone because of evapotranspiration. Plant damage is tied closely to an increase in soil salinity. Establishing a net downward flux of water and salts through the root zone is the only practical way to manage a salinity problem. Under such conditions, good drainage is essential to allow a continuous movement of water and salt below the root zone. Long-term use of reclaimed saline, drainage, and wastewater for irrigation is not possible without adequate drainage [7].

If more water is applied than the plant uses, the excess water will percolate below the root zone, carrying with it a portion of the accumulated salts. Consequently, the soil salinity will reach some constant value that depends on the leaching fraction, that is, the fraction of applied water that passes through and below the entire rooting depth:

$$\text{LF} = D_d/D_i = (D_i - ET_c)/D_i \tag{5.250}$$

where LF = leaching fraction (−), D_d = depth of water leached below the root zone (m), D_i = depth of water applied at the surface (m), and ET_c = crop evapotranspiration (m).

A high leaching fraction results in a lower salt accumulation in the root zone. If the salinity of irrigation water (EC_w) is known, the salinity of the drainage water percolating below the rooting depth can be evaluated by using

$$\text{EC}_{dw} = \text{EC}_w/\text{LR} \tag{5.251}$$

where EC_{dw} = salinity of drainage water percolating below the root zone (dS m^{-1}), EC_w = salinity of irrigation water (dS m^{-1}), and LR = leaching requirement (−).

The EC_{dw} value can be used to assess the potential effects of salinity on crop yield and on groundwater. For salinity management, it often is assumed that EC_{dw} is equal to the salinity of a saturated extract of soil, EC_e. This assumption is conservative, however, in that EC_{dw} occurs at the soil water potential of field capacity and EC_e occurs at a matrix potential of zero (by definition under the laboratory conditions for the test). As a rough approximation, the value of EC_{dw} can be estimated as twice the value of EC_e for most soils [6, 8, 9].

Water Infiltration Rate and Sodium

The adsorption of sodium by clays in the soil contributes to a reduction of physical fertility. Sodium can substitute other metals, such as potassium and other useful cations attracted to negatively charged exchange sites in the soil, thereby reducing the availability of essential elements for plant development. In addition, an excess of sodium produces a modification of clay structure with a consequent increase in the degree of soil impermeability and a progressive reduction of water availability for plants (deterioration of the physical condition of the soil with the formation of crusts, waterlogging, and reduced soil permeability).

If the infiltration rate is reduced greatly, it may be impossible to supply the crop with enough water for vigorous growth. In addition, irrigation systems using reclaimed wastewaters, saline water, or agricultural drainage waters often are located on less desirable soils, which already have low permeability and management problems. In these cases, it may be necessary to modify the soil profile by excavating and mixing soil in the affected areas.

The water-infiltration problem usually occurs within the top few centimeters of the soil and is related mainly to its structural stability. It has been demonstrated that this negative effect is directly linked to the concentration of sodium. It is inversely proportional to the presence of bivalent cations and is influenced by the carbonate–bicarbonate balance. For this reason, the SAR has been introduced:

$$\text{SAR} = \left(\text{Na}/\sqrt{(\text{Ca}_x + \text{Mg})/2}\right), \tag{5.252}$$

where the cation concentrations are expressed in meq L^{-1}.

In addition to the SAR index, a modified SAR index (adjSAR) has been proposed:

$$\text{adjSAR} = (\text{Na}/\sqrt{(\text{Ca}_x + \text{Mg})/2})[1 + (8.4 - \text{pH}_c)], \qquad (5.253)$$

where pH_c is the pH of the irrigation water in equilibrium with CO_2 in the soil and in contact with calcium carbonate.

It is possible to calculate by means of an appropriate table based on the concentrations of the ions Na^+, Mg^{2+}, Ca^{2+}, CO_3^{2-}, HCO_3^- as well as knowledge of hardness (the amount of calcium carbonate dissolved in water, usually expressed as parts of calcium carbonate per million parts of water), alkalinity (property of having excess hydroxide ions in solution determining the capacity of water to neutralize acids), and conductivity [6, 10, 11]. The adjSAR value is preferred in irrigation applications with reclaimed municipal wastewater because it more accurately reflects changes in calcium in the soil water. Values of pH less than 8.4 identify water that is supersaturated with carbonate, with an increase in the dangers associated with sodium. Values greater than pH 8.4 indicate a tendency to dissolve carbonates, which attenuates the sodium risk.

Another index that is used to express the phytotoxicity of sodium is the so-called exchangable sodium percentage (ESP), which can be calculated from the following relationship:

$$\text{ESP} = (\text{Na}^+ \times 100)/(\text{Na}^+ + \text{Ca}^{2+} + \text{Mg}^{2+} + \text{K}^+), \qquad (5.254)$$

where the concentrations of the ions usually are expressed in $\text{meq}\,\text{L}^{-1}$.

Residual Sodium Carbonate

The presence of carbonates corresponds to an increase in the concentration of sodium. The determination of the concentration of bicarbonate in water therefore represents another guideline in the evaluation of irrigation water quality.

Bicarbonate can be evaluated in terms of the sodium carbonate residue (SCR) defined by

$$\text{SCR} = \left(\text{CO}_3^{2-} + \text{HCO}_3^-\right) - (\text{Ca}^{2+} + \text{Mg}^{2+}), \qquad (5.255)$$

where the concentration of ions usually is expressed in $\text{meq}\,\text{L}^{-1}$ and where $\text{SCR} > 0$ for alkaline water.

Water containing a high percentage of HCO_3^- ions has the tendency to precipitate calcium and magnesium in the form of carbonates. This precipitation changes the SAR index. Sodium carbonate also can form. In this regard, useful waters have $Na_2CO_3 < 1.25\ \text{meq}\,\text{L}^{-1}$, waters that are not easily utilized have $Na_2CO_3 = 1.25\text{–}2.5\ \text{meq}\,\text{L}^{-1}$, unsuitable waters have $Na_2CO_3 > 2.5$ meq/L. Possible damages are caused by the deflocculation of soil colloids which, in turn, causes a definite reduction in soil permeability or clogs trickle irrigation equipment.

Boron and Other Specific Ions

When a decline in crop growth is due to excessive concentrations of specific ions rather than to osmotic effects alone, it is referred to as specific ion toxicity. The ions of most concern in wastewater are sodium, chloride, and boron (Table 5.27 and [6]). The most prevalent toxicity observed when using reclaimed municipal wastewater is from boron, which usually comes from household detergents or effluents from industrial plants. Boron is often present in small quantities in natural waters. When concentrations are less than 0.5 ppm, negative effects are not observed even on the most sensitive crops. Boron is, in fact, numbered among those micronutrients that are essential for plant growth. At low concentrations, it has a favorable influence on the activity of vegetation. In cases of boron deficiency (where there is a large demand for boron in crops), it may even be administered as a fertilizer. Boron is present in almost all irrigation waters at low concentrations (0.2–0.5 mg L^{-1}). Above these concentrations, however, it becomes a toxic element and can cause damage to and even destruction of crops (e.g., necrotic spots and burns at leaf edges). Water that contains this element in concentrations greater than 4 mg L^{-1} is unsuitable for almost all crops. Between the two extreme values of 0.3 mg L^{-1} (no effect) and 4 mg L^{-1} (maximum risk of damage) there are three ranges of boron concentration that can cause light to moderate damage and three plant categories that are differentiated on the basis of their tolerance to boron (see Table 5.27).

Heavy Metals and Other Trace Elements

The interaction of heavy metals with plants is, in many cases, complex. The presence of small quantities in plant-assimilable forms is essential for good plant growth because many of these metals are involved in different enzymatic complexes. In cases where the input of metals to soils is too high or is uncontrolled, however, they can have a phytotoxic effect [1, 12, 13].

The probability of exceeding the risk threshold is greater in acid soils with a low cation-exchange capacity (CEC), where the ability to hold metals is lower and, therefore, the metals are lost more easily to the soil solution and absorbed by the plants. Irrigation is not the only way in which metals are transferred to the soil. They can also derive from the administration of chemical fertilizers or biological sludge and from atmospheric deposition.

Metals are transferred to plants through the soil and the exchanges that occur therein. Contaminated water rarely can cause acute and immediate toxicity, but an excessive accumulation of some metals (cadmium, molybdenum, selenium, etc.) can cause an exposure risk to crop consumers. In this way, continued use of polluted water can lead to high soil concentrations up to and beyond the phytotoxic limit.

The concentration of metals is rarely very high in surface waters. Without doubt, treatment of industrial discharges (alone or mixed with urban discharges) results in notable reductions in concentrations. Such reductions occur through precipitation and absorption in biological sludge. Boron constitutes a particularly dangerous exception because of its anphoteric behavior.

The limiting values are lower in the case of coarse-textured soils because of a lower retention capacity of loose soils (greater availability for plants) and their greater permeability (greater percolation).

Plants take up water and nutrients from the soil through their root systems and they also take up undesirable elements. The root apparatus continuously extracts water from the soil, but the water content varies with rainfall and irrigation and influences the soil-chemical equilibria. In addition to the mechanical filtration of particles made from elementary compounds, other mechanisms (e.g., cation exchange, precipitation, adsorption, and complexion) make up ways of immobilizing trace elements of the soil. The CEC determines the potential for trace-element exchange in soil: The greater the CEC, the greater is the quantity of trace elements that can be tolerated in irrigation water. The lower limit corresponds to soils with a CEC < 5 meq/100 g and the maximum limit to soils with a CEC of 15 meq/100 g.

Nevertheless, to avoid metal accumulation in any soil, a total retention of the supplied quantity should be assumed. Then, it is possible to evaluate the maximum time period over which contaminated water can be used on that soil, on the basis of the maximum concentration that can be tolerated. For the average quantity of irrigation water applied, these values range from a few years for cadmium up to thousands of years for lead.

Plants vary considerably their tolerance to trace elements in soils; in many cases the chemical form has a marked effect on tolerance.

Effects of Trace Elements

The effects of the most common trace elements in irrigation water on plants are the following:

- *Aluminium.* Nonessential; typical soil concentrations much higher than those applied with water; toxicity only in acid soils; accumulates on and in the roots, causing plasmolysis; does not accumulate in leaves. Can reduce productivity in acid soils (pH < 5.5), but more alkaline soils at pH > 5.5 will precipitate the ion and eliminate any toxicity.
- *Arsenic.* Nonessential, crop tolerance to soil concentrations can vary, does not accumulate in roots, limits growth when dangerous concentrations are reached. Toxicity to plants varies widely, ranging from 12 mg L^{-1} for Sudan grass to less than 0.05 mg L^{-1} for rice.
- *Beryllium.* Nonessential; if present in excess, it causes a loss of production. Toxicity to plants varies widely, ranging from 5 mg L^{-1} for kale to 0.5 mg L^{-1} for bush beans.
- *Boron.* Essential. The margin between utility and toxicity is very narrow. Toxicity symptom is necrosis on leaf edges.
- *Cadmium.* Nonessential, plants show a large tolerance to soil concentrations, accumulates in plants to levels greater than those considered safe for animals and humans. It is toxic to beans, beets, and turnips at concentrations as low as 0.1 mg L^{-1} in nutrient solutions. Conservative limits recommended because of its potential for accumulation in plants and soils to concentrations that may be harmful to humans.
- *Chromium.* Nonessential; accumulates in and is toxic to plants if present in soils in the hexavalent form; not generally recognized as an essential growth element. Conservative limits are recommended because of a lack of knowledge on toxicity to plants.
- *Cobalt.* Toxic to tomato plants at 0.1 mg L^{-1} in nutrient solutions; tends to be inactivated by neutral and alkaline soils.
- *Copper.* Essential but toxic; particularly in acid soils, if present in high concentrations; can induce iron and phosphorus deficiencies. Sheep and lambs are sensitive

to elevated dietary concentrations. It accumulates in plants but usually not to levels greater than those considered safe for animals. Toxic to a number of plants at 0.1 to 1.0 mg L^{-1} in nutrient solutions.

- *Fluoride*. Nonessential. Toxicity is rare in plants from soils with high levels, but it does accumulate in the foliage of some crops, causing marginal necrosis and chlorosis. It is inactivated by neutral and alkaline soils.
- *Iron*. Essential, concentrations in soils much higher than those potentially applied, almost never toxic to crops. The presence in a solid amorphic form can cause phosphorus and molybdenum deficiency. It is not toxic to plants in aerated soils but can contribute to soil acidification and loss or reduced availability of essential phosphorus and molybdenum. Overhead sprinkling may result in unsightly deposits on plants, equipment, and buildings.
- *Lead*. Nonessential can inhibit plant-cell growth at very high concentrations. Concentrations are too low to be toxic to plants, but can be toxic if directly ingested by children or animals; concentrations can exceed safe limits.
- *Mercury*. Nonessential; low plant uptake and, therefore, low toxicity; toxic to humans and animals; lost by volatilization from soils.
- *Lithium*. Nonessential, mobile in soil, low plant toxicity. It is tolerated by most crops up to 5 mg L^{-1}, but is toxic to citrus at low levels (>0.075 mg L^{-1}). It acts in a way similar to boron.
- *Manganese*. Essential, soil concentrations well above those potentially applied, toxicity normally restricted to acid soils. Symptoms include dark spots on leaves, severe chlorosis, and brown coating of leaves. It is toxic to a number of crops at a few tenths of a milligram to a few milligrams per liter, but usually only in acid soils.
- *Molybdenum*. Essential, low plant toxicity. It can accumulate in crops to levels greater than those considered to be safe in soils with elevated concentrations; its solubility depends on soil acidity. It is not toxic to plants at normal concentrations in soil and water, but can be toxic to livestock if forage is grown in soils with high available levels.
- *Nickel*. Nonessential for plants. It accumulates and is toxic to growth in acid soils and with high concentrations, but is not absorbed by plants to levels dangerous for humans and animals. It is toxic to a number of plants at 0.5–1.0 mg L^{-1}; toxicity is reduced at neutral or alkaline pH.
- *Selenium*. Nonessential for plants; relatively low phytotoxicity, except in soils with high levels. It accumulates in plants if the soil concentration is greater than critical levels; in forage crops it can reach dangerous concentrations. It is toxic to plants at concentrations as low as 0.025 mg L^{-1} and toxic to livestock if forage is grown in soils with relatively high levels of added selenium. The availability for plants depends on the acidity of the soil. An essential element for animals but in very low concentrations.
- *Tin*. Effectively excluded by plants at relatively low concentrations.
- *Titanium*. Effectively excluded by plants at relatively low concentrations.
- *Tungsten*. Effectively excluded by plants at relatively low concentrations.

- *Vanadium*. Nonessential. Plant growth is reduced in highly contaminated soils, and it is toxic to many plants at relatively low concentrations.
- *Zinc*. Essential; accumulates in plants, but not to levels dangerous for humans and animals; is phytotoxic (sometimes acutely) in acid soils at high concentrations. It is toxic to many plants at widely varying concentrations; toxicity is reduced at pH > 6.0 and in fine-textured or organic soils.

Nutrients

Nutrients that are important to agriculture include nitrogen, phosphorus, and occasionally potassium, zinc, boron, and sulfur. They can be applied for crop production with fertilizers or reclaimed wastewater. However, nutrients can cause problems in certain circumstances when they are in excess of plant requirements. The most beneficial but also that which is most frequently present in excess in reclaimed municipal wastewater is nitrogen. The nitrogen in reclaimed wastewater can replace an equal amount of commercial fertilizer during the early to midseason of crop growth. Excessive nitrogen in the latter part of the growing season may be detrimental to many crops, causing excessive vegetative growth, delayed or uneven maturity, and/or reduced crop quality. If an alternative source of low-nitrogen water is available, a switch in water supplies or blending of reclaimed wastewater with other water supplies can be used to keep nitrogen levels acceptably low.

Reducing Substances

High levels of oxidizable substances can, in some cases, have a phytotoxic effect. Sulfides, in particular, can produce a precipitation of iron(II) sulfide in the root system, inducing symptoms of asphyxia in plants. Reducing water or water with low oxygen concentrations can be utilized by applying appropriate techniques that favor reoxygenation.

Organic Substances

Organic compounds in polluted water can be either natural or synthetic. Natural organic compounds principally derive from human settlements, livestock production, the food industry, paper mills, pesticides, and herbicides. Human and animal waste products and plant and animal residues are characterized by a high degree of biodegradability. They can be gradually degraded and mineralized when they are applied to soils. Alternatively, they can be treated efficiently using a biological plant.

Synthetic compounds, instead, principally derive from industrial activity and often consist of substances that are unknown in the natural environment. Many of these compounds are persistent or only degrade very slowly (e.g., hydrocarbons, detergents, and pesticides).

Pathogenic Agents

The presence of pathogenic microorganisms (metazoa, protozoa, bacteria, viruses, etc.) living, or in a resistant form (eggs, cysts, spores) in irrigation water is caused, principally, from untreated or partially treated sewage from urban settlements.

To a lesser but by no means negligible extent, pathogenic organisms also can be transferred to water by wild and domestic animals (e.g., rats). In time, transmission to people is possible, particularly in cases of epidemics (e.g., gastroenteritis), through fruit and vegetables irrigated with contaminated water, or by direct contact with the irrigation water itself. The presence of microorganisms in irrigation water does not limit the activity of the vegetation or the fertility of the soil because the prevailing presence of indigenous microflora is enough to eliminate most undesirable agents.

Pathogenic microbial species are not directly determinable, except by complex analytical methods, but the quantity of nonpathogenic microorganisms can be used as indicators of contamination from dirty water and the possible residual presence of pathogens present at the time of contamination [4, 7]. The qualitative requirements for irrigation water can be diverse and related to crop type, the growth phase, and the system selected for irrigation distribution. The environmental dispersion of these microbial species and the relative contact with people, in fact, can occur by the formation of aerosols with sprinkler-type irrigation systems. In general, a safety zone can be established 100–200 m from the source; the distance varies with the effects provoked by thermic shock, by variations in microclimatic conditions at the time of irrigation, and with different microbial species. A determination of the survival times of different species can be used to evaluate the pollution risk to the soil and the plants.

It is clear that untreated sewage should not be used for irrigation. Primary and secondary treatment are often sufficient to reduce the presence of pathogens considerably. Note, however, that, in many rivers with a torrential regime, the presence of coliforms, assumed as microbiological quality indices, is variable. The amounts are often comparable to levels in biologically treated urban waste. Note also that the strong recourse to chlorination as a means of effluent disinfection to indirectly reduce microbial loads in receiving water bodies appears to be unsuitable. Poor efficiency, in relation to the most dangerous (viral) species, and the inevitable formation of toxic halogenated compounds make such waters unsuitable. Halogenated compounds result when water with a high load of organic residues is treated with sodium hyperchlorite (normally used as the means of effluent disinfection).

Effects on Irrigation Equipment

Clogging problems with sprinkler- and drip-irrigation systems have been reported, particularly with primary and oxidation pond effluents. Biological growth (slimes) in the sprinkler head, emitter orifice, or supply pipes can cause blockages, as can heavy concentrations of algae and suspended solids in irrigation water. The most frequent clogging problems occur with drip irrigation. From a public-health standpoint, such systems often are considered ideal because they are totally enclosed and thus minimize any problems related to the exposure of operating personnel to reclaimed wastewater or drifting spray. In the case of treated wastewater, which has been chlorinated, residual chlorine concentrations of less than 1 mg L^{-1} do not affect plant foliage. Chlorine concentrations in excess of 5 mg L^{-1}, however, can cause severe plant damage when reclaimed wastewater is sprayed directly onto foliage.

Irrigation equipment generally is made of materials such as concrete, iron, steel, bricks, asphalt, plastic, and synthetic fibers. The cause of concrete degradation can be intrinsic (because of composition, manufacture, or aging) of an extrinsic chemical nature

(CO_2, mineral acids, pH, organic acids, bases, salts, NH_4^+, Mg, chlorides, sulfates), of a physical nature (elevated temperature), or of a mechanical nature (water velocity, abrasion, erosion). Ferrous materials can be subjected to physical and chemical degradation processes caused by the presence of pollutants in water, which can cause alterations and degradation by abrasion or erosion (suspended solids) or by corrosion (organic and inorganic chemical pollutants, pH). The use of water with high concentrations of suspended solids (e.g., algae) can cause blockages and/or loss of pressure in pipes, valves, and pumps. Filters and/or chlorination of eutrophic waters can be used to avoid these problems. Clay and silt deposits significantly affect irrigation canals in many parts of the world. Groundwater often is used for agriculture so as to avoid problems associated with quality and quantity. However, its exploitation is sometimes costly in terms of energy (pumping costs) and the environment (groundwater is often nonrenewable in the short term). In addition, it is often unsuitable for applications such as irrigation because of low temperature (thermic shock) or (occasionally) high salinity.

5.8.2 Effects of Agricultural Activities on Water Quality

Impacts on Water

Modern agriculture is characterized by the use of chemical substances (fertilizers and pesticides) for the protection of crops and to enable high levels of production. The environmental fate of these substances is determined by removal phenomena (adsorption onto the solid matrix, chemical and microbiological degradation, uptake by plants, photodecomposition) and transport phenomena. By air, transport phenomena include volatilization, dispersion, and aeolian (wind) erosion. In water, these processes are surface flow, solid transport and erosion, subsurface flow, and percolation.

It is, therefore, possible for agricultural pollution to affect water use. Some common examples are reported in the Table 5.28.

Point-Source and Diffuse Pollution

Agricultural activities typically generate pollution phenomena from diffuse sources (diffuse or nonpoint pollution), by which we mean the input of pollutant substances coming from sources that are distributed in space.

Point sources of pollution are rare in agriculture but do include some agroindustrial activities, the stocking and spreading of concentrated animal wastes on the soil, and the disposal of wastewater from washing out pesticide containers. Crop and animal production, harvest, preservation, food processing, and the sale of products (fruit, vegetables, oil, wine, milk, etc.) produce pollutant discharges (e.g., wastewater from hosing down animal excreta). The pollutant loads derived from these activities are characterized by high concentrations of organic substances in both dissolved and suspended forms. These loads vary, depending on the particular industry and/or season considered (e.g., BOD from 10 ppm to more than 100,000 ppm and suspended solids from 30 ppm to more than 10,000 ppm).

The final characteristics of waste products from animal production (especially where high densities of animals are raised in small areas) also are determined by the removal techniques adopted (mechanical or hydraulic systems). Some forms of diffuse pollution

Table 5.28. Effects of pollution from agricultural sources on water use

Pollutant Category	Form of Pollution	Impact: Primary	Impact: Secondary	Impact: Tertiary
Nutrients (phosphorus and nitrogen)	Total Nitrogen Nitrate Ammonia Total phosphorous Soluble phosphorous	Accelerated growth of algae and aquatic plants Eutrophication of slow-turnover water bodies Interference with potable and recreational uses	Decrease in oxygen Color changes Increased concentrations of NO_3 and NH_4	Death of fish in surface water bodies Reduction in recreational benefits Higher degree of treatment required for potable use Toxic effects of nitrate on infants
Sediments	Suspended solids	Blockage of channels, canals, and reservoirs Obstruction of outlets and filters Interference with recreational uses Increase in treatment costs	Reduction of light penetration Color changes	Negative effects on fish reproduction zone Adsorption/desorption phenomena of toxic substances
Salinity	Dissolved salts	Damage to fish community and aquatic vegetation Corrosion of pumps and turbines Conspicuous increase in treatment costs for potable use		Reduction of germinative capacity from downstream use Reduction in growth of some plant species
Herbicides, pesticides (insecticides, fungicides, nematodicides)	Organochlorine and organophosphorus pesticides Carbamate Triazine	Toxicity for fish and aquatic plants Heavy treatment for potable uses	Sublethal effects Synergistic effects	Biomagnification in the food chain Long-term mutagenesis
Animal wastes (nitrogen, phosphorus, oxidizable substances, pathogens)	Biological oxygen demand (BOD) Chemical oxygen demand Coliforms and streptococcus	Analogous to nutrients Reduction in fish populations Increase in level of treatment for potable use	Smell Color	Anaerobic transformations with methane, amine, and sulfide production

can ultimately derive from industrial activity. This includes chemical compounds released to the atmosphere by accident or on purpose (e.g., dioxin or radionuclides) and deposited (as fallout) by precipitation (wet deposition) or as particulates (dry deposition).

The most typical diffuse sources, however, come from agriculture and include nitrogen and phosphorus compounds present in fertilizers (both of which may be involved in the processes of eutrophication of water bodies) and the principal active agents present in pesticides and herbicides (which can be found in surface and deep waters or adsorbed to soil or sediments). Other substances, such as heavy metals, can come from spreading sewage sludge, wastewater, or compost obtained from solid urban wastes on the soil. Even sediments, salts, and organic material can be considered as "pollutants" of receiving water bodies if they are produced in an anomalous way from agricultural practices. The determination of the impact of substances used in agriculture on water bodies or on soil can be assessed by examining changes in natural loads (i.e., increases in the concentrations of single components), the introduction of extraneous substances, changes in the ratios of components present, and the occurrence of anomalous physical and/or biological phenomena.

The effects may include limitations in the direct and indirect use of the substance; increases in treatment costs (including maintenance of the equipment); the degradation of biological productivity; and the input of toxic, mutagenic, and teratogenic substances to the food chain.

The phenomenon of agricultural pollution is considered, in the majority of cases, to be unintentional, even if cultivation practices employed and the chemical substances used play a decisive role in the process. On nonirrigated land, pollution is linked to diffuse sources and regulated by the hydrological events that occur. On irrigated land, pollution due to irrigation waters depends on the transport of leached substances from the fields when excess water is applied.

An exact relationship between the use of a substance in agriculture and the consequent modifications in environmental quality is difficult to define properly because of the diffuse nature of the processes involved. Point-source contaminants have a uniquely localizable origin, such as effluent outfalls and dumps, for which it is possible to apply measurement techniques that allow a direct and reasonably accurate evaluation of concentrations and loads. Diffuse contaminants, on the other hand, come from nonconfinable sources, for which the determination of load consistency and the monitoring of pollution processes requires specific techniques. Such techniques can be applied in particular to the final receiving body, for example, taking into account the spatial variability of soil hydraulic properties.

Diffuse pollutants of agricultural origin can be transported to surface water bodies in solution by surface and subsurface runoff or as solids in erosion. They can reach groundwater bodies with water that infiltrates and percolates into aquifers. They can be adsorbed by soil particles. Finally, diffuse pollutants can be transferred to the atmosphere in significant amounts.

The different behaviors of the various pollutant substances is based on the prevalence of some fundamental factors for diffuse pollution processes. In the case of pesticides, for example, the degree of solubility and the degree of adsorption to soil particles can be highly variable from substance to substance. Nitrate, on the other hand, is characterized principally by transport in solution and its quantity depends on the soil-nitrogen cycle (mineralization, denitrification, and plant uptake processes). Phosphates and some radionuclides are, instead, adsorbed quite strongly to the soil and, therefore, they are associated mainly with the transport of suspended solids.

The transport of a pollutant to a surface water body from a nonpoint source is characterized by the response of the drainage area to a defined rainfall event. This has a limited duration, with high time variation in the initial phase. The hydrograph peak and that of the pollutant concentration may not coincide because the first part of a hydrological event is usually more polluting with respect to the remaining part because of the effect of leaching by the initial runoff.

The consistency of the nonpoint load depends on many factors, including the volume of rainfall, the intensity and the location of the storm event, the duration of the preceding dry period, the agricultural activity carried out in the area of interest and the phase in the cultivation cycle.

With the evolution of technology and a continued human requirement for water (especially for high-quality water), there has been a subsequent growth in the demands for abstractions from groundwater. To fulfill these demands, deeper and deeper wells are being drilled with larger diameters to continuously increase abstractions. The lack of adequate controls unfortunately, often has, lead to alterations in the deep stratigraphic structure and the opening up of interconnections between impermeable strata, thereby facilitating an increase in the transport of agrotechnological products, distributed at the surface, to aquifer systems.

Pollutants from Agricultural Activities

Pesticides

The generic term "pesticides" (or biocides) is applied to all chemical compounds used by people to inhibit or eliminate species that are considered damaging under particular circumstances (e.g., crop defense and water treatment for particular uses). They mainly consist of organic substances that control crop damage from fungi, insects, or other parasites and from weeds. Pesticides thus consist of fungicides, bactericides, herbicides, nematodicides, molluscicides, insecticides, and rodenticides (Table 5.29). They can be of an extremely variable nature both in terms of their chemical structure and in terms of their chemical and physical properties. They can be natural or synthetic, organic or inorganic. Organic pesticides represent the most common class of pesticide. Inorganic pesticides were used mainly in the past and now have limited use. Of the organometallic pesticides, the most frequently used contain mercury (such as ethyl mercuric chloride and phenylmercuric acetate) and tin (triphenyltin acetate, triphenyltin sulfide).

Carbon disulfide, carbon tetrachloride, cyanidric acid, sulfurous anhydride, and methyl bromide are employed as fumigants. Currently, the most commonly used pesticides in terms of quantity are, in decreasing order, organochlorines (chlorinated or halogenated organics), organophosphates (or phosphorated organics), and carbamates. Chlorinated pesticides are characterized by molecules with carbon-chlorine links (DDT and homologues, chlordane, heptachlor, aldrin, dieldrin). Phosphorated pesticides, in contrast,

Table 5.29. Percentage of world pesticide sales by major region, crop, and pollutant category, 1994–1995

Region	% 1995	Crop	% 1995	Category	% 1994
North America	29.5	Fruit and vegetables	25.7	Herbicides	47.0
East Asia	25.1	Cereals	15.0	Insecticides	28.8
West Europe	25.8	Rice	11.8	Fungicides	19.5
Latin America	10.8	Maize	10.7	Others	4.7
East Europe	3.4	Cotton	9.6		
Africa & rest of the world	5.4	Soybeans	8.2		
		Sugar beets	2.8		
		Rapeseed	1.6		
		Others	14.6		

Source: [20].

have molecules containing oxygen-phosphorus-sulfur links (parathion, methyl parathion, formothion, chlorpyrifos, malathion, phosdrin, thimet). Carbamates have molecules with nitrogen-carbon-oxygen links (carbaryl-baygon). Organophosphorus and carbamate compounds, normally, degrade rapidly into simpler substances. Organochlorines, on the other hand, remain unchanged in the environment for longer periods.

The most widely known herbicides are phenoxyalcanoics (2,4-D and MCPA), ammonium-quaternary compounds (diquat and paraquat), amid (metolachor and alachlor), linuron, and triazines (atrazine and simazine).

The phenomena that govern the behavior of pesticides in the soil are numerous and complex (see Fig. 5.77, [14]) with respect to their repartitioning (often expressed by the distribution coefficient K_d) among the liquid, solid, and gas phases; their interaction with living organisms; their transformation into residual products; and their chemical and biological degradation. The factors that influence the transport and the impact of a substance that falls within the wide category of pesticides include solubility, adsorption capacity, persistence, and other specific chemicophysical properties. Persistence is defined as the period of time necessary for complete degradation of the substance to other (reasonably innocuous) products. Degradation (often expressed by the dissipation coefficient K_s) is generally biochemical and therefore is influenced by factors that affect biological activity, such as low soil moisture content, temperature, organic-matter content, dissolved oxygen, and external pH.

The importance of each factor depends on the type of substance being degraded and on the characteristics of the site. Strongly adsorbed pesticides are relatively immobile in the soil profile and remain near the surface, where the greatest biological activity occurs and where the degradation rates are higher. Organochlorine substances are generally most persistent, whereas organophosphorus compounds and carbamates degrade more easily. For triazines, the degradation is essentially chemical and takes a long time. The retention of a pesticide on soil particles allows a prolonged degradation and a mitigation of the initial toxicity of the substance concerned.

Pesticide solubility, both in water and in fats, appears to be a fundamental property for the diffusion of pesticides in the environment and in the food chain. Pesticides migrate

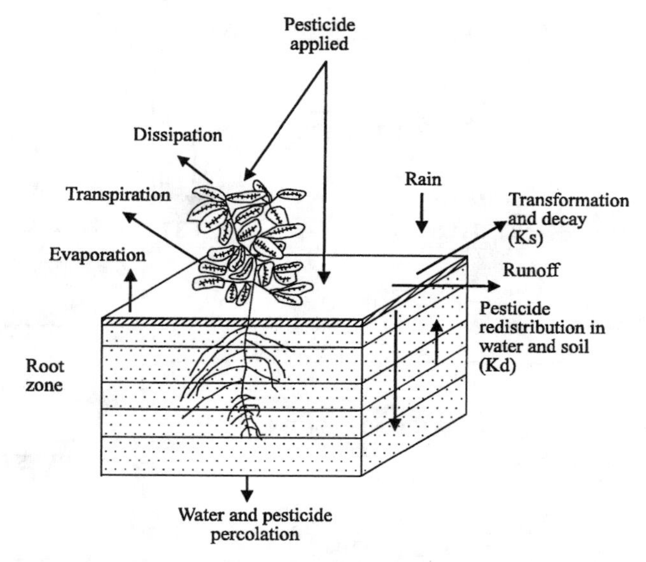

Figure 5.77. Interactions of pesticides, soil, and water (K_s = dissipation coefficient, K_d = redistribution coefficient) [13]. *Source:* [13].

from agricultural land toward surface-water bodies by volatilization and transport in a gaseous phase or by transport in solution or in suspension in runoff water. Transport toward groundwater occurs by molecular advection and diffusion. The adsorption characteristics, which determine the mode of transport, depend on numerous mechanisms based on molecular interactions of various types (including van der Waals forces and hydrophobic, electrostatic, hydrogen and ion exchange, and other bonds). Soluble and weakly adsorbed substances usually percolate through the soil profile or are lost in runoff during storm events and under certain hydraulic conditions (hydraulic conductivity, water content, slope, etc.). Moderately adsorbed substances are transported in runoff or percolating waters; those that are strongly adsorbed are transported on solid particles.

Fertilizers

Fertilization results in the input of nutrient elements to the soil, principally nitrogen and phosphorus. Some forms of these substances (nitrate and phosphate ions) also can be seen as exemplary illustrations of two main transport mechanisms (in solution and in suspension).

Nitrogen. This nutrient exists in soil principally in organic forms, which are slowly converted to more available inorganic forms (ammonium and nitrate). Nitrates are very soluble and easily leach below the crop rooting zone or can be removed in runoff water. Organic nitrogen and ammonia, on the other hand, are adsorbed onto soil particles and therefore tend to remain where they are, provided that they are not removed by erosion (Fig. 5.78).

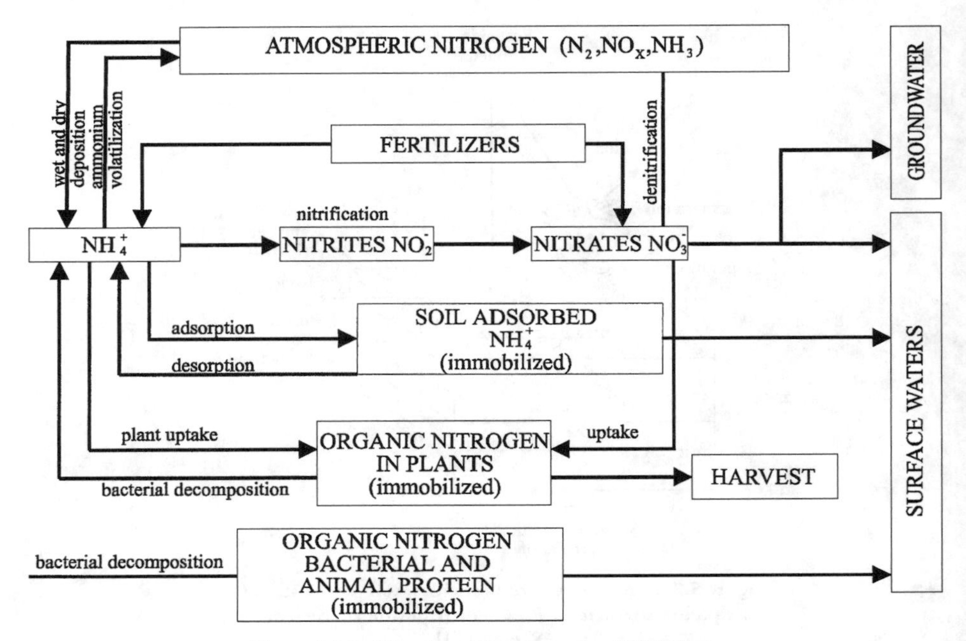

Figure 5.78. The nitrogen cycle, a simplified view.

Three microbially mediated transformation processes are involved in the balance of nitrogen. The first two (ammonification and nitrification) are part of the mineralization process that transforms organic nitrogen into inorganic forms, making it available for plants. The other (denitrification) results in a loss of nitrogen to the atmosphere and can reduce the quantity of nitrogen lost by leaching and runoff. Nitrogen also can be transferred to the atmosphere in a gaseous form by chemical processes, without requiring the presence of microorganisms. The latter process occurs in the case of heavy applications of urea- and ammonia-based fertilizers.

Practices that increase oxygenation or drainage also increase mineralization and therefore allow the formation of nitrate. Nitrate is available for plants, but also can be lost during runoff and leaching. The content of nitrogen compounds in the soil normally increases up to a depth of about 30 cm and then decreases because of the influence of leaching. Drainage favors nitrification but controls transport for groundwater. In general, migration of nitrate takes place at very low rates, with a subsequent increase in the concentration in groundwater. As a result, the problem of high nitrates in groundwater may get so bad that, even after the introduction of nonpolluting agricultural practices, it takes a long time to remove them. Migration rates are high when preferential and macropore flows dominate.

The period of fertilizer application is important in that, if it is close to the period of maximum vegetation development, it will favor a reduction in losses due to plant uptake. In particular for groundwater, it is recognized that, at least in some cases, a large part of the responsibility for the migration of nitrates is due to the abuse of fertilizers (mineral and organic), in terms of excessive and/or nonoptimal applications with respect to plant phenology. An additional factor that favors this migration is the mineralization of organic substances present in the soil, which is connected with unsuitable crop-cultivation practices.

Nevertheless, agriculture cannot be held responsible for all eutrophication phenomena in water or for all contamination by synthetic substances. Often the contribution of industrial or urban activities (sometimes unseen) are much more important (e.g., discharge and dispersion from sewers and wells).

Phosphorus. In natural systems phosphorus (P) is present as the orthophosphate anion (PO_4^{3-}), in both inorganic (insoluble calcium phosphate) and organic (plants and organic biomass) forms. Phosphorus is an important nutrient for aquatic ecosystems, but in excess it causes accelerated eutrophication manifested in an exaggerated production of organic material by algae, macrophytes, and other organisms containing chlorophyll. The rate of eutrophication in many water bodies can be exclusively regulated by the concentration of phosphorus, given that other nutrients are not limiting (in accordance with Liebig's law of the minimum).

The quantity of phosphorus coming from diffuse sources changes considerably with various pedological, meteorological, and environmental factors. In contrast to nitrogen, phosphorus is not particularly mobile in soils. The phosphate ion does not leach easily because it is fixed to clay and organic matter and it is largely lost by erosive processes (Fig. 5.79).

The principal factors (sometimes intercorrelated) that control the process of phosphorus fixation are

- Al and Fe oxides, which are responsible for retention in acid soils;

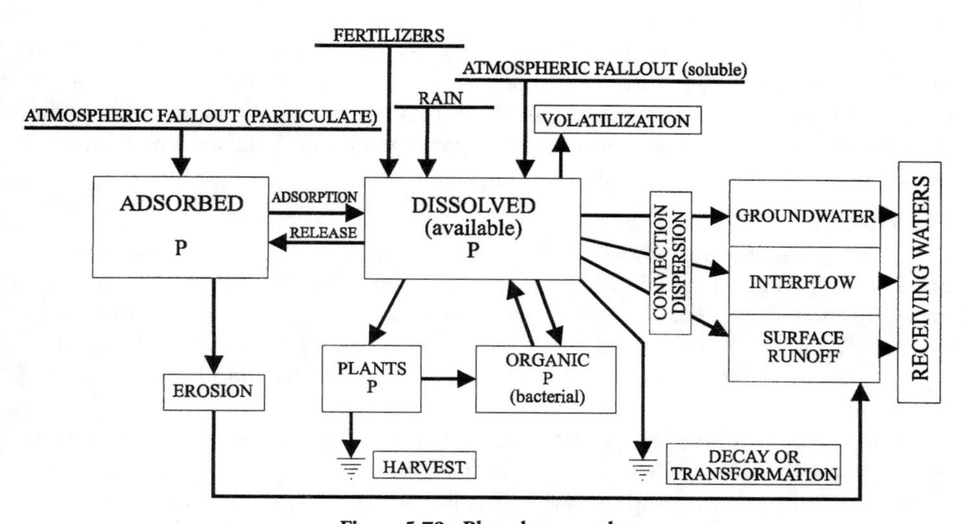

Figure 5.79. Phosphorus cycle.

- calcium compounds, which control solubility in calcareous soils; and
- the percentage of clay and organic matter that contribute to adsorption.

Adsorption is not an instantaneous process and occurs in an initial rapid phase followed by one that is slower. The adsorption characteristics are different in soils compared with transported sediments.

The greatest fraction of total phosphorus in most soils is not available for plants but remains adsorbed to solid particles. Part of the particulate inorganic phosphorus (the labile fraction) is the source of the dissolved form, but the transfer is very slow. In addition, continuous uptake by the root system means that the concentration of dissolved phosphorus is always lower than that estimated on the basis of the total phosphorus present. Phosphorus applied as a fertilizer also is retained and can be difficult to lose in a dissolved form. The transport of the element to water bodies may be largely in a particulate or adsorbed form. Runoff is the main process for transporting phosphorus to surface waters. Moreover, the balance between the labile particulate and dissolved forms also depends on the chemical and biological characteristics of the environment. Therefore, the quantity of dissolved phosphorus can vary during its transfer from agricultural soils to receiving water bodies and subsequently within these bodies. Analogously, a decrease in runoff can result in a reduced loss of total phosphorus, but may have little effect on the levels of available phosphorus in the receiving waters because of changes in the rate of transformation of the labile fraction into the dissolved form. The peculiar behavior of phosphorus is of specific interest in the degradation of the quality of surface water bodies and is the main factor controlling eutrophication, especially in freshwater.

Sediments

The effects of excessive sediment loading on receiving waters include a deterioration in aesthetic value, a loss of reservoir storage capacity, changes in aquatic populations and their food supplies, and an accumulation of bottom deposits that impose an additional oxygen demand and inhibit some advantageous benthic processes. Deposition of clay and silt in irrigation canals is a priority problem. Sediment is the most visible pollutant originating from nonpoint sources. Moreover, the finest sediment fractions are primary vehicles of other pollutants such as organic pesticides, metals, ammonium ions, phosphates, and other toxic materials.

Trace Metals

Generally, the metal content of soils is low, at least in soils that are not adjacent to main roads or zones of high industrial or mining activity. On agricultural land, however, the application of wastewater, sewage sludge, or compost (for disposal and/or fertilization) can introduce significant amounts of metals. In addition, the use of fungicides, insecticides, fuels, lubricants, and tires also can contribute to the introduction of metals to the soil.

Finally, chemical fertilizers can be the source of heavy metals derived from both primary materials and production processes. The rate of metal uptake by crops depends on the solubility of the metals in the soil solution. Many of them precipitate or are adsorbed onto soil particles. Those that are most immobilizable are Mn, Fe, Al, Cr, As,

Pb, and Hg, whereas those that can easily reach the plants are Cd, Cu, Mo, Ni, and Zn, especially under certain conditions of pH.

Microorganisms

Many pathogenic microorganisms (bacteria and viruses) can be present in the soil as a consequence of contamination by animal feces, as well as spreading of refuse and sewage. Their survival varies considerably under different environmental conditions.

Pollutant Transport Transformation Processes

The ways by which agricultural pollutants reach surface and groundwater bodies are essentially linked to the movement of water, which almost always represents the primary vector. The processes that most control transfer are erosion and sedimentation of soil particles, surface runoff, subsurface flow, infiltration, and percolation in the soil.

In the transfer of a pollutant to a water body, it is possible to identify several phases: supply, extraction, transport, and the impact on the receiving body. In each of these phases, the pollutants undergo physical, chemical and biological transformations that determine their method of transfer.

For substances that are strongly adsorbed onto soil particles, erosion and sedimentation processes are prevalent, even if they can subsequently detach from the particles and be released into solution. For moderately and weakly adsorbed substances, the prevalent transport processes take place in solution by surface runoff and deep percolation (Fig. 5.80).

The physical factors that influence pollutant transport in any specific situation are, therefore, numerous and include

- *precipitation characteristics*, that is,
  - frequency and season of the event,
  - intensity, depth, duration, and kinetic energy,
  - space-time distribution,
  - time after chemical treatment;
- *climatic characteristics*, that is,
  - temperature, wind velocity and direction, presence of ice, etc.,
  - solar radiation;
- *land characteristics*, that is,
  - pedology,
  - morphology;
- *planned crop cover*, that is,
  - crop type and degree of cover,
  - production, cultivation and irrigation practices rotation, and climatic conditions;
- *characteristics of herbicides*, that is,
  - formulation,
  - persistence, including microbial, chemical, or photochemical degradation, and volatilisation;
  - solubility,
  - adsorption.

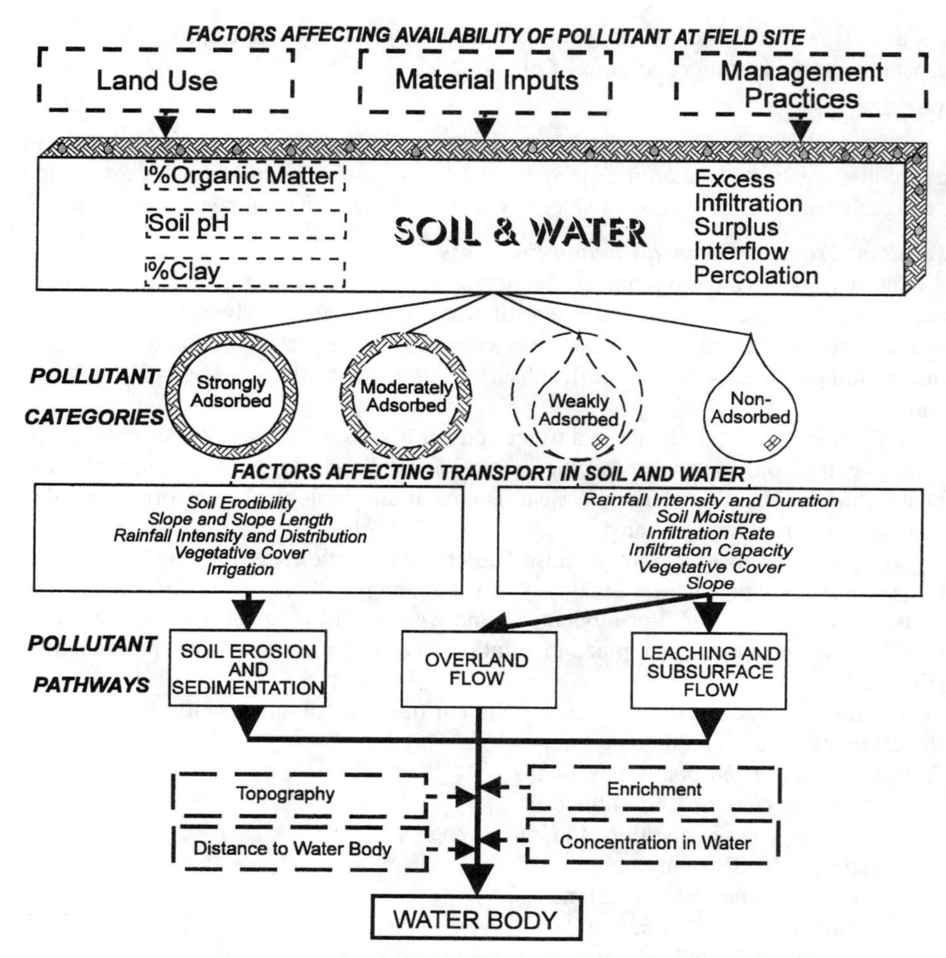

Figure 5.80. Elements and factors that intervene in the process of pollution transfer from agricultural nonpoint sources to receiving water bodies.

Meteorological conditions play a fundamental role in determining the hydrological and sedimentological behavior of a drainage basin. The temperature, moreover, also is linked with the quantity of nutrients that can be made available for transport because they are not used by plants. The greatest surface runoff and the greatest transport of particulate solids occurs during high-intensity rainfall events. Particles become detached from the surface because of the high kinetic energy of the raindrops. The erosion and transport capacity of water running over the soil increases with its velocity, which in turn increases with the angle of the surface slope.

The characteristics of the soil are also essential for the definition of the transport mode for agricultural contaminants. The texture and the mineralogical composition affect, for example, infiltration capacity, percolation, and the adsorption of pollutant substances in the lower layers. Coarse textures, in particular, favor high levels of infiltration and the transfer of pollutant substances toward lower layers. On the other hand, organic matter and clay particles are responsible for the consumption of high amounts of pollutant substances by adsorption mechanisms and chemical bonding. The fertility of the soil can, itself, influence the mobility of pollutants because, if fertility is high, there is no nutrient limitation on evapotranspiration. Increases in evapotranspiration can result in decreases in both the surface and the subsurface fluxes. Moreover, crop development can increase the infiltration of water and provide protection against erosion by rainfall and runoff. Finally, soil type can indirectly influence potential pollutants in that it affects the choice of crop organization and irrigation systems.

The method of irrigation determines the degree of impact that soil leaching and the return of water from irrigated fields to ditches and to groundwater can have on the receiving water bodies. Indeed the irrigation characteristics (volumes applied, number of applications, geometrical characteristics of the channels and sub-channels, etc.) influence runoff, percolation, extraction, and transport.

It is necessary to consider the chemical characteristics of the polluting substances along with the physical properties of the soil. These characteristics (especially the persistence, solubility, and adsorption capacity) determine pollutant availability and are responsible for the physical and biological transformations performed.

The factors that influence the distribution and the fate of agrochemicals in the environment arise from various factors that are related internally and externally. The main internal factors (linked to the substance itself) are solubility, polarity, volatility, molecular dimensions, and dissociation constant. The most significant external factors include adsorption capacity, the movement of the aqueous and air phases, temperature, pH, light, and biotic and abiotic activity.

The composition of these factors determines the occurrence and eventual prevalence of one of the following fundamental processes.

- adsorption from organomineral components with the possibility of subsequent release (e.g., according to Langmuir or Freundlich isotherm);
- transformation by chemical or enzymatic reactions into residual products with different nature and properties;
- removal from the system by surface runoff (overland flow), interflow, volatilization, leaching (or percolation), and chemical or biological degradation;
- uptake by living organisms, with eventual input to food chains.

The principal environmental compartments that are involved in the redistribution of agrochemicals are soil, water, and air. The contamination of surface waters (inland and marine) and groundwaters by pollutants of an agricultural nature is due to the natural process of percolation, the application of irrigation water, to direct spreading of pollutants, and to discharges from production or food processing units. Various chlorinated pesticides are present in river water but, because of their low solubility, they usually are associated with suspended material or bottom sediments.

Elevated concentrations of pesticides can be present in rainfall and snow and more often are associated with particles than in solution. It is estimated that about 50% of the pesticides from sprinkler systems could contaminate plants and animals near the target site because of transport by air currents.

In the study of the environmental fate of agrochemicals, it is often useful to refer to the hydrographic catchment system and principally to the soil system, which constitute the fundamental units in which the main transport and transformation processes of agricultural pollutants occur.

The fate of a substance in soil is governed by the solid phase as well as by the liquid and gas phases, which determine its concentration and mobility. The chemical and structural characteristics of a soil determine the availability of adsorption surfaces and fluid flow and these factors affect its behavior, e.g.,

- lower biological degradation, separation of pesticides and enzymes;
- facilitation of biological degradation, concentration of pesticides and enzymes;
- reduction in dispersion and volatilization;
- catalysis of nonbiological degradation;
- location of pesticide molecules in sites where microbial growth can occur;
- influence on phytotoxicity.

An examination of the data concerning diffuse agricultural pollution collected over the past 20 years [15] has revealed that most experiments can be grouped into long-term and short-term studies. Long- term studies are those that have been conducted under natural conditions (watersheds, fields, and plots) in order to assess effective losses over the crop year. Short-term studies are those that have been conducted under artificially controlled conditions (lysimeters, tilted beds, and small experimental plots) to assess the effect of individual factors and to simulate critical conditions (e.g., using rainfall simulators). The information obtained from short-term studies supplements that obtained from long-term studies regarding the maximum concentrations observed in runoff. The major problem with these studies is avoiding an overestimation of pollutant loads when the results are extrapolated to larger scales [16].

Effects of Agricultural Practices with Respect to Wastewater Uses

The agricultural utilization of refuse from anthropogenic activities can constitute a valid system of waste disposal [5, 17]. Solid refuse (urban waste compost and/or sewage sludge) can be used for fertilization (manure and conditioners) and liquid waste (wastewater) can be used for irrigation. There are several advantages associated with this type of waste use despite the environmental risks. The advantages include recycling of organic substances and plant nutrients in agroecosystems, reduction of spatial disposal requirements, and higher protection of aquatic ecosystems that otherwise would receive a major part of the wastewater (treated or not).

Application of Municipal Sewage, Livestock Manure, and Olive Oil Wastewaters to Land

Raw municipal sewage has always been used in agricultural areas surrounding human settlements until the development of large towns and the associated increase in pollution and sanitary risks resulted in its abandonment.

The problem of irrigation water quality is a current one, both in terms of the degraded state of a number of water bodies used for irrigation water supply and also in terms of the possibility of using more or less purified wastewater for irrigation. This solution has many potential environmental and agronomic advantages. Environmental advantages include the fact that the utilization of an alternative nonconventional water resource results in a saving of valuable primary resources and a reduction of pollution loads in receiving water bodies. Agronomic advantages include the availability of a new source of irrigation water and a contribution of nutrients (contained in the wastewaters) to agricultural soils. Nutrient inputs are often in the range of 10–1,000 mg L^{-1} for nitrogen, 5–25 mg L^{-1} for phosphorus, and 10–40 mg L^{-1} for potassium and organic compounds. Finally, the practice significantly simplifies the management of wastewater treatment.

In contrast to these positive aspects, however, wastewaters can carry toxic substances and pathogenic microorganisms that may be harmful to crops, soil, and people. The relative importance of the negative factors obviously depends on the provenance of the sewage and of the treatments received. In any case, sewage characteristics are highly variable (see Table 5.27 for an example). Thus, on one hand, the reuse of wastewater for irrigation is certainly an interesting prospect. On the other, however, a preliminary verification of water quality is still required to predict possible effects that pollutants may have and to define the methods and the doses for its use. The need for monitoring is also a need for a follow-up of impacts on plants, products, soil, groundwater, and irrigation systems. Irrigation with more or less purified wastewater can be applied using two techniques [7]:

1. Fertilizing irrigation, that is, extensive irrigation with a low load, the primary aim of which is to provide for agricultural development and to obtain maximum production, and secondary aim of which is the disposal and degradation of liquid sewage;
2. Purification irrigation or purification agriculture, that is, intensive irrigation with a high load, the principal aim of which is the disposal of wastewater, using the autopurification capacity of the soil-plant system to obtain the degradation of organic compounds present in liquid sewage and an absorption of other substances.

One of the biggest limitations to the application of wastewater to agricultural fields is the severity of current laws in some countries that prevents its use.

The suitability of wastewater for use in irrigation is based on a consideration of the qualitative aspects previously discussed.

One technical problem for the use of municipal wastewater for irrigation is the difficulty of organizing and synchronizing the continuous production of wastewater with periodic irrigation requirements. In some situations (e.g., summer holiday destinations, in which sewage discharge increases at the same time as major irrigation needs), it may be necessary to stock high volumes of irrigation water with consequent risks for stream water quality.

Land Disposal of Municipal Sewage Sludge and Solid Waste Compost

In the history of world agriculture, organic wastes have always been used as fertilizers of agricultural land because of the role of organic substrates as soil conditioners and fertilizers.

The use of sludges and composts as fertilizer would be desirable were it not for the presence of potentially harmful substances with negative effects on crops, human health, product quality, and the environment. The negative aspects to be considered are the following:

- *Bad odor*. Bad odors are produced in nonstabilized and poorly aerated sludges by the presence of substances, such as, hydrogen sulfide, ammonia, trimethylamine, indole, skatole, mercaptan, and volatile fatty acids.
- *Inert material*. The agricultural use of compost presupposes a preliminary screening of urban solid waste to remove inert (nondegradable) materials so that they do not constitute obstacles to operations, damage to plant operators, or eyesores on agricultural land.
- *Humidity*. The water content of sludges and composts, like their aeration, is an important aspect with respect to their agricultural utilization.
- *Pathogenic organisms*. Numerous pathogenic agents are present in sludge and compost (e.g., bacteria, viruses, yeasts, protozoa, and worms). The anaerobic digestion of sludge in a liquid phase can never permanently ensure the same level of hygienic safety that is achieved using aerobic stabilization in the solid phase or using innovative physical treatments.
- *Xenobiotic organic substances*. This includes many organic compounds unrelated to the biogeochemical cycles that can be found in municipal and (particularly) industrial sludges. Such sludges include oils, surfactants, aldehydes, aromatic amines, halogenated aromatic compounds (polychlorinated biphenyls), and polycyclic aromatic hydrocarbons (PAHs). On the basis of the mean concentrations found, it seems that an input to the soil of 10 $t\,ha^{-1}$ is acceptable.
- *Heavy metals*. The presence of heavy metals in sludge and compost, which accumulate in soils, represents the most serious limiting factor to their utilization in agriculture. The metals that are considered to pose the greatest environmental risk are Pb, Cd, Co, Cr, Cu, Hg, Ni, and Zn. Their principal origins are Hg, Cd, and Zn from batteries; Cr from leather tanneries; Cr, Pb, and Cd from paints; Cd from plastic and Pb from printed paper. The concentration of heavy metals is extremely variable and the limits for annual loads currently stipulated by laws vary from nation to nation.

The conclusions of most current studies is that the use of sewage sludge and solid urban waste compost appears to have some agronomic validity in terms of their potential use as fertilizers (assured by the input of plant nutrients and, to a lesser extent, organic substances) despite numerous doubts about the risk of contamination (viral, bacterial, or chemical) for stream water quality.

Pollution Problems in Water Bodies from Intensive Animal Production and Manuring

The first wastewaters used in agriculture were those produced from livestock. This practice enables the recovery of the water resource and any nutrients contained in the sewage or wastewater. However, it does pose some environmental risks linked principally to the current tendency to concentrate livestock rearing and to increasingly rely on feeds produced externally. This results in a high number of animals per unit area of farm and a consequent risk of excessive nutrient inputs when the animal wastes are applied to the soil, chemical and/or biological stream and lake pollution, eutrophication of slow-turnover water bodies, increase in foul odors, and proliferation of flies and other insects.

Other Relatively Specific Problems of Water Pollution in Agriculture

There are a number of other activities connected with agriculture that can cause bad pollution problems (e.g., of subsoil water), either accidentally or through poor practices. These additional activities include

- washing out and/or degradation of containers containing residues of pesticides, fertilizers, detergents, fuels, and lubricants for machinery;
- losses from stocks or during agricultural operations and/or other accidental losses of pesticides, fertilizers, detergents, fuels, and lubricants for machinery;
- incorrect disposal of residues from agrochemical companies;
- malfunctioning of septic tanks and subirrigation apparatus.

5.8.3 Methods of Identification, Evaluation, Management, and Control

Type of Intervention

The main intervention methods for avoiding the negative impact of water pollutants on agriculture or the impact of agricultural practices on water quality are

- monitoring and modeling for impact quantification, vulnerability evaluation, and watershed planning;
- treatment of wastewaters, sludge, and animal wastes in plants, lagoons, wetlands;
- implementation of management practices of pollution control.

Measures of agricultural pollution control are as diverse as the sources of pollution themselves and no single technology can be employed to control diffuse sources. The traditional means of pollution control (collection and treatment), which works well for most point sources, is prohibitively expensive for nonpoint pollution control. It can be used only in rare cases when all other methods of control at the source have failed.

Monitoring and Modelling for Impact Quantification and Vulnerability Evaluation

Mathematical models are used widely as predictive tools for assessing pollutant transport and transformation processes in the environment. In particular, the fate of agrochemicals in soil–water systems can be predicted using a number of models that vary in complexity, information requirements, data standards, data format, computer platform, and ultimate objectives (scientific research, governmental regulation, agricultural production, education, or planning).

One possible categorization of agricultural diffuse pollution models (ADPMs) [18] distinguishes between deterministic models, which assume that a system or process operates such that the occurrence of a given set of events leads to a uniquely definable outcome, and stochastic models, which assume that the outcome of any modeling exercise is uncertain and which therefore are constructed accordingly to account for this uncertainty. A second main distinction is between mechanistic and functional models. The term mechanistic is used for models that incorporate the most fundamental mechanisms of a process, as it is presently understood. Functional models, in contrast, incorporate simplified treatments of processes, such as water flow and solute transport and therefore usually require less input data and computing expertise for their use. In addition, it is necessary to distinguish between models that are primarily research tools, developed to aid the testing of hypotheses and the exposure of areas of incomplete understanding and models that are designed to be used as guides to management, as regulatory tools, or as teaching aids.

ADPMs generally have been developed to address agrochemical transport and transformation at specific spatial and temporal scales, according to the expected use. Such models have proved useful in designing water- and soil-quality management strategies and/or Best Management Practices (BMPs), which maintain agricultural productivity while minimizing adverse water- and soil-quality impacts [19]. In fact, the use of efficient and correctly calibrated models can highlight any significant effects that changes in management practices may have. It is necessary, however, to define sound protocols for model application. In any case, affordable simulation models are essential for obtaining information to assess the environmental impacts of agricultural activities, because they enable investigators to examine and compare scenarios with a wide range of agrochemical applications, soils, crops, management systems, hydrogeological settings, and meteorological conditions. Furthermore, models may be useful in extrapolating the results of experimental studies to larger scales or in analyzing the significance of changes in cultivation techniques on observed processes. Without the use of models, these steps would be more difficult, especially in the presence of stronger sources of variation, such as the rainfall pattern during the trial period. The principal modeling and monitoring approaches that can be adopted at various scales are shown in Table 5.30 [15, 18].

The transfer processes of compounds in flowing water and their transport to the basin outlet is, of course, rather complex. The first modeling tools developed for estimating the dynamics of diffuse pollution transport of agricultural origin were either point models (individual processes modeled at a single point) or aggregated-parameter models (lumped models) and generally only allowed the evaluation of mean event concentrations and total loads. These kinds of models can be used as land planning (screening) indices, in defining land vulnerability, and in planning water quality. They allow a quantitative evaluation of the effects of different control measures. They also can be used to examine the behavior of different chemical products (newly synthesized compounds or those that have never been used in certain environments). Data from experimental studies are more useful, according to one point of view, if they are used to calibrate models. Indeed, the loss of agricultural pollutants is influenced by various factors (e.g., the nature of the substance, the quantity and mode of application, the cultivation practices employed, the soil characteristics, and the meteorological and climatic conditions). Therefore, only a monitoring of final processes for use in model calibration permits an extension of the results obtained to other environmental systems. Because the monitoring of these phenomena is demanding from both technological and economic points of view, it is important to conduct “heavy” experimental studies to obtain really dependable data (even if it is site and contaminant specific). The collection of few data in many places should be limited to the initial (screening) phase of the study. Naturally, such models

Table 5.30. Categorization of monitoring and modeling approaches that can be adopted in agricultural diffuse-pollution studies

System	Spatial Scale	Example of Processes	Monitoring Approaches	Modeling Approaches
Province, region, county	$(i+4)$	contamination of groundwater	GIS, remote sensing	Mass balance functional models of fluxes combined with GIS ($i+1$ to $i+4$)
Soil regions (interacting watershed)	$(i+3)$	Ecosystem response to chemical inputs	Automatic monitoring network, remote sensing	Statistical models, compartment flux and pool models
Watershed with water bodies (rivers, lakes)	$(i+2)$ 1 ha–1,000 km^2	Delivery to surface water, groundwater recharge, solid transport (erosion)	Network of water-quality monitoring stations, soil sampling, meteorological gauges, remote control, remote sensing	Distributed (functional) or statistical hydrosedimentologic models; lumped parameter; functional, management models
Field, farm	$(i+1)$ >1,000 m^2	Soil-runoff extraction processes (enrichment and delivery), water transport in the soil, ammonification, erosion, lateral flow, atmospheric dispersal, etc.	Automatic monitoring systems, soil sampling	2- or 3- dimensional; lateral flow; *ad-hoc* stochastic use of deterministic models; functional, management models
Elementary cell, "pedon"	(i) 10 m^2–1 ha	$(i-1)$, $(i-2)$, $(i+1)$ processes combined with spatial variability; fluxes to hydrologic network, groundwater, and air	Experimental plots, tilted beds, lysimeters	Comprehensive deterministic models; $(i-1)$ models applied with knowledge of variability of water flow, sorption, degradation, volatilization
Profile, horizon	$(i-1)$ 0.1 m^2–10 m^2	Nonequilibrium physical and chemical transport; volatilization, degradation	Undisturbed lysimeters or monoliths	1- or 2-dimensional deterministic, mechanistic leaching with consideration of mediating processes, soil horizonation
Soil peds, aggregates	$(i-2)$ <0.1 m^2	Chemical–soil interactions during leaching; intraaggregate diffusion	Laboratory lysimeters or soil columns	Miscible displacement theory; macropore, preferential flow modeling
Soil–water–chemical mixtures	$(i-3)$	Sorption relationships; degradation kinetics	Laboratory (batch studies)	Langmuir, Freundlich isotherms, Michaelis-Menten, first-order degradation
Molecules (pore/particle)	$(i-4)$	Phase partitioning; polarity; hydrophobicity	Laboratory	Chemical reactions in solution; molecular structure, solubility

Source: [20].

have to be dynamic and take account of hydrological and sedimentological behavior and of biochemical transformations.

The concentrations of a contaminant transported by runoff water in liquid and solid phases depend on its concentration in the soil, which is related to the length of time from its date of application, its persistence, and its adsorption, as well as to the preceding precipitation events (dry period, moisture content, etc.).

Indices for screening or for evaluating vulnerability to agricultural pollution include attenuation factor (AF), retardation factor (RF), residual mass (RM), LEACH, LI, and groundwater ubiquity score (GUS). In existing models, the above-mentioned dynamics are schematized at various levels of detail. Available models include the following: PRT (1973), ACTMO (1975), Bruce (1975), ARM (1978), CNS/CPM (1979–1980), NONPT (1979), ANSWERS (1980), Haith (1980), PRS (1980), SWAM (1982), HSPF (1983), EPIC (1984), PRZM (1984) and RUSTIC (1989), GLEAMS (1986–1989), WRRB (1985) and SWRRBWQ (1991), LEACHMP (1986), SHE (1986), MRF (1987), AGNPS (1987), and others more recent such as PHYSIS (1993), RZWQM (USDA-ARS), OPUS (USDA-ARS).

Common limiting factors for the applicability of these tools include their mathematical complexity, difficulties of data retrieval, and the calibration of some parameters. The development of distributed models is, therefore, useful even if the mass of information necessary for a correct interpretation of the phenomena is large because they allow the effects and the space-time variation of a range of variables to be accounted for. Models that require the optimization of only a few physically based parameters (functional models [18]) are also useful in this respect. Obviously, the methods for monitoring and quantifying pollution processes from nonpoint pollution sources are not the same as those usually adopted for studying concentrated sources. It is the sampling method rather than the analytical techniques employed that most distinguishes the investigation of concentrated sources from that of nonpoint pollution sources.

With diffuse pollution, there is no precise point at which water can be sampled. Soil or groundwater must be sampled at different depths and locations in the drainage network.

The selection of sampling points must take into consideration the fact that the sampled water will probably be a mixture of water from a number of sources. For example, surface waters may interact with groundwater or with point sources of municipal, industrial, or agricultural sewage, even if the area under examination is of limited extent and is hydraulically confined. To compensate for this problem, many researchers have chosen to limit and control the study area. This usually means employing lysimeters and/or small plots. Limiting the size of the experimental area, in this way, resolves the problem of external interference. However, it also can produce artificial distortions in transport processes as a consequence of the experimental design. To resolve this problem, one must clearly try to understand all of the potential experimental difficulties and, if possible, avoid making use of only one scale of investigation (e.g., laboratory, plot, or watershed). Instead, a number of complementary scales should be used to accurately interpret the processes from the data collected. On the basis of experimental data and a theoretical knowledge of pollution phenomena, it is clear that the most effective monitoring method for both watersheds and plots is to sample more frequently during storm events when transport rates are higher. The ideal compromise could be to choose a system of research that allows information to be obtained at scales ranging from $(i-1)$ to $(i+2)$ in Table 5.30 and that allows for both a detailed analysis of processes at lower scales in the laboratory [from $(i-2)$ to $(i-4)$] and extrapolations to the higher scales [e.g., $(i+3)$ to $(i+4)$] using distributed modeling.

Treatment Systems

Potential pollutants from the waste products of the livestock sector and the agro-food industry are generally greater than those from municipal wastes (for which tried and tested treatment processes are more frequently employed). The ratios between the concentrations of pollutants in the two types of wastewater (livestock-agrofood and municipal) can vary between 2 and 35 for BOD_5, between 2 and 50 for nitrogen, and between 0.8 and 75 for phosphorus, with the highest values from the waste products of pig production. Local treatment of this type of waste is often complex in terms of digestion plant and it is, therefore, preferable to devise one or more systematic, simplified, and centralized solutions that allow the recuperation of nutrient elements and their agronomic use via spreading. These solutions should be designed to take account of both the pedological and crop characteristics of the area under examination. To calculate the area of land suitable for the spreading of waste on a livestock farm, one can base the distribution on phosphorus and then add extra mineral nitrogen or, alternatively, the area can be based on the available nitrogen, which inevitably involves accepting an excess of phosphate and potassium. The latter strategy usually results in a lower spreading area.

The purification of wastes can be carried out using biological or chemical processes or using a combination of both, according to the relative presence of organic (biodegradable) and/or chemical pollutants.

Usually, if it is possible to select the type of process, biological methods are preferred because of the lower production of sludge and the saving of reactive chemicals.

Biological sewage purification plants are systems for artificially accelerating the natural biological degradation of organic material, in a controlled environment, by establishing and maintaining an adequate mass of bacteria in the system. The biological processes, both aerobic and anaerobic, are based on the degradation of organic substances contained in the sewage via biochemical reactions mediated by the bacterial population and by the gradual formation of more stable products producing new cells. The process can be performed in treatment plants (Fig. 5.81), lagoons, wetlands, or other artificial or natural containers.

For example, the structures and plants necessary for the digestion of animal waste products are

- *solid-liquid separation*, comprising balancing vessel, separator, plate for the accumulation and maturation of the separated solid, pipes for the pumping of liquids;
- *stabilization*, comprising tanks and lagoons able to contain waste with a high water content (i.e., slurry), for at least 120–180 days, with pumps for aeration or homogenization (tanks may be covered for the recovery of the biogas produced and platforms are used for wastes with a low water content, i.e., farmyard manure);
- *field spreading*, comprising fixed or mobile plant or machinery;

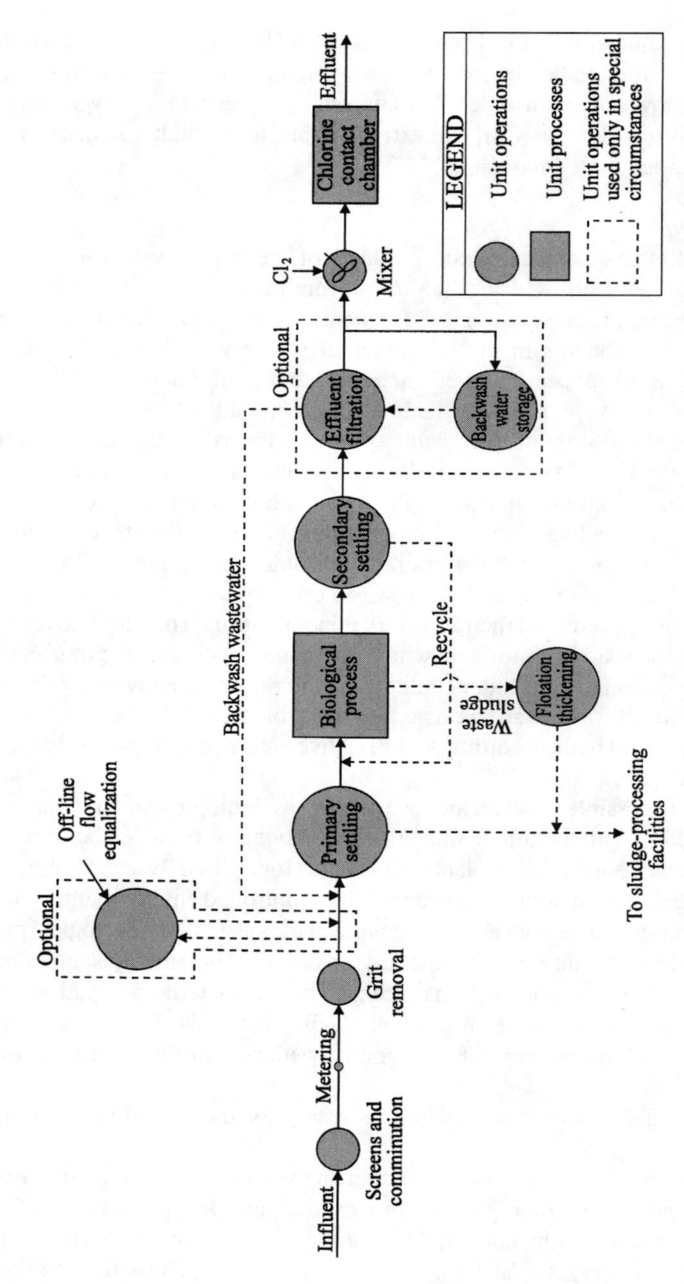

Figure 5.81. Location of physical unit operations in a wastewater treatment plant flow diagram. ***Source:*** **[9].**

- *purification*, In aerobic plants, the "active" biomass operates independently in the presence of oxygen, which is input via oxygenation or aeration equipment. In anaerobic plants, the microorganisms operate in the absence of oxygen and use different elements to oxygen as terminal electron acceptors in the respiratory chain.

Wastewater Processing. Purification plants consist of a series of treatments defined as primary, secondary, and tertiary, combined in different ways according to the slurry under treatment and according to the level of treatment required (Fig. 5.81).

Primary treatments consist of a grill (coarse or fine); a separator (cloth filtration on a rotating drum); sand-removal, horizontal-flow rectangular or circular flow tanks; sedimentation (with longitudinal or radial flow); and de-oiling (vessel with deflector).

Classical secondary treatments (Fig. 5.81) consist of an activated-sludge reactor (oxidation vessel), where agglomerated flocs of mixed bacteria form in the sewage and are aerated in such a way as to accelerate the processes of degradation. The mixture leaving the oxidation vessel reaches a sedimentor, where a solid–liquid separation takes place. The treated effluent is harvested in the upper part and the sedimented active sludges in the lower part are recirculated in the head of the oxidation vessel to maintain the optimal concentration of biomass (suspended solids or volatile suspended solids). In addition to supplying the oxygen necessary for bacterial metabolism, aeration systems (superficial with a vertical axis or turbine, with air or pure oxygen input), must adequately guarantee mixing and encourage contact between the bacteria and organic matter. Because BOD loads are very much higher in the case of pig slurry, a double phase of oxidation is considered.

In the case of lagoon treatment (biological ponds), the degradation of organic substances can occur aerobically, anaerobically, or facultativly (i.e., the operation of areobic and anaerobic pathways simultaneously), depending on the depth of the liquid mass. Aerobic lagoons have a medium depth (0.6–0.8 m) to allow an aerobic environment to be maintained (also assisted by the photosynthesis effect). Lagoons can be aerated artificially by floating turbines.

Attached-growth mass plants also exist. These are distinct from those previously described in that they have a *suspended-growth* biomass, with, for example, trickling filters or rotating contactors. Before clarification, but after sedimentation, slurry is percolated through supports of small stones or plastic material on which microbial biofilms (films of microorganisms) form. The biofilms feed on the organic substances present in the slurry and are oxygenated by the natural circulation of air through the interstitial spaces.Tertiary treatments are defined by the removal of nitrogen and phosphorus from the wastewater.

With respect to nitrogen, the activated-sludge process allows a fairly high level of nitrification to occur, depending on the load of the sludge applied in the oxidative phase (daily quantity of BOD per unit of active biomass present) and, therefore, it is not ideal when a combined removal of organic matter and nitrogen is required.

The biological removal of nitrogen takes place in two distinct processes: Nitrification (oxidation of ammonium to nitrate by autotrophic bacteria using an inorganic carbon substrate) and denitrification (reduction of nitrate to molecular nitrogen, which is released to the atmosphere by facultative heterotrophic bacteria, under anoxic conditions, using organic carbon substrates).

These processes can occur within a single heterogeneous population of bacteria (integrated systems with a single reaction volume subdivided into anoxic and aerobic zones or with denitrification reactors following those for nitrification). Alternatively, more specific bacterial flora can be developed (separate systems with a nitrification reactor and a denitrification reactor with the addition of a methane-type substrate). The reduction of phosphorus can be achieved using biological methods or with chemical additives (iron chloride, aluminium, iron sulfate, or lime) to obtain the precipitation of phosphate ions (before, after, or simultaneously with biological treatment). The biological process is based on the behavior of the bacterial flora which, when kept for a few hours under anaerobic conditions and then subjected to strong oxidative conditions, are able to assimilate phosphorus in much greater quantities than would be expected theoretically ($BOD_5/N/P = 100/5/1$).

Finally, disinfection is a refinement operation (chemical, physical, or mechanical) in which the majority of pathogenic organisms (bacteria and viruses) in effluents are eliminated.

Sludge Processing. The sludge produced from biological treatment plants requires additional treatments:

- *Stabilization and successive thickening.* Sludge from biological (anaerobic or aerobic) or chemical (elevation of pH with calcium carbonate or oxidation with chlorine) treatment processes have to be stabilized to avoid putrefaction.
- *Mechanical dewatering.* This can be obtained by natural (on desiccation beds) or mechanical methods (vacuum filter, solid bowl centrifuge, imperforate basket centrifuge, belt-filter press, or recessed-plate filter press).
- *Anaerobic or aerobic digestion.* Transformation of Organic substances or transformed and sludge residue is stabilized.
- *Thermic treatment.* Thermal reduction of sludge involves the total or partial conversion of organic solids to oxidized end products by incineration or wet-air oxidation, or the partial oxidation and volatilization of organic solids by pyrolysis or starved-air.

Sewage Treatment for Agronomic Use. The aim of sewage treatment for uses in agriculture is to stabilize the sewage (i.e., abatement of the $BOD_5 > 60\%$). This can be achieved rapidly using aerobic methods (air input into the mass) or, more slowly, using anaerobic methods (i.e., excluding the sewage from contact with air). Processes of *oligolysis*, dehydration and combustion, and composting also can be applied to animal slurries:

- *Aerobic treatments.* Atmospheric oxygen is input to the sewage using surface oxygenators (for large basins) or bottom oxygenators (with a submerged pump) in sufficient quantities to favor the development of aerobic microorganisms (low input for deodorization, medium input for stabilization, and intense input for purification). The efficiency is inversely proportional to the dimensions of the air bubbles and the expenditure of energy, which varies from 55 to 270 W·h/capita (15 to 75 mJ /capita).

- *Large anaerobic logoons*. This allows an anaerobic treatment at ambient temperature with one prolonged stocking and the production of biogas.
- *Controlled anaerobic digestion*. This is made in air-tight reservoirs in which the sewage reaches temperatures of 30°C–35°C (mesophilic conditions) and is degraded to inorganic compounds and biogas.
- *Oligolysis*. Sewage is deodorized via electrolysis. A good control of smell can be obtained with a very low expenditure of energy (2.4–18 W h day^{-1}).
- *Processes of desiccation and dehydration*. This type of treatment is used in poultry production to obtain a reduction of smell and a contraction of the bulk of waste material for distribution in fields.
- *Composting*. Aerobic fermentation of palatable materials is performed in towers or on plates. Organic substances are humified by biological activity, which also simultaneously reduces the load of pathogenic bacteria present in the detritus (the mass reaches temperatures of 70°C–75°C).
- *Combustion*. A chemical process of pyrolysis takes place in an *anoxic* environment at temperatures of 800°C–1000°C. A fraction of the organic substances is distilled, forming a combustible gas and an aqueous vapor.

Land-Use Management Practices for Pollution Control

The major pollutants associated with agriculture include sediments, nutrients (especially nitrogen and phosphorus), pesticides and other toxins, bacteria or pathogens, and salts (salinity). Different types of agricultural land use are more likely to contribute certain types of pollutants than others (crop production, pasture and rangeland, woodlands- silviculture, and intensive animal production). Runoff and subsurface water from agricultural and silvicultural operations are sources of several pollution types identified in Table 5.31.

BMPs, or Good Agricultural Practices (GAPs) refer to any methods, measures, practices, or combinations of practices that prevent or reduce nonpoint-source pollution to a level compatible with water quality goals. BMPs include, but are not limited to, structural and nonstructural controls, operations, and maintenance procedures. BMPs can be applied before, during, and after pollution-producing activities to reduce or eliminate the introduction of pollutants from diffuse sources into receiving waters.

BMPs can be classed as structural, vegetative, or management practices. When making a selection for the solution of a water quality problem (basically to "control a known or suspected type of pollution from a particular source or to prevent pollution from a category of land-use activity"), the following steps can be followed [19]:

1. Identify the water quality problem.
2. Identify the pollutants contributing to the problem and their probable sources.
3. Determine how each pollutant is delivered to the water system.
4. Set a reasonable water quality goal for the resource and determine the level of treatment needed to achieve this goal.
5. Evaluate feasible BMPs for water quality effectiveness, effects on groundwater, economic feasibility, and suitability of the practice for the site under consideration.

The application and selection of the BMPs should be based on

1. type of land-use activity (see Table 5.31),

Table 5.31. Agricultural and silvicultural operations as sources of pollution

Source	Pollutants[a]									
	Sediment	Nutrients	Pesticides, Herbicides	Metals	Salts	Bacteria	Pathogens	Viruses	Microoganisms	Organic Materials
Dry cropland	X	A/D	X		S					
Irrigated cropland[b]	S	A/D	A/D	T	X	S	S	S	S	
Pastureland	X	X	S			X				
Rangeland	S	X	O	O		X				S
Forestland	X	A								X
Confined animal feeding	X	A/D		X	X	X		X	X	X
Aquaculture[c]		D				X	X			
Orchards or nurseries[c]		D	X	T	X	X				X
Wildlife reserves[d]		X				X				

[a] X = most frequent occurrence or presence, S = sometimes; O = occasionally; A = adsorbed; D = dissolved; T = trace.
[b] In irrigation return flow (surface runoff and water percolation below the root zone).
[c] Special area.
[d] If wildlife populations become unbalanced.
Source: Adapted from [2].

2. physical conditions in the watershed,
3. pollutants to be controlled,
4. site-specific conditions.

When selecting BMPs for problem prevention, a more technology-based approach can be followed, by dividing agricultural land into land-use categories (e.g., irrigated cropland and rangeland) and specifying a minimum level of treatment necessary to protect the resource base.

Water Quality Policies and Institutional Issues

Contaminants in reclaimed wastewater that are relevant to health can be classified as either biological or chemical agents. Where reclaimed wastewater is used for irrigation, biological agents including bacterial pathogens, helminths, protozoa, and viruses pose the greatest health risks.

To protect public health, considerable efforts have been made to establish conditions and regulations that permit the safe use of reclaimed wastewater in irrigation.

Health-protection quality criteria for reclaimed water in developing countries often are established in relation to the limited resources available for public works, and other health delivery systems may yield greater health benefits for the funds spent: Confined wastewater collection systems and wastewater treatment are often nonexistent, and reclaimed wastewater often provides an essential source of irrigation water and fertilizer. For most developing countries, the greatest concern over the use of wastewater for irrigation is that, if it is untreated or inadequately treated, it often contains numerous enteric helminths (e.g., hookworm, ascaris, trichuris, and, in some circumstances, the beef tapeworm). These infectious agents, along with microbiological pathogens, are dangerous to crop consumers and also can harm farmworkers and their families.

The World Health Organization (WHO) has recommended that crops that are eaten raw should be irrigated with wastewater only after it has undergone biological treatment and disinfection so as to achieve a coliform level of not more than 100/100 mL for 80% of the samples. The criteria recommended by WHO for irrigation with reclaimed wastewater have been accepted as reasonable goals for the design of such facilities in many Mediterranean countries. In some Middle Eastern countries that have recently developed facilities for wastewater reuse, the tendency has been to adopt more stringent wastewater reuse criteria, along the lines of, for example, the Californian regulations. Adoption of more stringent regulations is made to protect an already high standard of public health by preventing the introduction of pathogens into human food, at any expense.

References

1. Giardini, L., M. Borin, and U. Grigolo. 1993. La qualità delle acque per l'irrigazione. L'Informatore Agrario, Verona, Italy.
2. Hoffman, G. J., J. D. Rhoades, J. Letey, and F. Sheng. 1990. Salinity management. *Management of Farm Irrigation Systems*, eds. Hoffman, G., *et al.*, pp. 667–715. St. Joseph, MI: American Society of Agricultural Engineers.
3. Kanwar, R. S. (s.a.). Agrochemicals and water management, s.l., pp. 367–387.

4. Hespanhol, I. 1996. Health impacts of agricultural development. *Sustainability of Irrigated Agriculture*, eds. Pereira, L. S., *et al.*, pp. 61–83. Dordrecht, The Netherlands: Kluwer Academic.
5. Kyritsis, S. 1996. Waste-water reuse. *Sustainability of Irrigated Agriculture*, eds. Pereira, L. S., *et al.*, pp. 417–428. Dordrecht, The Netherlands: Kluwer Academic.
6. Ayers, R. S., and D. W. Westcost. 1985. Water quality for agriculture. FAO Irrigation and Drainage Paper 29 Rev. 1, Rome: Food and Agriculture Organization.
7. Pescod, M. 1992. Wastewater treatment and use in agriculture. FAO Irrigation and Drainage Paper 47. Rome: Food and Agriculture Organization.
8. Rhoades, J. D., A. Kandiah, and A. M. Mashali. 1992. The use of saline waters for crop production. FAO Irrigation and Drainage Paper 48. Rome: Food and Agriculture Organization.
9. Metcalf, E. 1991. *Wastewater Engineering, Treatment, Disposal, and Reuse*. New York: McGraw-Hill.
10. Hagan, R. M., H. R. Haise, and T. W. Edminster. 1967. Irrigation of Agricultural Lands. Agronomy N° 11, American Society of Agronomy, Publisher Madison, Wisconsin.
11. Klute A. 1986. Methods of Soil Analysis. Agronomy N° 9. Part 1, American Society of Agronomy, Madison, WI: Soil Sci. Soc. Am.
12. Cavazza, L. (ed.). 1991. *Agricoltura e Ambiente*, Edagricole, Bologna: Accademia Nazionale di Agricoltura.
13. Moretti, L., G. Chiarelli, and A. Zavatti, (eds.). 1988. *Agricoltura Ambiente 1*, Bologna: Pitagora Editrice.
14. Cheng, H. H., G. W. Bailey, R. E. Green, and W. F. Spencer. 1990. *Pesticides in the Soil Enviroment: Processes, Impacts, and Modeling*, Madison, WI: Soil Sci. Soc. Am.
15. Preti, F. 1996. Main topics and recent experiences for environmental monitoring and modelling diffuse pollution. *Water Sci. Technol.* 33:4–5.
16. Novotny, V., and G. Chesters. 1981. *Handbook of Nonpoint Pollution: Sources and Management*. New York: Van Nostrand Reinhold.
17. Indelicato, S. 1996. Tutela dei corpi idrici: La riutilizzazione delle acque reflue in agricoltura, Convegno La Gestione dell'Acqua nell'Agricoltura Toscana, Alberese, Italy.
18. Wagenet, R. J. 1993. A review of pesticide leaching models and their application to field and laboratory data. *Proceedings of the IX Symposium on Pesticide Chemistry: Mobility and Degradation of Xenobiotics*, Edizioni G. Biagini Luca, Piacenza Italy, pp. 33–62.
19. Novotny, V., and H. Olem. 1993. Water Quality. Prevention, Identification and Management of Diffuse Pollution. New York: Van Nostrand Reinhold.
20. Preti, F., and G. Chester. 1996. Investigation on environmental rate of xenobiotic. Review Procedings X Symposium Pesticide Chemistry. The environmental rate of xenobiotics. La Goliardica Pavese s.r.l., Pavia, Italy, pp. 551–557.

Index